Proceedings of The International Forum on Climate Prediction, Agriculture and Development

April 26—28, 2000
Palisades, New York

International Research Institute for Climate Prediction
Linking Science to Society

IRI-CW/00/1
November, 2000

The IRI was established under a cooperative agreement between the NOAA Office of Global Programs and Columbia University.

Proceedings of The International Forum on Climate Prediction, Agriculture and Development, April 26—28, 2000, Palisades, New York

The International Forum on Climate Prediction, Agriculture and Development was the inaugural international conference for the International Research Institute for Climate Prediction's new home, the Monell Building, at Columbia University's Lamont Doherty Earth Observatory, Palisades, New York.

Supporting Sponsors
World Meteorological Organization
Global Change SysTems for Analysis, Research and Training (START)

Forum International Steering Committee
Cynthia Rosenzweig (Co-Chair), NASA, Goddard Institute for Space Studies, New York, NY, USA
Reid Basher (Co-Chair), International Research Institute for Climate Prediction (IRI), Palisades, NY, USA
Walter E. Baethgen, International Fertilizer Development Center (IFDC), Montevideo, Uruguay
Mohammed Boulahya, African Centre of Meteorological Applications for Development (ACMAD), Niamey, Niger
James Buizer, NOAA, Office of Global Programs (NOAA/OGP), Silver Spring, MD, USA
Mark Cane, Lamont-Doherty Earth Observatory (LDEO), Columbia University, New York, NY, USA
Michael Coughlan, WMO, World Climate Program (WCP), Geneva, Switzerland
Maxx Dilley, U.S. Agency for International Development (USAID), Washington, DC, USA
Paul Epstein, Harvard Medical School and Harvard School of Public Health, Boston, MA, USA
Graeme Hammer, Queensland Department of Primary Industries, CSIRO, Queensland, Australia
Ana Iglesias, Center for Climate Systems Research, Columbia University, New York, NY, USA
James W. Jones, Agricultural and Biological Engineering Department, University of Florida, Gainsville, FL, USA
Antonio Magalhaes, World Bank, Brasilia, DF, Brazil
Antonio Divino Moura, International Research Institute for Climate Prediction (IRI), Palisades, NY, USA
Rajul Pandya-Loch, International Food Policy Research Institute (IFPRI), Washington, DC, USA
Armando Rabuffetti, Inter American Institute for Global Change Research (IAI), Sao Jose dos Campos, SP, Brazil
Chet Ropelewski, International Research Institute for Climate Prediction (IRI), Palisades, NY, USA
Edward Sarachik, JISAO/Hays Center, University of Washington, Seattle, WA, USA
Hassan Virji, Global Change SysTems for Analysis, Research and Training (START), Washington, DC, USA

Local Organizing Committee
James Hansen (Chair) IRI
Ana Iglesias
Jennifer Phillips
Sandra Vitelli

Proceedings Committee
Ana Iglesias
James Hansen
Jennifer Phillips

Published by the International Research Institute for Climate Prediction (IRI), Palisades, New York, 10964, USA
IRI Publication IRI-CW/00/1

ISBN 0-9705907-0-9

Foreword

The International Forum on Climate Prediction, Agriculture and Development was organized to put a spotlight on how climate prediction information might be applied in agriculture to advance the development of societies, especially those vulnerable to climatic stresses. It aimed to develop better mutual understanding and networking among the research and user communities involved in climate, agriculture and development.

The International Research Institute for Climate Prediction (IRI) was established in late 1996 under a cooperative agreement between the National Oceanic and Atmospheric Administration's Office of Global Programs and Columbia University. The IRI's mission is to develop the science of climate prediction and its application for the benefit of societies worldwide. The concept for the institute arose out of the belief by many climate scientists that the advances in understanding the El Niño over the last few decades should be capitalized on—by developing the capabilities to predict climate fluctuations ahead of time and to act effectively on these predictions. It was recognized that this requires simultaneous advances on a wide front, including forecast design, information flow and use, social need and institutional response.

In his Dinner Address to the Forum on April 26, 2000, the Director of the NOAA Office of Global Programs, Dr J. Michael Hall, outlined his hopes for the future and his belief in the essential altruism underlying the IRI vision, i.e. that climate science can contribute to the alleviation of economic and environmental distress and that it is equitable for those who have a wealth of knowledge and technology to share this to benefit those who do not. While noting that it had been those of mature years who had brought the IRI to its present stage, he emphasized that the IRI was a young institute and that its future would primarily lie with the young people who were now engaging in its mission. He welcomed the large number of young scientists and professional who were participating in the Forum and challenged them to apply their altruism, creativeness, energy and commitment to further develop the vision and bring it to fruition through practical achievements.

It was natural for the IRI to focus on the key global issue of agriculture and development in this, its first large international conference, and the first to be held in its headquarters in the new Monell building. The Forum sought an integrative approach encompassing prediction science, agricultural systems science, social science and the development context, and to showcase the latest research and knowledge in each field. Recent experiences in the application of seasonal climate forecasts in many different regions of the world were reported. Specialist issues concerning the communication of uncertain climate information and the assessment of its economic value were considered. The event was also used to describe the IRI's strategy for fostering agricultural applications of climate prediction information and to seek feedback on this strategy.

The Forum and its associated travel expenses were largely funded from the IRI's NOAA base grant. I thank the World Meteorological Organization and the Global Change SysTems for Analysis, Research and Training office for their support to bring additional developing country participants. I am also thankful to the many speakers for the excellent presentations and the time put into them. Thanks also to the many organizations that supported their staff to attend the Forum. A large vote of thanks is due to all the members of the Forum committees (listed on page (ii) for their varied and significant efforts to steer and organize the meeting and to prepare this report. Special thanks go to Drs. Cynthia Rosenzweig and Reid Basher for their excellent work as Co-Chairs of the Forum Steering Committee.

Approximately 140 people from all over the world participated in the Forum. There were many opportunities for interaction and cross-disciplinary communication, both informally and in group sessions. By all accounts, the Forum was a great success. It gives me much pleasure to offer these proceedings to the reader.

Antonio Divino Moura
Director General
International Research Institute for Climate Prediction

Contents

THE GOALS OF THE FORUM

Reid Basher
International Research Institute, USA

Welcome to the International Forum on Climate Prediction, Agriculture and Development. It is wonderful to see you all and a big honor for the IRI to have such a constellation of expertise and experience here today.

So what are the goals of the Forum? In a broad sense, it is to address recent advances in *the application of seasonal-to-interannual climate predictions to agriculture and development.* There are few targets more important to climate science than agriculture, food, and development. We only need to think of the painful experience of Mozambique and Ethiopia in recent months to see this.

But this is not just another science meeting. It is not just about the latest research results. Of course you will learn from top experts on each of these topics, but there is more to why we are here today.

Let me call this the IRI difference. Recall that the IRI emerged from the scientific discovery of how the El Niño works and from the strong motivation of the leading scientists involved, many in this room today, to capitalize on this knowledge for the benefit of societies worldwide.

There are three key elements to the IRI difference.

- Firstly, the IRI has a mission: *The worldwide advancement of climate prediction and its application.* We do research for a purpose. We seek to learn how to achieve better predictions and more effective applications. And we aim to embed this learning in institutions worldwide, to build capacities, to achieve this mission.
- Secondly, the IRI community of interest extends far beyond this building—it includes you and many others. Networking, collaboration and cooperative programs are essential IRI activities to expand the whole worldwide effort.
- Thirdly, we take an integrated approach, what we call the *end-to-end* approach. Our mission demands we consider the whole picture, Concentrating on one factor won't do the job. Neglecting key issues will impede progress toward the mission.

By "end-to-end" we mean understanding and improving all the factors needed to actually achieve benefits for users and societies. Think of it as a value chain, where each step or each interaction can add value—or possibly detract value.

All factors must be researched and developed: data, predictions, impacts, communication, decision-making, users, institutions, infrastructure. Integrating physical, biological and social worlds.

This is the vision of the IRI. This is what we are struggling to create—an end-to-end mission.

So this meeting, the first major conference in our wonderful new building, is focused on advancing the mission of the broad IRI community. We will listen, learn, discuss, and network. We will review and identify issues and research directions. We will build cooperation and forge new paths. Your inputs will help the IRI clarify its strategies and its work programs.

The idea for the meeting was devised by a small group of advisers—particularly Cynthia Rosenzweig, Jim Jones, and Graeme Hammer—Cynthia seated here is Co-Chair of the Forum, and Jim and Graeme are also with us today. They wanted to accelerate the development of research in this field and help build the IRI's role. Given the newness of the IRI—most of us in the Applications Research group arrived on board late last year—we certainly have been accelerating!

We want to acknowledge the support of many people for making the Forum a reality. An International Forum Organizing Committee of experts provided us with valuable initial guidance.

The IRI through its supporting NOAA grant has funded most of the speakers and many participants from less developed countries. I specially want to thank two other sponsors, START, the SysTem for

Analysis Research and Training, and WMO, the World Meteorological Organization, for additional participant travel support. Those of you who paid registration fees have also contributed to assisting participants from less developed countries.

As I said before, this is our first major meeting, and it has been a sharp learning curve—there may be organizational glitches—let us know—we will do our best to fix things. This is a good moment to introduce the Local Organizing Committee:

Jim Hansen, the Chair of the committee and the IRI's lead scientist in agriculture; Ana Iglesias, who works with Cynthia at the Goddard Institute for Space Studies; Jennifer Phillips, IRI agricultural scientist; Sandy Vitelli and Ann Binder, supporting staff of the IRI. They have done a tremendous job and deserve all praise for their commitment and achievement.

Finally, let me say clearly that we want this to be your Forum, not the IRI's alone. On behalf of Co-Chair Cynthia Rosenzweig and myself, I invite you to engage fully in the planned activities and in individual interactions. We are looking forward to the exchange of information and ideas, and to the insights that surely will be generated over the course of the next three days.

KEYNOTE ADDRESS AND PLENARY PRESENTATIONS

Sustainable Development: Climate and Policy Linkages

Antonio R. Magalhaes
The World Bank, Brazil

WE LIVE IN AN UNSUSTAINABLE WORLD

In the world, 1.3 billion people are presently deprived of a sustainable livelihood. They survive under extremely unfavorable conditions with less than one dollar a day. They are mostly inhabitants of the developing world often affected by climatic variability. Poverty is a major indicator of unsustainability. Last year, the earth population reached 6 billion people. It doubled in about 40 years. It will reach 6.8 billion in 2010 and 8 billion in 2020. 90% of the population increase is occurring in the less developed and more vulnerable parts of the world(1).

Unsustainable human activities diminishes the capacity of the natural resource base to sustain life. As a consequence, land degradation, desertification, deforestation, and pollution are occurring in many regions(2). In countries like Brazil, part of the pressure to deforest the rainforest comes from migration originating from the drought-stricken Northeast(3). Excessive energy consumption and waste production that result from lifestyles in the affluent world are real challenges for the sustainability of the planet.

World water resources are limited and unequally distributed. Some countries already reached the limit to use their water resources(4). The problem of water quality has become serious in many parts of the world.

Food production distribution is highly unequal. Food deficits are foreseen in many countries, particularly in the third world ones that suffer from climatic uncertainty. Future food deficits in the developing world seem worse when taken into account the present nutritional deficits in poor countries of Africa, Asia and Latin America(5).

The unsustainable use of natural resources and the competition for water and cropland have resulted in conflicts in many parts of the world(6) Climate variability may increase unsustainability of life conditions everywhere. Adverse effects of droughts and floods are heavier on poor populations of Africa, Asia and Latin America. Millions of people are affected by droughts in Northeast Brazil, Sub-Saharan countries, South Asia. In poor countries, extreme climatic events result in serious social problems.

Governments and the international community have responded to extreme climatic events in many ways. The international community has helped poor countries with food supplies, medicines and other forms of social assistance. In Brazil, workfare programs have created millions of jobs for the drought displaced rural workers in the semi-arid Northeast(7). The World Bank has helped countries to recuperate from extreme events like the 1988 floods in Rio de Janeiro and the 1998 El Niño related floods in Ecuador and Peru. The international community is still shocked with the images of the recent floods in Venezuela(8) and Moçambique.

However, responses so far have looked into effects, not causes. Effects may be dramatic because populations and ecosystems are vulnerable. To reduce vulnerability, sustainable development should be promoted. This was the message from the participants of the ICID Conference held in Fortaleza in 1992. "The environmentally sustainable economic and social development of the semi-arid regions must be pursued as the ultimate goal"[(9)].

SUSTAINABLE DEVELOPMENT: AN OPERATIONAL VIEW

For planning and decision making purposes, a simple definition of sustainable development is sufficient: "It is development that lasts"[(10)]. Whenever a development initiative is taken, some questions should be answered: will the results last? will natural resources be used in a sustainable way? will benefits continue in the long run?

To respond to these questions, various dimensions of sustainability should be considered. In the Aridas exercise that followed the ICID conference[(11)], four dimensions of sustainability were defined: economic, social, environmental, and political dimensions(12). Economic sustainability has to do with he capacity of the initiative to be durable from an economic and financial point of view. Initiatives that depend permanently on government incentives or market distortions are not economically sustainable. At the macro level, economic sustainability requires a stable fiscal situation. Social sustainability is related to the capacity of the social system to provide social stability. High income inequalities, poverty, and exclusion are indicators of social unsustainability. Initiatives that foster or maintain such social problems may prove unsustainable in the future. Political sustainability is an important dimension that has not often been considered. It has to do with the capacity of a society to define rules of the game to govern their relationships and that are accepted by all. Political sustainability is better attained in a democratic system in which opportunity of participation is shared equally by all groups in society. Minority or underprivileged groups must have equal opportunities to political participation. Forms of participation vary. Some are related to the way the state itself is organized, as well as the relationships among states, and within states between the state and its citizens. Sustainable institutions are key for the success of any sustainable development effort. At the micro level, the participation of the beneficiaries and other stakeholders in any development initiative is a condition for the political sustainability of that initiative. One problem is related to the fact that government mandates are limited to a certain number of years. While this is important to assure the right of citizens to change their governments, new governments tend to ignore initiatives of previous governments and interrupt programs or projects that are trademarks of their antecessors. If a broad participation process is undertaken, societal ownership will be translated into political demand for the continuation of good policies.

Environmental sustainability has to do with the way natural resources and the environment are used. The idea of sustainable development grew up from the preoccupation with environmental sustainability. Mankind has tended to consider earth resources as inexhaustible. Economic development theorists and practitioners have not included in their calculations the cost of the natural resources used in production processes. Development banks used to prioritize projects based on local natural resources, even if such resources would eventually be depleted. Environmental sustainability has to do with the preservation or enhancement of the productivity of the natural resources combined with the preservation of habitats, biodiversity, and landscape. The sustainable use of natural resources should aim at maximizing "the net benefits of economic and social development, subject to maintaining the services from, and stock of natural resources over time"(13). The rules for environmental sustainability are: (a) renewable resources should be utilized at rates less than, or equal to their natural regeneration capacity; (b) the utilization of non-renewable resources should be subject to the substitutability between these resources and technological progress; (c) waste should be generated in a rate less than, or equal to the assimilative capacity of ecosystems to absorb it; (d) intra and inter-generational rights should be protected.

Traditional economic development theory has focused on the role of physical man made capital. Economic growth is a function of investment in the creation of physical capital. The capital/output ratio would determine the ratio of economic growth. This theory has been improved along time, to include other variables such as education (human capital) and technology. The permanence of major parts of the globe under conditions of underdevelopment, irrespective of efforts done so far, raises the issue of inadequacy or insufficiency of this approach. We now think that sustainable development may result from the creation and utilization of four forms of capital:

- Man made capital, including infrastructure of transport, energy, communications, urban infrastructure, water reservoirs, housing, factories, agriculture projects etc;
- Human capital, where education is the major component, but also health and nutrition. This is probably the single most important form of capital. Access to human capital increases the access of people to opportunities;
- Natural capital, including the natural resource base, soils, water, vegetation, biodiversity, climate;
- Social capital, in terms of human institutions and social links that facilitate collective action in favor of the community well-being. Social capital is distinguished from human capital. Also, like cholesterol, there is good and bad social capital. Where the good social capital prevails—for example, community associations or voluntary services related to the community's well-being, sustainable development is more easily achieved[(14)].

There is a direct relationship among the four forms of capital and sustainable development. Some forms of physical capital, like war tanks, may not lead to sustainable development. There may be substitutability between forms of capital. For instance, the use of non-renewable natural resources like oil could be compensated by the increase in human or social capital. Therefore, the wealth or poverty of nations should take into account not only one form of capital, but all four. The World Bank has done some estimations of the wealth of nations using these criteria[(15)]. Depletion of natural resources, then, may represent an impoverishment of the respective country even if its GNP, according to standard national accounts, is seeing to be increasing.

One issue is how to measure progress and sustainable development. In particular, how to value environmental goods. It is easy to value a machine or a factory. It is difficult to measure the value of a watershed, an ecosystem, a biodiversity reserve, or the local climate. Natural capital embodies different kinds of environmental assets[(16)]. The total economic value of such assets depends on whether they have an use value or not. If they have an economic value, it may be a direct use value or an indirect use value. Examples of direct use value include outputs that can be used directly such as food or vegetal fibers. Indirect use value includes some benefits that are not consumed but are useful, such as flood control. Non-use values include certain things like the value of knowledge from the continued existence of a certain ecosystem even if humans do not use it. Finally, option values include certain things that may have a future direct or indirect use value, but not presently, such as biodiversity. Now, it may be easy to give a value to an environmental asset from which food is produced, but practically impossible to value an endangered species.

Notwithstanding such difficulties, progress in environmental accounting has been made. In exercises sponsored by the World Bank and the United Nations, interesting results came out. When deducting the costs of natural resources that were depleted in the production process, GDP growth turns out to be lower than originally calculated. Further, if other costs like degradation of environmental quality, or the loss of some environmental services, are also included, a positive GDP growth rate may be reduced or even become negative[(17)].

CLIMATE, VARIABILITY, AND SUSTAINABLE DEVELOPMENT

Climate is a component of natural capital. It is different, however, from other environmental assets such as soils, water, biodiversity. In principle, climate cannot be depleted or changed, except perhaps in the long run. However, climatic phenomena like rains may be affected by air pollution, as in the case of acid rains that impact negatively on forests and crops. Local microclimates may also be affected, as in the case of cities and deforested areas where temperatures tend to increase[18]. In the long run, the build up of greenhouse gases in the atmosphere may cause drastic changes in the global and local climates. Such alterations of the climate may have economic and human impacts, but it is difficult to value them. It is possible however to take action in the elimination of causes that alter the climate, through sustainable development practices that reduce greenhouse gas emissions or increase sinks.

Also different from other environmental assets, climate is subject to a great degree of natural variation. The normal pattern of climate in a certain region includes intra and inter-annual variations of different magnitudes. Such variations often cause climatic events like droughts, floods, temperature peaks, freezes. These events impact on society, on the economy and on the environment[19]. Climatic events cannot be changed: droughts or heavy rains will continue to occur in many regions, though it is possible that their pattern will change as a consequence of global warming. Statistical and probabilistic studies show that droughts and floods are recurrent, but can not inform how the climate of the next year will be. Recent scientific development by the climate community has made impressive progress with regards to climate prediction, so that it is now possible to have an idea about the characteristics of the climate in the next season. This information may be highly valuable in regions that are more vulnerable to such events.

Impacts of climatic variability depend on the intensity of the events. Scientific information that predicts with high probability the occurrence of certain events is thus very useful. On the other hand, the intensity of impacts depends on the vulnerability of the region, natural resources, populations, and economic and social activities. Poor people are more vulnerable than well off people. Marginal regions are more vulnerable than non-marginal regions. Natural resources under stress become more vulnerable than preserved ecosystems. The issue of vulnerability is consequently a key element in a sustainable development strategy that aims to reduce or neutralize the adverse effects of climatic events, as well as to increase the benefits of climate.

Vulnerability is "the risk that the household's entitlements will fail to buffer against hunger, famine, dislocation or other losses" [20]. Poor people do not have the entitlements—for instance, food that the household can obtain through production, exchange or other means such as reciprocal relations or kinship obligations[21]—that will provide them with some capacity to resist to crises. When a drought comes, they have no protection. The understanding of vulnerability is thus very important for defining societal responses to the impacts of extreme climatic events. As Ribot puts it: "each instance in which a household's production, investments, stores and claims are insufficient to sustain them—chains of causality can be traced out. From an understanding of causality, we can move toward policies to reduce the vulnerability which is located at the confluence of these causal chains"[22] So, the policy issue becomes one of how to foster sustainable households. By the same token, how to foster sustainable economic activities, sustainable environment, sustainable communities in order to reduce vulnerability to crises, including those arising from climatic variations.

PLANNING FOR SUSTAINABLE DEVELOPMENT: WHERE THE CLIMATE VARIABILITY COMES IN

A process of sustainable development planning and implementation involves: (a) a holistic approach, where all dimensions of sustainability come together; (b) a long term perspective, at least one generation (20-30 years); (c) a combination of scientific and technical expertise with a comprehensive process of stakeholders participation and community empowerment; (d) a sense of strategy. The Aridas methodology suggested a seven-steps approach to planning sustainable development[23]: (a) an assessment of present sustainability—social, economic, environmental, political—of the planning subject, starting with on the identification of key indicators of sustainability, including climate indicators; (b) an assessment of the present vulnerability to crises, like those resulting from droughts and floods; (c) a business-as-usual scenario for the future, based on the projection of key indicators—including climate indicators to capture for the possibility of climate change in the region—and the best expert knowledge; (d) again, an assessment of future vulnerability to crises; (e) a desired scenario, based on extensive stakeholders participation; (e) an analysis of lessons learned with regards to past and present policies; (f) the definition and negotiation of a sustainable development strategy—for instance, focusing on the creation of human capital, rational utilization and preservation of the natural capital, and the fostering of good social capital, targeting of the poor, reduction of vulnerability to crises—aiming at deviating the path of the business as usual scenario towards the desired scenario; (g) the seventh step is the implementation of the strategy, a complex process that involves changes in legislation, building or strengthening of institutions, adoption of market and non-market instruments for sustainability, continuous evaluation, and feed-back.

The climate variable is treated as a component of the natural capital endowment of the region and as a factor that may trigger crises that impact people, economic and social activities, and the environment. Climate prediction information can be introduced in the planning process as an input to the design of adaptation/mitigation plans. This information will be most valuable during the implementation phases: decision-makers must be ready to act as soon as they receive reliable information on the climate of the next season. It is important that in the planning phases some triggers be agreed upon, based on which the contingency plans will be put into action. The issue of the communication between scientists and decision-makers becomes one of the utmost importance, as both communities normally speak different languages.

CLIMATE INFORMATION AND POLICY IMPLEMENTATION

Three kinds of climatic information are necessary for sustainable development planning and implementation. First, the knowledge of the regional climate. Climate information is necessary to the diagnosis of the natural capital endowment of a given region. Second, the knowledge of the impacts of climatic variability: social, economic, environmental vulnerabilities, how the impacts occur, causality chains. Third, the information on climate prediction with regards to the next season, to allow decision makers to prompt themselves for action in order to reduce adverse impacts or enhance positive effects of climate variations.

The experience of development planning so far has not given due attention to climate information, either as a component of the natural capital base or as trigger of socioeconomic crises. In order to deal properly with climate in the process of sustainable development planning and implementation, the knowledge base should be expanded to the regional climate (including climate prediction), impacts, adaptation and mitigation policies.

At the local level, particularly in the less developed regions, there is still much to do to improve climate information. Climate is a component of the natural capital and as such is an important element of sustainable development. There is a need for better information on climatic patterns, precipitation, temperatures, seasonal and inter-annual variations, teleconnections, possibilities of climate prediction.

In face of adverse climatic events, policy makers and producers have to decide on how to act. Though the local climate may be well known, there is much uncertainty on when and how a drought or a flood will happen and how intense it will be. Climate prediction helps to reduce this uncertainty. Information on climate prediction facilitates the decision making process for the next season[24]. If prediction points to a drought year, decision-makers (governments, companies or individuals) may prepare adaptation and mitigation plans in advance and not wait until the initial impacts have occurred. In case of floods, it is not difficult to identify the vulnerable areas to floods, like urban slums located on the low plains or on the hills. Once these areas are identified, preventive actions may be taken, and civil defense services may be prompted to respond. In the medium and long run, sustainable land use policies should be adopted to reduce vulnerability of people and areas affected.

The assessment of climatic impacts aims at understanding the effects of climate on the vulnerable population, on the social and economic activities, on the soils and water resources, on biodiversity. To understand the causality chains between a given climatic event and its final impact is crucial for the identification of sustainable policy responses. A climatic extreme event must be seen as the trigger of a crisis that affects the most vulnerable. Because of differences in vulnerability levels, the same climatic event can be harmless in one place and a social or environmental disaster in another place. Hence, it is essential to understand the nature of such vulnerabilities and how impacts are transmitted from the initial physical event to the final outcomes(25). Adaptation and mitigation policies aim at (a) reducing vulnerability of peoples, activities, and regions and increasing resiliency to climatic events, and (b) assisting populations hit by climatic events such as droughts and floods. Adaptation policies may be more effective if adopted within the context of a broader sustainable development strategy that takes into account all dimensions of sustainability. One of the common mistakes of many adaptation policies has been a narrow view of the nature of the impacts. In the drought prone Northeast of Brazil, for instance, responses to drought have been for the most part restricted to building water infrastructure. In other words, if drought was seen as a shortage of water, the right response should be to accumulate water. This kind of response does not address the complex issue of the social and economic—as well as political and environmental—vulnerabilities. A more appropriate response along the lines of the Aridas project, for instance, emphasizes the building of human capital and education, a coordinated water policy focused on institutional building and creation of social capital, more attention to the conservation of the natural capital of the region, and market incentives to foster sustainable economic initiatives(26). In lieu of only building dams or digging wells, adaptation policies should aim at increasing households entitlements, changing rural and urban land-use patterns, changing the composition of economic activities towards less climate-dependent activities and so on.

The problem of communication between scientists and policy makers needs to be addressed. Scientists and policy makers speak different languages. Scientists are usually not able to communicate their findings in a language that will be properly understood by policy makers. Policy makers, in turn, want clear cut information. The probabilistic nature of climate prediction is not easily understood. It may happen that scientists may be tempted to offer policy makers the kind of certainty the latter are demanding. This is dangerous, as it may erode the credibility of science if the prediction is not confirmed by the facts. Strategic communication should be adopted to bridge the gap of information between scientists and policy makers.

CONCLUDING REMARKS

There has been remarkable progress in the science of climate and climate prediction in the last few decades. While this progress will continue, it is necessary to mainstream the climate variable into the development planning and implementation processes. This requires an understanding of how climate variability impacts on society in a country, region, or community. Successful sustainable development strategies must embody the elements of adaptation and mitigation to climate variations.

NOTES

(1) For population and poverty data, see The World Bank World Development Indicators, 1997; Atlas of the Environment, Lean and Hinrichsen, 1992.
(2) See, for instance, Hassan and Dregne, 1997.
(3) On the case of Brazil's Northeast and the Amazon, see Bitun et al., 1996.
(4) See, for instance, Steer, 1997, for projections of key indicators like food, population, fertilizer use, fish stocks, water, urban growth. Steer presents his projections in the context of analyzing causes of conflict in the World. See also Brown, 1994.
(5) See OECD, 1998, p.27: Chronic under-nutrition will decrease, but will still be affecting 12% of the population of the developing countries in 2010, circa 680 million people. See also Rosenzweig and Hillel, 1998, for a discussion on food security and vulnerability to famine, p. 234-244.
(6) See Magalhães, 1997, for a discussion of the relationship between sustainable development and peace, based on the proceedings of the seminar on this topic held at the University of Texas at Austin, in 1996. See also Homer-Dixon, 1999.
(7) See Magalhães and Glantz, 1992.
(8) See Kreimer and Walker, 2000. It is estimated that 50,000 people died during the floods in Venezuela in January 2000.

(9) The ICID-International Conference on Climate Variation and Sustainable Development in Semiarid Regions was held in Fortaleza, Brazil, in January-February of 1992, as a contribution to UNCED. See ICID Declaration in Ribot, 1996.

(10) This definition was adopted by the Aridas project (see Magalhães et al, 1994). The Aridas project successfully attempted incorporate the idea of sustainability into the regional development planning process in the Northeast of Brazil. The experience was later extended to several states and other regions of Brazil.

(11) See Ribot et al, 1996.

(12) See Magalhães et al, 1994. In the World Bank literature, three dimensions are considered: Economic, Social, and Environmental. However, the social dimension here encompasses the social, political and institutional dimensions. See Munasinghe, 1993; Serageldin, 1996; Steer, 1997.

(13) Munasinghe, 1993.

(14) See Putnam, 1993; Serageldin, 1996.

(15) The wealth of nations is not only in produced physical goods, but mainly in human and natural capital. See Serageldin, 1996; Steer, 1997.

(16) See Munasinghe, 1993; Serageldin, 1996.

(17) See Munasinghe, 1993

(18) Recent research by Daniel Rosenfeld, Hebrew University of Jerusalem, suggested that air pollution particles impede the formation of raindrops, thus affecting rainfall at the local level. See Gerhardt, 2000.

(19) There is an extensive literature on climate impacts, especially after the IPCC was launched in 1989. See, for instance, Parry et all, 1988; Magalhães and Glantz, 1992; IPCC, 1995.

(20) Cf. Ribot, 1996, p. 2. The concept of vulnerability and entitlements is discussed by Sen, 1981 Downing, 1992 and Ribot, 1996, among others.

(21) See Drèze and Sen, 1989, cited by Ribot, 1996.

(22) See Ribot, 1996, p. 3; Jessop, 1990, cited by Ribot, 1996.

(23) See Magalhães, 1992, 1998; Magalhães et al, 1994; McKaughan, 1997.

(24) The state of Ceará, in Brazil, for instance, has a program called "time to plant", when it advises farmers to initiate the planting of their crops based on the information on climate prediction provided by its meteorological agency (Funceme). See Stern and Easterling, 1999, for a comprehensive discussion on climate forecasts.

(25) See Note 18. On methodologies of climate impact studies, see UNEP, 1999; Parry and Carter, 1998; Glantz, 1988; Kates, 1985; Smith at all, 1996.

(26) See, for instance, the proposals of the Aridas project in: Projeto Aridas, 1995.

REFERENCES

Bitoun, J., L. Guimarães and T. B. Araujo, 1996. Amazonia and the Northeast: The Brazilian Tropics and Sustainable Development, p.129-146. In: J.C.Ribot et al. Climate Variability, Climate Change and Social Vulnerability in the Semi-Arid Tropics. Cambridge University Press, Cambridge.

Brown, L. R. and H. Kane, 1994. Full House: Reassessing the Earth's Population Carrying Capacity. The Worldwatch Environmental Alert Series.W.W. Norton & Company, London..

Downing, T. E., 1992. Vulnerability and Global Environmental Change in the Semi-Arid Tropics: Modeling Regional and Household Agricultural Impacts and Responses. ICID Conference, Fortaleza-Ceará, Brazil, 27 January-1 February.

Dreze, J. and A. Sen, 1989. Hunger and Public Action. Clarendon Press, Oxford.

Gerhardt, I., 2000. Air Pollution Reduces Frequency of Rains, p. 11. In: Folha de São Paulo, Mundo, São Paulo. (In Portuguese).

Glantz, M. H. (Ed.), 1988. Societal Responses to Regional Climatic Change: Forecasting by Analogy. Westview Press, Boulder.

Hassan, H. and H. E. Dregne, 1997. Natural Habitats and Ecosystems Management in Drylands: An Overview. Environment Department Papers nr. 51, ESD. The World Bank, Washington.

Homer-Dixon, T. F., 1999. Environment, Scarcity, and Violence. Princeton University Press. Princeton.

ICID—International Conference on Climate Variations and Sustainable Development in Semi-Arid Regions, 1992. "Declaration of Fortaleza", p.167-169. In: J. C. Ribot et al. (ed), 1996. Climate Variability, Climate Change and Social Vulnerability in the Semi-Arid Tropics. Cambridge University Press, Cambridge.

IPCC, 1996. Climate Change 1995: Impacts, Adaptations and Mitigations of Climate Change: Scientific-Technical Analysis. In: R. T. Watson, M.C. Zinyowera, and R.H. Moss (eds). Intergovernmental Panel on climate Change, Cambridge University Press, Cambridge.

Jessop, B., 1990. State Theory: Putting the Capitalist States in Their Place. University Park. The Pennsylvania State University Press.

Kates, R. W., J. H. Ausubel, and M. Berberian, 1985. Climate Impact Assessment. (Scope 27) John Wiley & Sons, NY.

Kreimer, A. and P. Walker, 2000. Reducing the Misery of Disasters. In: Bank's World, Vol. 4, no. 2, Feb. 2000. The World Bank.

Lean, G. and D. Hinrichsen, 1992. Atlas of the Environment. WWF. Harper Collins Publishers, NY.

Magalhães, A. R. (1992).Understanding the Implications of Global Warming in Developing Regions: The Case of Northeast Brazil, p.237-256. In: J. Schmandt and J. Clarkson. The Regions and Global Warming: Impacts & Strategies. Oxford University Press, Oxford.

Magalhães, A. R, 1998. Planning for sustainable development in the context of global change. Global Environmental Change (London), Vol. 8, Nr. 1, 1-30.

Magalhães, A. R., E. B. Neto and S. S. Panagides, 1994. Project Aridas: A Strategy for the Sustainable Development of the Northeast. Basic Document. (In Portuguese). Secretariat of Planning of the Presidency of the Republic, Brazil. Brasília.

Magalhães, A.R. and M. H. Glantz (Ed), 1992. Socioeconomic Impacts of Climate Variations and Policy Responses in Brazil. UNEP, Gov. of Ceará, Esquel Brasil Foundation. Brasília.

Magalhães, A.R. (Ed), 1997. Sustainable Development: Implications for World Peace. LBJ School of Public Affairs, University of Texas at Austin, Austin.

McKaughan, S. E, 1997. A Sustainable Development Guidebook for Planners. Ministry of the Environment, Brazil. 1997 (manuscript, to be published).

Munasinghe, M., 1993. Environmental Economics and Sustainable Development. World Bank Environment Paper Number 3. The World Bank, Washington.

OECD, 1998. The Future of Food: Long-term Prospects for the Agro-food Sector. Paris.

Parry, M. and T. Carter, 1998. Climate Impact and Adaptation Assessment. Earthscan, London.

Parry, M.L., T.R. Carter, N.T. Konijn (Ed), 1988. The Impact of Climate Variations on Agriculture. Vol 1: Assessments in Cool Temperate and Cold Regions. Vol. 2: Assessments in Semi-Arid Regions. UNEP/IIASA. Kluwer Academic Press, Dordrecht, The Netherlands.

Projeto Aridas, 1995. A Strategy for Sustainable Development in Brazil's Northeast. Ministry of Planning and the Budget. Brasília.

Putnam, R. D., with R. L. and R. Y. Nanetti, 1993. Making Democracy Work: Civic Traditions I Modern Italy. N. J. Princeton University Press, Princeton.

Ribot, J.C., A.R. Magalhães and S.S. Panagides (Ed), 1996. Climate Variability, Climate Change and Social Vulnerability in the Semi-Arid Tropics. Cambridge University Press, Cambridge.

Ribot, J.C., 1996. Introduction. Climate Variability, Climate Change and Vulnerability: Moving Forward by Looking Back, p. 1-10. In: J.C. Ribot et al (Ed). Climate Variability, Climate Change and Social Vulnerability in the Semi-Arid Tropics. Cambridge University Press, Cambridge.

Rosenzweig, C. and D. Hillel, 1998. Climate Change and the Global Harvest. Potential Impacts of the Greenhouse Effect on Agriculture. Oxford University Press, NY.

Sen, A., 1981. Poverty and Famines: An Essay on Entitlements and Deprivation. Oxford University Press, Oxford.

Serageldin, I., 1996. Sustainability and the Wealth of Nations. First Steps in an Ongoing Journey. Environmentally Sustainable Development Studies and Monographs Series No. 5, The World Bank. Washington.

Smith, J. B. et al (Ed), 1996. Adapting to Climate Change: Assessments and Issues. Springer, NY.

Steer, A., 1997. A Global View, p. 143-171. In: A. R. Magalhães (Ed). Sustainable Development: Implications for World Peace. LBJ School of Public Affairs, University of Texas at Austin, Austin.

Stern, P. C. and W. Easterling (ed), 1999. Making Climate Forecasts Matter. National Academy Press, London.

UNEP-United Nations Environment Programme, 1999. Handbook on Methods for Climate Change Impact Assessment and Adaptation Strategies. UNEP, Institute for Environmental Studies, Nairobi and Amsterdam.

Wilhite, D. A. (Ed), 1993. Drought Assessment, Management, and Planning: Theory and Case Studies. Kluwer Academic Press, Dordrecht.

The World Bank, 1997. World Development Indicators, Washington.

The World Bank, 1992. World Development Report: Development and the Environment. Washington.

Society and Climate

Steve Rayner
International Research Institute for Climate Prediction, USA

Before I begin, I should offer a couple of disclaimers. I have no expertise in fields of climate prediction or agriculture. My work on the human dimensions of climate change, until very recently, has focused on long-term secular climate change, so I can claim very little expertise on issues of climate variability. On the other hand, I have spent some considerable portion of my professional lifetime pondering issues of sustainable development, sustainable livelihoods, and institutional aspects of information use and decision making under risk, uncertainty, and indeterminacy. I want to signal some of these issues in the hope of provoking discussion over the next couple of days.

I will start with a highly abbreviated intellectual history of work in climate and society. From the time of Herodotus and Hippocrates to the early years of the 20th Century, the relationship between climate and society was viewed largely deterministically. For example, the climatic determinists of the turn of the last century generally equated hot weather with the tendency of a culture towards indolence and low productivity. Temperate zones led nations to be industrious and productive. Of course, most writers thought that the climate of their own location was the one most likely to promote the ideal level of intellectual activity! This chauvinistic approach led scholars throughout most of the 20th century to ignore the climate-society relationship altogether. As a result, there is an 80-year gap in the social science research record on this topic. When attention returned to the issue during the 1970s and 80s, in the context of concerns about global warming, the result was a tendency towards a static view of climate that assumed that the earth's current state is an optimum state that ought to be preserved.

Since the 1970s, two major research traditions that ought to inform our thinking about probabilistic approaches to climate and society have developed separately and have barely touched. These are the field of natural hazards research and the field of technological risk research. The natural hazards tradition, led by geographers, has focused on disaster warning, coping, and recovery. Its principal sponsor in the United States have been the Federal Emergency Management Agency (FEMA). On the other hand risk analysis, led by engineers but subverted by various social sciences, has focused on disaster prevention through anticipation and risk reduction. Its principal US sponsors have been the Department of Energy (DOE) and the Environmental Protection Agency (EPA). Hopefully, this conference will represent one step in bringing these research traditions into closer collaboration.

The standard equation for risk in the 1970s was R = PM (Risk equals the probability times the magnitude of an unwanted event). During the following decade, social scientists sought to introduce a variety of additional factors into understanding of risk, summarized by the expanded equation R = PM + TLC (where the novel terms indicate the social arrangements of trust, liability, and consent respectively). In the field of climate change, natural hazards research has only partially informed the thinking of the Intergovernmental Panel on Climate Change (IPCC) about climate impacts. The influence of the expanded view of risk analysis on the IPCC has been even less.

This brief history suggests that, from the point of view of social decision making, it is important to recognize that the kind of information appropriate to decision making depends on a combination of (1) the degree of uncertainty and ignorance about the behavior of a system and (2) the decision stakes that are involved in acting under any particular degree of uncertainty. By mapping systems uncertainty and decision stakes as orthogonal dimensions, philosophers of science Silvio Funtowicz and Jerome Ravetz have generated a conceptual model of 3 kinds of science and policy discourse corresponding to low, medium, or high levels either variable (Figure 1). Where both systems uncertainty and decision stakes are low, decisions are dominated by decision rules or routines. Where either systems

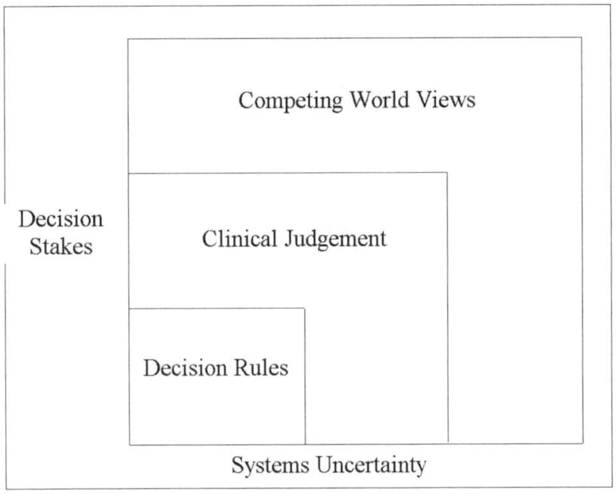

uncertainty or decision stakes increase, decisions are made through the exercise of clinical judgment in which established routines begin to be displaced by the exercise of craft skills based on practitioner experience. Where either systems uncertainty or decision stakes are high, decision making is dominated by discourse among competing world views or core beliefs about how the world works.

The Funtowicz and Ravetz framework also emphasizes the social justice argument that risk cannot be defined simply in terms of the probability and magnitude of an event, but in ordinary language is intrinsically bound up with arrangements for TLC.—In other words, in the real world, fairness in distribution cannot be separated from assessment of consequences. For this reason, scholars at the International Research Institute for Climate Prediction have been paying particular attention to the equity implications of improving humanity's ability to predict seasonal to interannual variability of climate.

The existence of different worldviews and variability in the experience upon which clinical judgment is based both indicate that the *hydraulic* concept of information, popular in economic theory, is of limited usefulness in a pluralistic world. The hydraulic concept assumes that information, like water, naturally flows "downhill" from source to recipient. Where it does not, it is due to the presence of "barriers" (such as incorrect prices, or institutional failures) that must be eliminated to restore the natural flow. The hydraulic concept is also one that sees information as being like water transferred in a bucket chain, remaining largely unchanged as it passes from one bearer to another. Exceptions to this expectation are regarded as distortions.

In contrast to the hydraulic concept, psychologists, sociologists, and anthropologists tend to converge on the idea that both individuals and institutions utilize a *briccolage* approach to information. Briccolage is the process of cobbling together useful devices out of found objects. Applied to information processing, it suggests that upon encountering new information people tend to dismantle it into components and then reassemble the components in such a way that they fit with their existing mental and cultural models of how the world works. In contrast with the hydraulic concept, this is a model of active filtering of information by recipients to admit what is consistent with their prior expectations.

This process of active filtering calls attention to a phenomenon that is best described as the social construction of ignorance—the tendency of institutions to exclude information that is threatening to their operation especially with regard to their legitimacy. For example, research on decision making among water resource managers indicates that they judge their success by their ability to remain invisible to policy makers and the general public. Hence they seek to minimize risk of interruptions to supply through redundancy in capacity and consequently tend to be resistant to information about improvements in efficiency that they perceive as also having the potential to increase the possibility of supply interruptions.

Another important characteristic of information is that it is only information if it can be acted upon—otherwise it is at best noise. For instance, in wake of the 1973 drought, Mickey Glantz interviewed a cross section of Sahelian decision makers who reported that even perfect foresight about the drought would not have changed their behavior as they were operating under inflexible institutional and resource constraints.

The issue of prior expectations strongly suggests that local knowledge about resource management in the light of ecosystem and climate behavior merits serious understanding before attempts are made to intervene in existing resource management regimes. This relates to both how scientific climate forecast information fits into local resource management systems as well as how local knowledge can contribute to scientific forecasting, as exemplified by the investigation by Ben Orlove and his colleagues into the scientific basis for the Andean farmers' tradition of predicting El Niño according to the visibility of the Pleiades constellation at a certain time of the year.

The issue of incorporating local knowledge can provoke scientific opposition in some quarters, particularly in less-industrialized societies where scientific organizations and institutions have remained insulated from postmodernist impulses and see themselves as in competition with superstitious belief systems.

However, the sustainable livelihoods literature points out the dangers of displacing traditional techniques of risk spreading through strategy switching, for example among crops or from agriculture to fishing and back. This is particularly so on the pulsing margins of the global economy such as described by George Bond in parts of Africa where crop diversification and traditional grain stores were displaced by market oriented monoculture, supported by fertilizers, all of which collapsed in what was, in global terms, a barely detectable contraction of the global economy.

We also need to be aware that the application of scientific capabilities can increase global productivity at the expense of local welfare. A salutary story being that of the Mahaweli irrigation project in Sri Lanka that led to a shift in production from millet to higher value grains that also led to a shift from female to male control over cultivation. Because women spend more of their income on the household than men, the increase in productivity was actually accompanied by an increase in child malnutrition.

The obvious source of value in the ability to predict climatic variation is the ability to warn of extreme events. However, it may well be the case that in both less-industrialized and industrialized societies, incremental changes in routine resource management behavior may yield larger societal dividends than more visible improvements in emergency management. For example, the development of weather derivatives and the efflorescence of consulting climatologists and private weather forecasting services serving agricultural and private utility companies in the US.

The case of consulting climatologists and private forecasting services highlights another important issue: the importance of addressing intermediate information users. A good example arose in a recent study, in which I was engaged, looking at dam operations in the US. It soon became clear to the research team that a potent way to change the management of water impoundments would be to influence the standard operation procedures of the US Corps of Engineers.

I would like to close these remarks with two final thoughts. First, we should always be aware that technological change may have a significant impact on the value of climate forecast information. When frost protection technology in Pacific Northwest apple orchards was provided by kerosene heaters, farmers would have to rise in the dead of night to light them. The economic value of frost warnings were estimated at millions of dollars. However, when the kerosene technology was replaced by water misters that were controlled automatically by thermostat, the frost warning became virtually valueless to apple growers.

Second, we should consider the possibility that uncertainty about socioeconomic development may be more significant to decision makers than climate uncertainty—hence it makes sense to think about how to incorporate climate information into mainstream policy agendas such as economic, trade, or agricultural, policy, rather than promote climate forecasting as a stand-out activity.

I hope that these somewhat telegraphic remarks will help to provide some broad context on climate and society and the use of probabilistic forecasts of climate variation in the more specific setting of agricultural activity.

REFERENCES

Bond, G. 2000. Paper presented in panel session "Predicting Rain: The Science and Politics of Forecasting", Conference on The Politics of Climate Change in Africa: Farming, Food and Forecasting, Columbia University, March 24th 2000

Funtowicz, S. & J.R. Ravetz 1985. "Three Types of Risk Assesment: A Methodological Analysis." In C. Whipple & V. Covello (eds.) Risk Analysis in the Private Sector, Plenum, New York

Glantz, M. 1977. "The Value of a Long-Range Weather Forecast for the West African Sahel." BAMS 58: 150-158

Katz, R. personal communication

Orlove, B.S., J.C.H. Chiang, and M.A. Cane. 2000. Forecasting Andean rainfall and crop yield from the influence of El Niño on Pleiades visibility. Nature 403:68-71.

Climate Variability, Seasonal Prediction, and Agriculture

Cynthia Rosenzweig
NASA/Goddard Institute for Space Studies, USA

INTRODUCTION

Since agriculture began, some 10,000 years ago, farmers have known that variations in productivity are linked to variations in climate. Some form of seasonal climate prediction has probably been around for nearly as long. An oft-cited example is Joseph's prediction for Egyptian agriculture, made over 3,000 years ago, on the basis of dream interpretation: "Behold, there come seven years of great plenty throughout all the land of Egypt: And there shall arise after them seven years of famine; and all the plenty shall be forgotten in the land of Egypt; and the famine shall consume the land;" (Genesis 41:29-30).

Seasonal forecasts for agriculture through the millennia have been made primarily on the basis of signs and portents. Some of these methods may be based on indicators of large-scale forcing mechanisms such as the El Niño Southern Oscillation phenomenon. Recent work by Ben Orlove and colleagues relates that potato farmers in drought-prone regions of the Andes have for more than four centuries made observations of changes in the apparent brightness of stars in the Pleiades at the time of the southern winter solstice in order to forecast interannual variations in summer rainfall and autumn harvests (Orlove et al., 2000). Poor visibility of the Pleiades in June—caused by an increase in subvisual high cirrus clouds—is indicative of an El Niño year, which is usually linked to reduced rainfall during the growing season several months later (Figure 1).

In recent decades, there's been a dawning of hope in "scientific" seasonal climate prediction based on improvements in the understanding of large-scale forcing mechanisms and in the skill of

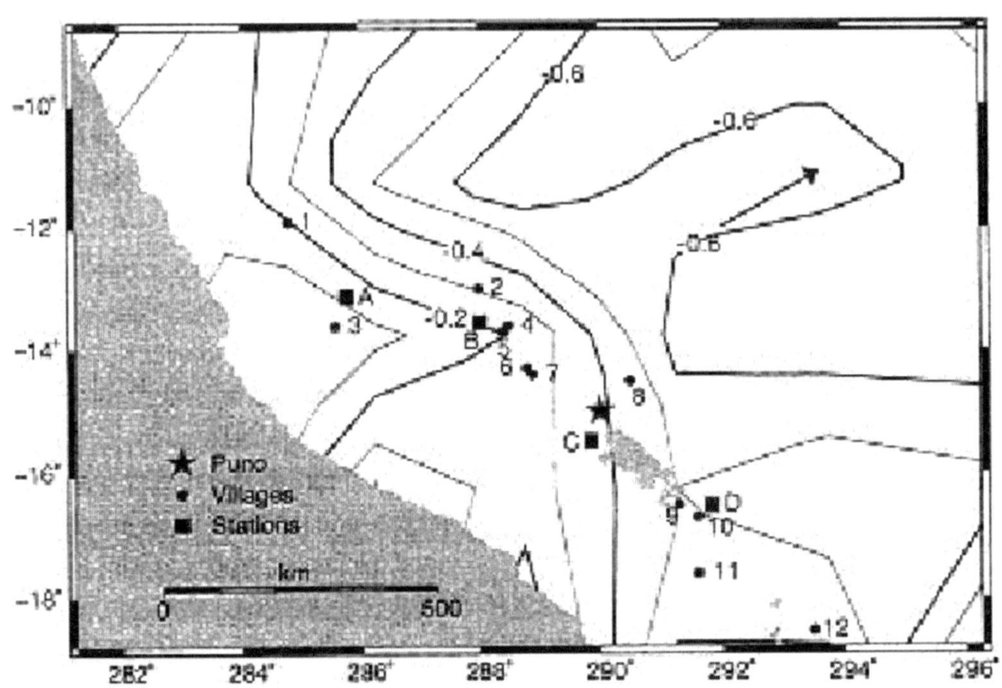

***Figure 1.** Correlation of high cloud in late June with potato yield in the central Andes. (Orlove et al., 2000).*

models that predict them. In particular, prediction of the strongest of these mechanisms, the ocean-atmosphere interactions associated with the El Niño cycle, offers potential for agriculture because of the widespread "teleconnections" to many regions where farming activities are vital (Figure 2) (Ropelewski and Halpert, 1987). Strongly affected regions are NE and SE South America, Australia and SE Asia, the Indian subcontinent, and SE Africa. Lesser but still significant effects are found in the SE U.S.

These teleconnections affect crops within growing seasons and offer potential understanding and foreknowledge of year-to-year variability. In regions such as Zimbabwe with strong correlations between crop yields and the ENSO cycle, Pacific SSTs explain up to 60% of yield variation (r = 78) (Figure 3) (Cane et al., 1994). In other regions, ENSO effects are weaker; in the U.S. Corn Belt, for example, Pacific SSTs explain only about 15% of crop yield variation (Rosenzweig and Wannebo, 2000; Phillips et al., 1998). Analyses have shown that the El Niño cycle can cause significant economic consequences in many regions (Adams et al., 1998; Heal and Yun Lin, pers. com.), and that improved climate forecasts can have economic value (Fox et al., 1999; Mjelde et al., 1998; Solow et al., 1998) (Table 1).

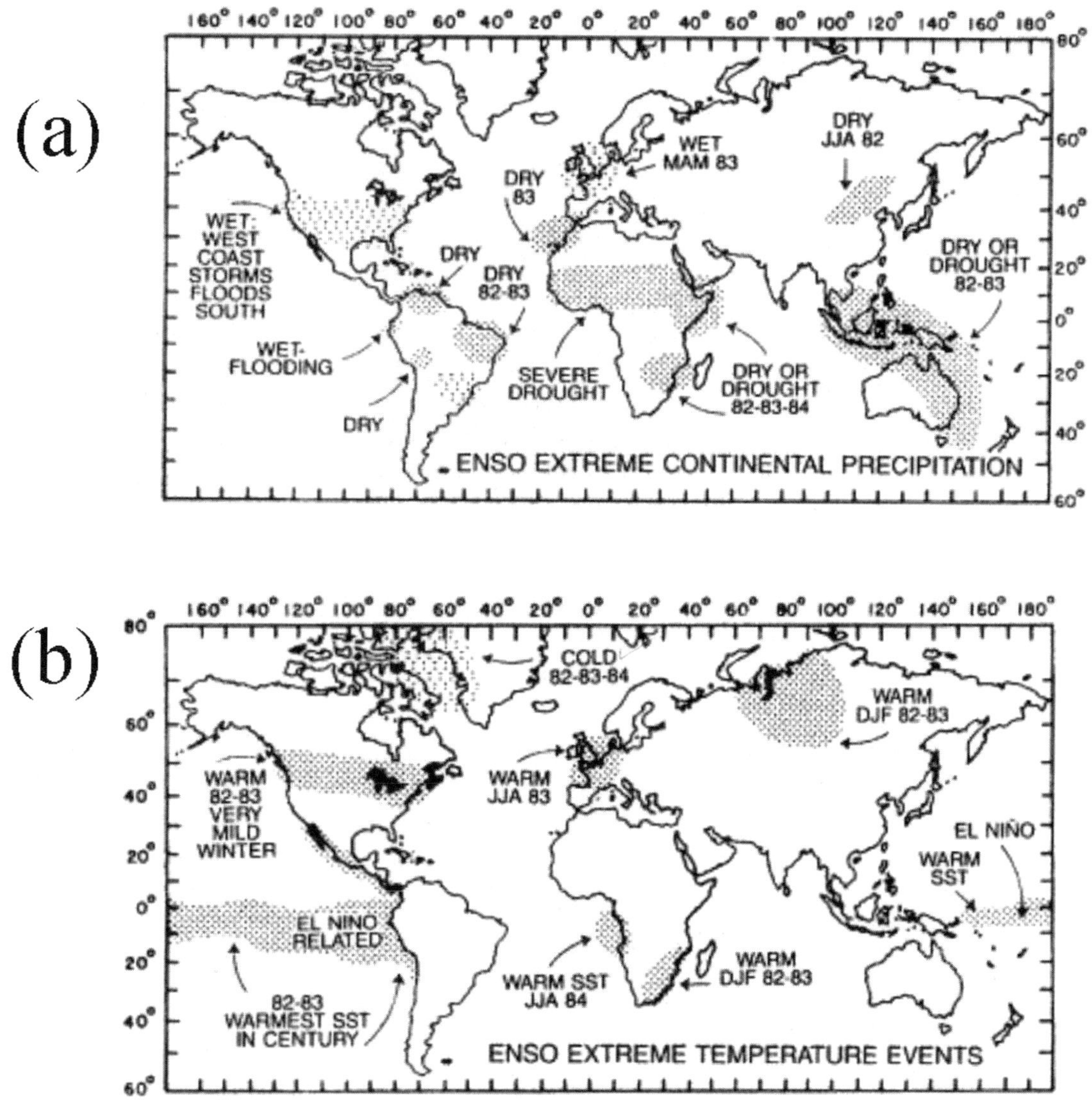

***Figure 2.** (a) Typical rainfall impacts from El Niño (Ropelewski and Halpert, 1987); (b) typical temperature impacts from El Niño (Halpert and Ropelewski, 1992).*

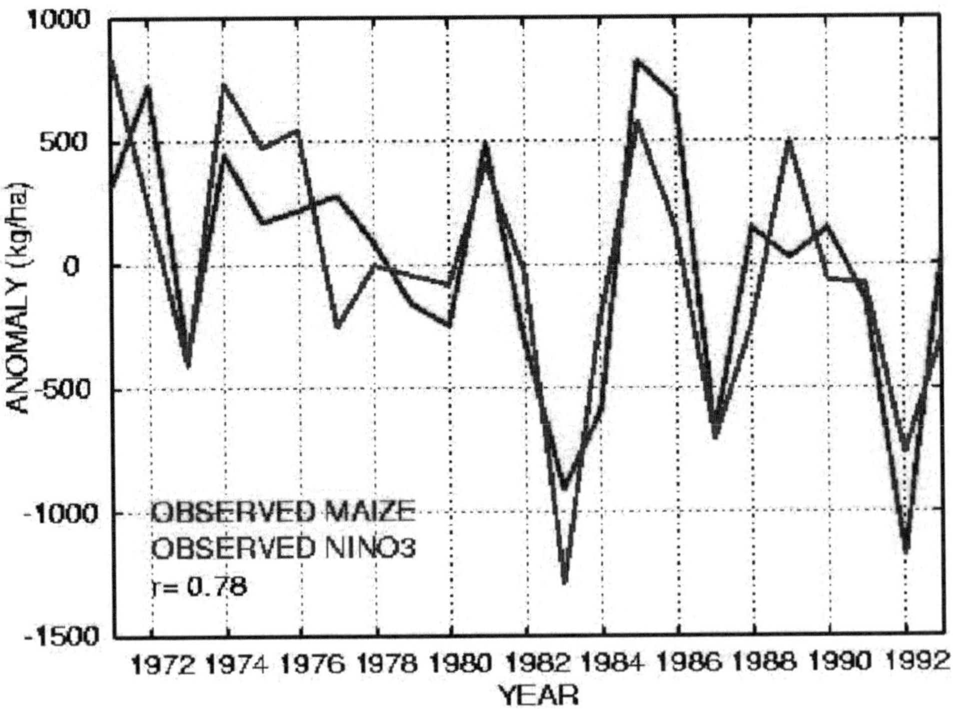

Figure 3. *Correlations between crop yields in Zimbabwe and Pacific SSTs in the NINO3 region, 1971-1993 (Cane et al., 1994).*

This is a tremendously exciting development for agriculture. To have some notion of probability of occurrence of key climate variables, especially precipitation, but also temperature, before the season begins allows for development of agricultural strategies in advance, and can help to inform mid-season and post-season decisions as well. Key actors are farmers, industries, and regional planners (Table 2).

At the farm level, pre-season decisions that can take changed probabilities of seasonal climate into account include crop type (e.g., drought-tolerant sorghum vs. more moisture dependent maize), variety (early-maturing for drought avoidance vs. late-maturing for wetter seasons), planting density (low density if rainfall may be scarce vs. high density if rainfall is predicted to be more plentiful), and inputs (lower levels of fertilizer application in drought-predicted seasons vs. higher levels to take advantage of good seasons). Other farm-level decisions can be affected by seasonal climate predictions during and after the growing season. Irrigation regimes are planned before the start of the growing season and carried out and modified during the growing season. Mid-season corrections include thinning of a densely planted crop if soil moisture becomes limiting. After the growing season, decisions can also be modified in terms of marketing and storage.

Agricultural industries can utilize ENSO predictions to plan effective input production, prepare for the provision of credit and crop insurance, and improve transportation of both inputs and products. There are also many decisions that are made by regional and national planners on regional scales that may benefit from ENSO predictions. Water resource planning is primarily carried out on

Table 1. *Economic impacts of ENSO on agriculture in selected countries.*

	Argentina	Brazil	USA	Zimbabwe
GDP in agriculture US$ Million*	3,956	121,283	197,645	244
La Nina Change US$ Million	-352	-848	-2,852	22
El Nino Change US$ Million	-226	-1,053	-4,838	-88
La Nina Change %	-9	-5	-5	-5
El Nino Change %	-6	-7	-9	-19

*US$ 1987

***Table 2.** Seasonal decisions affected by ENSO predictions by farmers, industry, and regional planning.*

ON FARM		
Pre Season	*Mid Season*	*Post Season*
Crop		
Variety	Irrigation	Irrigation
Plating density	Harvest	Storage
Inputs		
INDUSTRY		
Input production Credit Transportation Insurance		
REGIONAL PLANNING		
Water resources Emergency management		

regional scales as is emergency management; examples are responding to drought-related famine and providing relief from major floods.

There has been a surge of research and programmatic initiatives aimed at understanding how best to make use of seasonal climate forecasts for agriculture in the larger context of development. Much of this work has been done on a region-by-region basis. The objective here is to identify and assess major research issues, drawing from, and comparing and contrasting, results from across the regional activities.

MAJOR ISSUES

Research activities in the field of seasonal climate prediction for agriculture may be grouped according to several major conceptual, scientific, and technical issues (Table 3), each with associated questions.

Conceptual models

What conceptual models are useful to encourage the interdisciplinary, combined basic and applied research needed to enhance the use of seasonal climate forecasts in agriculture? How can physical, biophysical, and social scientists work together effectively? How do we nurture effective interactions between researchers and users of seasonal climate forecasts?

***Table 3.** Major issues in seasonal climate prediction for agriculture.*

1. Conceptual models
2. Impediments to use of seasonal climate forecasts
3. Transfer of regional assessment methods
4. Tools, linkages, and scales
5. Communicating probabilities
6. Global assessment and forecasting

Impediments to use of seasonal climate forecasts

What are barriers are in place that limit the effective use of seasonal climate forecasts in agriculture? Does farming system pre-determine the usefulness of climate prediction?

Transfer of regional assessment methods

Are regional assessment methods transferable? Would development and dissemination of a set of guidelines and/or an assessment template be effective in expanding the usefulness of regional climate forecasts?

Tools, linkages, and scales

What are the scientific and technical tools and linkages needed to improve forecasts for agriculture? What temporal and spatial questions need to be resolved?

Communicating probabilities

What is the most effective way to express climatic prediction and agricultural outcomes in probabilistic terms? What climate variables and agricultural outcomes are most relevant?

Global assessment and forecasting

Is there a role for the global scale in seasonal climate prediction for agriculture? How can global agricultural assessments be devised?

CONCEPTUAL MODELS

Conceptual models are important because they guide our research and application projects, help-

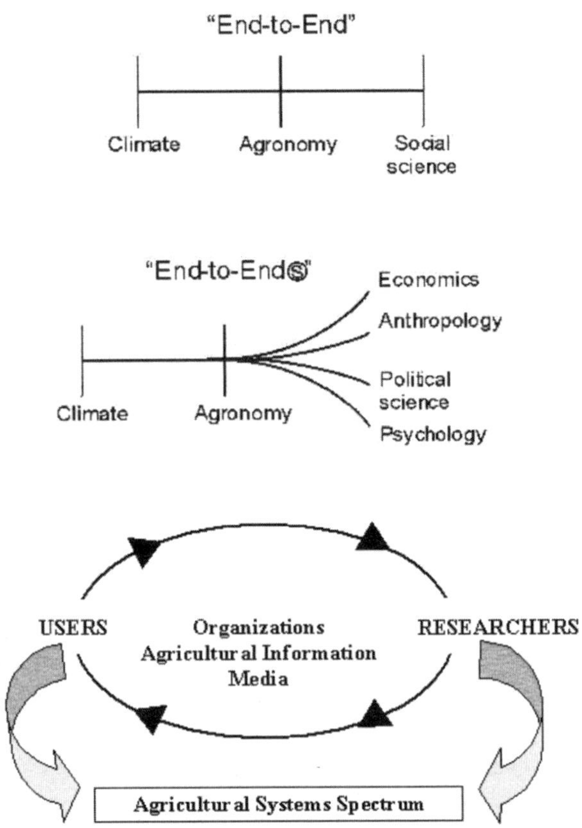

Figure 4. (a) "End-to-end" conceptual models for the study of seasonal climate forecasts and agriculture. (b) "Feedback process" conceptual model for the study of seasonal climate forecasts and agriculture.

ing physical, biophysical, and social scientists to work together effectively across disciplinary divides. Equally and possibly more important, they help to nurture effective interactions between researchers and agricultural stakeholders. An early conceptual model for enhancing the usefulness of seasonal climate forecasts is known as the "end-to-end" approach (Figure 4a). This model consists of a uni-directional linear trajectory from the physical science, through agronomy, to social science—primarily economics. This model quickly became the "end-to-multiple-ends" approach, since social science consists of many disciplines besides economics. Outcomes and insights regarding the use of seasonal climate forecasts may be very different, depending on whether the disciplines of economics, anthropology, political science, or psychology were involved. A weakness of both of these conceptual models is the absence of users in the research process.

At the recent CLIMAG Workshop in Geneva, held in September, 1999, a new conceptual model was proposed: the "feedback process" (Figure 4b). Here, users are included explicitly, and researchers from both the natural and social sciences are represented as working together. Communication between the users and researchers flows in both directions. The important role of information transfer is highlighted, through public and private organizations, agricultural information providers, and the media. Both users and researchers are located on a spectrum of agricultural systems ranging from small-holder, food-based systems, to commercial, market-based systems.

IMPEDIMENTS TO USE OF SEASONAL CLIMATE FORECASTS

Climate forecasts in and of themselves are not useful, and may even be harmful. Societal factors are critical, especially in evaluating vulnerability to climate variability. Describing the drought period in the Sahel of 1968-1973, Glantz wrote: " . . . Absence of meteorological forecasts was not the main constraint on enhancing food production or in averting food shortages in Africa, but societal factors dominate . . . " (Glantz, 1996, p. 125).

It is clearly essential to identify the farming system that one is targeting for study and delivery of climate forecast information. Current research seeks to identify if and how systems may differ in effective use of forecasts.

TRANSFER OF REGIONAL ASSESSMENT METHODS

Since ENSO climate effects on agriculture are particular to regions, researchers in places with strong impacts have been developing and testing methods for the primary assessment of regional agricultural ENSO impacts and for enhancing the effectiveness of seasonal climate predictions. Key foundation work has been done in Australia, Argentina, Uruguay, India, and elsewhere.

What has been learned from these regional experiences? Methods of analysis have been suggested as well as key interactions with regional researchers and agricultural stakeholders needed to ground assessments in whatever agricultural system is targeted (Table 4). Developing a set of guidelines for regional seasonal climate forecast and agriculture would be useful to encourage the extension of efforts beyond the regions in which work has already been started. Steps in an ENSO-agriculture regional analysis include: identifying ENSO signal on regional agricultural productivity; choosing the target agricultural system(s); assessing the temporal and spatial patterns of ENSO events; characterizing regional climate-yield relationships; developing seasonal climate prediction scenarios; testing crop management strategies; and analyzing the value of forecasts and effects on regional and national production.

At each step in the analysis, engagement with agricultural stakeholders, such as farmers, cooperatives, and extension agents is essential. Interactions with associated support industries and regional agricultural and water resources planners will also facilitate effective use of seasonal forecasts.

TOOLS, LINKAGES, AND SCALES

Linkage tools under development and testing include: climate scenarios for driving agricultural simulation models; regional Geographic Information Systems (GIS) with soil, weather, management, yield, and economic data; and crop and pasture simulation models. There are two main approaches for creating climate scenarios appropri-

Table 4. *Technical guidelines for regional ENSO-agriculture analysis.*

Step	Methods	Interactions
(1) Identify agriculture-ENSO signals and choose agricultural system(s) for analysis.	• Conduct SST-climate-yield statistical tests at national and regional levels.	Meet with researchers, extension specialists, farmers' groups, associated support industries, etc.
(2) Assess temporal and spatial patterns of ENSO effects.	• Create SST and crop calendar. • Analyze remote sensing observations (including ground truth).	Survey farmers for timing of operations.
(3) Characterize regional climate-yield relationships.	• Develop regional GIS. • Calibrate and validate crop and pasture models.	Collect soil, weather, management, and yield data from agronomic research experiments and farmers' fields.
(4) Develop seasonal climate prediction scenarios.	• Characterize major sources of regional climate variability (ENSO, NAO, etc.) • Downscale forecasts (GCMs, regional climate models, weather generators).	Interact with meteorologists and climatologists.
(5) Test crop management strategies.	• Use crop models and GIS to identify management practices that optimize the use of climate forecasts.	Survey farmers and other users to identify constraints and opportunities.
(6) Analyze forecast value and effects on regional and national production.	• Conduct economic analyses on value of forecasts and effects on supply.	Collaborate with economists.

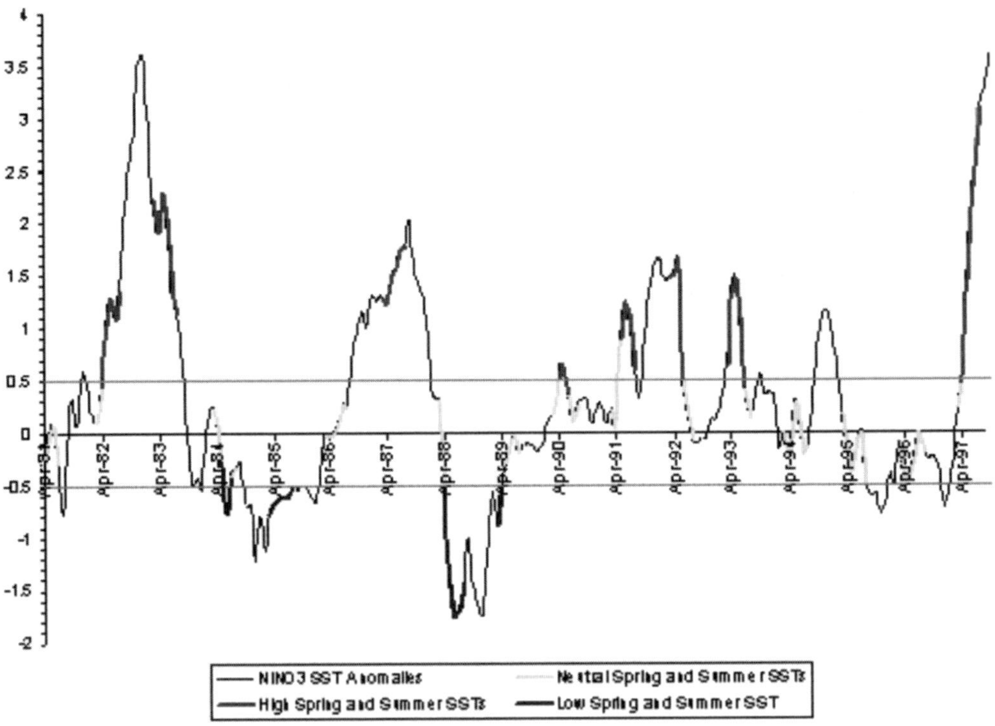

Figure 5. *NINO3 SST anomalies and U.S. Corn Belt crop calendar (Wannebo and Rosenzweig, 2000).*

ate for driving site-based crop simulation models at a daily time-step: (1) forcing a regional climate model with GCM results (e.g., Druyan et al., 1999b); and (3) training a weather generator with observed weather data and GCM anomalies (Wilks, 1999; Rajagopalan and Lall, 1999). Geographic information systems are used to link on-farm and agronomic experiment databases, economic data, climate forecasts, and remote sensing data to produce strategy optimizations and risk analyses. Crop and pasture simulations models include the DSSAT family of crop models (Ritchie; Jones et al.) and the CENTURY pasture and soil carbon model (Parton et al.)

Temporal and spatial questions include: how best to characterize ENSO events in relation to regional agricultural outcomes (Figure 5); what are the spatial patterns of ENSO agricultural effects and do they persist; and how do differing land covers and uses affect vulnerability to ENSO effects. Satellite imagery is a helpful tool for characterizing patterns and persistence of these effects on continental and regional scales and for distinguishing the role of land-cover in responses to climate variability (Figure 6) (Myneni et al., 1996; Wannebo and Rosenzweig, 2000; Archer, 2000).

COMMUNICATING PROBABILITIES

This is perhaps the most crucial question we need to answer in order to enhance the use of seasonal climate prediction for agriculture. We know that large-scale mechanisms of climate variability account for only a portion of the variance, first of regional climate, and then of crop yields. Even in Zimbabwe, where connections are very strong, ENSO accounts for only 60% of maize yield variability (Cane et al., 1994). Thus, even with perfect predictions of, in the case of ENSO, Pacific sea-surface temperatures and then perfect predictions of regional climate teleconnections, seasonal climate predictions for agriculture must be expressed in terms of shifting probabilities. How to do this effectively is a challenge. Several methods are currently employed for presenting precipitation probabilities. Probabilities of key agricultural vari-

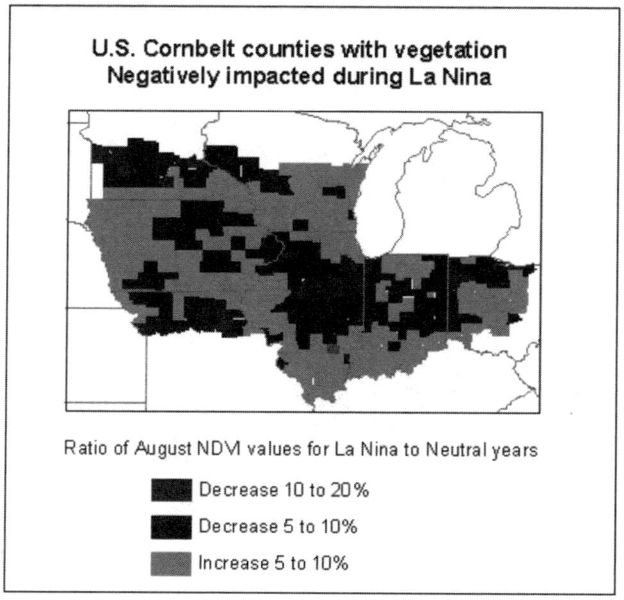

Figure 6. *U.S. Corn Belt counties with vegetation negatively impacted during La Niña (Wannebo and Rosenzweig, 2000).*

ables and outcomes, such as soil moisture, yields, and area harvested, are needed as well. Presentation methods for agricultural outcomes are currently being tested in several regions (Figure 7a and b).

GLOBAL ASSESSMENT AND FORECASTING

Research on ENSO-related impacts on agriculture has been done primarily on a country-by-country or even region-by-region basis. Although such studies are rich in regional detail, they lack a common context; results are often anecdotal. A more rigorous framework may be required, so that knowledge gained in one location (for example, what form and timing of ENSO predictions are most effective) can be compared and applied in others. Many climatological, agronomic, and economic datasets and models are already available but need to be marshaled for this purpose. A global perspective may help to bring these fragmented elements into a coordinated whole (Figure 8). A global perspective could also be embodied in the creation of a global agricultural forecasting product. Such an effort might to contribute a useful component to food security forecasting efforts.

CONCLUSION

We face many challenges as we seek to enhance the exciting prospect of bringing scientific seasonal climate forecasts to bear on agricultural systems around the world. To advance the development of agricultural applications of climate predictions requires a continuing strengthening of research capacities, interdisciplinary communication, and international cooperation.

(a)

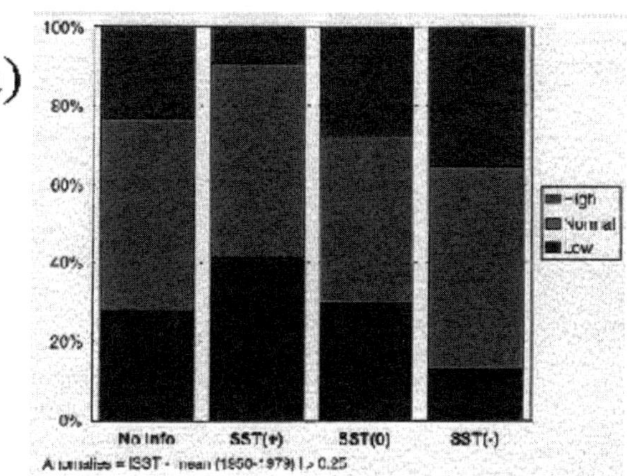

(b)

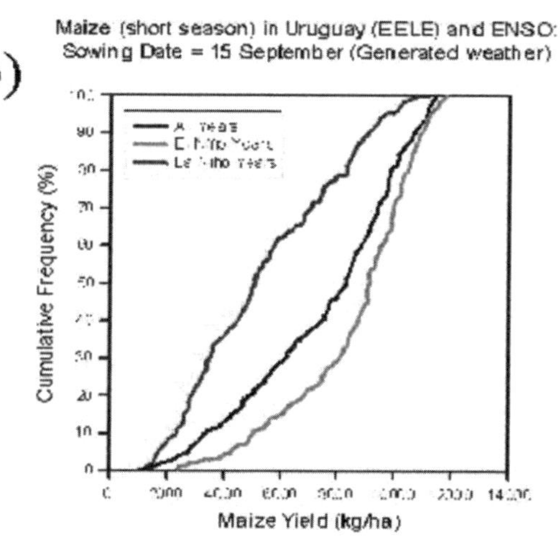

Figure 7. *(a) Maize yields and SST anomalies in the U.S. Southeast (Baethgen). (b) Cumulative frequency of short-season maize yields in all years, in El Niño years, and in La Niña years in Uruguay; planting date September 15 (Baethgen).*

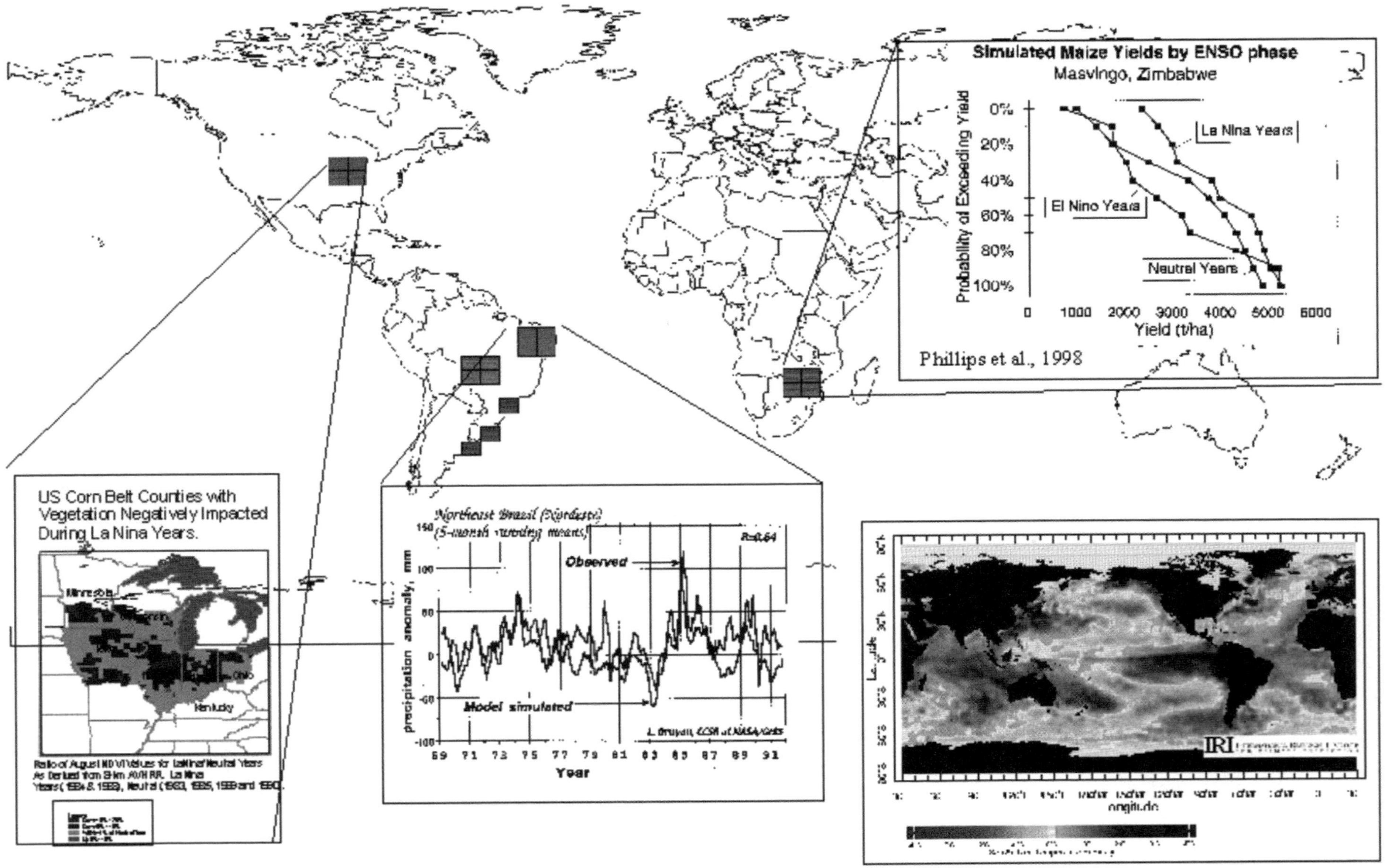

Figure 8. *Contributions to global assessment and forecasting (Rosenzweig, 2000).*

REFERENCES

Adams, R.M., C.-C. Chen, B.A. McCarl, and R. Weiher. 1998. The economic consequences of ENSO events: The 1997-98 El Niño and the 1998-99 La Niña. Faculty Paper Series 99-2. Depart of Agricultural Economics. Texas A&M University. College Station, TX.

Archer, E. 2000. Fragile Calculus: Climate, Political Economy and Vegetation Change in the Semi-arid Karoo, South Africa. Ph.D. Thesis. Graduate School of Geography. Clark University. Worcester, MA.

Cane, M.A., G. Eshel, and R.W. Buckland, 1994. Forecasting Zimbabwean maize yield using eastern equatorial Pacific sea surface temperature. Nature 370:204-205.

Druyan, L. and M. Fulakeza. 2000. Use of a Regional Climate Model (RCM) in Zimbabwe for ENSO teleconnections. (in preparation).

Glantz, M.H. 1996. Currents of Change: El Niño's impact on climate and society. Cambridge University Press. Cambridge. 194 pp.

Halpert, M.S. and C.E. Ropelewski. 1992. Surface temperature patterns associated with the Southern Oscillation. Journal of Climate 5:577-593.

Fox, G., J. Turner, and T. Gillespie. 1999. The value of precipitation forecast information in winter wheat production. Agricultural and Forest Meteorology 95:99-111.

Jones, J.W., G.Y. Tsuji, G. Hoogenboom, L.A. Hunt, P.K. Thornton, P.W. Wilkens, D.T. Imamura, W.T. Bowen, and U. Singh. 1998. Decision support system for agrotechnology transfer: DSSAT v3. In Tsuji, G.Y., G. Hoogenboom, and P.K. Thornton (eds.).

Understanding Options for Agricultural Production. Kluwer Academic Publishers. Dordrecht, The Netherlands. pp. 157-177.

Mjelde, J.W., H.S.J. Hill, and J.F. Griffiths. 1998. A review of current evidence on climate forecasts and their economic effects in agriculture. American Journal of Agricultural Economics 80(5):1089-1095.

Myneni, R.B., S.O. Los, and C.J. Tucker. 1996. Satellite-based identification of linked vegetation index and sea surface temperature anomaly areas from 1982-1990 for Africa, Australia and South America. Geophys. Res. Letters. 23:729-732.

Orlove, B.S., J.C.H. Chiang, and M.A. Cane. 2000. Forecasting Andean rainfall and crop yield from the influence of El Niño on Pleiades visibility. Nature 403:68-71.

Parton, W.J., D.S. Schimel, C.V. Cole, and D.S. Ojima. 1987. Analysis of factors controlling soil organic matter levels in Great Plains grasslands. Soil Sci. Soc. Am. J. 51:1173-1179.

Phillips, J., B. Rajagopalan, M. Cane, and C. Rosenzweig. 1999. The role of ENSO in determining climate and maize yield variability in the U.S. Cornbelt. International Journal of Climatology 19:877-888.

Rajagopalan, B. and U. Lall. 1999. A k-nearest neighbor simulator for daily precipitation and other weather variables. Water Resources Research (in press).

Ritchie, J.T. 1985. A user-oriented model of the soil water balance in wheat. In W. Day and R.K. Arkin (eds.). Wheat Growth and Modeling. Plenum Press. New York.

Ropelewski, C.E. and M.S. Halpert. 1987. Global and regional scale precipitation and temperature patterns associated with the El Niño/Southern Oscillation. Monthly Weather Review 115:1606-1626.

Rosenzweig, C. 2000. Impacts of the El Niño-Southern Oscillation on agriculture: Guidelines for regional analysis. In Rosenzweig, C. et al. (eds.). 2000. Impacts of El Niño and Climate Variability on Agriculture. Special Publication. American Society of Agronomy. Madison, WI. (in press).

Solow, A.R., R.F. Adams, K.J. Bryant, D.M. Legler, J.J. O'Brien, B.A. McCarl, W. Nayda, and R. Weiher. The value of improved ENSO prediction to U.S. agriculture. Climatic Change 39:47-60.

Wannebo, A. and C. Rosenzweig. 2000. Using remote sensing to identify agricultural areas in the US Cornbelt sensitive to the El Niño-Southern Oscillation phenomenon. International Journal of Remote Sensing (submitted).

Wilks, D. 1999. Multisite downscaling of daily precipitation with a stochastic weather generator. Climate Research 11:125-136.

Climate Prediction

Mark Cane
Columbia University, USA

INTRODUCTION

The most dramatic, most energetic and best defined pattern of interannual variability is the global set of climatic anomalies referred to as ENSO, an acronym derived from its oceanographic component, El Niño, and its atmospheric component, the Southern Oscillation. Historically, El Niño refers to a massive warming of the coastal waters off Ecuador and Peru. This warming leads to widespread mortality of fish and guano birds, crippling the local economy. The heavy rainfall associated with the event results in catastrophic flooding in coastal land areas.

The atmospheric component of ENSO, the Southern Oscillation, is a more recent discovery. The seminal figure in delineating it was Sir Gilbert Walker, the Director-General of Observatories in India. Walker assumed his post in 1904, shortly after the famine resulting from the monsoon failure in 1899 (an El Niño year); he set out to predict the monsoon fluctuations, an activity begun by his predecessors after the disastrous monsoon of 1877 (also an El Niño year). Walker was aware of work indicating swings of sea level pressure from South America to the Indian-Australian region and back over a period of several years. In the next 30 years he added correlates from all over the globe to this primary manifestation of the Southern Oscillation. For example, he found that heavy rainfall in the central equatorial Pacific, drought in India, warm winters in southwestern Canada and a cold one in the southeastern U.S. characterize periods of low Southern Oscillation Index. However, Walker did not consider El Niño, and although both El Niño and the Southern Oscillation had been known at the turn of the century, it was only in the 1960's that the close connection between the two was finally appreciated, principally through the work of Jacob Bjerknes.

Bjerknes did more than point out the empirical relation between the two; he also proposed an explanation. Bjerknes suggested a tropical coupling between El Niño and the Southern Oscillation; he also proposed that the changes in atmospheric heating associated with tropical Pacific SST anomalies cause changes in mid-latitude circulation patterns. This teleconnection idea is consistent with the global nature of Walker's Southern Oscillation. However, in his theory the causes of ENSO are rooted solely in the coupling of the atmosphere and ocean in the tropical Pacific. They are entirely *internal* to the climate system, not responses to volcanic eruptions, solar variations, biological activity, etc.

Work in the past two decades, especially that under the auspices of the international TOGA (Tropical Ocean Global Atmosphere) Program, has provided theoretical and observational support for Bjerknes' concept. An essential addition, equatorial ocean dynamics, was introduced by Klaus Wyrtki in the 1970s on the basis of data from a network of Pacific island tide gauges. The first model to successfully simulate ENSO, that of Zebiak and Cane (described in *Mon. Wea. Rev.,* 115, p 2262, 1987), was based explicitly on the Bjerknes-Wyrtki hypothesis.

This numerical ENSO model depicts in a simplified manner the evolution of the tropical Pacific Ocean and overlying atmosphere. It is a dynamical model, rather than a statistical one; that is, it relies on the governing physical equation rather than simply a sequence of observations. Such dynamical models also provide a means for physical interpretation and understanding of whatever they simulate. One of the most significant results of the model simulations was the recurrence of ENSO at irregular intervals as a result of strictly internal processes; that is, without any imposed perturbations. Analysis of the model helped in developing a now widely accepted theory that treats ENSO as an internal mode of oscillation of the coupled atmosphere-ocean system, perpetuated by a continuous imbalance between the tightly coupled surface winds and temperatures on the one hand, and the more sluggish subsurface heat reservoir on the other.

PREDICTION

Starting with experimental studies in 1985, our group at Lamont has used the Zebiak-Cane model not only to simulate, but also to predict El Niño. As noted above, theory argued for a deterministic origin of ENSO; that is, a systematic evolution throughout the cycle rather than a sequence of random events. However, even deterministic systems that are chaotic have limited predictability, and in this case the situation was made worse by a very poor observational base over vast regions of the tropical Pacific. By making predictions based on past periods we could compare forecasts directly with reality. This we did, starting with 1970. The results clearly demonstrated predictive skill at lead times longer than one year. This set the stage for the first predictions of the future, made in early 1986, which called unambiguously for an El Niño occurrence later that year. Differences in timing and other details showed that the prediction scheme was far from perfect, while the mere appearance of El Niño proved it successful. Prior to the 1997 event, the model was generally been successful in forecasting the major events (1972, 76, 82, 86 and 91) a year or more ahead. But it has little skill in predicting smaller fluctuations, some of which may influence climate elsewhere on the globe. Early 1993 was a notable failure. Moreover, some periods (e.g. the 1970s) are harder to predict than others (e.g. the 1980s).

Though the reaction to the first ENSO forecasts may have been surprise that it could be done at all, at this point the natural question to ask is why it isn't being done better. The factors limiting the current skill of forecasts are: inherent limits to predictability; flaws in the models; gaps in the observing system; flaws in the data assimilation systems used to introduce the data into the models.

There is no question that the predictability of ENSO is limited. The ENSO system exhibits "sensitivity to initial conditions" so that small inaccuracies in the estimate of the initial state of the climate system can grow into substantial errors. Our best guess, based on numerical experimentation with models such as that of Zebiak and Cane (1987) is that only in a small percentage of cases will the forecasts at a lead time less than a year be substantially impacted. At present the primary problems lie elsewhere. Model shortcomings are surely a major limitation on forecast skill. Though the observing system was certainly inadequate in the past, part of the legacy of the TOGA program is an array of highly instrumented buoys in the tropical Pacific. The experience of others (Leetmaa, private communication) confirms our own estimates that the present observing system provides adequate initial conditions. Two caveats apply. First, the 1997 experience may show us that this optimistic assessment is wrong (see below). Second, maintenance of the observing system at an adequate level is not guaranteed, so interested parties must exercise due vigilance.

However, there are important shortcomings in the way the observational data is used. In the decades of experience with numerical weather prediction, data assimilation has been developed into a sophisticated and effective technology. There is only a decade of experience with coupled model ENSO forecasting and our understanding of data assimilation for this purpose is in its infancy.

The 1997 Experience

The Lamont forecasting system utterly failed to forecast the 1997 event. Subsequent work showed that the forecast is much improved by initializing the forecasts using wind fields derived from the space borne NSCAT scatterometer rather than the FSU (Florida State University) wind fields used in the standard procedure (Chen et al 1998). Also, the forecast could be greatly improved by assimilating ocean thermal data, something not done in the original procedure. It should be added that many, perhaps most, of the models, did a credible job of predicting the 1997 event, although few, if any, of the forecasts made before the event became evident in early 1997 predicted the outsized nature of the event to come (viz. NOAA's *Experimental Long Lead Forecasting Bulletin* for late 1996 and early 1997). Future analysis of this event should identify the reasons for this, leading to improvements in the forecasting procedures.

There is certainly room for improvement, and this analysis of the limits to ENSO forecast skill is hopeful. The inherent limits to predictability and the gaps in the observing systems are, respectively, the most intractable and most expensive of the limiting factors. The best estimate is that these are not now the major limiting factors, though they will have to be faced if the other factors are reduced. The major problems at the moment are shortcomings in the simulation models and in the systems for introducing data into these models. Both should diminish in the face of concerted efforts by forecasting centers such as IRI and NCEP. The difficulty of the work is increased by the data scarcity for past events, which restricts the time period suitable for evaluating putative improvements in the model and data assimilation system. Rapid gains in data assimilation should result by borrowing techniques developed for numerical weather prediction. Fixing the model flaws appears to be more difficult, though progress should be expected.

GLOBAL IMPACTS

As a general rule, the effects of an ENSO event are strongest and most reliable in the tropical Pacific genesis region and contiguous continents. When there is a warm event one can be fairly certain of heavy rains in Peru, drought in Australia and fewer typhoons there. Typical consequences are somewhat less reliable in the global tropics, but still highly likely. Thus, there is frequent concurrence of ENSO warm events and a poor monsoon in India, poor maize yield in Zimbabwe, drought in the Nordeste of Brazil, and reduced numbers of hurricanes in the Atlantic.

There is no question that ENSO has an influence in extra tropical latitudes, but the response is less certain than in the tropics. Other factors may intervene, and, characteristically, the extra tropical climate system is more chaotic—less predictable. Thus in these latitudes an ENSO event should be thought of as putting a bias in the system rather than as a certain cause. With warm (El Niño) event heavy rains in the Great Basin region of the U.S. are more likely and with a cold event drought (e.g. as in 1988) and lower corn yields are more likely in the Midwest. Certain patterns are more likely to persist, altering the paths of hurricanes, typhoons, and winter storms. However, there is some evidence that long relied on patterns may be changing. For example, the Indian monsoon has not failed in the past 15 years, despite many strong El Niños (cf. Kumar et al.,1999 Science, 284, 2,156-2,159). The1997 monsoon was quite normal despite the strong El Niño.

FUTURE PROSPECTS

It is unquestionably a great advance to have demonstrated the possibility of long range climate prediction. The forecasts are certainly far from perfect, especially so for the connections to local conditions with the greatest human consequences. We know that the nature of the climate system does not allow unlimited predictability, so that even a perfect forecasting system would not be able to deliver precise forecasts. Thus to be correct a forecast must be phrased as a probabilistic statement. Some aspects of the climate are more predictable than others. For example, given an ENSO event for some climate features the probability distribution will differ greatly from the climatological expectation, while for others it will not. For the former knowing that there is an ENSO event adds information. The forecast should be useful, all the more if the probability conditioned on ENSO is sharply peaked so that a rather specific outcome may be expected. Our current knowledge of the predictability of different features of the global climate is quite incomplete, but research over the next few years will add to it rapidly.

Farming Strategies for a Variable Climate—An Indian Case Study

Sulochana Gadgil
Indian Institute of Science, India

INTRODUCTION

Variation of rainfall on interannual and intra-seasonal time-scales has a large impact on agricultural production of tropical regions such as India, because a large fraction of the cultivated area is rainfed. While the production of irrigated regions increased rapidly during the green revolution, relatively little progress has been achieved in the rainfed regions. The phenomenal growth of production during the green revolution was achieved by the introduction of new fertilizer responsive, dwarf varieties. The search for appropriate varieties for rainfed regions i.e. varieties which are tailored to the climate variability, has not been as successful. It has been suggested that for the complex rainfed ecologies, greater enhancement of production will result from tailoring of management strategies to the climate variability rather than from varietal changes (Siddiq 1999).

Even the strategies suggested by the research so far, are generally not adopted by the farmers in the rainfed regions and the yield gap between the agricultural research stations and the farmers' fields is very large (e.g. Figure 1). Swaminathan (1998) suggests that the major reasons are "The research farms' programmes have mostly been scientist oriented and not farmer or user centered. These were perceived, planned, implemented, supervised and evaluated by scientists. The transfer of results followed a top down approach. In this "take it or leave it approach", the farmer was at best a passive participant. Scientific findings which became the so-called 'technologies' were born from small plots and short-term research and were invariably not associated with critical cost-benefit studies."

It is seen from Figure 1 that for years with low rainfall, the yield gap is small. In other words, the farmers appear to be doing almost as well as possible in years with deficit rainfall. However, for years with high seasonal rainfall, the yields of the farmers' fields are much lower than what could be achieved with the available know-how, and the yield gap is very large. One of the major differences between the management practices at agricultural stations and that on the farmers fields is in the degree to which fertilizers and pesticides are applied. It is important to note that over rainfed regions, the benefit of measures such as application of fertilizers, in terms of the yield enhancement, are not the same every year. In years with reasonable rainfall in which good yields are expected, such measures lead to a large enhancement in the yields. In years with poor rainfall, the yields are low irrespective of such applications. Hence in

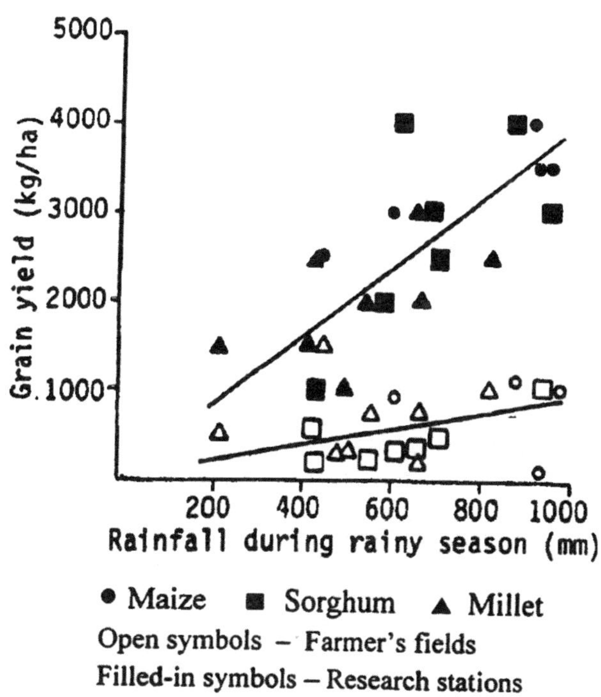

Figure 1. Relationship between rainfall during the rainy season and yield of maize, sorghum, and millet at 15 dryland locations in India (after Sivakumar et al., 1983).

such years the farmers' yield is comparable to that at the field stations.

Since the yields over rainfed regions tend to be low and fluctuate considerably from year to year, the farmers have less resources available than those in the irrigated regions with assured yields. The farm-level decision of whether to invest in the recommended measures, depends upon the ratio of the associated benefit to the additional cost. When the yield (and hence the benefit associated with the recommended measures) fluctuates from year to year, the decision depends upon the expected benefit over a period of few years. Given the low levels of resource availability, in the absence of estimates of the expected benefit in variable climate, the farmers tend to minimize additional investments. On the other hand, adequate investment at the agricultural stations leads to enhancement of yields in good rainfall years. Hence the yield gap is very large in years of good rainfall. It should be noted that some of these farmers do make such investments over irrigated parts of their lands where the yields are assured (Rao and Gadgil 1999). So it is not the lack of know-how, but rather the lack of estimates of the expected benefit in variable climate, that results in the large yield gap in years with good rainfall.

Several other management strategies, such as water/soil conservation measures or farm ponds for supplementary irrigation, have been recommended for different regions in the rainfed belt. However, the farmers tend not to make additional investments for such measures because in their estimate, the expected benefit is not commensurate with the cost. The relative lack of success in promoting strategies derived at agricultural stations in rainfed areas has led to an increasing realization of the need for identifying farmer-acceptable strategies (Katyal 1998).

It is clear that for a strategy to become acceptable to farmers, it is necessary to derive estimates of the expected benefit in the face of rainfall variability of the type experienced over the specific region. For this, the variation of the yield in response to rainfall variability under adoption of different management practices has to be investigated. For understanding the impact of climate variability on yield, the results of short term experiments at agricultural stations have to be complemented with analysis of the nature of rainfall variability from the available meteorological data and investigation of the impact of such variability on the yield simulated by crop models. This in turn requires the development of crop models which can realistically simulate the observed variations in yield in response to rainfall variability.

If for a specific year, meteorological predictions either on the seasonal scale (i.e. of the seasonal rainfall and its distribution) or on the sub seasonal scale (wet/dry spells in the next few days) are available, strategies appropriate for such predictions can also be derived. However, the skill of such predictions i.e. the probability of correct prediction, needs to be higher than the ratio of the additional cost to the expected benefit associated with the strategy for it to result in achieving the goal (Katz and Murphy 1987, Gadgil et. al. 1995).

It is important to ensure that the optimum strategies are sought from amongst the options available to the farmers. Furthermore, the constraints of the farmers due to aspects other than production, such as availability of markets, facilities and cost of transportation etc. have to be taken into account. Hence, the approach for seeking appropriate, farmer-acceptable crops/varieties and management practices for the complex heterogeneous rainfed ecosystems, has to be farmer-centered and radically different from the 'lab to land' approach hitherto adopted successfully for enhancing the production in the irrigated regions.

Here I present an approach for identifying appropriate strategies for rainfed regions based primarily on information and prediction of climate variability. The methodology is general but applied to the specific case of production of rainfed groundnut (peanut) in a semi-arid part of the Indian peninsula-the Anantapur region (Figure 2). In the next section, the farmers' perspective is elucidated. The essential tools/knowledge are briefly discussed in Section 3. Use of knowledge and/or predictions of interannual and intraseasonal variation in farm-level decision making is discussed in below.

FARMERS' PERSPECTIVE AND FARMING STRATEGIES

A distinguishing attribute of agricultural production in rainfed regions is the large variation in yield from year to year in response to the variation in rainfall (e.g. Figure 3 for Anantapur district). In

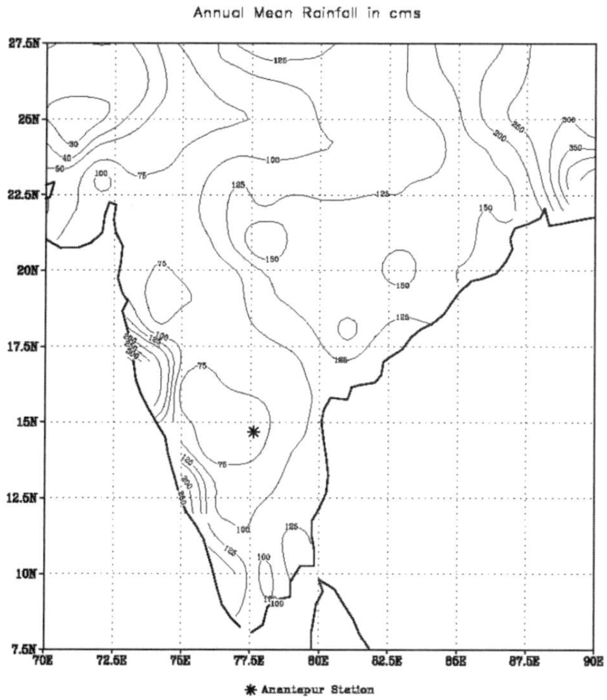

Figure 2. *Annual mean rainfall in India (cms); * Anantapur Station.*

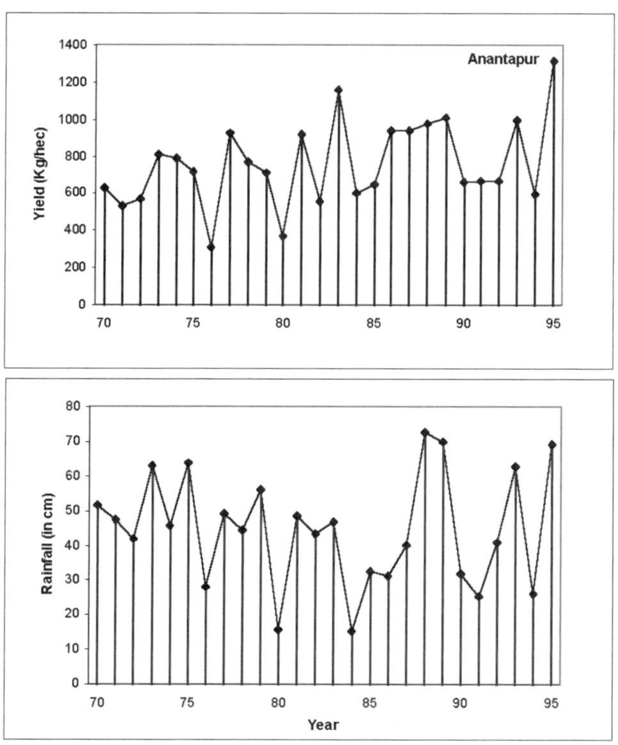

Figure 3. *(a) District yield, and (b) seasonal rainfall at Anantapur Station.*

some years the production is so low (i.e. less than 500 kg/ha) that even the cost of cultivation is not met. Such crop failures have a very large impact on marginal farmers (with landholding less than 2 hectares). However, the farmers with large landholding can tolerate a few such years, provided that they can make adequate profits in other years. Generally, the goal for the large farmers is maximizing the profit over periods of a few years, while for the marginal farmers minimizing the risk of crop failure is as important (Rao and Gadgil 1999). The constraints faced by the farmers also differ for the different categories. The available options and hence the appropriate strategies for attaining the specific goal necessarily depend on the socio-economic factors. A decision support system must, therefore, encompass all the possible scenarios and goals. Farmers' input and participation is central to the development of such a system.

We set up a farmers network in the Anantapur area to elucidate farmers' perspective on important facets of cultivation of rainfed groundnut. For validation of the existing models, data were collected during the kharif season (July-November) of 1998 on the important meteorological variables, soil moisture, plant growth, development, yield and incidence of pests/diseases/weeds. Detailed information on the farming situation, farming practices and farm-level decision making was obtained through discussions with the farmers. In particular, information was collected on the basis for decision about the cropping pattern, the management options available to the farmer, the farmers' assessment of the associated costs, benefit and the nature of meteorological information/prediction required for the decisions at each stage (Rao and Gadgil 1999).

Some of the farm-level decisions along with the farmers' estimates of the additional cost, what the benefit depends on and the range of benefit are given in Table 1 as a percentage of the basic cost of cultivation. It is seen that the additional cost is higher than the minimum benefit, except for the decision about seed treatment and hence in the absence of estimates of expected benefit it is not considered worthwhile to incur the additional costs. Note that the benefit of application of soil amendments depends upon the yield levels in four successive years. In fact, the benefit (in terms of yield enhancement) associated with many of the farm-level decisions depends upon the expected yield in one or a series of few years and hence the input required for making these decisions is the expected yield.

Some of the decisions such as timing of harvest, depend upon prediction of dry/wet spells (Table 1). There are some decisions, such as the choice of the sowing opportunity, which do not involve additional costs. However, a wrong decision can lead to a loss, as when sowing is postponed to be nearer the recommended time-slot and a sowing opportunity does not occur later in that season.

Our discussions with farmers revealed that there were some important farm-level decisions, which

Table 1. *Farm level decision making*

Soil & Water Management				
Decision about investing in	Timing of the decision	Benefit depends on	Benefit* range	Cost* per year
Apply soil amendments e.g. farm yard manure 12 t/ha once in 4 years	Before land preparation	Expected yield in next four years; benefit increases with yield	10-45%	20%
Increase of fertilizer application to recommended dosage N.P.K 25:50:25	Two days prior to sowing	Expected yield; benefit increases with yield	10-40%	17%
	Plant protection measures			
Seed treatment fungicide at 5g/kg	Before sowing	Seed rot incidence; benefit increases with severity of incidence	10-20%	2%
Late leaf spot Spray fungicide	7 days after incidence	i) severity of incidence ii) expected yield; benefit increases with both	10-50%	10-19%
	Farming operations			
Decision about whether to	Timing of the decision	Benefit depends on	Benefit* range	Penalty for wrong decision
Harvest immediately (Soil dry, no wet spell predicted) or (ii) Postpone harvest (soil dry, wet spell predicted)	90-100 days	Occurrence of wet spell in the following 3-10 days	0 10%	10% 25%
Harvest immediately (soil wet, no dry spell predicted) or (ii) Postpone harvest (soil wet, dry spell predicted)		Occurrence of dry spell in the following 3-10 days	0 10%	10% 25%

depend on the nature of rainfall variability of the region, regarding which they wanted our advice. One of these is the choice of the sowing window i.e. period within which sowing is done when the soil moisture becomes adequate. The results of our investigation of this problem (Section 4), are now being tested in field trials by the farmers. This suggests that the results of a farmer-centered approach in which farmers contribute right from the stage of posing the problems, will yield strategies that become readily acceptable to the farmers.

ENHANCEMENT OF RAINFED PRODUCTION—THE AVAILABLE KNOWLEDGE AND TOOLS

It must be noted that farmers have considerable knowledge of the complex rainfed ecosystems because rainfed agriculture has been carried out over such semi-arid regions, for a very long time. During this period, the farmers have been making decisions regarding the choice of cropping patterns and various management options such as timing of sowing, application of fertilizers etc. In fact some of the crops grown today such as Sorghum and Pigeonpea have been cultivated for over two thousand years in our country (Achayya 1994). Even the relatively recently introduced crops like groundnut (peanut) have been cultivated for over hundred years. Hence we expect the farmers of the rainfed regions to have considerable knowledge of the rainfall variability of their region as well as the impact of the rainfall variability on the different stages of such traditional crops. The traditional cropping patterns and farming practices probably evolved by trial and error to be appropriate to the climate variability of the region.

However, the present cropping pattern over a large part of the rainfed belt is markedly different from the traditional pattern. Thus, in large areas of the Anantapur region, the complex multi-cropping traditional system, is replaced largely by groundnut (variety TMV-2, which is not the traditional one) intercropped with one of the traditional crops like pigeonpea (Rao and Gadgil 1999). Hence, while there is considerable traditional knowledge of rainfall variability, the detailed impact on the specific crops/varieties grown now, is not known since the experience of one or two decades is not adequate. It is therefore necessary to combine traditional knowledge with modern scientific tools to gain an insight into the links between rainfall variability and production of crops/varieties cultivated at present.

Rainfall Variability and Prediction for Groundnut Farmers

The traditional knowledge of rainfall variability of farmers all over the country, is in time-units, not weeks or months, but so called 'nakshatras' which are 13 or 14 day periods, also based on the solar calendar. The appropriate time for farming operations is worked out in terms of these time periods. Hence the knowledge of rainfall variability derived from analysis of quantitative data collected over a century can be discussed with the farmers most effectively if worked out for each nakshatra rather than month or week (e.g. Figure 4). We found that using this 'language that connects' made a major difference in our interaction with the farmers.

It is seen that although the rainy season extends from May to November, rainfall of 1 cm or more occurs with over 75% probability only during the nakshtras Uttara (13–26 September) and Hasta (27 September–9 October). In fact there is a local proverb which says that if Uttara rains fail, the time has come to move off from the region. The probability of no rain at all in a nakshatra decreases from over 50% in the first part of the rainy season to be the minimum at 20% in Uttara and picks up to above 50% from the last week of October. The farming strategies have to be tailored to this kind of variability.

Modeling Impact of Climate Variability

One of the most important tools is a model which can simulate the impact of climate variability on yield realistically. Such a model has to perform reasonably well in simulating the yield under different farming situations (e.g. different soils), for the different management options available (such as to spray or not to spray pesticides) and for the different varieties that are available.

The PNUTGRO model for the growth, development and yield of groundnut developed by Boote et al (1986) has been validated for the prevalent variety TMV-2 at the Anantapur agricultural station by Singh et al (1994). They showed that the year to

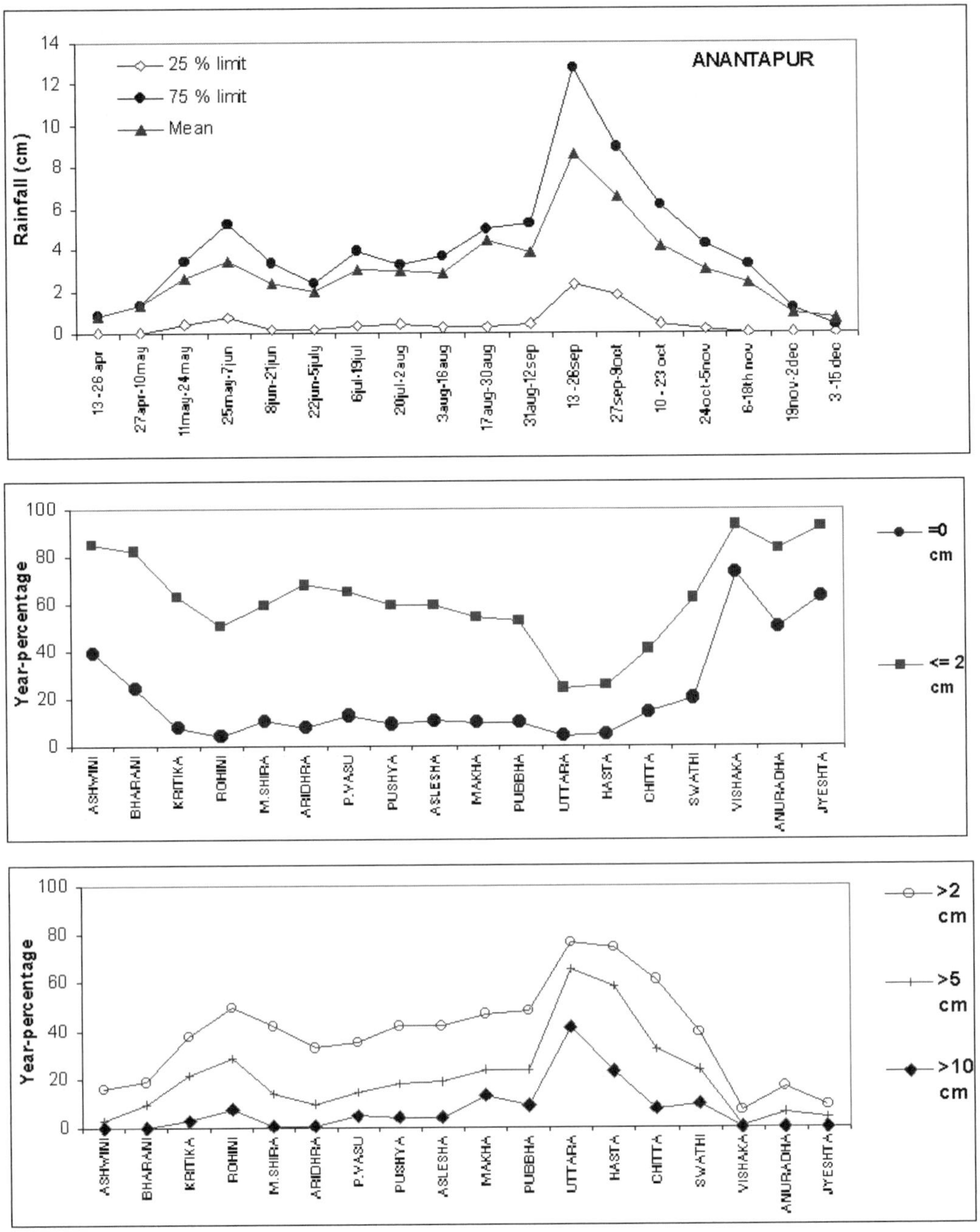

Figure 4. (a) Seasonal rainfall distribution by 'nakshatras', or 13 to 14 day periods recognized by farmers in Anantapur; and (b and c) Percentage of years with rainfall above various thresholds at 18 sites in the region.

year variation of the observed yield (in response to the variation in rainfall) is well captured in the simulation. Thus under pest-disease free conditions, akin to those at the agricultural station, the PNUTGRO model can be used to simulate the yield and its interannual variation in response to that of the rainfall. However, generally there is a large gap between the PNUTGRO yield and the district-average yield. One of the reasons is that PNUTGRO does not incorporate losses due to pests and diseases which can be quite large on the farmers fields.

Climate variability can also have an impact on the productivity via triggering of pests and diseases. A heuristic model, which simulates triggering of pests/diseases by wet/dry, spells and also the timing of the farming operations such as ploughing, sowing etc has been developed on the basis of the farming practices of the region (Gadgil et al 1999). This model as well as PNUTGRO have been validated with the data collected from the farmers network (Rao and Gadgil 1999).

We used the heuristic model for indirect impact in conjunction with the PNUTGRO model for the available data at the Anantapur agricultural station to simulate the variation in yield during 1970-1990. The yield thus simulated is found to be close to the observed district yield (Gadgil et. al. 1999).

CLIMATE VARIABILITY AND PREDICTION FOR FARM LEVEL DECISIONS

Information on climate variability of the region can be used to derive the probability of yield/benefit associated with each management option in variable climate. This can lead to an identification of strategies which can maximize benefit over a period of few years. Identification of the optimum seed rate (i.e. plant population density) or the optimum sowing window (period during which seeds are sown when the soil moisture is adequate) are examples of strategies which are tailored to the climate variability of the region. If, in addition, prediction of seasonal rainfall and its distribution within a season is available for a specific year, strategies which can maximize benefit for that specific rainfall pattern can be identified. However, the kind of prediction required for the specific season is different for different decisions. For example, the decision whether to apply a chemical fertilizer, depends upon the expected yield. On the other hand, the choice of the sowing opportunity depends upon a prediction of the detailed rainfall pattern including the timing of wet and dry spells several weeks ahead.

Data on observed yield are available for the prevalent variety of groundnut, only for about twenty-five years. It is therefore difficult to generate reliable estimates of the probabilities of high/low yields with these data. The data on observed variation of yield with sowing date are even more meager and not adequate for determining optimum sowing date for each of the types of rainfall patterns that occur over the region. However, daily data on rainfall which is the most important meteorological variable (Rao et. al. 2000) are available for a long period (1911-98). For each of these years, the sowing date as per the existing farming practices can be determined by the heuristic model and the yield simulated by the PNUTGRO model. Thus the variation of simulated yield can be analyzed for this long period and probabilities (a priori as well as conditional) of low, high yield estimated. Also, the impact of the variation of sowing date on yield can be readily analyzed with the PNUTGRO model. However, in applying the results of analysis of simulated yield to the farm-level/district yield, the fact PNUTGRO yield generally higher, has to be incorporated.

Optimum Sowing Window

One of the problems the farmers wanted us to investigate is identification of the optimum sowing window for this region. Generally, in any season there are several opportunities, when the soil moisture is adequate for sowing. At present, the farmers sow at the first opportunity in the sowing window (22 June-17 August), which comprises the latter part of the recommended sowing window (UAS 1994) of May to July and 17 days of August (during which sowing is not recommended). The farmers do not sow in the earlier part of recommended window. Furthermore, if no opportunity arises till the end of July, they do sow in August, despite the recommendation to the contrary. This is because in their experience there have been years with good yield for such late sowing. Hence they sought our advice on the optimum sowing window for this agroclimatic regime.

Analysis of the variation of the PNUTGRO yield with sowing date for 1911-1998 (Rao et. al. 1999) showed that the probability of crop failure is very high for sowing dates before mid-June (Figure 5). This is consistent with the choice of the commencement of the sowing window used by the farmers. An unexpected result is that the yield is enhanced considerably if the sowing is postponed to after mid-July. An analysis of the moisture stress in the model showed that the most critical period is the pod filling stage, which is 60-80 days after

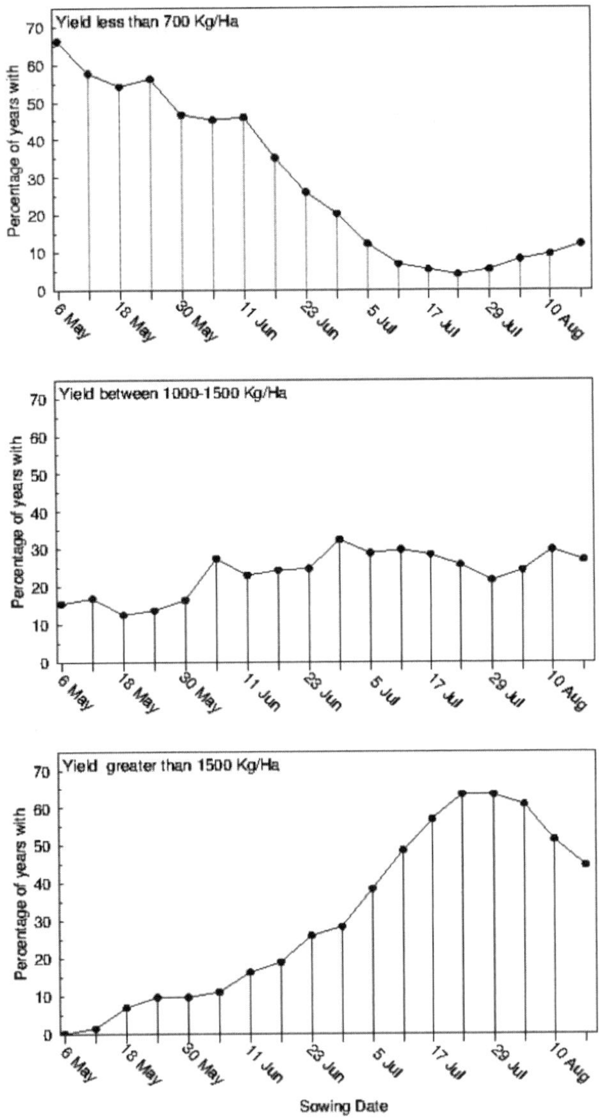

Figure 5. Simulated peanut yield as a function of sowing date.

sowing. A wet spell has a high probability of occurrence at this stage if coincides with Uttara (13-26 September) and Hasta (27 September—9 October) i.e. only when sowing is done after mid-July. Hence, such a sowing window is optimal for the rainfall variability of this region.

In this region of high rainfall variability the farmers are reluctant to postpone sowing, because they fear that another sowing opportunity may not occur. Using the heuristic model, we showed that giving up a sowing opportunity before mid-July does not involve much risk. This is because for every year during 1911-98 for which such an opportunity occurred before mid July, another occurred after mid-July. The response of the farmers to these results was very positive. In the kharif season of 1999 several farmers have conducted field trials with late sowing to test this finding.

Prediction of Seasonal Rainfall, Yield

We find that the observed district yield as well as the yield simulated by PNUTGRO model are well correlated with the rainfall during July-December (Figure 6). The yield depends upon the seasonal total as well as the distribution of rainfall within the season. Hence, for a given value of seasonal rainfall, there is considerable variation in the yield. There appears to be a threshold of 50 centimeters for seasonal rainfall. For seasons with less rainfall, the yield ranges from zero to well above 2 tons/ha. In such seasons, we expect the range of the district yield to be also large from below 500 kg/ha i.e. crop failure to good yields. However, if the seasonal rainfall exceeds 50 centimeteres, the lower limit of PNUTGRO yield increases to 700 kg/ha.

It turns out that the probability of the seasonal rainfall being less than 50 centimeteres increases significantly in El Niño years (Figure 7). Out of the 23 El Niño years during 1911-98, 21 were charac-

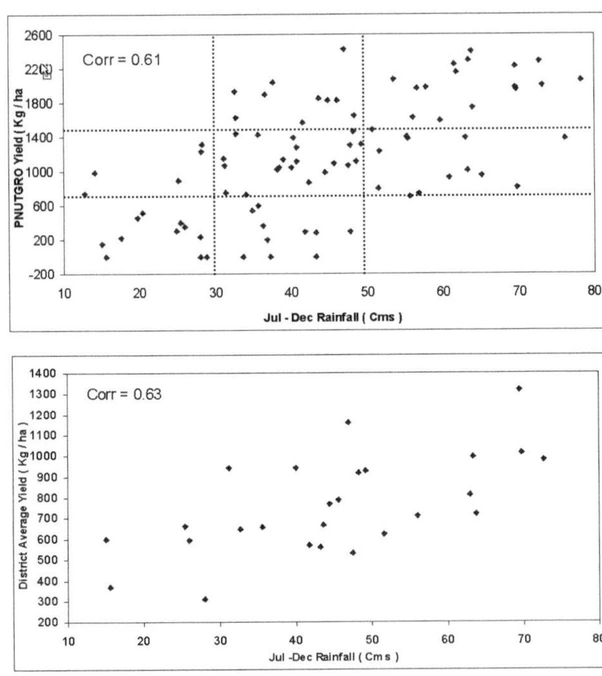

Figure 6. Relationship between July to December rainfall and (a) simulated peanut yield, and (b) observed district-level peanut yield.

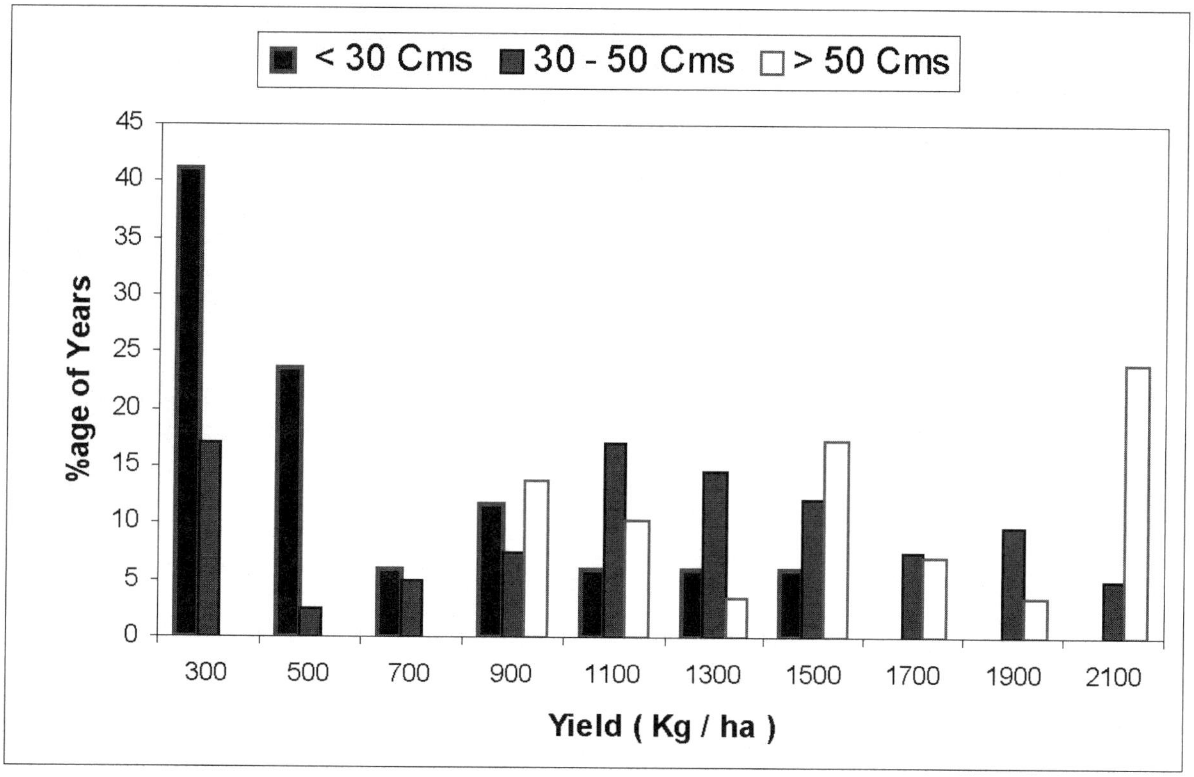

Figure 7. *Deviation of rainfall from (a) 50 cm for July to December at Anantapur, (b) annual mean for June to September for all-India, (c) rainfall distribution and (d) simulated peanut yield as a function of ENSO years.*

terized with seasonal rainfall less than the threshold implying an increase in probability from 0.67 (for all years) to 0.92 (for El Niño years). In fact the relationship of the seasonal rainfall at Anantapur is stronger than that of the all-India summer monsoon rainfall to El Niño (Figure 7). The probability of low PNUTGRO yield (<700 kg/ha) is considerably higher and of good yield (>1500 kg/ha) considerably smaller for El Niño years (Figure 7). On the other hand, in the season after an El Niño, the probability of rainfall greater than 50 centimeteres increases significantly i.e. to 0.62 in comparison to 0.33 for all years (Figure 7). The probability of low yield in such years is much lower (0.05 in comparison to 0.18) and good yield is also higher (0.48 in comparison to 0.33). This suggests that investments in fertilizers, pesticides etc. will be highly worthwhile in El Niño + 1 years, whereas during an El Niño, additional investments should be minimized. It may be worthwhile to explore alternative cropping patterns for El Niño years.

The interannual variation of seasonal rainfall as well as PNUTGRO yield contain a strong biennial component. This can lead to useful conditional predictions, particularly for the year following one with good yield. The probability of a low yield in a year succeeding a season with good yields is 0.5 which is twice that of the a priori probability of low yields of 0.25 (Table 2). The farmers tend to invest more in the season following a good harvest because of the high level of resource availability. Given the increased chance of low yield in such seasons, clearly such investments should only be made for measures such soil amendments which have an impact for 3-4 years and not in measures such as application of fertilizers which can enhance yield in the following year only if the yield level is reasonable. In fact, many marginal farmers used the profit made in the good season of 1998 to lease land in the season of 1999. The season of 1999 turned out to be a poor one with deficit rainfall and very low yields and hence the farmers did not get any return on the additional investment. Had the farmers invested in acquiring more sheep/goats instead, they would have been better off.

Thus it appears that even the patterns readily derived from analysis of the impact of rainfall variability so far, can lead to decisions which imply

Table 2. *Probability of different levels of PNUTGRO and district average yields.*

PNUTGRO yield	Low <700 kg/ha	Medium	High >1500 kg/ha
a priori probability	0.25	0.45	0.3
Conditional probability p (yield/L)	0.18	0.5	0.32
p (yield/H)	0.5	0.27	0.23
District yield	Low <600 kg/ha	Medium	High >900 kg/ha
a priori probability	0.23	0.42	0.32

enhanced yields or minimum losses. Further investigations such as identification of the optimum cropping for this region, in El Niño, El Niño + 1 and other years are required for substantive gains.

ACKNOWLEDGEMENTS

It is a pleasure to thank my collaborators, P. Seshagiri Rao, Narahari Rao and the enthusiastic farmers of the region. I am grateful for financial support from START and the Ministry of Environment and Forests, Government of India.

Stimulating discussions with colleagues at the Center for Atmospheric and Oceanic Sciences, particularly Professors R. Narasimha, J. Srinivasan at the Center for Ecological Sciences, Professors Madhav Gadgil and N. V. Joshi on all facets of climate variability and agriculture contributed immensely to this work. Our work in this area over the last five years has been actively encouraged by several scientists involved in WCRP, START and IGBP. I would like to thank, particularly Drs. R Fuchs, W. L. Gates, H. Grassl, M. Manton, P. Tyson and H. Virji for criticism, help and advice.

REFERENCES

Achayya, K. T. 1994: Indian food: a historical companion. Oxford University Press, Delhi, pages 322.

Boote, K. J., J. W. Jones., J. W. Mishoe, and G. G. Wilkerson, 1986: Modeling growth and yield of groundnut. In: Agrometeorology of Groundnut. Proceedings of an International Symposium, ICRISAT Sahelian Centre, Niamey, Niger, 21-26 August 1985. International Crops Research Institute for the Semi-Arid Tropics, Patancheru, India, pp. 243-254.

Katyal, J. C., 1998: Soil, Water and Environmental Sciences, paper presented at Brainstorming Session on 'Scientists Perception for Agriculture—2020' held at NBPGR Auditorium, 4-5 June 1998, New Delhi, India.

Katz, C. W and A. H. Murphy, 1987: Am. Statist., 40, 187-189.

Gadgil, Sulochana, P. R. Sheshagiri Rao, N. V. Joshi, and S. Sridhar, 1995: Forecasting rain for groundnut farmers—How good is good enough? *Current Science,* 68 (3), 301-309.

Gadgil Sulochana, P. R. Seshagiri Rao and S. Sridhar., 1999 : Modeling impact of climate variability on rainfed groundnut, *Current Science,* Vol. 76, No. 4, p557-569.

Rao P. R. Sheshagiri, and Sulochana Gadgil., 1999: Groundnut cultivation in Variable Climate—Farmers' Perspective, CAOS report 99AS 7.

Rao, K. Narahari., Sulochana Gadgil, P. R. Sheshagiri Rao, and K. Savithri., 2000: Tailoring strategies to rainfall variability—I : the choice of the sowing window (accepted for publication in Current Science).

Siddiq, E. A., 1999: Rice in a variable climate, pp 107-123, Eds, Y P Abrol, Sulochana Gadgil, APC Publications, Delhi.

Singh, P., K. J. Boote., A. Yogeswara Rao., M. R. Iruthayaraj., A. M. Sheikh., S. S. Hundal., R. S. Narang and Phool Singh, 1994: Evaluation of the groundnut model PNUTGRO for crop response to water availability, sowing dates, and seasons. *Field Crops Research,* 39, 147 –162.

Swaminathan, M. S., 1998: Climate and Sustainable food security, Vayu Mandal, Bulletin of the India Meteorological Society, Vol. 28, No. 1-2, p 3-10.

UAS 1994: Package of practices for high yield for central dry zone, publication of University of Agricultural Sciences, G.K.V.K, Bangalore.

Applying Climate Forecasts in the Agricultural Sector: The Experience of South East South America

***Walter Baethgen*[1] *and Graciela O. Magrin*[2]**
[1]International Fertilizer Development Center, Uruguay
[2]INTA, Argentina

INTRODUCTION

Climate variability measured in seasonal and interannual scales is a key factor affecting human development. Societies, cultures and economies in the world's history have successfully developed by mastering their abilities to adapt to climatic conditions. Consequently, temperature or precipitation anomalies have often resulted in catastrophic consequences, particularly if such anomalies were not predicted. Although often public or media attention focused on extreme events, climate variability can also have subtler consequences, but with significant economic impacts.

Countries are usually unprepared to cope with climatic anomalies in an effective way. Governments typically react to a climatic extreme event through "crisis management" rather than through the formulation and implementation of anticipatory measures commonly referred to as "risk management". A typical reason mentioned by decision makers for the lack of such risk management policies, has been the lack of means to predict climate conditions (e.g., precipitation, temperature) with sufficient skill and lead time. In selected regions of the world, this situation has changed dramatically due to recent advances in the capacity to predict climate anomalies linked to the onset and intensity of a warm or cold event as part of the El Niño/ Southern Oscillation (ENSO) phenomenon. ENSO is the main source of interannual climate variability in many parts of the world (Baethgen, 1999).

Although the advances in ENSO prediction are a significant scientific achievement, there are still limitations in our ability to establish effective risk management policies or, in general, to take advantage of the improved forecasting capabilities. Since the early 1990's several International and National organizations started efforts to establish research projects in Latin America oriented to develop effective applications of climate forecasts in the agricultural sector of the region.

The Inter-American Institute for Global Change Research (IAI) has been playing a leading role in these activities by both, facilitating the communication between researchers and by funding multinational research projects. Other organizations such as NASA, NOAA, IDB, etc., have also started funding research activities in this subject in Latin America. More recently, National governments have established research activities oriented to develop effective applications of climate forecasts in different production sectors, including agriculture.

In this article we describe the experience of South East South America (SESA) as an example of applications of climate forecasts in the agricultural sector that are currently being developed in Latin America, and to propose research needs for the near future.

THE SOUTH EAST SOUTH AMERICA REGION (SESA)

The Southeastern South America (SESA) region is one of the largest food producers (cereals, soybeans and livestock) of Latin America. The economy of the region is largely based, directly or indirectly, on agriculture (crops and livestock). Agricultural production in SESA is largely based on the highly fertile soils of the Pampas, an ecosystem in which vast areas of native temperate and subtropical grasslands have been converted to croplands. Interannual and interseasonal climatic fluctuations in the region result in high variability in crop production with negative consequences on food supply and economy at the regional and national level. The uncertainty related with weather variability is a continuous challenge for the farmers in the SESA region. For example, mean annual grain production in the Pampas (1988-

1998) was, on average, 37 million tons (Mt) but it varied from 23 Mt in the worst years to 59 Mt in the best ones. Climatic fluctuations also result in similar variability in the animal-based production (beef, wool and milk).

APPLICATIONS OF CLIMATE FORECASTS IN THE AGRICULTURAL SECTOR OF SESA: ONGOING AND PROJECTED RESEARCH ACTIVITIES

In December 1997 Uruguay hosted the first Regional Climate Outlook Forum for Southeast South America. This first forum was organized by Uruguayan institutions including the Rural Association (ARU), National Agricultural Research Center (INIA), with support from International organizations such as IAI, the Federation of Rural Associations of Mercosur (FARM), the Office of Global Programs (OGP-NOAA) and the International Research Institute for Climate Prediction (IRI). The forum was attended by several climate scientists, agricultural researchers, hydrologists, specialists in natural disasters and representatives from the communities of users (agriculture, disasters, hydrology).

This forum can be viewed as a cornerstone for the research activities in both, climate forecasts and their applications in agriculture in SESA. For the first time climate scientists from Argentina, Brazil, Paraguay and Uruguay, as well as from International organizations (NOAA, IRI) met to discuss methodologies and results of their research in climate forecasting, and produced a consensual regional climate outlook for the following three months. Also, researchers working in agricultural sciences and hydrology discussed possible applications of climate forecasts and defined collaborative research projects that needed to be established in the region to improve applications of climate forecasts. Finally, scientists met with representatives of the community of users and explained the importance of climate outlooks and the potential scope of their applications in different sectors.

The strong 1997/98 El Niño had been stimulating vast media coverage around the world, and therefore this forum was also attended by several members of the regional and International press. Since that first forum of December 1997, organizations from SESA have continued to organize regional climate outlook meetings, and producing consensual climate forecasts every three months.

Although many projects are currently being conducted throughout Latin America, in this article we will describe some of the research activities that are being conducted in Argentina and Uruguay on applications of climate forecasts for the agricultural sector.

Researchers in these two countries have been using similar research approaches to apply climate forecasts in agriculture. In both countries climate scientists are developing forecasts at the National scale mainly based on the impacts of El Niño/Southern Oscillation (ENSO) on rainfall and temperature, although more recently they have also started to consider other ocean anomalies (e.g., Atlantic). These climate forecasts are developed using different combinations of mesoscale circulation model runs and statistical methods. The research on applications of climate forecasts is oriented to develop and improve tools to incorporate the climate forecast information in the planning and decision-making processes of the agricultural sector. In the following sections we summarize the methods being used and present some of the results being obtained in the research on applications of climate forecasts in the agricultural sector of Argentina and Uruguay.

RESEARCH APPROACH

Agricultural scientists in the Argentina and Uruguay are following similar research steps in their activities to develop applications of climate forecasts. These steps can be divided into three main classes of activities: (1) risk assessment, (2) risk management and (3) evaluation.

Research methods used in these countries include simulations of individual crops and pastures, crop mixes and sequences, and field experiments conducted in demonstration farms. Also, workshops are periodically organized in which researchers and users discuss climate forecasts (variables and months being considered, format in which forecasts are expressed, etc.); define potential applications; and discuss recommended practices suggested by the research results and by farmers.

Risk Assessment

Risk assessment involves quantifying the agricultural impacts (biophysical and socioeconomic) of climate variability and its association with ENSO. A set of activities are used to help identify the agricultural sectors that are most affected by this variability and therefore have the greatest potential to benefit from climate forecast applications. These activities measure the degree of vulnerability of agriculture to weather variability, quantify the strength of the influence of ENSO activity on weather and agriculture, and provide information needed for the risk management activities.

ENSO Signal

The first published studies conducted on the impact of ENSO phases on rainfall and temperature in southeastern South America were those by Ropelewski and Halpert (1987 and 1989). Their work defined a region that included Uruguay, southern Brazil, southern Paraguay and eastern Argentina. These authors concluded that precipitation during November-February in this region tended to be above normal during years with warm ENSO events (Ropelewski and Halpert, 1987). They also found that years with high values of the Southern Oscillation Index (typically, La Niña years) presented negative rainfall anomalies during June-December (Ropelewski and Halpert, 1989).

Researchers from the SESA region have conducted detailed studies on the precipitation anomalies associated with ENSO phases. Working in southern Brazil, Grimm et al. (1998) identified 4 sub-regions with homogeneous behavior with respect to rainfall anomalies during warm phases of ENSO. Large positive precipitation anomalies were found in all sub-regions during the austral spring months of El Niño years. Positive rainfall anomalies were also found in the fall-winter months of the years following El Niño in 3 sub-regions covering southern Paraná, and the states of Santa Catarina and Rio Grande do Sul. The authors also reported a tendency to dryness during the year previous to the warm event and the first half of the warm event year.

Regarding the impact of cold phases, these authors identified three coherent sub-regions, with a common period of strong negative rainfall anomalies during the spring of the La Niña years, mainly during October-November. Further analysis revealed a minimum of precipitation in the second semester of La Niña years throughout southern Brazil, with magnitudes that were stronger than the corresponding to the El Niño years. In the Pampean region of Argentina Grondona et al. (1998) also found that the negative precipitation anomalies during November-December of La Niña years were clearer than the positive anomalies during the El Niño years. In a more detailed study for the same region, Magrin et al., 1998 found that the negative precipitation anomalies during La Niña start in Sep-Oct, they progress during Oct-Nov, and are strongest in Nov-Dec. Then the signal practically disappears during Jan-Feb, restarting in Mar-Apr and disappearing again in May-Jun . During El Niño years the positive precipitation anomalies begin later (Nov-Dec) and are weaker than in La Niña, specially when it restarts in the Fall.

Díaz et al. (1998) studied the relationship between rainfall anomalies in Uruguay and Rio Grande do Sul (URSOL) and anomalies in the SSTs of the Pacific and Atlantic Oceans. They found positive precipitation anomalies during November-February of years with warm anomalies in the tropical Pacific, which were specially clear in central and western URSOL. These authors also found positive rainfall anomalies during October-March in years with cold SSTs in the Southern Pacific Convergence Zone and warm SST in the El Niño region. These results are consistent with those of Pisciottano et al. (1994) who reported negative rainfall anomalies in Uruguay during October-December in high-SOI years. Finally, Díaz et al. (1998) found that precipitation anomalies during the fall-winter months in URSOL were better correlated with SST anomalies in the Atlantic Ocean than in the Pacific. Thus, years with warm SSTs in the Southern Atlantic Convergence Zone (SACZ) tended to have positive rainfall in Uruguay and Rio Grande do Sul during April-July. The authors explained that the anomalies in SACZ may be the result of a remote forcing by anomalies in the equatorial Pacific or may be independent from ENSO.

Studies conducted in the Argentinean Pampas region indicate that during "El Niño" years, precipitation tends to be higher than normal between November and January, but in La Niña years rainfall tends to be lower than normal between October and December (Tanco and Berri, 1996; Vila and Grondona, 1996; Moschini et al., 1996; Barros et al., 1996).

Although all these studies indicate the existence of significant spatial variability in southeastern South America, they also reveal a general pattern with positive rainfall anomalies during October-March in El Niño years, and negative rainfall anomalies during the austral late winter and spring months in La Niña years.

The impact of ENSO phases on temperatures (maximum, minimum and mean) has been subject of very few studies in the region. Bidegain and Krecl (1998) have recently found that during El Niño years mean temperatures in northern Uruguay tend to be higher than normal during the winter months, and lower than normal during the summer months. On the other hand during La Niña years all the weather stations considered in their study showed negative anomalies in the mean temperature of all months with a few exceptions for some summer months. In the Pampas region of Argentina, maximum temperature showed a seasonal pattern with lower (higher) values in the cold season in La Niña (El Niño) years, being the behavior opposite in the warm season. While minimum temperature showed the same trend all over the year, with higher (lower) values in El Niño (La Niña) years (Messina, 1999, Magrin et al., 1999). Considering the known correlation between rainfall and temperatures in the region, these results are also consistent with the ones discussed above on the impact of ENSO phases on precipitation in the region.

In summary, the first detailed studies conducted in southeastern South America reveal the existence of a near symmetry between impacts of El Niño and La Niña on precipitation as well as on crop productivity. Positive rainfall anomalies prevail in El Niño years, and negative rainfall anomalies prevail in La Niña years, during the austral Spring and/or Summer months. Some research results also suggest that the impacts of La Niña are stronger and/or less variable in both, rainfall and crop yields than the impacts of El Niño. Finally, the work conducted by Díaz et al. (1998) suggest the need to consider the connection between SST anomalies in the Atlantic ocean and precipitation anomalies in southeastern South America.

Observed Impacts of ENSO Phases on Crop Productivity

Often the annual variability observed in crop and pasture productivity is an excellent indicator of the impacts of large scale phenomena (e.g., ENSO) on regional climatic conditions. Crops and pastures are "integrators" of the general climatic conditions (rainfall, temperatures) during their growing seasons. In addition, most crops present growth stages in which they are specially sensitive to environmental conditions (e.g., flowering, grain filling, etc.). When ENSO phases affect the climatic conditions during those critical crop growth stages, the impacts are very evident. For example, Myneni et al. (1996) found strong negative anomalies in the NDVI's of the rangelands of Uruguay during the 1988/89 La Niña, and much weaker positive NDVI anomalies in that same region during the 1982/83 El Niño.

Baethgen (1997; 1998) found stronger negative maize yield anomalies in Uruguay during La Niña years than positive maize yield anomalies during El Niño years. Similarly, Podestá et al. (1998) reported that in Argentina the depression of maize and sorghum yields during La Niña years were larger and less variable than the yield increases during El Niño years. The typical correlation found between maize and sorghum yields and rainfall during the flowering months (December-February) reinforce the findings of positive rainfall anomalies during El Niño years and (possibly stronger) negative anomalies during La Niña years.

More recent studies conducted in the Argentinean Pampas revealed that in general, maize and soybean yields tend to be higher (lower) than normal during El Niño (La Niña) years (Magrin et al., 1998, 1999). The same trend was detected in wheat but only in the southern portion of the region, while sunflower response was opposite and weaker (Magrin et al., 1998, 1999). Maize yield reductions in La Niña years were associated with negative precipitation anomalies during November, while yield increments in El Niño were positively related to December increments in precipitation (Travasso et al., 1999).

Risk Management

These are activities which provide information aimed to help decision makers apply climate forecasts to minimize negative consequences of climate variability and to take advantage of favorable conditions. In order to benefit from climate forecasts, decision makers must possess flexibility to change their management practices in response to the improved information (Sonka et al., 1986). For that reason, activities in the region aim to specify alternative management options that are feasible and reasonable from the perspective of stakeholders. Research is focused on crop production decisions that are sensitive to future weather conditions. Simulation models are being used to identify optimal management with and without climate forecasts. The selected optimal options allow estimating the efficacy and value of climate forecasts, and identifying options to present to decision makers

Following this approach, a number of activities were carried out to evaluate the acceptance and value of ENSO-based climate forecasts for agricultural decision-making.

In a preliminary study, Hansen et al. (1996) found that predicted benefits of tailoring soybean planting dates to forecasted ENSO phases ranged from US$ 2.4 to 32.4, according to location and soybean prices.

Messina et al. (1999) found "optimal crop combinations" for each ENSO phase in different Pampas' locations which depend on location, risk aversion and initial wealth. On the other hand, Magrin et al. (1999) concluded that the best management option in "La Niña" years for both maize and soybean was to delay the sowing date. Inversely, early sowings were more likely to optimize yields and incomes in "El Niño" years. These researchers also found that nitrogen rates in maize should be higher during "El Niño" events for maximize expected maize yield and profit. However, due to important intra-phase climate variability (e.g., rainfall during flowering is not always lower in La Niña years), making crop management recommendations based on mean values may still lead to negative results (Magrin et al., 1999).

Evaluation

These are the activities aimed to obtain user responses, identify additional information needs, and evaluate the potential for practical routine application of climate forecasts. Results of the risk management analyses are probabilistic. Improved management options and their estimated value apply to the long-term use of climate forecasts, but, given current forecast skill, there are no guarantees that the economic outcome in any particular season will be favorable (Hammer et al., 1996).

Researchers in Argentina and Uruguay are working with focus groups of agricultural decision makers to obtain end user responses and identify additional information needs. This ongoing activity has several requirements. First, it requires the characterization of decision makers' attitudes toward climatic and other forms of risk. Second, the characterization of attitudes toward climate forecasts, including an assessment of perceived impediments (e.g., insufficient forecast skill or lead time). Third, define projected uses of climate forecasts. Fourth, provide decision support tools so decision makers can explore outcomes of alternative management options in response to forecasted climate. Feedback from the focus groups will enable researchers to improve the content and format of climate forecasts and production advice based on forecasts.

RESEARCH NEEDS AND PROPOSED FUTURE STEPS

Research achievements in these three classes of activities in the region is variable, some are very advanced and others are only starting. Most of these on-going activities are now being conducted uncoordinatedly in the region and even within each country. There is a clear need to develop a research framework to coordinate these efforts, and to discuss methodologies, share results, and define common research protocols. Such coordination should be oriented to expand this type of activities to other countries in the region, starting with neighboring southern Brazil and Paraguay where ENSO signals are similar to those of Argentina and Uruguay.

A key common element in the research activities in Argentina and Uruguay is that the applications are intended to be end-to-end. Thus, the methods and tools developed in the research activities must reach decision makers acting at different levels: from governmental agencies to individual farmers.

A fundamental concept that must be considered when exploring applications of climate forecasts in agriculture is that the climate information is only one of a large number of variables that stakeholders need to consider in the planning and decision making processes. Effective application of climate forecasts must frame the climate information in broader decision support tools that include also data on prices, land use feasibility, evaluation of technologies, etc.

With this challenge in mind, IFDC has been collaborating with National agricultural research centers in Latin America (e.g., INIA in Uruguay, IAPAR in Brazil) to develop Information and Decision Support System (IDSS) for the agricultural sector. The main idea behind these IDSSs is to take advantage of modern information tools, process large and complex databases, and produce information which can be easily understandable by stakeholders acting in the agricultural sector.

The most recently developed IDSS (Uruguay) includes information generated in experimental climate forecasts. The system uses climate outlooks generated at three scales: (a) regional forecasts such as those developed by IRI, NOAA, ECMWF, etc.; (b) regional forecasts produced in the Regional Outlook Fora since December 1997; and (c) National experimental climate forecasts generated by the local scientific community.

The IDSSs are constructed by linking: (a) maps (soils, weather, land use, political divisions, market centers, routes, rivers), (b) regional statistics; (c) prices at the farmers gate; (d) remotely sensed information on vegetation and weather; (e) climate forecasts; (f) crop simulation models; (g) a decision support system specifically designed for agrotechnology transfer; (h) an automated system for land use evaluation; (i) simulation models for studying soil organic matter and nutrient dynamics and soil erosion losses.

The IDSS links maps and databases through a Geographic Information System which allows spatial analyses at the regional and at the farmer level. Examples of applications of the IDSS include: (a) develop crop and pasture production forecasts; (b) issue drought and flood alert reports; (c) conduct impact studies of interannual and long-term climatic variability; (d) perform economical and environmental impact studies including risk assessment; (e) establish land use feasibility classes for different agricultural production systems.

Information generated in the climate forecasts has become an essential component of the new IDSSs. The climate outlooks allow for better crop and pasture production forecasts and more precise drought alert systems. This in turn results in the information required for adequate planning and decision making from governments to farmers, and from commercial enterprises to famine early warning systems (FEWS).

REFERENCES

Baethgen, W.E. 1997. Relaciones entre la temperatura superficial del Pacífico tropical y los rendimientos de cultivos en Uruguay. Proceedings of the Workshop and Conference on the 1997-98 El Niño: Impacts and Potential Applications of Climate Prediction in Southeast South America. December 1997. Montevideo, Uruguay.

Baethgen, W.E. 1998. El Niño and La Niña Impacts in Southeastern South America. In: M. Glantz (ed.), Review on the causes and consequences of cold events: A la Niña Summit. NCAR, Boulder, CO.

Baethgen, W.E. 1999. Aplicaciones de Pronósticos Climáticos en el Sector Agropecuario. CREA, 204:22-24.

Barros, V.; Castañeda, M. E. y Doyle, M. (1996) Variabilidad ineranual de la precipitación: Señales del ENSO y del gradiente meridional hemisférico de temperatura, en Impacto de las variaciones climáticas en el desarrollo regional un análisis interdisciplinario—VII Congreso Latinoamericano e Ibérico de Meteorología, p.321-322

Bidegain, M. and P. Krecl. 1998. Comportamiento de la temperatura en el sudeste de Sudamerica (Uruguay) asociado al fenómeno ENSO. Congreso Iberico—Latinoamericano de Meteorologia. Brasilia, Brazil. September 1998.

Díaz, A.; C.D. Studzinski and C.R. Mechoso. 1998. Relationships between precipitation anomalies in Uruguay and southern Brazil and sea surface temperature in the Pacific and Atlantic oceans. J. Climate, 11:251-271

Grimm, A.M.; S.E.T. Ferraz and J. Gomes. 1999. Precipitation anomalies in southern Brazil associated with El Niño and La Niña events. J. Climate (in press).

Grondona, M.O.; G.O. Magrin; M.I. Travasso; R.C. Moschini; G.R. Rodriguez; C. Messina; D.R. Boullon; G. Podestá and J. Jones. Impacto del fenomeno "El Niño" sobre la producción de trigo y maíz en la region Pampeana Argentina. Workshop and Conference on the 1997-98 El Niño: Impacts and Potential Applications of Climate Prediction in Southeast South America. December 1997. Montevideo, Uruguay.

Hammer, G.L.; D.P. Holzworth and R. Stone. 1996. The value of skill in seasonal climate forecasting to wheat crop management in a region with high climatic variability. Aust. J. Agric. Res. 47:717-737.

Hansen J.W., Jones J.W., Magrin G.O., Meira S.G., Guevara E.R., Travasso M.I., Díaz R.A., Marino M., Hordij R., Harwell C., Podestá G. 1996. ENSO effects on yields and economic returns of wheat, corn and soybean in Argentina. Proceedings of CONGREMET VII-CLIMET VII. (pp 89-90).

Magrin G.O, Grondona M.O, Travasso M.I., Boullón D.R., Rodriguez C.D., Messina C.D. 1998. Impacto del Fenómeno "ENSO" sobre la Producción de Cultivos en la Región Pampeana Argentina. Boletín Técnico. INTA..

Magrin, G.O., M.O.Grondona, M.I.Travasso, D.R. Boullón, G.R.Rodríguez and C.D.Messina.1999. ENSO impacts on crop production in the Argentina's Pampas Region Proc. 10th Symposium on Global Change Studies, 10-15 Jan. 1999, Dallas, TX, by the AMS, Boston, MA. pp. 65-66.

Magrin G.O., Travasso M.I., Jones J.W., Rodriguez G.R., Boullón D.R. 1999. Using Climate forecast in Agriculture: A pilot application in Argentina. The third International Symposium on Systems Approaches for Agricultural Development (SAADI-II). 8-10 Nov, Lima, Perú.

Messina C.D. 1999. El Niño-Southern Oscillation y la productividad de cultivos en la zona Pampeana.: Evaluación de estrategias para mitigar el riesgo climático. Tesis Magister Scientiae, Producción Vegetal. Escuela para Graduados, Facultad de Agronomía, UBA. 132 pp.

Myneni, R.B.; S.O. Los and C.J. Tucker. 1996. Satellite-based identification of linked vegetation index and sea surface temperature anomaly areas from 1982-1990 for Africa, Australia and South America. Geophys. Res. Letters. 23:729-732.

Moschini, R. C.; Grondona, M. O. y Vila, D. A. (1996) Influencia del ENSO sobre la distribución de las probabilidades condicionales diarias de precipitacion en algunas estaciones de la Región Pampeana, en Impacto de las variaciones climáticas en el desarrollo regional un análisis interdisciplinario—VII Congreso Latinoamericano e Ibérico de Meteorología, p.307-308

Nuñez, Mario N. y W. M. Vargas (1998): El Niño 1997—1998: Un factor económico y social. Boletín Informativo Techint Nº 295, 9-36

Pisciottano, G.J; A.F. Díaz and C.R. Mechoso. 1994. El Niño-Southern Oscillation impact on rainfall in Uruguay. J. Climate, 7:1286-1302.

Podestá, G.R.; C.D. Messina; M.O. Grondona and G.O. Magrin. 1999. Associations between grain crop yields in central-eastern Argentina and El Niño-Southern Oscillation. J.Applied Meteorology 38:1488-1498.

Ropelewski, C.F. and M.S. Halpert. 1987. Global and regional scale precipitation patterns associated with El Niño/Southern Oscillation. Mon. Wea. Rev. 115:1606-1626.

Ropelewski, C.F. and M.S. Halpert. 1989. Precipitation patterns associated with high index phase of Southern Oscillation. J. Climate, 2:268-284.

Sonka, S.T., P.J. Lamb, S.E. Hollinger, and J.W. Mjelde. 1986. Economic use of weather and climate information: Concepts and an agricultural example. J. Climatol. 6:447 457.

Tanco, R. And G.J. Berri. 1996. Acerca del efecto del fenómeno El Niño sobre la precipitación en la Pampa húmeda Argentina. Pp. 319 320. In Proceedings, VII Congreso Argentino de Meteorología y VII Congreso Latinoamericano e Ibérico de Meteorología, 2-6 Sep. 1996.

Travasso, M.I., Magrin, G.O. and Grondona, M.O. 1999. Relations between climatic variability related to ENSO and maize production in Argentina. Proceedings 10th AMS Symposium on Global Change Studies, 10-15 Jan. 1999, Dallas, TX. Boston, MA. pp.67-68.

Vila, D. A. y Grondona, M. O. (1996) Estudio preliminar sobre las relaciones del ENSO y la frecuencia de días con lluvia en la Pampa Húmeda, en Impacto de las variaciones climáticas en el desarrollo regional un análisis interdisciplinario—VII Congreso Latinoamericano e Ibérico de Meteorología, p.309-310.

Application of Long-Range Rainfall Forecasts in Agricultural Management: A Review of Africa's Experiences

Leonard Unganai
Zimbabwe Department of Meteorology

INTRODUCTION

Inter-annual rainfall variability represents one of the biggest challenges facing agricultural production across Africa. The continent's high dependence on rain-fed agricultural production, low input production systems and degraded soils increase the farmers' vulnerability to vagaries in weather. In drought years, production in the small-holder farming sector may drop to zero whereas in wet seasons production never rises beyond even half the potential yield level.

It has been widely suggested that if society could have advance information on the most probable outcome of the weather pattern during the crop growing season, then the adverse effects associated with extreme weather conditions could be alleviated or minimized. Following recent advances in the understanding of climate variability particularly in those regions affected by the El Niño/Southern Oscillation (ENSO) phenomena, there has been a number of initiatives across parts of Africa to generate and disseminate long-range rainfall forecasts and promote their application in farm management, particularly during the 1990s. Enhancing the ability of resource poor communities to capture the potential benefits of applying seasonal climate forecasts in farm decision making remains an important developmental issue. However, translating seasonal climate information into appropriate actions by farmers and other resource managers remains a significant problem across the whole of Africa.

The object of this paper is to review progress across the African continent in the areas of long-range forecast generation, dissemination and application. Literature on climate information application in agricultural management across Africa is very scarce. In this paper results from case studies in Mozambique and Zimbabwe are used to highlight some of the challenges facing both farmers and forecasters in the effective application of seasonal climate forecasts in farm management. The paper also highlights weaknesses in the current long range forecast products which if addressed could enhance the potential benefits of applying the products in farm management. The main focus of the paper is the small-holder farming sector, which represents the bulk of agricultural producers across much of Africa.

SYSTEM DESCRIPTION

Climate of Africa

Africa experiences a wide variety of climatic regimes. The poleward extremes of the continent get some winter rainfall associated with the passage of middle-latitude westerly waves. Across the Kalahari and Sahara deserts precipitation is inhibited by subsidence virtually throughout the year. Moderate to heavy precipitation associated with the Inter-Tropical Convergence Zone (ITCZ) characterizes the equatorial and tropical areas. The mean climate of Africa is further modified by topography. Inter-annual rainfall variability is controlled by a number of factors, of which the ENSO is the most dominant over eastern and southern Africa, whereas the North Atlantic Oscillation (NAO) is a key factor in North Africa.

Small-holder Farming Systems

The small-holder sector across much of Africa is characterized by mainly family run small scale farming operations. In most cases agricultural production is primarily for subsistence purposes. Plots for arable farming are generally small (below five hectares) and pastures are communally owned. Most land preparation is done by donkey or ox-drawn plough or hand. Planting, weeding, harvesting and threshing are generally by hand. Use of hybrid seed and fertilizer is now quite common within the small-holder farming sector across the continent. Family members provide most of the farm labor and credit for purchasing inputs.

Yields in the smallholder sector are generally low compared to the large-scale commercial farming sector. Figure 1 shows national average maize yield per hectare in the smallholder and large scale-farming sector respectively over the period 1969 to 1998 for Zimbabwe. The seasonal rainfall amount is superimposed. It is quite clear from the figure that rainfall variability has a strong influence on production. For the communal sector, the correlation between rainfall and maize yield after 1980 is about 0.80, and only 0.42 if the whole period is used. This distortion arises from the trend in production which is quite evident from the graph. Production in the small holder sector is about five times lower than the large scale sector, a reflection of differences in farming skills, access to resources and the quality of farming environment.

After a major drought, draft power for land preparation can become a major constraint, unless a household has resources to hire a tractor. Since most of the smallholder farmers are in marginal areas, crop mixes that have evolved over the years reflect an attempt to adapt to climate variability in those areas (Figure 2). From Figure 2, it appears that farmers choose crops and make capital investments assuming the *unpredictability of climate variations or at least they have some prior probabilities.* Skillful forecasts can lead to a change in this paradigm.

Decisions on what to plant, when to plant, how to plant and how much area to plant appear to be largely conditioned by the family's preferences (these will have evolved over time), prior commitment, precedents, recent events, historical rainfall distribution, cost and government policy. Indigenous knowledge systems for seasonal climate forecasting are quite common, although these do not appear to be used in formulating farm management strategies. Small-holder farmers wait for every cropping season with optimism, for they strongly believe that through rain-making ceremonies, the character of any rainy season can be changed for the better. Allocation of plots on arable land is such that every household has a mix of soil types to cater for differences in inherent soil fertility and water holding capacity.

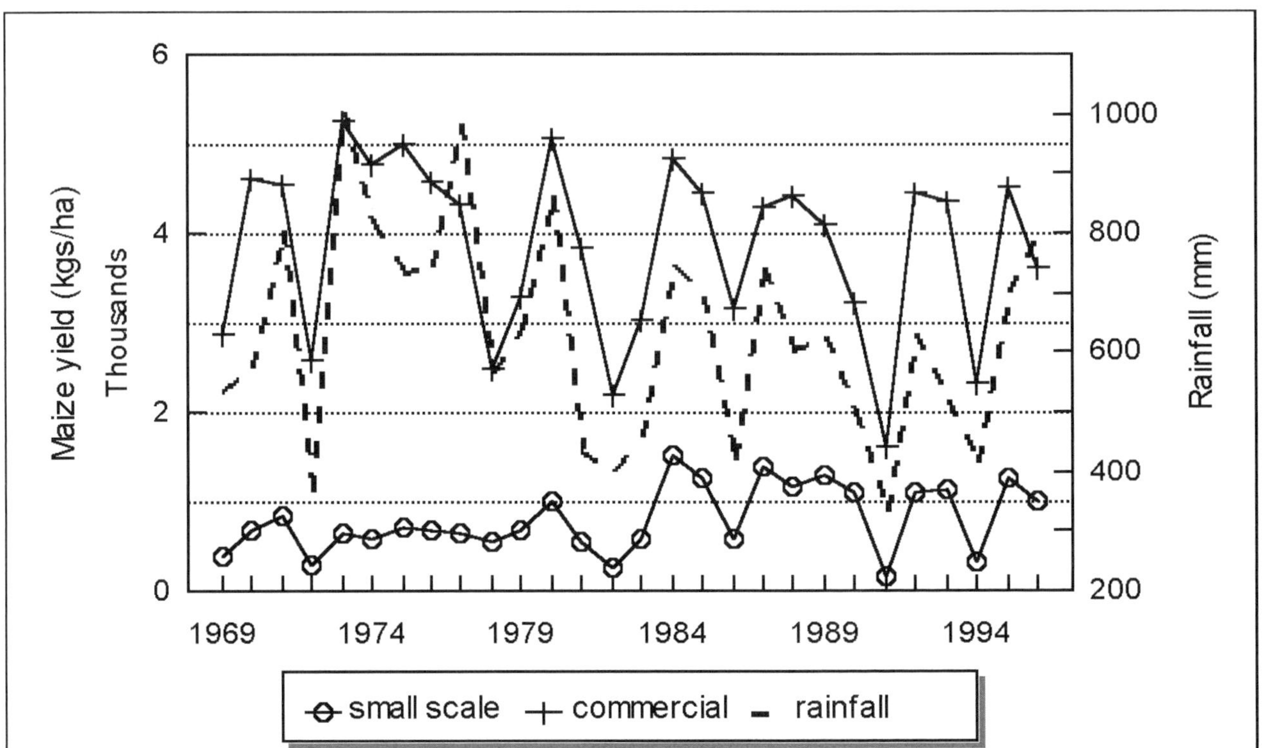

Figure 1. *Time series of average maize yield in the smallholder and commercial farming sectors in Zimbabwe. The national average rainfall is superimposed.*

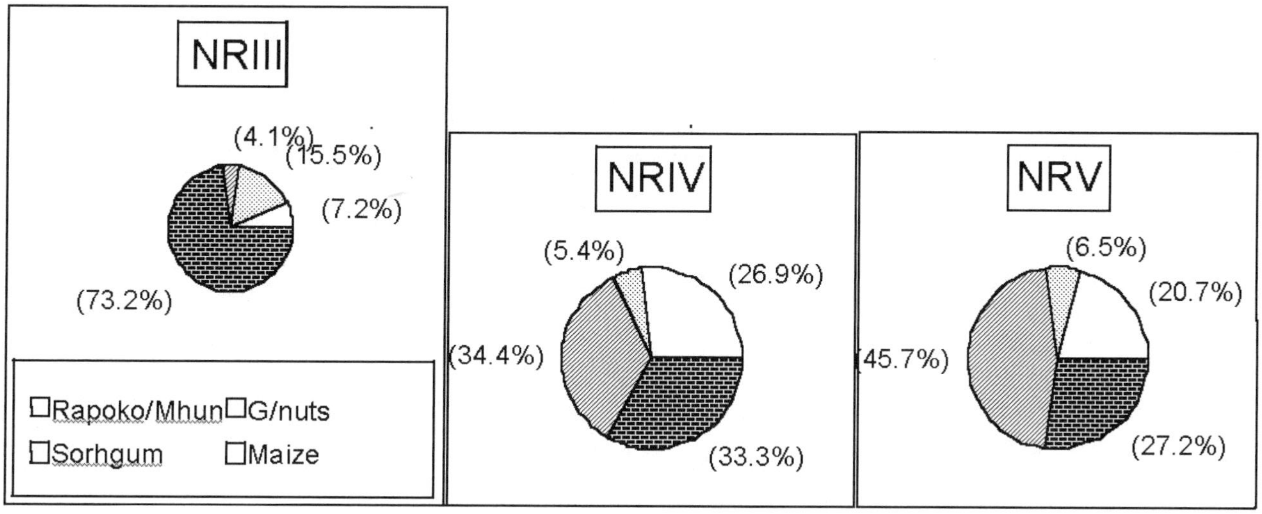

Figure 2. *Crop mixes in three contrasting farming regions in Zimbabwe. Natural Region III (NRIII) gets more rain than NRIV and V. NRV is the driest. Source of Data: Zimbabwe National Early Warning Unit.*

Current State of Long Range Forecasts

Current long range forecasts generated on the African continent are largely based on empirical-diagnostic analyses with Global SST anomaly patterns, the state of ENSO, NAO and upper level winds being the main predictors. Available forecast schemes explain up to about 60% of the variance. The forecasts are issued in terciles, below normal, normal and above normal with the probability of rainfall being in each of the three categories being stated. In addition these forecasts are issued for relatively large homogeneous rainfall regions extending over three month periods. The forecast parameter is rainfall amount. An example of a typical seasonal rainfall forecast is shown in Figure 3 for OND of 1999 and JFM of 2000. However, the probabilities are not disseminated for public consumption.

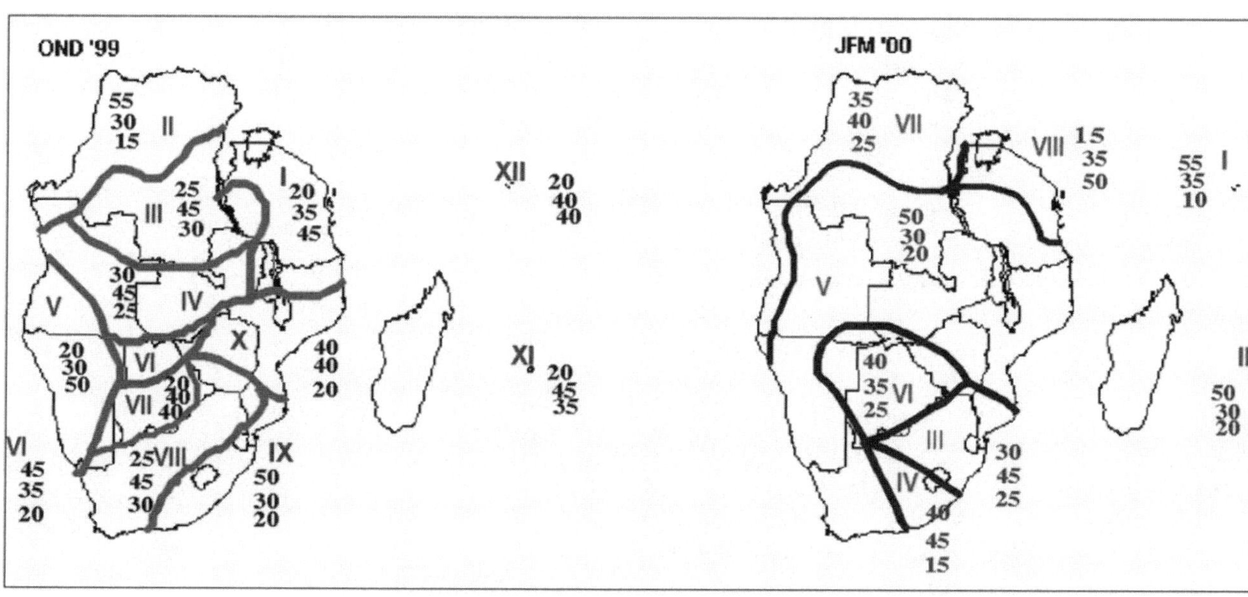

Figure 3. *Example of consensus regional seasonal climate outlook generated by forecasters for southern Africa. The numbers represent the probability of rainfall amount being in a particular tercile. Source: Drought Monitoring Centre, Harare.*

Farmer's Climate Information Requirements

Survey results and feedback from workshops suggest that for farmers to capture the potential benefits of long range forecasts, the forecast has to contain the following information, with respect to rainfall:

- date of onset of rainy season
- quality of rainy season (rainfall amount)
- date of end of rainy season
- chance, frequency and timing of adverse weather events (e.g. dry spells) within the season
- temporal and spatial distribution of rainfall
- interpretation of above information in terms of which crops and varieties to plant and when to plant.

Above all, the forecast needs to be stated in a language and terms the user understands, and has to be issued in good time.

EXAMPLES OF APPLICATIONS

1997-98 Rainy Season

Following the evolution of one the strongest El Niño event in recorded history during 1997, a forecast for a high probability of low rainfall was issued for the whole of eastern and southern Africa as early as September 1997. Memories of the devastating droughts associated with the 1982-83 and 1991-92 warm ENSO events resulted in most people preparing for the worst possible drought across southern Africa. 1997. Using survey results from a pilot study implemented in Zimbabwe from 1997, the effectiveness of forecast dissemination during that year is shown in Figure 4. It can be seen from the figure that more than 90% of the interviewed farmers received the forecast and more than 80% of those that got the forecast believed it. The survey also revealed that most of the farmers got the forecast through radio.

Table 1 shows survey results from Mozambique comparing small-holder farm management strategies in response to the forecast with and without external advice. From the Table it can be seen that there is no difference in the top two drought mitigation strategies with or without external advice.

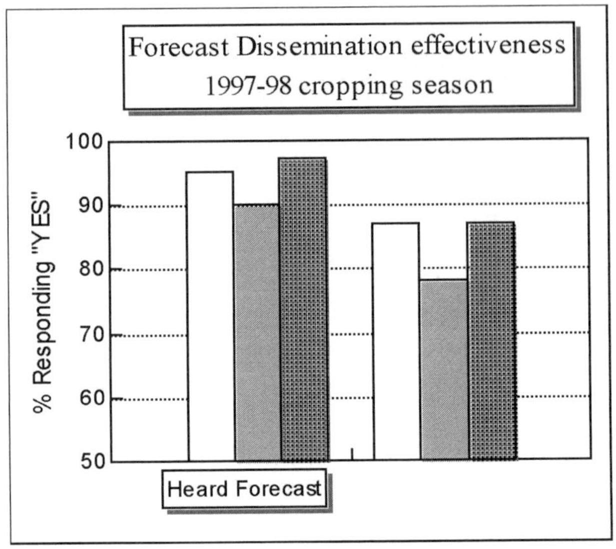

Figure 4. *Forecast dissemination effectiveness in a random sample surveyed in Zimbabwe. The second set of bars shows the percentage respondents that believed the forecast. Source: Phillips, et. al., 1999.*

This suggests that given an accurate forecast, most of the farmers already know how to respond accordingly.

Mozambique. The 1997-98 turned out to be wet and most crops that had been planted in lowlands got flooded.

Zimbabwe. In Zimbabwe the following advice is always given to farmers for crop management if low rainfall is anticipated, as was the case during 1997-98.

- shift crop mix to more drought-tolerant and short season varieties and crops,
- reduce area planted, initially, then increase area planted gradually, depending on nature of season,
- stagger planting dates,
- increase plant spacing,
- maximize use of clay soils where these are available (these have a high water holding capacity)
- implement soil-water conservation techniques (pot-holing, weeding)
- adjust level and timing of fertilizer applications

Table 1. *Traditional and external drought mitigation strategies during the 1997-98 rainfall season in Mozambique.*

Management strategy	Percentage
A. Traditional strategy	
1. *Use drought tolerant varieties and diversify production*	57
2. *Increase use of lowlands*	37
3. *Judicious management and use of local water sources*	3
4. *Early planting*	6
5. *Appropriate seed and food stocks management*	3
6. *Traditional ceremonies*	22
B. External advice	
1. *Use drought tolerant varieties and diversify production*	64
2. *Increase use of lowlands*	45
3. *Judicious management and use of local water sources*	4
4. *Early planting*	24
5. *Appropriate seed and food stocks management*	22
6. *Traditional ceremonies*	0

Source: AEDES, Mozambique.

Some of these strategies worked for the southern parts of the country where rainfall was low during the 1997-98 rainfall season. However, for the northern areas, the season turned out average to above average and some opportunities were missed.

1998-99 Rainy Season

During 1998, a forecast for an average to above average rainfall season was issued. During January and February, it rained nearly every day in some parts of southern Africa. The following crop management strategies had been issued to farmers in Zimbabwe:

- plant more high yielding varieties if planting is done early,
- Optimize use of inorganic fertilizers—split fertilizer applications as risk of leaching will be high,
- Diversify crop mix with some commercial crops,
- Increase plant population,
- Improve field drainage,

The 1998-99 rainy season turned out excessively wet across much of the country, adversely affecting crop production. Loss in maize production as a result of excess wetness was estimated at 30%.

DISCUSSION AND CONCLUSIONS

Experience across much of Africa suggests that connecting seasonal climate forecasts into applications remains an area that needs attention. The on-farm decision making environment is complex. As much as the seasonal weather conditions are an important factor of production, availability of labor, inputs, government policy, farmer skills, recent events, historical rainfall distribution and market trends all have an important influence on final decisions. Users of forecasts have different goals, which in most cases may be conflicting, creating the possibility of winners and losers. Many seasonal climate forecasts that are currently available only provide general predictions and farmers argue that this makes it difficult or impossible to use them in formulating farm specific management strategies.

Feedback from farmers following the 1997-98 decision errors shows that there is intense regret for a loss incurred because one changed decisions as a result of the forecast. Most smallholder farmers prefer to fail conventionally rather than unconventionally. In most cases there is little interest in adopting new production strategies since government or donor food aid is virtually assured in the event of widespread crop failure. To overcome some of these impediments in the use of long range forecasts, there is need to avoid or reduce over-confidence in the forecast, whilst at the same time seeking ways to ensure that decision makers co-own the forecasts. Through research, there is need to establish which areas the forecast is usable, quantify the economic benefits of using the forecasts and have feedback to seasonal forecast design and content. Getting the message out correctly and getting the farmers to use the forecast correctly is still a big problem.

As was quite evident during the El Niño of 1997-98 and the La Niña event of 1998-99, a forecast may produce or relieve anxiety about extreme weather events, especially in those sectors with little or no resources with which to mitigate or cope.

When one looks at the gap between farmers' climate information requirements and what is currently available, the question is then raised:

- How does forecast information available in an irrelevant format, get relevant and incorporated into individual decision makers information requirements, which differ considerably from one decision maker to the next?.
- Whereas larger operations may benefit by spreading the fixed costs of interpreting and applying forecast information, how can smaller and less economically successful operators also benefit. It is thus important to estimate the value of climate forecasts both throughout the economy and disaggregated by sector, region and type of actor.
- How can the utility of forecasts can be systematically increased to bring scientific output and users' needs closer together.
- How should the forecasts be delivered to ensure that they reach the intended target and are used for an intended purpose.
- How do we manage pre-existing mental modes (in individuals) and pre-existing routines (in organizations) that may affect response. Responses to past forecasts also affect the use of new ones.
- How do we handle conflicting goals between different stakeholders?. Climate forecasts have different effects on different regions, sectors and actors. One actor may use the forecast in a way that may be harmful to the goals and aspirations of other actors.

Whereas the potential benefits of applying seasonal climate forecasts to improve farm management are unquestioned, the following areas will also need to be attended to before the full benefits can be captured by the farmers:

- Improve spatial and temporal resolution of the forecast
- Invest in extension services for climate risk management
- Improve information dissemination pathways
- Skill of the forecast has to be always established and communicated to the user,
- Establish the economic benefit of changing decisions based on the forecasts
- The forecast has to have appropriate content and issued in good time.

Given the fact that the radio is the most effective means of disseminating seasonal climate forecast information in Africa, one wonders how easy it will be to package the forecasts so as to include all the relevant information for decision makers. It is also concluded that whereas long-range forecasts enable broad production decisions to be made, medium range forecasts can enable more specific management decisions to be implemented during the growing seasons. Therefore, reliable medium range forecasts, issued regularly will enable the potential benefits of long-range forecasts to be captured by most farmers.

REFERENCES

AEDES, 1998. Traditional techniques in climate forecasting and subsequent alternatives in farm management in Mozambique, *Unpublished Survey Report,* Maputo.

Cane, M.A., G. Eshel and R. W. Buckland, 1994. Forecasting Zimbabwe maize yields using eastern equatorial Pacific sea surface temperatures, *Nature,* 370, 204-205.

Hammer, G (1999). A general systems approach to applying seasonal climate forecasts, In, G.L. Hammer, N. Nicholls, and C. Mitchell (eds.), *Applications of Seasonal Climate Forecasting in Agri-*

cultural and Natural Ecosystems—The Australian Experience. Kluwer Academic, The Netherlands.

Janowiak, J.E., 1988. An investigation into interannual rainfall variability in Africa. J. *Climate,* 1, 240-255.

Mazzocco, M.A., B. J. Sherrick, S. T. Sonka and P. J. Lamb, 1992. The role of decision maker expectations in valuing climate information. In Proceedings of the 5th International Meeting on Statistical Climatology, pp. j115-119.

Meinke, H. and Zvi Hochman, 1999. Using seasonal climate forecasts to manage dryland crops in Northern Australia- Experiences from the 1997/98 seasons. In, G.L. Hammer, N. Nicholls, and C. Mitchell (eds.), *Applications of Seasonal Climate Forecasting in Agricultural and Natural Ecosystems*—The Australian Experience. Kluwer Academic, The Netherlands.

Nicholson, S.E., J. Kim and J. Hoopingarner, 1998. Atlas of African Rainfall and its interannual variability, Dept. of Meteorology, The Florida State University, Tallahassee, Fl, USA.

Phillips, J.G., E. Makaudze and L.S. Unganai, 1999. Current and potential use of climate forecasts for farm management in Zimbabwe. *Unpublished report on USAID/NOAA Pilot project in Zimbabwe.*

Tyson, P. D., 1986. Climatic change and variability in southern Africa. Oxford University Press, Cape Town.

Experiences in Agricultural Applications of Climate Predictions: Australasia

Holger Meinke and Graeme Hammer
Queensland Department of Primary Industries, Australia

INTRODUCTION

Australia and the neighboring countries in Southeast Asia and the Pacific experience the highest rainfall variability in the world (Nicholls et al., 1997). This exposes farm managers to considerable production risks, but also results in degradation of the resource base. The nature of these risks and possible management responses to elevate them depend strongly on the socio-economic context within which farmers operate: In less developed nations food security is often of overwriting importance. In developed countries financial viability and environmental impacts need to be reconciled.

The impact of climatic variability—although most dramatic at the farm level—is apparent throughout the entire Australian economy and can even affect macroeconomic indicators such as international wheat prices (Chapman et al., 2000b), employment or the exchange rate (White, 2000). Our current understanding of the causes of climatic variability and our increasing ability to predict these cycles and their likely consequences has made 'managing for climate variability' an important feature of Australian farming system.

CLIMATE

Significant, physically based lag-relationships exist between an index of the ocean/atmosphere El Niño/Southern Oscillation phenomenon (ENSO) and future rainfall amount and temporal distribution in Australia and many other areas across the globe. Stone et al. (1996) have shown how phases of the Southern Oscillation Index (SOI) are related to rainfall variability and can be used for rainfall forecasting for many locations. For large parts of Eastern Australia a rapid rise in SOI over a two months period is related to a high probability of above long-term median rainfall at certain times of the year. Conversely, a consistently negative or rapidly falling SOI pattern is related to a high probability of below median rainfall for many regions at certain times of the year. As the SOI pattern tends to be 'phase-locked' into the annual cycle, the SOI phase analysis provides skill in assessing future rainfall probabilities for the season ahead. This is an objective and scientifically sound approach of assessing the regional effect of ocean/atmospheric conditions on future rainfall patterns and does not depend on expert classification of either El Niño or La Niña events. Probabilistic forecasts based on this system for the next three to nine months are available as early as late May each year (Meinke and Hochman, 2000).

THE ROLE OF CROP MODELING

Climate signals such as the SOI translate via rainfall variability into associated production variability. However, rainfall anomalies are not the only determinant of yield and factors such as starting soil moisture, planting dates and timeliness of rainfall strongly impact on final yields. Simulation models integrate all these effects and SOI phases can be used to directly estimate seasonal production (Meinke and Hochman, 2000). Although rainfall and yield are strongly correlated, consequences of rainfall variability will differ from season to season due to these other influences on yield formation.

Figure 1 clearly shows this effect: The area around Dalby in Queensland is strongly influenced by ENSO. In this environment, wheat is grown on clay soils that can store a substantial amount of water during fallow periods for use by subsequent crops. The long-term median rainfall during the wheat-growing season from June to October is 190 mm. However, in years when the April/May SOI phase is consistently negative, this winter rainfall is often drastically reduced (median = 114 mm). These years do often, but not always, correspond with official 'El Niño' years. However, there is still

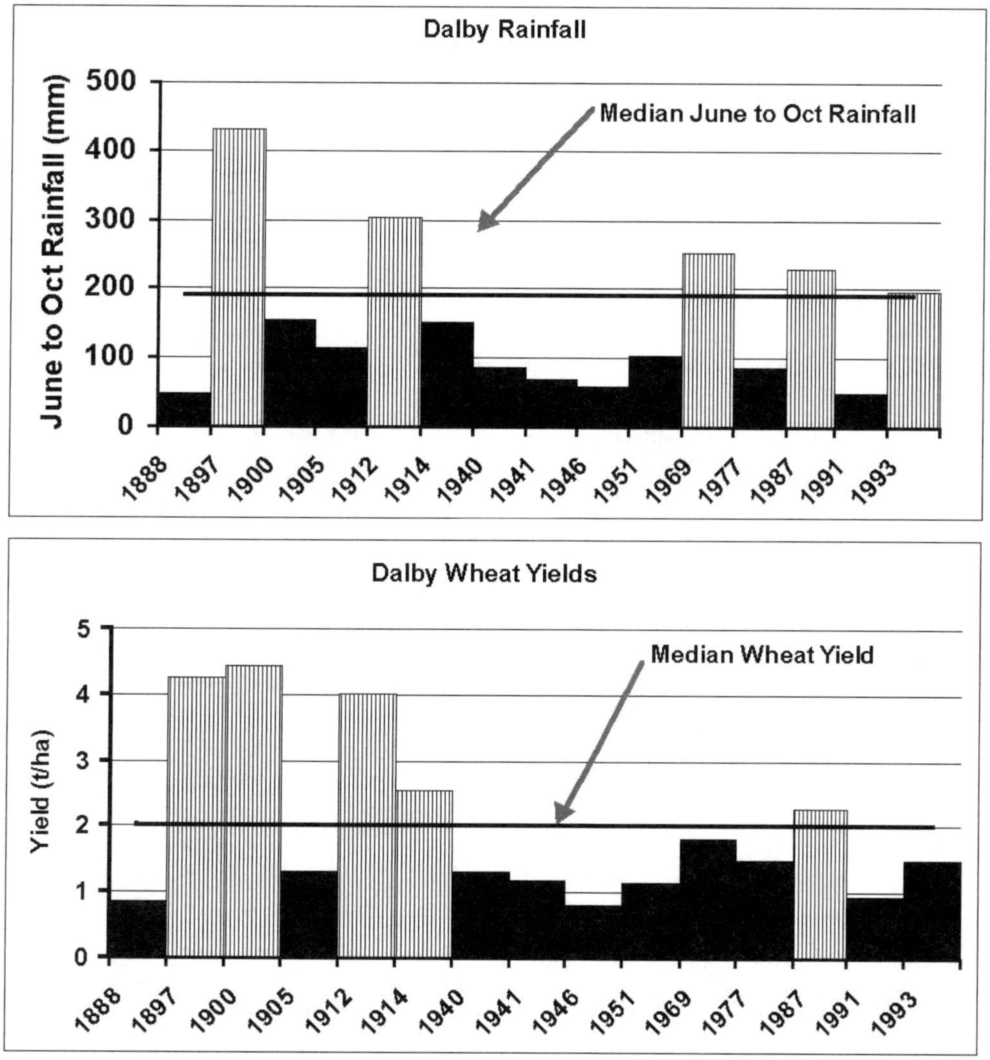

Figure 1. *Seasonal rainfall and corresponding simulated wheat yields for Dalby (northeastern Australia) in years when the April/May SOI was consistently negative. The median was derrived from the entire 107-year rainfall and yield record.*

a distinct possibility that the long-term median rainfall might be exceeded even under these conditions (5 out of 15 years with historical records going back to 1888, Figure 1). Using a well-tested wheat model to simulate wheat yields for the same years shows the impact of this reduced winter rain. Long-term simulated median wheat yields were reduced from 2 t ha^{-1} to 1.5 t ha^{-1} in seasons with a negative SOI phase. However, as with rainfall, 5 of these 15 years resulted in wheat yields that exceeded the 2 t ha^{-1} long-term median. Importantly, these were not always the years during which rainfall records exceeded the long-term median! This highlights the importance of estimating the future impact of seasonal rainfall variability rather than just quantifying the likely effect.

O'Meagher et al. (1998) also reported the significant impact of droughts in Australia and South Africa and on the associated policy frameworks that deal with such climatic impacts on agricultural production. This raises the question: How can prior knowledge of significant shifts in rainfall probabilities and our understanding of their impact on production be translated into better farm outcomes? A recent publication (Hammer et al., 2000) provides many examples of such successful climate applications. Central to all of them are the use of simulation models and the probabilistic nature of the information provided to the decision makers.

With European settlement of Australia about 200 years ago, farmers of the 'new' continent were exposed to an environment that differed fundamentally from their experience. They had no means of assessing the land's suitability to cropping other than by trial and error. Their hard-won experiences, often featured in Australian contemporary art and folklore, were passed on and led to today's manifestation of diverse, regional cropping systems. Analyzing and improving these systems requires sound understanding of physical, chemical, physiological and climatic processes but also tools to evaluate their interactions. Effects of management strategies need to be assessed and quantified in terms of productivity and their impact on the resource base. High rainfall variability means that often even one lifetime of cropping experience can be insufficient to sample the underlying variability adequately (Meinke and Hammer, 1995). Cropping systems models are an obvious choice to address such issues and have many potential applications ranging from environmental issues and policy matters to farm optimization and variety adaptation (e.g. Reyenga et al., 1999; Keating and Meinke, 1998; Hammer et al., 1996a). Generation of probabilistic information useful to systems managers (e.g. What if? When? How often?) can help to identify gaps in current knowledge. By simulating the production system, the state of the system at any point in time is known and alternative management options and their long-term impact on sustainability and productivity can be evaluated.

The Agricultural Production systems SIMulator, APSIM, is such a simulation environment and provides a common communication platform for the many scientific disciplines (McCown et al., 1996; http://www.apsru.gov.au/Products/apsim.htm). APSIM can predict and evaluate the dynamics of soil condition and crop production while allowing management intervention through tillage, irrigation, or fertilization as well as choice, timing and sequencing of crops either in fixed or flexible rotations. Examples of APSIM applications can be found in Hammer et al. (1999).

To obtain real world benefits from this approach requires an integrated approach to seasonal climate forecasting and modeling. Such an approach engages decision-makers in the research process and uses modeling to integrate the newly gained insights. The model is used analytically by scientists and decision-makers to gain further knowledge and insight. In other words, systems approaches in a problem solving context always require on-going connections between decision-makers, advisers, modelers and researchers for effective outcomes (Hammer, 2000).

GURUS OR PROBABILITIES?

The ENSO phenomenon provides a scientific explanation for the significant shifts in rainfall and yield distributions (Figure 1). Actual rainfall received in any given season can be anywhere on this distribution and, in fact, even outside the distribution whenever an event more extreme than previously recorded occurs. Hence management decisions based on knowledge of the state of ENSO will have positive outcomes in some years and negative outcomes in others. This must not be regarded as either a 'win' or a 'failure' of the strategy employed, since each season only represents a sample of one from a not very well defined distribution of possible outcomes. To assess the true value of such probabilistic information requires comparison of results in each season against outcomes that would have been achieved in the absence of such information. A sufficient number of seasons must be sampled so that statistical methods can be employed to assess the probability space of each management option (de Maia and Meinke, 1999). Often we are told that people want simple answers, they 'don't like' probabilities, they require certainty for decision making. This gives rise to the 'guru' who is prepared to give such a categorical forecast. If gurus understand the climate system and interpret the atmospheric and oceanographic signals correctly the law of probabilities dictates that they will be 'right' more often than not. Still, such categorical forecasts do not belong to the realm of science. Although our understanding of the mechanisms responsible for climatic variability is constantly increasing, there will always be a considerable element of chaos or chance involved. We have to be blunt and acknowledge this fact freely. We live in an uncertain world, but we survive by weighing up the odds. All science can reasonably provide are such probabilities.

AUSTRALASIAN CASE STUDIES

To use this technology effectively by actively influencing management decisions is a great challenge. Decision making in agriculture happens at many levels and involves a wide range of possible users. To provide these clients with the most appropriate tools for decision making requires clear focus on their specific requirements and needs. Although farmers are one obvious client group they are not necessarily the ones most responsive to a forecast. This responsiveness depends very much on the socio-economic and political circumstances, local infrastructure and the agricultural system in question. To clearly identify clients and their decision points it is helpful to classify them according to geographic scale and information needs (Hammer, 2000). Such a conceptual framework not only helps to identify the information needs of decision-makers, it also assists in the identification of the most appropriate and efficient tools to use. Although modeling approaches are frequently the tools of choice, the type of model required will differ and depends on geographic scale, required inputs and information needs. In cases where uncertainty is high and model input parameters are sparse a plethora of modeling approaches can add weight to the credibility of the results (e.g. drought policy, its assessment and declaration, O'Meagher et al., 1998; Keating and Meinke, 1998; de Jager et al., 1998). These issues need to be considered before 'the Australian approach' can be successfully employed under different socio-economic circumstances.

Farmers as Decision-Makers

Good farm managers understand the interaction of agricultural systems components (e.g. the relative importance of stored soil moisture versus in-season rainfall) and are skilled at incorporating new information into the decision making process. A simulation approach can help such farmers to replace "gut feeling" about their complex system with (a) hard data about the current state of their system and (b) with probabilistic information about the way in which the unknown future rainfall will affect the outcome of alternative management decisions. Effective decision making ultimately has to improve the long-term economic performance of the farming enterprise either by increasing profits, improving sustainability indicators (e.g. erosion, soil organic matter) or reducing risk. Using the 1997/98 El Niño and the 1998 La Niña event as a case studies, Meinke and Hochman (2000) demonstrated how seasonal forecasting can aid farm managers in their tactical decision making processes. Specifically, they profile one particular wheat/cotton grower who uses climate-related information in most of his management decisions. These decisions range from benchmarking for fertilizer applications, crop protection and choice of cultivar (or crop) to marketing.

Agricultural decisions are made within the context of crop rotations and whole farm management. Carberry et al. (2000) point out that the common thread across a range of scale in decision-making is exposure to the chance of making a loss, or risk (financial risk, land degradation and other environmental risks). Climate variability generates these risks because outcomes of decisions cannot be predicted with certainty. This also creates opportunities, as there is exposure to the chance of making financial and environmental gains. In their study, they compared a standard, fixed dryland cropping rotation of long fallowing from sorghum to cotton with two alternative fixed rotations and with a flexible rotation influenced by an SOI forecast. The decision point is October of the second year where the manager can choose to proceed with the standard summer fallow, plant sorghum or plant cotton with the intention of definitely planting cotton in the following summer. The various scenarios were assessed using APSIM. The simulation analysis showed that by changing between rotations based on the SOI phase in the August-September period preceding the decision point, average gross margins per year increased by 14%. At the same time, simulated soil loss from erosion was reduced by 24%. The SOI strategy did however increase the risk of an economic loss from 5% of years for the standard fallow-cotton rotation to 9%, but this risk was considerably less than for the two alternative fixed rotations.

Scientists as Decision-Makers

Better quantification of genotype x environment interactions (GxE) can significantly speed up plant breeding programs and help to develop better regionally adapted cultivars. Shorter et al. (1991) and Hammer et al. (1996) showed how physiologically based crop simulation models can assist in this process. For sorghum, Chapman et al. (2000a) showed that using a simulation model to quantify the frequencies of certain types of water stress (early, mid or late season stress) allows plant breeders to select germoplasm that is better adapted to these stresses by categorizing breeding trails according to stress type rather than location. Once this new germoplasm has been incorporated into improved cultivars, seasonal forecasts will allow producers to select the most adapted cultivar for the likely season ahead, reducing risk and using the scares water resources more efficiently.

Policy Decisions

The determination of appropriate drought policies and the assessment and declaration of drought exceptional circumstances require objective information regarding the impact of climate variability on agricultural production. Seasonal forecasting and simulation models allow not only an assessment of the severity and frequency of drought events, but can be used to evaluate management responses that result in a higher level of drought preparedness at the farm, regional and institutional level. A wide range of studies have shown how these concepts can be operational (e.g. de Jager et al., 1998; Keating and Meinke, 1998; Stafford Smith and McKeon, 1998). The approach puts short-term climate perturbations into a long-term historical context. Targeted drought alerts based on seasonal forecasts have led to a much higher level of drought preparedness in Australia and increased the capacity to better absorb negative impacts of drought.

Historical increases in atmospheric CO2 concentrations are well documented. Increasing atmospheric CO2 affects agricultural production both directly through the stimulation of photosynthesis and improved water use efficiency and indirectly as the increased concentration of CO2 and other greenhouse gases in the atmosphere may induce other climate changes. The evidence suggests a continuing trend in temperature increases (Nicholls, 1997). Our current understanding of the plant physiological consequences of climate change (Morison and Gifford, 1984), combined with evidence of the impact on yields allows us to use simulation models for scenario analyses (Reyenga et al., 1999).

For wheat, Howden et al. (1999) investigated key adaptation options such as choice of cultivars and sowing windows and found significant regional differences for 10 sites throughout the Australian wheat belt. Specifically, they found likely impacts not only on production but also on grain quality characteristics such as protein content. Their findings imply that nitrogen fertilization rates need to be increased in future if current grain quality levels are to be maintained. Using the same modeling approach, Reyenga et al. (1999) found that by 2100 changes in temperature, CO2 levels and rainfall patterns could lead to a movement of the 'cropping frontier' in eastern Australia by about 100 km to the west. Such studies are likely to influence future land use policy decisions.

Climate varies at a range of temporal scales with ENSO and global change at the high and low end of the frequency spectrum, respectively. Decadal and interdecadal climate variability are the new buzzwords in climate science (Allan, 1999; Navarra, 1999; Power et al., 1999). Such fluctuations can have considerable consequences for crop production. While there have been many examples where farming practices in Australia were originally developed during periods of above-average rainfall, research has found that in later decades there were periods when reliable rainfall patterns did not persist. Meinke et al. (2000) demonstrated the impact of decadal climate variability on Australian crops and cropping systems and showed that simulated historic wheat yields varied strongly according to these decadal signals. Simulations of a range of cropping systems management options also showed considerable potential to increase profitability and sustainability by strategically altering cropping intensity and crop type based on knowledge of such long-term fluctuations.

EXISTING SERVICES AND FUTURE DEVELOPMENTS

The previous sections described a wide range of R & D activities in Australia that are related to climate prediction. Information generated by these activities is disseminated via a range of services that provide seasonal outlooks for climate, crops and pastures via rural press, feature articles, fax services, TV and radio. WEB sites such as the 'Long Paddock' (www.dnr.qld.gov.au/longpdk) or the SILO project (www.bom.gov.au/silo) provide up-to-date and in some cases even customized information. Software products such as 'Rainman' and 'Whopper Cropper' have been developed to cater for specific needs. Coordinated R, D & E teams such as APSRU and QCCA have been established and have links to other research organizations and client groups across the country and around the world. Client response and support have been encouraging but there are several areas that require future investment.

Enhanced Connection to Capacity-Building

So far, the integrated approach to systems research advocated earlier has not been very effective in linking with University-based teaching and research. University courses rarely teach the systems analytical skills required and it is often difficult to attract appropriately trained staff. This has been recognized and steps are currently being taken to address this issue.

Improving Farm Scale Applications

A greater emphasis on whole-farm economics is required to make the information more relevant. This should also allow the evaluation of new policy decisions on farm performances. Economists should be encouraged to become active team members and help to shape project development and design. A more holistic form of choice assessment must also account for the human dimension and evaluate options within a given socio-economic context.

Further Improving Seasonal Climate Forecasting

Further climate research that is well targeted towards relevant applications is required. This should include research into longer lead times of the forecasts and the investigation of prediction schemes with timescales beyond ENSO. Sensibly interfacing climate (GCMs) and agricultural simulation models should be perused via multidisciplinary research projects and teams.

Internationalization of the Systems Approach to Applications

In this paper we briefly outlined some of the R, D & E activities in Australia. Other groups around the world have also shown considerable interest and skill in operationalizing climate predictions in order to develop effective climate applications (e.g. groups in Uruguay, USA, Zimbabwe and India to name a few). There is an urgent need to further enhance existing links between these groups and to build new partnerships. Organizations such as IRI, ICASA, APN, ENRICH, IAI and START are already playing a vital role in facilitating such partnerships. Active participation in such international efforts should be strongly supported by all players so that the emerging scientific breakthroughs are captured quickly to the benefit of the global community. This requires true global collaboration for notable and effective local delivery of benefits.

REFERENCES

Allan, R.J. 1999: ENSO and climatic variability in the last 150 years. In Diaz, H.F. and V. Markgraf (eds), El Niño and the Southern Oscillation: Multiscale variability and its impacts on natural ecosystems and society. Cambridge University Press, Cambridge, UK, in press.

Carberry, P., Hammer, G.L., Meinke, H. and Bange, M. 2000. The potential value of seasonal climate forecasting in managing cropping systems. In: G.L. Hammer, N. Nicholls, and C. Mitchell (eds.), Applications of Seasonal Climate Forecasting in Agricultural and Natural Ecosystems—The Australian Experience. Kluwer Academic, The Netherlands, in press.

Chapman, S.C., Cooper, M., Hammer, G.L., and Butler, D. 2000a. Genotype by environment interactions affecting yield of grain sorghum. II. Frequencies of different seasonal patterns of drought stress are related to location effects on hybrid yields. Aust. J. Agric. Res., in press.

Chapman, S.C., Imray, R. and Hammer, G.L. 2000b. Relationship between seasonal forecasts and grain prices. In: G.L. Hammer, N. Nicholls, and C. Mitchell (eds.), Applications of Seasonal Climate Forecasting in Agricultural and Natural Ecosystems—The Australian Experience. Kluwer Academic, The Netherlands, in press.

de Jager, J.M., Potgieter, A.B. and van den Berg, W.J. 1998. Framework for forecasting the extent and severity of drought in maize in the Free State Province of South Africa. Agric. Syst., 57: 351-366.

de H.N. Maia, A. and Meinke, H. 1999. Non-parametrical survival analysis as a statistical tool in agricultural decision making. International Symposium, Modeling Cropping Systems, Proceedings European Soc. For Agronomy, Div. Agroclimatology and Agronomic Modeling, Lleida, Spain, 21-23 June 1999, 103-104.

Hammer, G.L., Butler, D., Muchow, R.C. and Meinke, H. 1996. Integrating physiological understanding and plant breeding via crop modeling and optimization. In: Cooper, M. and Hammer, G.L. (eds). Plant adaptation and crop improvement. CAB International, ICRISAT and IRRI, pp. 419-441.

Hammer, G.L., Keating, B.A., Meinke, H., Carberry, P.S., Freebairn, D.M., Probert, M. and Holzworth, D.P. 1999. An integrated systems approach to improving agricultural systems using the agricultural production systems simulator, APSIM. International Symposium, Modeling Cropping Systems, Proceedings European Soc. For Agronomy, Div. Agroclimatology and Agronomic Modeling, Lleida, Spain, 21-23 June 1999, 31-37.

Hammer, G.L. 2000. A general systems approach to applying seasonal climate forecasts. In: G.L. Hammer, N. Nicholls, and C. Mitchell (eds.), Applications of Seasonal Climate Forecasting in Agricultural and Natural Ecosystems—The Australian Experience. Kluwer Academic, The Netherlands, in press.

Hammer, G.L., Nichols, N. and Mitchell, C. (eds) 2000. Applications of Seasonal Climate Forecasting in Agricultural and Natural Ecosystems—The Australian Experience. Kluwer Academic, The Netherlands, in press.

Howden, S.M., Reyenga, P.J. Meinke, H. and McKeon, G.M. 1999. Integrated global change impact assessment on Australian terrestrial ecosystems. Report to the Australian Greenhouse Office, CSIRO Wildlife and Ecology Working Paper 99/14, Canberra, pp 51.

Keating, B.A. and Meinke, H. 1998. Assessing exceptional drought with a cropping systems simulator: a case study for grain production in north-east Australia. Agric. Syst., 57: 315-332.

McCown, R.L., Hammer, G.L., Hargreaves, J.N.G., Holzworth, D.P., and Freebairn, D.M. 1996. APSIM: A novel software system for model development, model testing, and simulation in agricultural systems research. Agric. Syst., 50: 255-271.

Meinke, H. and Hammer, G.L. 1995. Climatic risk to peanut production: a simulation study for northern Australia. Aust. J. Exp. Agric., 35: 777-780.

Meinke, H. and Hochman, Z. 2000. Using seasonal climate forecasts to manage dryland crops in northern Australia. In: G.L. Hammer, N. Nicholls, and C. Mitchell (eds.), Applications of Seasonal Climate Forecasting in Agricultural and Natural Ecosystems—The Australian Experience. Kluwer Academic, The Netherlands, in press.

Meinke, H., Power, S., Allan, R., Potgieter, A., Stone, R. and deVoil, P. 2000. Impact of decadal to multi-decadal climate variability on Australian agriculture. 7th National Aust. Meteorological and Oceanographic Society Conference, AMOS Conference Committee (eds.), AMOS Publication No 16, Melbourne, January 2000, 85.

Morison, J.I.L. and Gifford, R.M. 1984. Plant growth and water use with limited water supply in high CO2 concentrations. II. Plant dry weight, partitioning and water use efficiency. Aust. J. Plant Physiol., 11: 375-384.

Navarra, A (editor) 1999. Beyond El Niño: Decadal and Interdecadal Climate Variability. Springer, Berlin, 1999. 374 pp.

Nicholls, N. 1997. Increased Australian wheat yield due to recent climate trends. Nature 387: 484-485.

O'Meagher, B., du Pisani, L.G. and White, D.H. 1998. Evolution of drought policy and related science in Australia and South Africa. Agric. Syst., 57: 231-258.

Power, S. Tseitkin, F., Mehat, V., Lavery, B., Torok, S., and Holbrook, N. 1999. Decadal Climate variability in Australia during the 20th century. International J. Climatol., 19: 169-184.

Reyenga, R.J., Howden, S.M., Meinke, H., McKeon, G.M. 1999. Modeling global change impacts on wheat cropping in south-east Queensland, Australia. Environmental Software, 14: 297-306.

Shorter, R., Lawn, R.J. and Hammer, G.L. 1991. Improving genotypic adaptation in crops—a role for breeders, physiologists and modellers. J. Exp. Agric., 27: 155-175.

Stafford Smith, D.M. and McKeon, G.M. 1998. Assessing the historical frequency of drought events on grazing properties in Australian Rangelands. Agricultural Systems, 57: 271-299.

Stone, R.C., Hammer, G.L. and Marcussen, T. 1996. Prediction of global rainfall probabilities using phases of the Southern Oscillation Index. Nature 384: 252-255.

White, B. 2000. The importance of climate variability and seasonal forecasting to the Australian economy. In: G.L. Hammer, N. Nicholls, and C. Mitchell (eds.), Applications of Seasonal Climate Forecasting in Agricultural and Natural Ecosystems—The Australian Experience. Kluwer Academic, The Netherlands, in press.

Agricultural Applications of Climate Predictions: Bridging the Gap between Research and its Application in the SE USA

James W. Jones[1], J.W. Hansen[2], J. J. O'Brien[3], G. Podestá[4], and F. Zazueta[1]
[1]University of Florida, USA
[2]International Research Institute for Climate Prediction, USA
[3]Florida State University, USA
[4]University of Miami, USA

INTRODUCTION

Agriculture is one of the most important economic activities in the Southeast USA, generating tens of billions of dollars of revenue each year. Florida ranks fifth in the nation and first in the Southeast in the value of crops produced ($5.0 billion in 1997) (NASS). It is second only to California in the categories of fruits and nuts, and commercial vegetables. The region is diverse in terms of its climate and agriculture. Although most of the Southeast is temperate, tropical fruit and winter foliage and vegetable production are feasible in subtropical southern Florida. Agricultural production systems are highly diverse in terms of enterprises, scale and resource endowment, particularly in Florida. The El Niño-Southern Oscillation (ENSO) exerts a substantial influence on both climate and agriculture in this region.

Vulnerability of the region's agriculture and economy to climate fluctuations and weather extremes prompted formation of a consortium of Florida universities (Florida State University, University of Florida and University of Miami) to seek to capitalize on the potential predictability of climate impacts associated with ENSO. The goal of our research in the Southeast US is to reduce economic risks and improve social and economic well being by facilitating the routine and effective use of climate forecasts for agricultural decision making. Our objectives, aimed at bridging the gap between those who supply climate forecasts and those who make agricultural decisions or policies, are to: (i) characterize stakeholder needs, uses and perceptions of climate forecasts to guide product design and approaches for routine agricultural uses, (ii) evaluate the usefulness and limitations of climate forecasts to help stakeholders understand risks for specific applications and gain confidence in their use across the Southeastern USA, (iii) adapt and enhance research tools, methods, and data products for translating climate forecasts into information required to support agricultural decision/policy making, and (iv) demonstrate successful applications of climate forecasts with targeted stakeholders. This paper provides an overview of the approach we have used to address these objectives and lessons that we learned in the process, with particular emphasis on the successful transition from research to application in Florida.

ENSO IMPACTS

Approach

Much of our early effort in the Southeast US was devoted to understanding and quantifying impacts of ENSO on the region's climate and agriculture. Through statistical analysis of time-series data, we supplemented what was already known about climate response, and documented crop response to ENSO in the region.

What We Learned

Climatic impacts of ENSO vary substantially in space and with time of year. In the winter and early spring, El Niño is associated with lower temperatures and enhanced precipitation in most of the region (Green et al., 1997). Summer impacts of El Niño are more localized, including drier conditions along the Atlantic Coast and from North Texas to northern Alabama. Evidence indicates that regional climate anomalies associated with very strong El Niño events are not simply amplifications of normal El Niño conditions (Rosenberg et al., 1997).

With some exceptions, La Niña shows effects that are opposite those of El Niño. Although La Niña winters are usually dryer than normal, precipitation tends to increase in the Gulf Coast region in the spring. Effects in the subsequent summer are weaker and spatially more variable, and include reduced temperatures in the extreme South and reduced precipitation in parts of Louisiana. We have seen some evidence that the summer rainy season is shifted earlier in southern Georgia and Alabama. El Niño conditions significantly reduce Atlantic hurricane landfall frequency in the US whereas La Niña conditions have a smaller, positive influence (Bove et al., 1998). Florida's highly profitable citrus and fresh winter vegetable industries are particularly vulnerable to low temperature extremes. Unfortunately, ENSO apparently does not influence the probability of low temperature extremes (Hansen et al., 1999) or agriculturally-important freezes (Downton and Miller, 1993) in Florida in a consistent or predictable manner.

Regional impacts of ENSO on agriculture, and the potential value of ENSO-based climate forecasts to agriculture, are substantial. In North America, the main summer cropping season does not coincide with the strongest ENSO signal. However, we found that ENSO significantly influenced maize, wheat, cotton, tomato, rice, sugarcane and hay in eight southeastern states (Hansen et al., in press). ENSO phases explained a shift of $212 million, or 26% of the long-term average, inflation adjusted value of maize and $133 million, or 18% for soybean in a four-state (Alabama, Florida, Georgia and South Carolina) region (Hansen et al., 1998a). Field crop yield response in the Southeast to the two strongest El Niño events analyzed (1982-83 and 1997-98) was generally opposite in direction to response to weak-to-moderate events (Hansen et al., in press). ENSO influences yields of high-value crops such as some citrus species and winter vegetables (tomato, bell pepper, snap beans and sweet corn), and prices of bell pepper and snap bean in Florida (Hansen et al., 1998b, 1999). The Florida Department of Agriculture attributed about $165 million of losses in economic activity to agriculture and forestry in Florida due to the 1997-98 El Niño event. Adams et al. (1995) estimated the potential annual benefit of ENSO-based climate forecasts to agriculture in the Southeast US at around $100 million.

DECISION MAKERS

Approach

Our interactions with decision makers in the agricultural sector have taken several forms, including weather and climate workshops with Florida farmers; farmer surveys in northern Florida and southern Georgia and Alabama, open-ended surveys of Florida Agricultural Extension service personnel; District Extension meetings; contributing to training workshops; and various interactions with agri-business (input and information suppliers, marketing, and commodity organizations). Surveys of extension personnel, who daily interact with and advise farmers, proved to be a particularly effective means of obtaining information about farmer perspectives and opportunities for using climate prediction within agricultural production.

What We Learned

Our interactions with farmers revealed consistent interest in forecasts of weather events, particularly hurricanes, freezes, floods and abnormally high temperatures. However, they showed a wide spectrum of attitudes toward seasonal climate prediction, ranging strong skepticism to strong optimism. Reasons for skepticism included lack of understanding of the nature of ENSO influence on climate, and inadequate information about degree of uncertainty of climate forecasts at particular locations and time. Some tended to view and evaluate climate forecasts as if they were deterministic. Awareness of spatial variability of local weather events was manifest in interest in climate forecasts for their specific locations. Many farmers were sensitive to price impacts of increasing globalization, and wanted forecasts for their competitors' regions. While growers of rainfed field crops were concerned about climate fluctuations, market variations tended to dominate decisions for high-value crops. Perceived flexibility to adjust management in response to climate expectations varied among farm types and sizes. Farmers showed consistent interest in learning more about ENSO impacts and climate prediction independent of their degree of confidence in, or perceived flexibility to respond to, such information.

One of the practical lessons from our decision maker interactions was that farmers in the region tend to evaluate the credibility of information and advice based on its source. While they tend to be skeptical or at least cautious about "experts" from outside the community, they generally trust the state extension services and their county agents. A second lesson was the difficulties in understanding the potential value and risk associated with agricultural uses of climate forecasts in this region, mainly due to the diversity in agriculture and farming situations in this region and the extensive nature of agriculture. The extension services have offices in communities throughout the region and they have agents and specialists who understand the agricultural systems as well as the farmers. It became clear that our research effort had to work closely with the extension services in order to achieve widespread understanding and use of climate forecasts by the farming community in this region.

MANAGEMENT OPTIONS

Approach

We followed a two-fold strategy for identifying management options for taking advantage of climate forecasts. Interactions with decision makers (discussed above) included assessment of actual and potential uses of climate prediction for adjusting decisions. We also used crop simulation models and systems analysis to explore the potential for tailoring field-scale crop management and farm-level crop mix to expected climate associated with ENSO phases for a typical farm in Southern Georgia that grows. Analyses considered the probabilistic nature of forecasts within an economic risk framework.

What We Learned

The most common decisions that farmers and extension personnel indicated might be adjusted in response to credible climate forecasts were: field preparation, marketing, livestock stocking rates, crop choice or area of each crop, choice of crop varieties, and crop management. North Florida farmers did not reporting changing specific management in response to ENSO information. In South Florida, however, farmers cleared state-operated drainage canals and reformed their potato fields for better drainage when they learned about the expected 1997-98 El Niño. One South Florida strawberry grower reported using climate forecasts to decide on what variety to plant.

Analytical studies to date have focused on two of these strategies for rainfed field crop production: crop selection and crop management. Analysis of optimal allocation of farm land among crops integrated crop simulation results with, whole-farm budgeting, climate risk and risk aversion in an expected utility framework (Messina et al., 1999). Results indicated that conditioning crop mix on ENSO phases in small to medium field crop farms could increase profits an average of \$4 to \$6 ha^{-1} y^{-1} (Jones et al, submitted). A global optimization algorithm linked to crop simulation models and enterprise budgets (Royce et al., submitted) was used to explore maize and wheat management strategies for increasing average profits. Results indicate that potential average benefits of tailoring planting date, variety, plant population, and N fertilizer management to ENSO phase are on the order of \$5 to \$30 ha^{-1} y^{-1} (Hansen et al., in press; Jones et al., submitted). Results of these studies are, however, still quite tentative. They have not been conducted at enough locations to establish their generality. More importantly, they have not yet been evaluated under on-farm conditions. Plans for on-farm evaluation of model-derived crop management strategies by cooperating farmers have been postponed in part because of concerns about unrealistic predictions of maize response to planting date.

Results of interactions with agricultural decision makers and analytical studies lead us to conclude that practical, acceptable options exist for using climate forecasts to improve farm decision making. The range of options has not been fully explored for the field crop study, and there are many more types of agriculture in this region that we have not even attempted to study. Those options that have been identified need further evaluation before they can be recommended broadly with confidence.

TOOLS AND METHODS

Approach

Use of climate prediction in agriculture is a new technology, and does not yet have an array of established and tested methods. Much of our effort has been devoted to developing, implementing and testing tools and methodology, including methods for processing and analyzing time-series data; use of crop models, weather generators, optimization procedures, and economic and risk analysis; and interactions with stakeholders. We developed a field-scale management optimization tool (Royce et al., submitted), a farm-scale land allocation optimization tool that considers risk attitudes (Messina et al., 1999), stochastic weather generators for translating monthly forecasts or ENSO phases (Grondona et al., submitted) into daily weather realizations, a nested regional dynamic climate model (building on prior efforts) (LaRow and Krishnamurti, 1998), and a framework for evaluating alternative forms of climate forecasts in agriculturally-relevant terms.

What We Learned

Analytical tools are necessary but not sufficient for guiding use of climate prediction information. Dynamic crop models have proven highly useful to researchers for understanding agricultural impacts of climate variability, and for assessing both the potential value and risks of applying climate forecasts to adjust management. Farmers are generally open to using analytical tools for decision support in the context of a broad range of information from multiple sources. They want access to analyses, forecasts and decision support tools that are tailored to their local conditions. This will require additional work on methods for downscaling forecasts and other relevant information. Further development of user-oriented decision support tools is needed and should be guided by stakeholder needs, but not at the exclusion of strategic innovation from researchers.

FROM RESEARCH TO APPLICATION

Approach

The confidence that farmers place on their county extension agents and the lessons learned in our initial research encouraged us to investigate the possibilities of a joint research and extension program on climate forecast applications in Florida. We first obtained a strong endorsement from the Extension administration of the University of Florida's Institute of Food and Agricultural Sciences (IFAS) through a series of meetings with the Dean for Extension as well as Extension District Directors. The administration encouraged us to work closely with an existing extension weather program (the Florida Agricultural Weather Network, or FAWN). FAWN provides real-time (15 minute) weather data through a system of automated weather stations distributed throughout Florida. It was created in response to demand from the agricultural community after the National Weather Service discontinued support for agricultural weather services in 1996). An unexpected freeze in Florida the next winter caused major losses to citrus and vegetable industries.

A pilot project was funded in 1999 by IFAS research and extension administration to integrate research and extension efforts to expand the mission of FAWN. One goal of the new research and extension team is to evaluate the potential acceptance and use of ENSO and climate prediction information on a sustained basis by clients in our state. One objective of this cooperative effort is to determine the most effective methods for delivering climate and weather information for different Extension Districts and commodities. The FAWN website has been redesigned to accommodate climate forecast information and to provide both weather and climate decision aids to the agricultural community in Florida (http://fawn.ifas.ufl.edu).

The initial implementation of the new FAWN (June, 2000) contains background information on ENSO oriented toward farmers, and a variety of figures that show ENSO impacts on climate in the region (O'Brien et al., 1999). Farmers have expressed interest in knowing how climatic impacts of ENSO vary both in space and in time through the growing season. Maps of mean seasonal temperature and precipitation anomalies associated with El Niño and La Niña show impacts at a relatively high spatial resolution (Figure 1). Monthly graphs show the seasonal patterns of mean climate and anomalies for each ENSO phase for any of the 88 sites selected by the user (Figure 2). Finally, users are provided with probabilistic presentations of monthly climate variables, expressed as smoothed histograms or time series plots, segregated by ENSO phase (Figure 3).

In addition to providing climate information and related decision aids, the new FAWN will have tac-

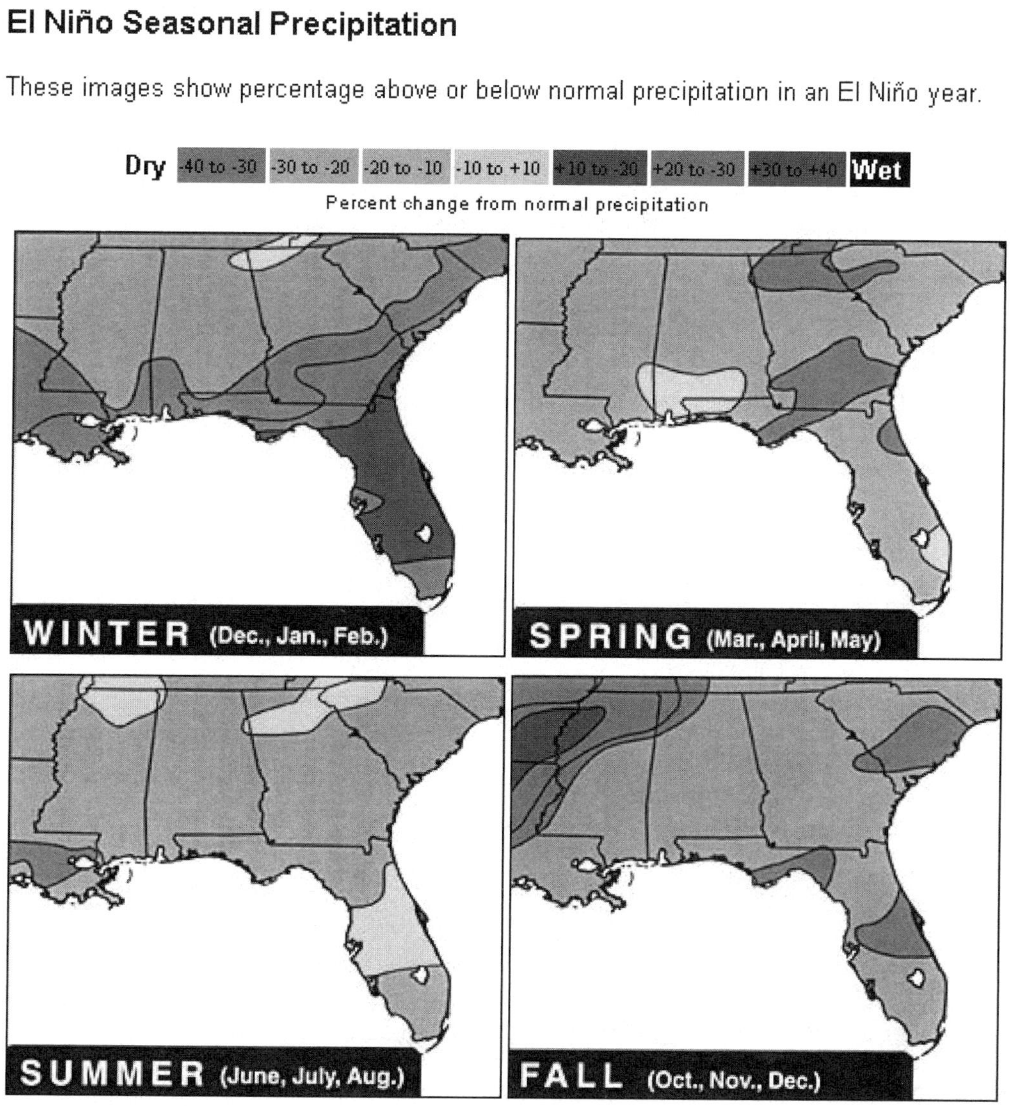

Figure 1. *Spatial distribution of seasonal precipitation response to El Niño events in the Southeast USA (source: http://fawn.ifas.ufl.edu).*

tical decision aids that use real time weather from the FAWN weather station network. Two decision aids have been completed. The first is a model to predict minimum night time temperatures to help citrus and vegetable growers decide whether or not to implement frost protection measures during nights when freezing temperatures may occur. This decision aid is now available for use. The second example is a model to predict the severity of the late blight disease on potato. An existing model (Krause et al., 1975) was programmed as a web application to allow users to select a location and planting date to determine a fungicide application schedule recommendation (Figure 4). This decision aid is ready for implementation (Koo, personal communication).

Having an Internet climate information system for agriculture in Florida was only the first step. A Statewide Major Program (SMP) is now being established on weather and climate in Florida. An SMP is the official mechanism for engaging extension specialists and county agents throughout the state. This will allow extension to develop in-service training and other educational programs for its clients, to assess impacts of their programs, and to contribute to the continued evolution of the new FAWN by providing feedback from the broad agricultural community.

The strategy of the research team was to first develop an application program in Florida, then expand into other states if successful, taking into account lessons learned in this first effort. Because

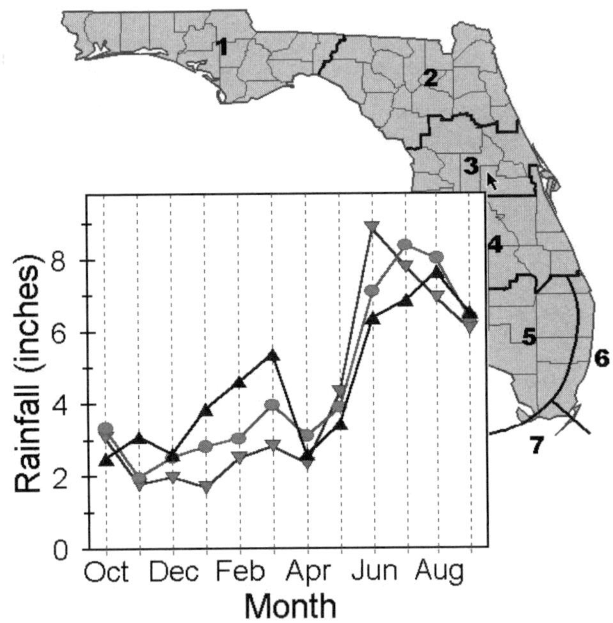

Figure 2. Mean monthly precipitation total for each ENSO phase in Florida climate district 2 USA (source: http://fawn.ifas.ufl.edu).

of our initial success and growing interest in climate in this region, partly due to the extreme drought during early months of year 2000, cooperators have been identified in Georgia. A research proposal was written to obtain funds needed to expand into Georgia and beyond.

What We Learned

Time is required to understand mechanisms of predictability of climate fluctuations and impacts on agricultural systems, to identify and evaluate opportunities to use that information to improve decisions, to understand decision maker perspectives, and to understand the roles of institutions that influence those decision makers. In addition to the efforts of a research team, successful and sustained application of climate prediction requires both interest of individual beneficiaries and commitment on the part of relevant institutions. In Florida, the agricultural extension service proved to be an ideal mechanism for fostering sustainable application of climate prediction. Given the lack of distinction in the minds of area farmers between weather and climate, we found that a climate prediction information system can complement the state's existing weather programs. The SMP will provide emphasis on climate and weather program throughout the state and through all agricultural enterprises. It will provide information on the use of this information and on the impacts of its use.

We have found that close cooperation with both existing institutions and individual decision makers greatly influences research priorities, and is essential for success. Close cooperation with extension has led our team to reconsider its research priorities. User demand for information and decision aids for implementation of a useful climate information system for agriculture requires investment of time and resources beyond what our research team has. Demand for research aimed at delivering information needed by clients must be balanced

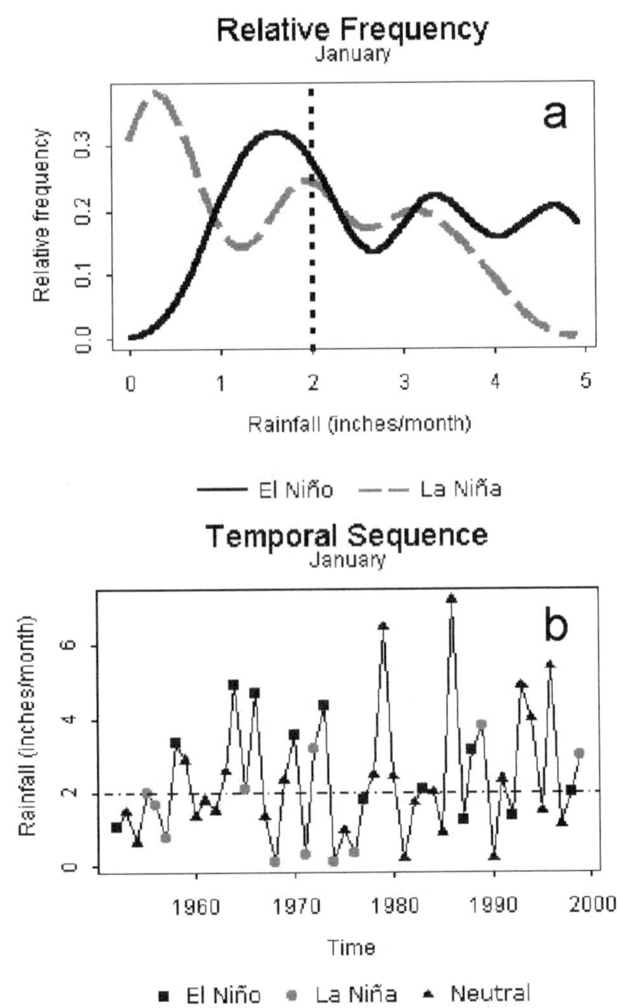

Figure 3. Influence of ENSO on January precipitation total in Orlando, Florida, presented probabilistically as (a) smoothed histogram and (b) time series USA (source: http://fawn.ifas.ufl.edu).

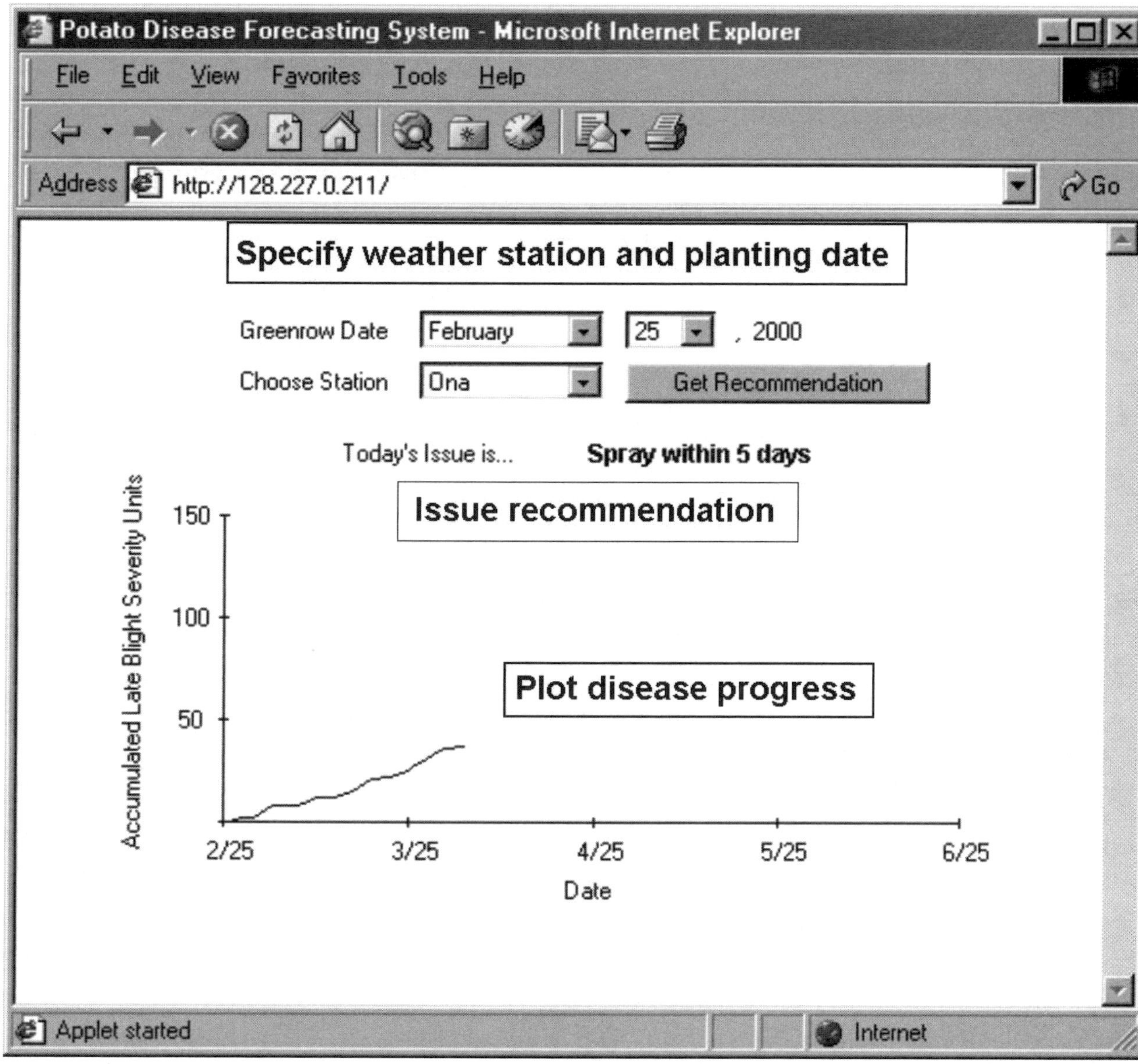

Figure 4. *Blitecast potato late blight tactical decision aid adapted for internet access by J. Koo, University of Florida.*

with the goal of strategic research to investigate new innovative methods and approaches. Nevertheless, we have learned that this "research on demand" is essential for bridging the gap between our specific studies and needs of the agricultural community. By combining our knowledge and methods with the resources of the extension service, considerably more can potentially be accomplished. Because of the SMP, we anticipate that Extension Specialists will help define and develop specific information and decision aids for specific commodities and decisions.

CONCLUSIONS

Opportunities for using predictability of climate fluctuations and agricultural impacts associated with ENSO in the Southeast US offers considerable economic potential. However, that full potential will be difficult to realize. An exceptional effort is needed to put our present understanding of impacts and opportunities into practice. Sustainable application of climate prediction information to management of agriculture is possible. The strong link between stakeholders and extension services provides a mechanism. There is a need and opportunity to develop an operational climate prediction application program by linking work done by the Florida Consortium with state agricultural extension programs.

REFERENCES

Adams, R.M., Bryant, K.J., McCarl, B.A, Legler, D.M., O'Brien, J., Solow, A., Weiher, R., 1995. Value of improved long-range weather information. Contemporary Economic Policy 13, 10-19.

Bove, M.C., J.B. Elsner, C.W. Landsea, X. Niu, and J.J. O'Brien. 1998. Effect of El Niño on U.S. landfalling hurricanes, revisited, Bull. Am. Meteorol. Soc. 79:2477-2482.

Downton, M.W. and K.A. Miller. 1993. The freeze risk to Florida citrus. Part II: Temperature variability and circulation patterns. J. Climate 6:364-372.

Green, P.M., Legler, D.M., Miranda V, C.J.M., O'Brien, J.J. 1997. The North American climate patterns associated with El Niño-Southern Oscillation. Report 97-1, Center for Ocean-Atmospheric Prediction Studies, Tallahassee, FL 32306-2840.

Grondona, M.O., G.P. Podestá, M. Bidegain, M. Marino, and H. Hordij. In press. A stochastic precipitation generator conditioned on ENSO phase: A case study in southeastern South America. J. Climate.

Hansen, J.W., A.H. Hodges, and J.W. Jones. 1998a. ENSO influences on agriculture in the southeastern United States. J. Climate 11:404-411.

Hansen, J.W., Irmak, A., and Jones, J.W. 1998b. El Niño—Southern Oscillation influences on Florida crop yields. Soil and Crop Science Society of Florida Proc. 57:12-16.

Hansen, J.W., J.W. Jones, A. Irmak, and F.S. Royce. In press. ENSO impacts on crop production in the Southeast US. In Impacts of Climate Variability on Agriculture. ASA Special Publication. American Society of Agronomy, Madison, Wis., USA.

Hansen, J.W., J.W. Jones, C.F. Kiker, and A.H. Hodges. 1999. El Niño-Southern Oscillation impacts on winter vegetable production in Florida. J. Climate 12:92-102.

Jones, J.W., J.W. Hansen, F.S. Royce, and C.D. Messina. Potential benefits of climate forecasting to agriculture. Submitted to Agric. Ecosys. Env.

Krause, R.A., L.B. Massie, and R.A. Hyre. 1975. Blightcast: A computerized forecast of potato late blight. Plant Dis. Reptr. 59:95-98.

LaRow, T.E., and T.N. Krishnamurti. 1998. Initial conditions and ENSO prediction using a coupled ocean-atmosphere model. Tellus, 50A, 76-90.

Messina, C.D., Hansen, J.W., and Hall, A.J. 1999. Land allocation conditioned on El Niño-Southern Oscillation phases in the Pampas of Argentina. Agric. Syst. 60:197-212.

O'Brien, J.J., Zierden, D.F., Legler, D., Hansen, J.W., Jones, J.W., Smajstrla, A.G., Podestá, G., and Letson, D. 1999. El Niño, La Niña and Florida's Climate: Effects on Agriculture and Forestry. The Florida Consortium (Florida State Univ., Univ. of Florida and Univ. of Miami).

Rosenberg, N.J., R.C. Izaurralde, M. Tiscareño-Lopéz, D. Legler, R. Srinivasan, R.A. Brown, and R.D. Sands. 1997. Sensitivity of North American agriculture to ENSO-based climate scenarios and their socio-economic consequences: Modeling in an integrated assessment framework. PNL-11699/UC-0000, Pacific Northwest National Laboratory, Washington, DC.

Royce, F.S., J.W. Jones, and J.W. Hansen. Model-based optimization of crop management for climate forecast applications. Submitted to Trans. ASAE.

IRI Strategy for Agricultural Applications

James Hansen
International Research Institute for Climate Prediction, USA

INTRODUCTION

Agriculture—"probably the most weather-dependent of all human activities" (Oram, 1989)—is the endeavor most critical for human survival and the most important sector of most national economies. Humanity is particularly vulnerable to the impacts of climate variability on agriculture in regions that are climatically marginal, strategically important, or socio-economically vulnerable. This vulnerability is expected to increase as the world's population increases and more marginal lands are brought into production. One of the reasons that climate variability is often so devastating to agriculture is that we do not know when to expect either favorable or unfavorable weather conditions, and we are thus unprepared.

Many critical agricultural decisions from farm to policy level have complex interactions with the climate, but must be made several months or seasons before the impacts of weather are realized. If the agriculture sector had better predictions of climate three to six months ahead of time, it may be possible to adjust decisions to decrease unwanted impacts and take advantage of expected favorable conditions. Although the potential benefits are quite promising, the agricultural sector is complex. Effective application of climate prediction information to agricultural production requires thorough understanding of both the physical and biological mechanisms by which climate fluctuations influence the agro-ecosystems that support production, and the set of social, economic and institutional factors that can constrain or enhance the use of that information (Jones et al., submitted; Hammer et al., submitted).

The International Research Institute for Climate Prediction (IRI) was established in 1996 with the mission to "continually assess and develop seasonal-to-interannual climate forecasts, and to foster the application of such climate forecasts to the explicit benefit of societies." Recent planning activities have shifted the IRI's emphasis toward enhancing society's capability to understand, predict and manage the impacts of seasonal climate fluctuations. Within the IRI, an Applications Research Group is charged with conducting research and fostering efforts outside the IRI directed toward apply climate forecasts within climate-sensitive sectors of society in a manner that will benefit society. Agriculture is one of three priority sectors that the IRI has targeted. Early in 1998, the IRI enlisted an Advisory Group to guide development of an IRI Agricultural Applications Strategy. The group outlined objectives, elements of a framework and initial areas of activity; proposed guidelines for external review and guidance; and identified priority staff positions. Building on the work of the Advisory Group, this paper presents the IRI perspective on purpose, guiding principles and appropriate strategy for agricultural applications efforts. Although it represents the current position of the IRI, it is an intermediate product of an ongoing planning process intended to elicit constructive feedback.

GUIDING PRINCIPLES

The purpose of the IRI's Agricultural Applications Strategy is to enhance society's capability to apply predictions of seasonal climate fluctuations to improve stability and sustainability of agricultural production, distribution and consumption systems. In addition to the IRI mission and this statement of purpose, refinement and evaluation of emphases, projects and activities must consider other issues such as the needs and expectations of clients and intended beneficiaries relative to the strengths and resources of the IRI.

Clients and Beneficiaries

We sometimes refer to climate prediction application as a comprehensive "end-to-end" process that spans the gap between those who produce and those who use climate forecasts. Although farmers or, from a food security perspective, consumers (including subsistence producers) are typically regarded as the ultimate end users in the agricultural sector, a network of intermediaries—researchers, extension and private-sector advisors, input suppliers, lenders, marketing and distribution channels, policy makers and insurers—potentially act both as users and disseminators of climate prediction information as they process, interpret and act on that information in a manner that influences or constrains farmer decisions. Intermediaries concerned with food security include policy makers, institutions responsible for food security alerts, and food relief donor agencies.

IRI agricultural applications efforts will seek to benefit target populations of producers or consumers in a manner that does not degrade the natural resource base, or threaten future food security or livelihood. Evaluation of particular projects and activities will consider the vulnerability of producer or consumer populations to impacts of predictable components of climate variability on agriculture. However, the IRI recognizes the many intermediate actors who have the potential and mandate to use climate prediction information to benefit farmers and consumers as potential clients.

Climate forecasts can be made more useful when the characteristics (e.g., timing, climatic variables, spatial and temporal resolution, presentation of uncertainty) of the forecasts are tailored to the requirements of the various users. Agricultural applications research efforts will therefore regard the climate prediction community at the IRI and elsewhere as an appropriate client.

What Resources and Strengths Does the IRI Offer to the Agriculture Community?

Agriculture in any given region is supported by research and developments efforts of an enormous network of public, NGO and private institutions. Resources of the IRI will always be small relative to the magnitude of the challenge of managing impacts of climate variability on agriculture. Yet the IRI is developing a broad base of expertise that complements the resources and expertise of existing agricultural research and development institutions. The unique contribution of the IRI is its integrated capacity to address both production and application of climate forecasts. Specific strengths that the IRI currently offers include:

- Climate prediction expertise, tools and products.
- A quantitative systems framework and tools for understanding the interface between climate variability and crop production systems at field, farm and regional levels.
- Expertise related to the institutional factors that influence both vulnerability to climate variability impacts and adoption of improved strategies conditioned on climate forecasts.
- An evolving framework for global-scale synthesis and interpretation of regional impacts of climate variability on food production.
- Ability to engage and coordinate expertise of broader research communities related to climate prediction and agriculture.

The IRI will seek to continue to enlist additional relevant expertise both internally and through its partners.

Facilitating "End-to-End" Applications

Prerequisites of beneficial application of climate prediction information include understanding of potentially complex interactions among human decision makers, climate variability and ecosystems; unimpeded and undistorted flow of appropriate information; and political and institutional commitment to promote and sustain beneficial application of the information. We use "end-to-end" to describe the entire process needed to support and sustain application of climate prediction information; it does not imply that producers and end users are the only relevant actors in the process. A breach at any point in the "end-to-end" process can impede application and constrain the value of the forecast information. In line with its evolving mission, the IRI has adopted the role of facilitating understanding and addressing gaps in the entire process that can enhance or obstruct potential benefits of climate prediction information. Although details of this role are evolving, it will include:

- Identifying and analyzing key knowledge and methodological gaps, and policy and institutional impediments to the successful application of climate prediction to agriculture for the benefit of society,

- Studying ongoing efforts to apply climate prediction information to agriculture in order to identify lessons that can be transferred and applied elsewhere, and
- Actively seeking to influence the agendas of relevant institutions responsible for information, research and implementation in response to identified knowledge gaps and institutional impediments.

All IRI program activities, including those focused on agriculture, will be assessed on their potential to enhance, and address gaps and impediments in, the entire "end-to-end" process of generation, dissemination, application and evaluation of climate prediction information.

Principles and Criteria

IRI involvement in projects and activities related to agricultural application of seasonal climate forecasts are based on several principles and criteria:

- The IRI will give priority to projects, locations, agricultural systems and populations that show the highest likelihood of the greatest societal benefit. Potential societal benefit is related to the degree of vulnerability of human populations to variability of agricultural production or food availability associated with predictable components of climate fluctuations. Likelihood of success is influenced by a range of factors including availability of data, identified IRI expertise, regional collaborators and institutional support.
- The IRI will seek to enhance, and address gaps in, the entire "end-to-end" process of generation, dissemination, application and evaluation of climate prediction information. Although a particular project might not address every aspect of that process, the portfolio of combined projects of the IRI and collaborators in a region should address the process in a comprehensive manner. The IRI will also consider more limited exploratory projects that have the potential to contribute to the development of a more comprehensive program.
- Clear opportunities to address key scientific questions, refine climate prediction research and products, develop or enhance tools and methods, or glean lessons that are general or transferable to other regions will favor IRI participation in particular projects.
- The potential for leveraging other organizations' efforts toward the IRI's mission will favor IRI participation or facilitation. In most regions of the world, a network of local, national and international research, extension, development and aid organizations support agriculture. Many are already researching and promoting applications of seasonal climate prediction. The IRI recognizes the contribution of such organizations, and their essential role in any sustainable efforts to apply climate prediction to agriculture.
- The IRI will seek to build local and regional technical and institutional capacity to understand, predict and manage the impacts of seasonal climate fluctuations on agriculture. Capacity will be enhanced through several means including project collaboration, visitor programs and training. The IRI will avoid competing with capable local or regional institutions.
- Serendipity and political concern are legitimate considerations in project selection where relevant to maintaining a portfolio that is well balanced in terms of, e.g., geographic region, project scope, spatial scale, or stakeholder interest.

RESEARCH THEMES

Most of the effort currently devoted to applying climate prediction to agriculture is in the context of local-to-regional projects oriented toward food production. The Agricultural Applications Advisory Group recognized that, although the IRI must foster and participate in such efforts, it should not and will not have the resources to compete with regional efforts led by universities or agricultural research institutes. With this in mind, the Advisory Group proposed two themes for the IRI Agricultural Applications strategy: application of climate prediction to global food security issues, and to improve management of agricultural production at local-to-regional scales. Two complementary areas of research—development of tools and methods, and refinement of IRI forecast products—are necessary to support the food security and regional agricultural application efforts.

Refinement of IRI Forecast Research and Products

The Applications Research Group has the responsibility to make sure that forecast products are relevant to the needs of users in the targeted sectors. Two general needs have already been identified within the agricultural sector. The first is objective, quantitative, probabilistic characterization of the uncertainties associated with climate forecasts. From a research and policy perspective, characterization through hindcast analysis facilitates economic valuation of climate forecasts. From a communication and decision perspective, farmers have been known to err both from too little and too much confidence in forecasts in the absence of objective information on their uncertainties communicated in a relevant format. Second, crop simulation models are now used widely to understand biological response and tailor management strategies to predicted climate fluctuations creates. Mismatch between the spatial scale and temporal characteristics of model-based climate forecasts and the requirements of crop models creates a need for approaches to downscale climate forecasts in space and time. The downscaling issue relates both to refinement of IRI forecast products and methodological research.

The type and timing of climate prediction information required by agricultural decision makers will vary among regions. As the IRI works with agricultural decision makers in various regions, it will identify and prioritize their information needs in order to guide prediction research and refine forecast products.

Development of Tools and Methods

As the IRI studies and participates in the end-to-end agricultural application process, we expect to find gaps where we can make substantial contributions. Contribution to tools and methods that can be applied widely and adopted by relevant institutions have good potential to advance the IRI mission. Therefore the IRI will promote and, when appropriate, conduct research on methods and tools when they clearly address gaps in the end-to-end application process. Relevant methodological research questions are likely to include appropriate downscaling of model-based forecasts in space and time (discussed below), evaluation of climate forecasts in agricultural terms, appropriate probabilistic presentation of forecast uncertainty for various agricultural decision makers, and more robust and realistic quantitative representation of farm decision making and risk management strategies in a systems analysis, simulation and optimization context.

Regional-to-Local Application to Food Production

A number of independent initiatives are seeking to apply seasonal climate forecasts to management of crop production systems to improve secure and sustainable producer livelihoods. Participation in such efforts is consistent with the IRI mission and a good match for current staff expertise. In order to foster regional agricultural application efforts without competing with capable local or regional institutions, the IRI will consider three different levels of participation in regional projects:

- Catalytic: In a catalytic role, the IRI will foster the development of new projects with assistance that might include identifying potential research partners, guidance in securing external funding, providing requested guidance in our areas of expertise and, in a few cases, providing seed funding to facilitate project development.
- Participatory: Where appropriate and feasible, IRI personnel will participate actively as investigators in project development and implementation. We will identify areas of IRI participation where either (a) expertise of IRI investigators matches needs identified within regional project teams, or (b) IRI investigators can address strategic research issues that transfer to other regions or projects. Such projects are likely to include funding for postdoctoral associates or scientific exchanges supervised by permanent IRI staff. The IRI is a participa-

tory partner in agricultural projects in southern India, Mali and Zimbabwe.

- Ownership: In a few cases, the IRI will initiate and take a leadership role in regional projects, while actively seeking and enlisting collaborators. These are likely to be projects in which building regional climate prediction capacity is the initial objective. Work in the Greater Horn of Africa is an example of IRI ownership of a climate prediction capacity building project that will include agricultural application efforts.

Selection of local-to-regional projects and the level of IRI participation will be governed by the criteria discussed earlier.

Conceptual and methodological frameworks adopted by the various regional efforts to apply climate prediction to agriculture show many common elements. The IRI has a potential role in coordinating such regional programs to facilitate exchange of approaches and tools, and synthesis of lessons learned. The IRI will promote a voluntary network of regional programs and projects that will meet periodically; share approaches, tools, data and lessons learned; identify key research gaps; and perhaps foster new initiatives. The IRI may also provide small grants to a smaller set of regional projects to facilitate sharing of information. We will ask these projects to commit to providing specific information that the IRI will then synthesize and interpret at a global scale as part of the global food security theme.

Global-to-Regional Food Security Alert and Response

Intermittent food shortages and distribution problems occur in many regions. Global population growth, declining rates of increasing yields of some important staple food crops, expected climate change and degradation of the natural resource base are expected to increase the frequency, scope and severity of food shortages in the future. Several organizations have food security monitoring and alert systems in place in vulnerable regions, particularly in sub-Sahara Africa. Proper use of seasonal climate forecasts may improve the lead time and refine the accuracy of food production forecasts and food shortage alerts, thereby improving the timeliness of interventions.

The initial plan advanced by the Advisory Group called for the development of a computer-based Global Food Alert System that will identify countries and regions of the world likely to experience food production shortages with a three-to-six month lead time. Based on further evaluation by IRI personnel and external program review, the comprehensive Global Food Alert System outlined in that plan appears to be beyond the present resources of the IRI, and redundant with other efforts. Instead, the IRI will contribute to the global food security emphasis in at least two ways. First, it will explore current and potential use of climate forecasts with organizations (e.g., FAO, USAID) that already issue food production forecasts and security alerts at continental-to-global scales. A separate strategy document will detail this effort. Second, it will collect information from other sources, particularly from a network of regional programs that use climate forecasts for predicting agricultural impacts (discussed above). That information will be synthesized, interpreted and disseminated through the IRI web site. In the short term, statistical analysis will be used to examine ENSO influence on historic country-level crop production and to explore the potential for using ENSO to predict food production anomalies. Both thrusts are likely to identify new avenues for the IRI to enhance the use of climate prediction information for global food security forecasting and mitigation.

REFERENCES

Hammer, G., J.W. Hansen,, J. Phillips, J.W., Mjelde, H.S.J. Hill, and A. Potgieter. Submitted Advances in application of climate prediction in agriculture. Paper presented at Systems Approaches for Sustainable Agricultural Development: Methodologies for Interdisciplinary, Multi-scale Perspectives, Lima, Peru, 8-10 November 1999.

Jones, J.W., J.W. Hansen, F.S. Royce, and C.D. Messina. Submitted. Potential benefits of climate forecasting to agriculture. Agriculture, Ecosystems and Environment.

Oram, P.A. 1989. Sensitivity of agricultural production to climatic change, an update. p. 25-44. In Climate and Food Security. IRRI, Manila, The Philippines.

Linking Climate, Agriculture, and Decision-Making: Experiences and Lessons for Improving Applications of Climate Forecasts in Agriculture

Graeme Hammer and Holger Meinke
Queensland Department of Primary Industries, Australia

INTRODUCTION

No one can predict what the next season will bring. Farmers, input suppliers, traders, marketers, and governments would all like to know because it is critical to their decision-making. Instead they face uncertainty that delivers risk (i.e. a chance of incurring loss, environmental degradation, loss of markets) to their business and livelihood. However, this uncertainty also generates opportunities as decision-makers can make substantial gains in good years. Strategies to cope with risk and capitalize on opportunities generated by climate variability are often critical to survival of the farmer, agribusiness, or even the government. At the very least, strategies to cope with risk due to climate variability significantly affect the economic, environmental and social stability of agricultural systems and associated communities.

Skill in climate prediction offers considerable opportunities to managers via its potential to realize system improvements (i.e. increased food production and profit and/or reduced risks). Realizing these opportunities, however, is not straightforward as the forecasting skill is imperfect and approaches to applying the existing skill to management issues have not been developed and tested extensively. While there has been much written about impacts of climate variability, there has been relatively little done in relation to applying knowledge of inherently imprecise climate predictions to modify actions ahead of likely impacts, i.e. applications of climate prediction. An effective application of a seasonal climate forecast is defined as use of forecast information leading to a change in a decision that generates improved outcomes in the system of interest. In other words, effective applications either increase profits and/or decrease risks or involve trade-offs between the two that realize improvements.

A considerable body of effort in various parts of the world is now being focused on this issue of applying climate predictions to improve agricultural systems (see other contributions to this forum) and valuable experience has been gained. These experiences have generated key lessons and insight about the approach needed to apply climate prediction in agriculture effectively. This paper will focus on the general lessons and insights rather than specific examples of applications. The material is drawn from the synthesis of contributions to a recent book, which cover a diverse range of examples and experiences in this field (see acknowledgments). The reader is referred to the extensive literature cited throughout that book. No attempt is made here to cite supporting material.

THE LESSONS

Each of six lessons will be discussed. The lessons are –

- Climate impacts are not simple transformations of climate
- Applications are about pre-emptive decisions
- Applications are about managing risk in target systems
- Targeted information is better than a general climate forecast
- Wisdom is more robust than information
- Effective application is not a linear process—it requires an integrated systems approach with co-learning

Lesson 1—Climate impacts are not simple transformations of climate

There is a common misconception that an agricultural system will respond in the same way as climate. In particular, in semi-arid environments, it is often believed that a high rainfall season will be a high yield season and a low rainfall season will be a low yield season. This is not the case. While there is often a good correlation between rainfall and yield, there are many system attributes and interactions (such as soil moisture storage and temporal patterns of crop water demand in relation to moisture supply) that work in more complex ways. Figure 1 shows seasonal rainfall and the associated simulated seasonal wheat yield for years in a particular ENSO phase for a site in NE Australia. There are many above/below median rainfall years that correspond to below/above median yield.

It is important to realize that the dynamic complexity of the target agricultural system often ensures that the climate-impact link is usually not simple.

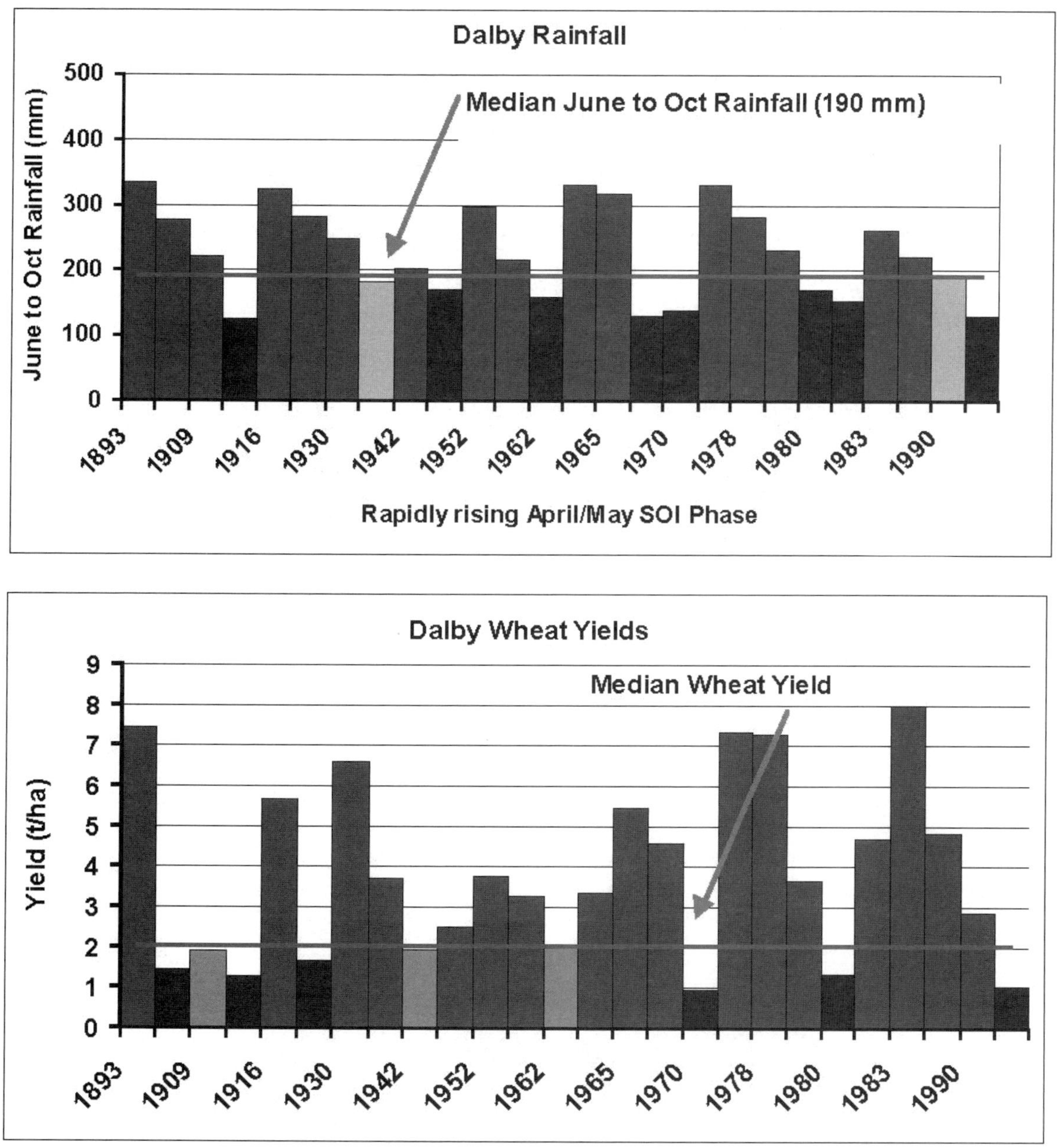

Figure 1. *Winter (June to October) rainfall at Dalby, Queensland (top panel) and simulated wheat yield (lower panel) for years with rapidly rising SOI phase in April-May (see Stone, R.C. Hammer, G.L. and Marcussen, T. (1996). Prediction of global rainfall probabilities using phases of the Southern Oscillation Index. Nature. 384: 252-255)*

Lesson 2—Applications are about pre-emptive decisions

This point has been made in the introduction. Applications use knowledge of imprecise (probabilistic) climate forecasts to consider opportunities to modify actions ahead of likely climate impacts. The focus is on decisions and likely outcomes—not the climate forecast. Agricultural system managers are most interested in the likely outcomes from their business/livelihood critical decisions.

Lesson 3—Applications are about managing risk in target systems

To be credible and scientifically honest, a seasonal climate forecast must be a probabilistic statement. Hence, applying seasonal climate forecasts must be done in a risk management framework. The importance of risk implications emphasizes an often poorly grasped point—the skill in capturing the unexplained variability in the imprecise (probabilistic) forecast is at least as important as the skill in capturing shift in the median or mean. Further, as climate impacts in agricultural systems are mostly not simple transformations of climate, it is also important to convert credibly the climate risks to system outcome risks. This provides a key role for agricultural systems modeling (see Figure 2 and Lesson 4).

Whether or not to modify a decision depends on the expected change this brings in the outcome and any change in the associated risk. The attitude of decision-makers to risk is crucial in their decision-

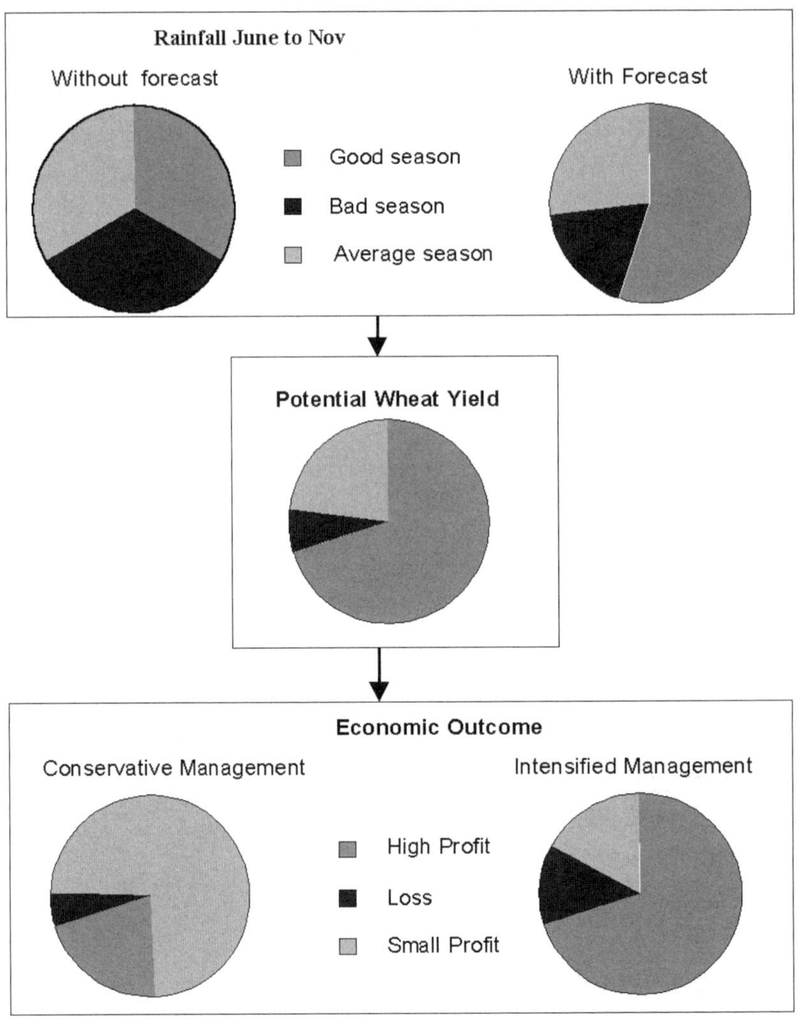

Figure 2. Yield and economic risks associated with conservative and intensified management applied in seasons with a climate forecast with high chance of good rainfall.

making process. Hence, a significant increase in expected production or profit by changing management may not be realized because any associated increase in risk (or perceived increase) is unacceptable to the decision-maker. Awareness of these issues among researchers, analysts, and decision-makers is highly variable and this variability in awareness generates many of the difficulties observed in effectively applying climate forecasts in agriculture.

It is imperative to move beyond categorical statements of forecasts and system outcomes to ensure effective applications of seasonal climate forecasts. It is simply unacceptable to use communication difficulties as an excuse for falling into categorical language. Some existing skill measures for climate forecasts, which focus on "hits" and "misses" are no help in this regard. Overcoming this major impediment may need an educative approach for both decision-makers and forecasters.

The example shown in Figure 2 builds on the rainfall-crop yield association shown in Figure 1. The first panel shows the pie chart for climatology (on the left) and the pie chart for the seasonal forecast based on SOI phase analogues. The potential wheat yield pie chart associated with the probabilistic climate forecast shows an amplified chance of a good-yielding season. There is still some chance of a poorly-yielding season, but it is much reduced from the one-third that would be associated with climatology. To achieve potential yield, however, requires high (and costly) inputs. An intensified management regime approximates this and achieves a high chance of high profit (last panel), but there is a significant chance of loss. The conservative management regime is not so risky, but has much reduced chance of high profit. The decision will depend on attitude to risk and the scale of the risk this decision brings in relation to the business. It is noteworthy that the pie charts showing the economic outcomes of the decision options look nothing like the forecast!

Lesson 4—Targeted information is better than a general climate forecast

What is the critical piece of information a decision-maker needs in making a risky decision? Often this is information on expected system outcomes and risks associated with feasible management options (see Figure 2). Such targeted information is directly relevant to the decision-maker and is more important than the general climate forecast, which on its own has little, or no, intrinsic value.

There is a critical linking role for agricultural modeling and systems analysis in generating this information on likely system outcomes. The models provide the means to explore "what if" questions on use of climate forecasts in key decisions in farming systems—what if I plant crop A rather than crop B?; what if I plant earlier than normal?; what if I intensify inputs?. Simulation studies of such questions with credible models provide estimates of the full distribution of likely outcomes and enable open discussion and evaluation of risk implications of decision options. Use of targeted information in discussions of this nature can minimize communication issues associated with probabilistic information.

Lesson 5—Wisdom is more robust than information

Knowing how to use appropriate information engenders resilience to agricultural system managers. Knowing what is important to agricultural system managers engenders insight to system scientists. Both are aspects of wisdom that can only be learnt effectively by interaction. Effective applications of seasonal climate forecasts focus on developing knowledge and wisdom systems via interaction among key players rather than by remote one-way information flow systems. No one player has a comprehensive understanding and view of the entire system, or a full awareness of any other player's understanding and view. Hence, it is critical to use a process that facilitates interaction and learning among the players in the applications milieu (see lesson 6).

Lesson 6—Effective application requires an integrated systems approach with co-learning

It is argued that an integrated systems approach (Figure 3), which connects decision-makers, analysts, modelers and researchers in a participatory manner, is the most likely means to foster learning and overcome awareness limitations in applying seasonal climate forecasts in agriculture. The sys-

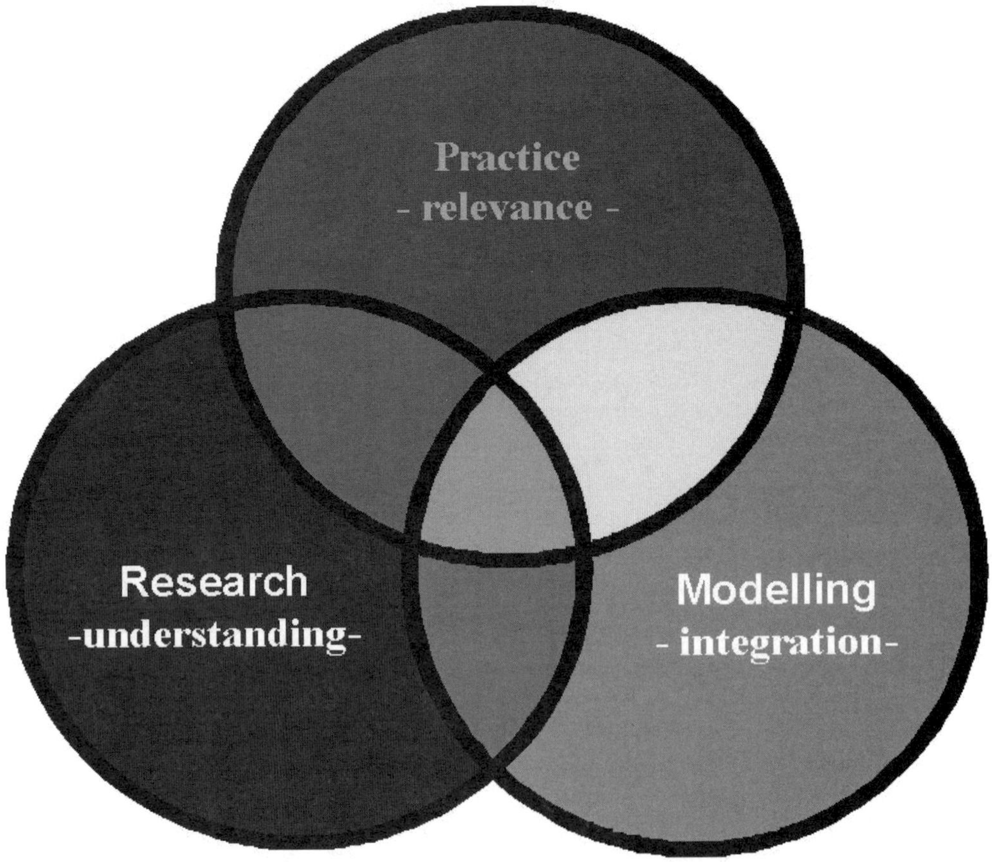

Figure 3. *Integrated systems approach to applying seasonal climate forecasts showing overlapping roles of practitioners, analysts/modelers, and researchers.*

tems approach involves practitioners in identifying opportunities for relevant system improvements in conjunction with researchers. This approach often includes scenario analysis based on simulation with credible agricultural system models as a key aspect of the learning process. The models rely on research to understand and predict responses in the target system. In turn, research can be targeted by analysts and practitioners at the issues most likely to give best payoff. Predicting system responses to management using models is a means to explore risk implications of decision options in that system. However, while these models have a considerable degree of utility in their own right, it is argued that the dialogue and learning they generate among the disparate players is equally important to effective applications of climate forecasting. The modeling provides the means to build the trust and effective relationships needed among the disparate players.

ACKNOWLEDGMENTS

The material in this paper is drawn from the synthesis of contributions to Applications of Seasonal Climate Forecasting in Agricultural and Natural Ecosystems, edited by G.L. Hammer, N. Nicholls, and C. Mitchell, Atmospheric and Oceanographic Sciences Library—Volume 21, Kluwer Academic, The Netherlands, 2000.

Climate Variability and Climate Change: Implications for Agriculture

Richard M. Adams
Oregon State University, USA

SUMMARY

Crop yield variability is a defining characteristic of agriculture. Variations in yield and production are strongly influenced by fluctuations in weather, both in terms of overall seasonal weather characteristics and extreme events. Variability also arises because of the influence of changing production practices, development of new diseases and pests, and government policy. Most climate research to date has focused on the aggregate, long-term affects of climatic change on crop yields, and the economic consequences of these changes in mean yields. This presentation instead focuses on the impact of short -term phenomena of climatic variability on crop yields. The discussion is organized around three questions: 1) what hypotheses have been advanced about climate change and variability? 2) do current data on yields suggest climate change will increase variability? and 3) what are the economic consequences of extreme events becoming more common? The exploration of these three questions draws on research and literature reviews from both the topic of long-term climate change, and the issue of shorter-term climatic variations.

A number of arguments have been made which support various hypotheses that relate climate change to changes in weather that agriculture is likely to face in the future. First, an increase in mean (maximum and minimum) temperature will increase the likelihood of extreme daily temperature events. A small change in mean will result in a large change in probability of extremes. Second, simulations performed with General Circulation Models (GCM's) show seasonal weather patterns change in selected regions or latitude bands. Daily temperature variance increases in summer, but decreases in winter. Consequently, the frequency of extreme temperature events rises with a mean shift in temperature and greater variance. Decreases in the frequency of low temperature extremes in winter are due to overall rises in mean temperature and a resulting decrease in variance. Third, GCM model simulations support an increase in precipitation intensities.

The climate literature suggests that weather patterns may become more variable under a changing (warming) climate. However, how crop yields will be affected by climatic variability depends on both biophysical and economic (human adaptation) responses to climate change. Biophysical factors related to crop response include temperature, rainfall, CO2 levels, climate variability and extreme events, and other indirect effects. Some of these factors are expected to increase yields, while others will have the opposite effect. For example, findings show as temperature rises, crop yields decrease. Rainfall is forecast to be increasing under warming, though not uniformly. When combined with temperature, some areas will have greater productivity, some areas less. Increased CO2 in the atmosphere will increase photosynthesis and plant-water use efficiency resulting in increased production. Conversely, extreme events are expected to increase thus decreasing crop yields. And indirect effects could include rising sea levels, increases in pest and pathogens, soil erosion and degradation, change in stratospheric ozone levels and ultra violet radiation, and changes in air pollution, as well as in groundwater and surface runoff.

The influence of human response and adaptation to climate variability focuses on two levels of response, under the assumption that humans possess significant potential to adapt to climate change. These levels are the farm (producer) level and the market level. At the farm level, producers adapt to new plant and harvest dates, crop rotation, crop selection, water use, fertilizer type and rate of application, and tillage practices. At the market level, trade reallocates commodities from areas of surplus to areas of scarcity. Price signals arising from activity at both levels are anticipated to spur development of new crop varieties under future climates and investment in irrigation and other forms of management. Overall, crop yields are expected to increase overtime with or without climate change, due to technological change. However, some concern is directed at the nature and speed of adaptations, relative to the speed of climate change. In addition, changes will generate increased costs in technology (necessary research and development) and may affect local/regional agricultural economies. Further, it is suspected that adaptation will be more difficult in developing countries.

To determine the consequences of climate change on actual crop yields, this paper draws on the results of a recent study that examined how year-to-year and region-to-region climate variation alters the distribution of crop yields. Data analysis included state level yields and acreage harvested, drawn from USDA-NASS, state level climate data from the National Oceanic and Atmospheric Administration (NOAA), based on time series observations for thousands of weather stations. The data were then analyzed using established production functions forms to estimate the result of both the average effect of climate on yield, and on the variance of yield. Results suggest that temperature and precipitation changes affect both the mean and variances of crop yield. On average, crop yields increase with more rainfall and decrease with higher temperatures. In terms of variability, increases in rainfall result in decreased variability of crop yield, and an increase in temperature results in increased variability of crop yield. When looking at specific crops, it was found that with increasing temperature, corn yields decrease and yield variance increases. This suggests that a warmer future climate in corn growing regions could result in reduced yields and greater year-to-year fluctuations in corn yields and total production. The yield functions were evaluated using a different source of climate change data to compute the projected yield changes. The results showed uniform decreases in corn and cotton yield variability, and mixed results of yield variability for sorghum, soybean and wheat.

A number of atmospheric scientists have hypothesized that global climate change may alter the frequency and strength of extreme events. Focusing on one such extreme event, the El Niño Southern Oscillation (ENSO), it is suggested that global change would cause the tropical Pacific region to change towards a state corresponding to present day El Niño conditions. In addition, stronger and more skewed inter-annual variability will result, with more extreme year-to-year climate variations, and strong cold events becoming more frequent. Changes in frequency of ENSO events will carry substantial consequences in term of U.S. and world economic welfare. It is further expected that the frequency of the extreme phases of El Niño and La Niña will increase while the neutral phase will decrease. An economic model of the U.S. agricultural sector was used to explore the economic consequences of such an occurrence. The model results produced three major insights into aggregate economic welfare projections under shifts in ENSO frequencies. First, ENSO events cause economic damage (losses) and more frequent events will lead to greater economic losses. Second, at the farm level, operating with ENSO frequency information forecasts will be useful to individual producers, though not enough to offset losses. . Third, in the aggregate, U.S. producers and foreign exporters will experience a net loss in economic welfare, while U.S. consumers will experience benefits from the use of forecasts.

In conclusion, research to date is beginning to suggest some likely patterns of response of agriculture to climate change and variability, although knowledge regarding possible shifts in the frequencies of extreme events with a new climate regime is limited. It is also difficult to incorporate certain types of weather events into climate change scenarios for use in economic assessments. For example, crop models can simulate the effects of a seasonal drought, but they cannot adequately simulate responses to very short-term events. Finally, caution must be exercised in interpreting the results of any particular climate and economic model forecasts; policy makers need to resist the temptation to rely on specific numerical estimates of economic damages or gains.

Practical & Conceptual Challenges to Climate Forecast Applications

Kenneth Broad
International Research Institute, USA

Over the past several years there have been increased efforts by the climate forecast community to demonstrate the utility of seasonal-to-interannual climate forecasts for decision makers in climate sensitive sectors. These efforts have been facilitated by a public awareness of the socioeconomic implications of seasonal climate variability following the extreme 1997-98 El Niño event. The use of climate information during the recent El Niño event, as well as experiences of organizations such as the International Research Institute for Climate Prediction (IRI) which are studying the potential societal applications of climate information, have revealed a number of challenges to achieving societal benefit from climate forecast information (1). These challenges include distributional and participatory equity; unintended negative consequences of forecast provision; and the political and ethical implications of various forecast dissemination strategies (2).

Difficulty anticipating the use of climate forecasts arises from the combination of the uncertain, or probabilistic characteristics of the forecasts, combined with a number of societal factors. Characteristics of forecasts that contribute to the difficulties adopting this information into decision making schemes include: 1) multiplicity of forecasts (i.e., there are a number of forecasts—often contradictory—available from different sources); 2) very low spatial and temporal resolution of forecasts; and, 3) difficulty linking forecast model outputs to parameters relevant for decision making (e.g., stream flow or crop yields). In addition to these limitations of the forecasts, every ENSO event presents itself differently which makes boilerplate planning for the next event difficult. This essay will focus on the societal factors which influence the use of information and will make suggestions as to how to approach these challenges.

"Objective" scientific information, such as a climate forecast, is interpreted and acted upon in light of a number of sociocultural factors. Some of these factors may be a result of societal structural (e.g., de jure and de facto regulations, availability of credit) some may be more culturally influenced (e.g., perceptions of the resilience of nature, tradition, trust in authority). Below are some of the recurrent issues which have been observed to influence the use of climate information. Grouped together, they demonstrate the challenges of equity, unintended consequences, and ethical dilemmas mentioned above.

Common Property Issues

Neither climate impacts nor forecasts adhere to national boundaries. Further, many resources (e.g., fish stocks) which are important to various (and often competitive) sets of stakeholders are effected by climate variability. The use of information by one group may adversely impact another group. For instance, during the 1997- 98 El Niño, in order to protect their environmentally stressed fish stocks, the Peruvian government attempted to place a fishing ban in the south of the country. This fish stock is shared with Chile, but nationalistic opposition in Peru argued that they should not protect their fish if they are going to be caught by Chile. Thus, the most "rational" use of climate information—from a resource sustainability standpoint—did not occur due to geopolitical concerns over common property.

Legal Structure & Economic Philosophy

A country's legal processes will affect the ability to respond to climate forecasts. For instance, the maximum lead time in which we have confidence in forecasts is approximately 6 months (3). This is a relatively short time frame in which to make legislative moves to free resources to take advantage of the anticipated conditions (e.g., subsidies for switching crops, changing production process, or identifying new markets) (4). Further, as many countries are adopting an extreme free market economic approach, there is a decreasing tendency for government intervention to facilitate reactions to climate information. In most sectors, this leaves it completely to the private sector to finance proactive measures, which tends to exclude small scale producers due to their limited access to credit.

Media

The majority of the public in most parts of the world receive their information via the media. The media has consistently displayed a tendency to distort climate related information, either due to misinterpretation of facts, or in order to sensationalize information to attract an audience. The media filter can thus thwart even the most carefully constructed scientific statements about the likelihood of certain events occurring. Presenting probabilistic information in deterministic headlines may prompt irresponsible actions by decision makers. In the case of subsistence level farmers, a lack of awareness of the uncertainty in forecasts, which may be masked by the media's interpretation, may result in planting decisions which have potentially catastrophic effects.

Tradition

Many groups have adapted to the rhythms of seasonal climate variability and have built-in mechanisms for coping with climate uncertainty (e.g., crop mixing, occupational multiplicity, communal grain reserves). Further, it is likely that there are a number of traditional indicators used to anticipate the seasons that may conflict with these "scientific" forecasts. Thus, there lies a danger in displacing such mechanisms by imposing the use of computer generated, or "scientific" climate forecasts on these groups. Again, small scale producers can not afford the risk of crop failure for the marginal gains offered by the use of climate forecasts.

Memory & Semantics

The few strong ENSO events which occurred prior to the 1997-98 event resulted in impacts which left lasting impressions on effected populations. Thus there was an expectation of similar impacts for the forecasted 1997-98 event, which in many cases (such as drought conditions in parts of southern Africa) did not occur. The question arises, rightly so, as to how much trust decision makers will have in the next "El Niño" forecast. Part of the problem in conveying the probabilistic nature of the forecasts lies in the fact that when scientists make a forecast of ENSO, they are referring to sea surface temperature in a specific area of the Pacific Ocean, and not to impacts in the teleconnected regions. However, to many decision makers, the term El Niño evokes images of drought or floods in their particular region (based on their individual and statistically insignificant experience).

What are some of the implication of the above observations?

Multidisciplinary approach to forecast applications

In order to understand what use, if any, climate forecasts can have in solving complex societal problems, there is a need for multidisciplinary collaboration on identifying the impediments to forecast use. In some cases these barriers can be overcome and in other they may require a complete restructuring of sectors (e.g., urban zoning, water resource management).

Transparency & Education

In order to minimize the unintended consequences of forecast use, forecast providers must strive to honestly convey the skill of their models in terms comprehensible to a lay user. Further, basic education allowing the user to understand the historical climate variability in their region may allow a better intuitive understanding of the probabilistic nature of climate forecasts. Forecast providers must work closely with those intermediary organizations (e.g., agriculture extension services, nongovernmental organizations) who are best situated to convey information in the proper form to decision makers (5).

Winners & Losers

Forecast providers must recognize that their information may not necessarily help all stakeholders and, in fact, even a very good forecast can result in harm to certain groups. Though it is a difficult political and ethical decision, forecast providers should decide which groups they want to help, and adopt the optimum strategies to achieve benefit for that group. Research should be done to in order to build scenarios which anticipate which groups will be affected in prior to beginning dissemination efforts (6).

In summary, forecast providers should take a research approach prior to an intervention approach. In carrying out this research, there should be recognition that:

- a forecast provider's actions inevitably de facto endorse a view of societal benefit;
- a forecast provider must define the societal benefit it wishes to confer and also acquire some socioeconomic understanding of a forecast's target area to choose optimum dissemination strategies;
- retrospective studies of how forecasts were used in real-world decisions would permit more informed future choices by forecast providers.

If a decision to undertake intervention is made, we ought to learn from past experiences and:

- take small steps;
- favor reversibility;
- plan on surprises;
- plan on human inventiveness (7).

NOTES

1 For an historical overview of the scientific and societal aspects of El Niño, see Glantz, 1996: *Currents of Change: El Niño's impact on climate and society.* Cambridge University Press, New York.

2 For discussion of uneven benefits and the implications for forecast dissemination strategy, see Pfaff, A., K. Broad, and M. H. Glantz, 1999: Who benefits from climate forecasts? *Nature,* 397, 645-646.

3 For discussion of temporal constraints on responding to climate information, see Orlove, B.S. and J. Tosteson, 1998: The Application of Seasonal to Interannual Climate Forecasts Based on El Niño-Southern Oscillation (ENSO) Events: Lessons from Australia, Brazil, Ethiopia, Peru, and Zimbabwe. Berkeley, Institute of International Studies, University of California, Berkeley: 1-60.

4 For an overview of basic economic and decision theories applied to weather and climate forecasting cases, see Katz, R.W. and A. Murphy, 1997: Economic Value of Weather and Climate Forecasts. Cambridge University Press, Cambridge. See also Mjelde, J. W., T. N. Thompson, et al., 1997: Utilizing a farm-level decision model to help prioritize future climate prediction research needs." *Meteorol. Appl.* 4: 161-170.

5 For a discussion of these points based on the 2000 Ethiopia famine, see Broad, K., and S. Agrawala, 2000: The Ethiopia Famine: Uses and Limits of Seasonal Climate forecasts. Science (in press).

6 For a detailed argument for this approach see Broad, K., A. S. P. Pfaff, and M. H. Glantz, 2000: Effective and Equitable Dissemination of Seasonal-to-Interannual Climate Forecasts: policy implications from the Peruvian fishery during El Niño 1997-98. *Journal of Policy Analysis and Management.* (in review)

7 Scott, J., 1998: *Seeing like a state: how certain schemes to improve the human condition have failed.* Yale University Press, New Haven.

EXPERIENCES OF AN AGRICULTURAL INPUT SUPPLIER

Percy M Malusalila
Seed Co, Limited, Zimbabwe

BACKGROUND

Seed Co Limited

Seed Co Limited is an agricultural seed producer and supplier to the sub-region of southern Africa, whose existence spans over 60 years of dedicated service. This is a Zimbabwe registered company, operating within the region through seed distribution agencies and strategic partners, as well as subsidiary companies in a number of countries.

Seed Co has an established research programme to service the target clientele, with crop breeding programmes in maize, soybeans, wheat and groundnuts. Extensive crop screening programmes exist. Seed produced is mainly of the field crop group, covering in the main, maize, sorghum, millet, sunflower, groundnuts, soybeans, wheat, barley, cowpeas and beans. Seed Co has distribution agency on a range of vegetable seeds.

Countries to which seed business is conducted include all countries in Africa south of the equator. With crop seed demand on the increase within Africa, Seed Co has established itself with a mandate 'to be the leading supplier of world class crop seeds in sub-Sahara Africa'(Seed Co, 2000).

Agriculture in Zimbabwe

Agriculture in Zimbabwe is confined to a single cropping season, with the rainy season being November to March/April. The country is divided into agricultural zones, based in the main on soil type, rainfall regime, and altitude. The growing season ranges from under 90 days on one end, to over 160 days on the other extreme. Agriculture is one of the main contributors to the Zimbabwe economy. Farmers in Zimbabwe make an integral part of the community. About 60% of the crop-producing farmers, largely small holder farmers, live in the fragile environments, with rainfall less that 700mm per annum.

IMPACTS OF SEASONAL FORECASTS

Current operational framework

Seed Co has to embark on extensive pre-agricultural season preparations, to ensure that the farming clientele needs are fully met. In this preparation phase, correct seed types related to the anticipated and predicted seasonal forecast, correct maturity ranges, and correct pack requirements in the right locations, are all part of the targeted programme.

The farmers on one hand are also in a planning phase. Maturity groups suitable for their areas, the crop type as advised by the Government and input supplier-driven extension messages, disease tolerant varieties, the right pack size as determined by the disposable income at that purchasing moment, have to be available.

The preparatory and planning phase is crucial, to the extent were any delay in the programme will impact negatively on the overall supply chain. No quick-fix alternatives are available in the seed supply chain. Some farmer reactions have to be anticipated in the farmer scenario. A prediction of a good season is likely to result in the farmers purchasing high yielding varieties. A bad season prediction would in turn result in short season varieties demand, increase in small pack requirements, low yield projections, with a strong possibility of possible national government and non-governmental seed aid programmes to vulnerable communities in the following season.

Short Term impact on planning activities

Given in volume terms, about 70% of the seed produced and sold by Seed Co Limited on an annual basis is maize. About 23 varieties are available on the market (Seed Co, 1999), with maturity grouping ranging from very early {110days} to very late {160days}. Added to the seed provision equation are smaller volumes of other crops that will have to be provided.

A simplified range of product supply for the current season, the short-term schedule of activities, would be as follows:

ACTIVITY	MARCH APRIL	MAY JUNE	JULY AUGUST	SEPTEMBER OCTOBER	NOVEMBER DECEMBER	JANUARY FEBRUARY	MARCH APRIL
Rainfall season					XXXXXX	XXXXXX	XXXXXX
Field production estimates [seed]	XXXXXX						
Crop packing requirements estimates		XXXXXX					
Order of packaging		XXXXXX					
Packing exercise			XXXXXX				
Design season marketing and sales programme			XXXXXX				
Target seed to holding depots				XXXXXX			
Seed sales programme				XXXXXX	XXXXXX		
After sales programmes						XXXXXX	
Alternate crops sales programme						XXXXXX	
When is seasonal forecast released?				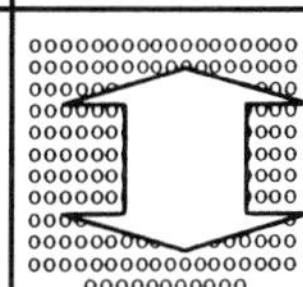			

The seasonal forecast, noting from the Zimbabwe experience, has been released in the second half of September (Zimbabwe Department of Meteorology, 1999). Given the timing, all activities conducted from March to August of the current season will have been conducted with limited information for the planning of the following season. Increase in lead-time would be very beneficial.

Medium to long term impact on planning activities

Seed production is a seasonal programme. The crop cycle has to be subjected to the full developmental cycle. At best in southern Africa, two cropping seasons per annum can be realised, utilising the warm environments in winter, with full irrigation. In the medium to long term, quarterly reviews of planning programmes can be summed up as below. Note that this example starts from the first quarter year 2000, up to the third quarter year 2001. The determined activities from January 2000 would directly impact on the quality of the seed sales in the selling season of 2001.

QUARTERS IN THE YEAR	JFM 2000	AMJ 2000	JAS 2000	OND 2000	JFM 2001	AMJ 2001	JAS 2001
Review of carry over seed stock.	XXXXXX						
Production programme review, type and variety		XXXXXX					
Research meeting review, setting priorities		XXXXXX					
Summer season production plan instituted			XXXXXX				
Review of achieved targets				XXXXXX			
Winter production plan to meet set targets					XXXXXX		
Estimates of production targets						XXXXXX	
Estimate of sales plan							XXXXXX

The sales plan instituted in the third quarter 2001 is planned with no information what so ever on the projected demand regime, or the predicted climatic conditions. Sales trends are taken as the norm, information forming the broad basis of planning.

The alternative information sources

Seed production planning and marketing and sales programme execution cannot wait for the official release of the climatic predictions from the Department of Meteorological Services. Information from the Met Services is used once data becomes available, and there on, but previous to that time, there are alternative sources of climate information. Sources of data on climatic information for planning purposes are:

- Department of Meteorological Services. Annual stakeholder planning seminar. The official government bulletin in released. Widely disseminated with extensive exposure.
- Department of Meteorological Services. Mid season update, about December or thereabout. Limited dissemination and exposure.
- The University of Natal Regional Climatic Prediction, useful only as an informal source, as it can not be quoted in official publications.
- The traditional indicators, which offer discussion indicators, cannot be quoted, quantified and qualified, but offering the other side of the coin—the farmers' perspective of the season ahead.

CLARITY AND USEFULNESS OF CLIMATIC DATA

The conventional presentation methodology

The conventional way of expressing climatic data is the three boxes we are all familiar with now, each showing the probability of above normal, normal, and below normal outcomes. Although this is technical and sound, the interpretation leaves too wide a range for effective qualification. The climatic predictors leave it to the users to interpret, but a more definite system would be appreciated from the user perspective.

Farmer perceptive and receptiveness to data

The farmer has only one opportunity for a source of income in the season. He is very guarded, cautious in approach, and would like to be sure before he employs the released technologies. On the other hand, the farmer prefers to spread the risk in his operations given the range of seed products available on the market. Farmer planning processes are done very early in the season, to the extent that most advice which comes to the farmer does not affect the immediate operations, but are documented as planning tools for post season evaluation. His latitude to adapt at very short notices is small. This is particularly so with the small holder farmers. A different scenario could be equally developed, which is the avenue being explored. If small holder farmers where provided climatic prediction data in a simplified form, in a timely manner, with specific recipes for agricultural survival, what would be the farmer reaction? This needs to be explored, to avoid assumptions!

CHALLENGES ON THE INTERPRETATION AND USEFULNESS OF DATA

Market Innovations

In an environment where there is market liberalisation, where there are several competitors looking at the same market base, the same clientele, addressing similar product attributes, survival in that market has a direct relationship to innovativeness. The company has to excel in the farmers' minds, has to be distinct in approach, offering services with distinct advantages to the benefit of the farmer. There is no limit in innovativeness, save for the fact that innovativeness has to be creative, exciting, easy to follow, and educative.

Data Presentaiton and Interpretation

Seed Co Limited has adopted an easier methodology for farmers to master the range of seed products. Mention has been made of the number of maize seed products available on the market. All these products exhibit different attributes, and the farmer has to look for the products which satisfy his demand and taste. With revamped technologies, revamped research, and the quest to improve services to the farmers, Seed Co Limited has designed a system which addresses the need for the market to be provided with accurate information in the easiest and innovative, yet informative manner.

Looking at the animals of the universe, they do have and exhibit natural characteristics, which can be taken as synonyms to the attributes of the maize seed varieties. Qualification of the attributes then realises farmer description of varieties as animal attributes.

ANIMAL	ATTRIBUTES	VARIETY GROUPING	ACCURATE ATTRIBUTES
Monkey	Very quick, active and clever.	SC401, SC403, SC405, SC407.	Very early maturity, 90-130 days, adaptable within the group to tolerances of drought, diseases, with medium to fast drying rates.
Zebra	Hardy animal, graceful to look at, heavy in stature.	SC501, SC513, SC515.	Hardy. Early maturity, up to 138 days, strong adaptation to drought, with stable good yields.
Lion	Fierce.	SC601, SC621, SC625, SC627, SC629	Very high yield potential under rain-fed conditions.
Elephant	Big, slow in movement.	SC701, SC709, SC711, SC715.	Slow growing, high yielding, ideal for high management regimes, including irrigation.

This format, if presented on text in colour, makes very interesting observation. This system of addressing Seed Co Limited released varieties has been registered and copy right protected.

Information Communication

The farmer has to keep abreast with technology. He may retain specific variety names for discussion, or look at memorising the group attributes, hence look at the memory of animal maturity groupings. Drier environments will require the farmer to plant 'monkey hybrids' only. Disease attributes can be super-imposed.

With climatic prediction available, the farmers are clearly advised on the varieties to grow looking at the animal attributes. Should it be a dry season prediction, earliness is the way to advise. Monkey group and zebra group would be ideal. In wet years, zebra varieties are suitable. You have to superimpose the cob disease tolerance aspect, standability tolerances, as well as grain-drying rate. To spread the risk, plant the lion varieties first, followed by the zebra group, with the monkey group suitable for late plantings, early green meallie harvest, and an insurance against hunger. Applications and interpretations are as varied.

CONCLUSION

Climate predictions are very useful in the decision making processes of the farmers, but they have to be timely. They also have to be put in such a language that the farmers can interpret and trust. From an input supplier perspective, forecasts have to assist in planning. Timeliness and accuracy are of the essence. To date, many decisions are still made based on past trends, with room for change.

ACKNOWLEDGEMENTS

For exposure and practical awareness, we acknowledge the Department of Meteorological Services in Zimbabwe, Dr. L. Unganai. Discussions on the applicability of forecasts have varied widely, and the exchange of ideas and concepts with Dr Jennifer Phillips and Dr Ben Orlove need special mention. Funding for the opportunity to exchange and share views at the International Forum is attributed to the IRI, to which a word of appreciation is extended. Editing of this submission is acknowledged from the Seed Co General Manager, S.B. McCarter.

REFERENCES

Department of Meteorological Services. Sept 1999. Proceedings from the 1999-2000 Seasonal Rainfall Forecast Workshop.

Seed Co Limited Annual Report. 2000. Reaching out to the Region.

Seed Co Limited 1999. Millennium Seed Co seed manual. A farmer advisory and reference manual.

The Role of Seasonal Climate Prediction in the Development and Utilization of Water Resources for Agriculture

Daniel Hillel
Center for Environmental Studies, Israel

As this forum is aimed at exploring the range of issues involved in the application of seasonal climate forecasts to the development of agriculture, it is appropriate for us, in the first place, to consider the term "development" so as to clarify what it means to us in the present context. In human physiology, development refers to the process of growth from infancy to maturity. More broadly, in evolutionary biology it refers to elaboration from relatively simple to increasingly complex and specialized organisms in mutually interactive living communities. As appropriated by economists, the same term is used to describe the mobilization and application of human and physical resources to the production and utilization of goods and services, aimed at improving human welfare. Nowadays that is often taken to imply a market economy linked to world trade.

In recent decades we have become increasingly and—I might add—painfully aware of the potential dangers posed by hasty economic development. By "hasty" I mean the type of economic development that serves its purpose in the short run while eroding the resource base in the long run. Such development is, in the current use of the term, unsustainable. A typical example is the exploitation of an irreplaceable resource, or the practically irreversible degradation of a resource, for short-term gain. Such exploitative "development" may take place knowingly. More often, however, it occurs unwittingly, if only because those who practice it are unaware of the gradual processes induced by their actions, which all too often lead to the eventual loss of productivity. Acting out of ignorance is positively dangerous. Obviously, we cannot reliably manage a system we do not understand and monitor. True development must be based on improvement, or at the very least maintenance, of the productive base.

Nowhere is this principle truer than in the development and efficient use of water resources. The ultimate source of fresh water is precipitation, which is a function of climate. Even in relatively humid regions, where precipitation is abundant and dependable, water must be managed to regulate and distribute its supply and to safeguard its quality. The need for management is especially crucial in semiarid and arid regions, where, by some accident of history and geography, nearly one-third of humanity lives.

In arid regions, the supply of water by precipitation is least even while the requirement of crops for water, imposed by the evaporative demand of the dry atmosphere, is greatest. The dilemma thus posed by the imbalance between supply and demand is compounded by the fragility of the environment in arid regions, which is all too prone to such processes of degradation as denudation of vegetation, erosion and salination of soil, and depletion as well as pollution of water resources. Arid regions, moreover, are typically marked by climatic instability, with wide variations in the relative quantity and intensity of rainfall from season to season and from location to location. No wonder therefore that the arid regions are where we find the world's poorest nations, most vulnerable to droughts and floods and being in the greatest need for economic development. The populations of these countries is growing rapidly even as their ability to feed themselves is under threat. Their situation is even further complicated by the specter of global warming, which may cause the climate to be even less stable by intensifying the frequency and severity of extreme weather events.

The *sine qua non* of economic development in arid regions, most especially in the agricultural sector, is the supply of fresh water and its efficient utilization. Water resources exist in two principal forms: surface waters and ground waters.

Surface waters occur in springs, streams, and lakes, and their variable quantities depend directly on the vagaries of annual precipitation. Harnessing such waters generally requires the construction of expensive dams and storage reservoirs, as well as extensive distribution systems consisting of canals, pipelines, and pumping stations. The potential advantages of such works are that they can serve to control floods, regulate the supply of water, generate hydroelectric power, and allow the development of aquaculture and of recreation facilities. The potential problems involved with such development are the submergence of valleys, the destruction of habitats and cultural assets, the displacement of people, the loss of water by evaporation and unrecoverable seepage, the loss of storage capacity by sedimentation of reservoirs, the hazard of dam failure, as well as the contamination of water and the spread of water-borne diseases affecting public health. For these and other reasons, there is increasing opposition to the further construction of large dams. Moreover, there are fewer sites yet remaining where additional dams can be safely and economically located.

The major alternative source of fresh water is groundwater, obtainable from aquifers by mean of wells. It major advantages are the relative stability of supply (being less dependent of the seasonal variation of rainfall, though it is dependent on long-term trends such as climate change). Aquifers generally provide natural protection against evaporative losses. The problems, however, are the energy and capital costs associated with pumping the water from depth, the danger of aquifer contamination by the percolation of toxic residues or intrusion of seawater, and possible depletion by overpumping (i.e., withdrawals exceeding recharge rates over an extended period), which may also cause land subsidence. Especially vulnerable to depletion and contamination are aquifers in arid regions, where the annual recharge rate is low. Aquifers in such regions may be brackish from the outset, due to dissolution of salts from incompletely leached geologic strata.

All these problems require that the development and management of water resources be done with great care and with attention to long-term sustainability and economic efficiency. The essential criterion in this regard is the maintenance of a proper balance between the demand for water and the supply, the latter being dependent not only on mean climate but also on climate variability in time and space. The prospect of climate change, which may well exacerbate variability, makes that criterion all the more crucial. The construction of storage, pumping, and conveyance facilities and their operation in practice must therefore be based on timely information regarding climate-affected patterns of supply and demand.

As one example among many, let us focus on the case of the Nile River, which is the principal water resource for Egypt and the Sudan, and to which Ethiopia as well lays claim (Figure 1). The combined populations of those countries already exceeds 160 million, and may well exceed 250 million within twenty years.

The total flow of the united Nile, measured at Khartoum, averages some 84 billion cubic meters per year. About 90 percent of that flow is con-

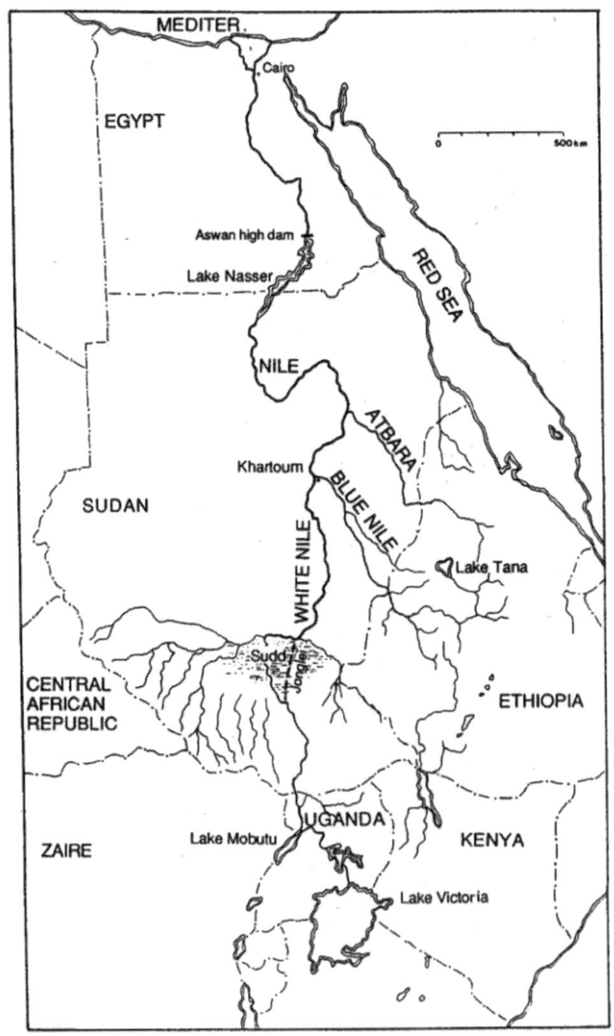

Figure 1. *The Nile River and its sources. (Hillel, 1994).*

tributed by the Sobat, Atbara, and—especially—the Blue Nile, which convey runoff from the highlands of western Ethiopia. Those highlands are visited annually by the monsoonal rains that roll in from the Indian Ocean during the summer months.

Although the timing of the monsoonal rains is fairly regular, the quantity is variable from season to season. The ancient Egyptians, who depended utterly on the annual flooding of the river, had no idea from whence the flood originated, and imagined it to arise by divine power from the bowels of the dry Sahara. Hence they could hardly make contingency plans ahead of time. (That is what makes the story of Joseph's forecast, of seven good years to be followed by seven bad ones, so unusual). The Egyptians did, however, make regular measurements of the level of the water in the river, in order to decide annually whether they could divert some of the water and spread it over larger tracts of land. To this end, they set up gauging devices known as Nilometers, the recorded readings of which extend back thousands of years. Prior to the construction and expansion of upstream regulatory facilities during the last century, the maximum river level at Cairo (just above where the river fans out over the delta) registered the effects of summer (June to September) rainfall over the highlands of Ethiopia. Though not completely continuous, those records were assembled and analyzed in recent years, and were found to correlate with the sequence of ENSO events (Quinn, 1994; Glantz, 1996).

The explanation for this correlation is that rainfall over the Ethiopian portion of the Nile catchment represents the western end of the Southern Oscillation (Figure 2). The large-scale feature known as ENSO, which extends over the lower latitudes from South America westward to East Africa manifests itself as a "seesaw" in ocean-atmosphere activity between the areas in and around the tropical Pacific Ocean and the tropical Indian Ocean.

When the Southern Oscillation Index (SOI, defined as the sea-level atmospheric pressure difference between sites in the southeastern Pacific high-pressure region and sites in the Indonesian equatorial low-pressure region) is high, it indicates that the large low pressure system extending over India and the Arabian Sea is well developed and the summer monsoon rainfall is likely to be heavy. On the other hand, when the SOI is low, it indicates that the low pressure system over that area is not well developed or is displaced to the east, and that the summer monsoon rainfall is likely to be deficient.

Hence a season characterized by a strongly positive SOI is likely to be one of high summer rain-

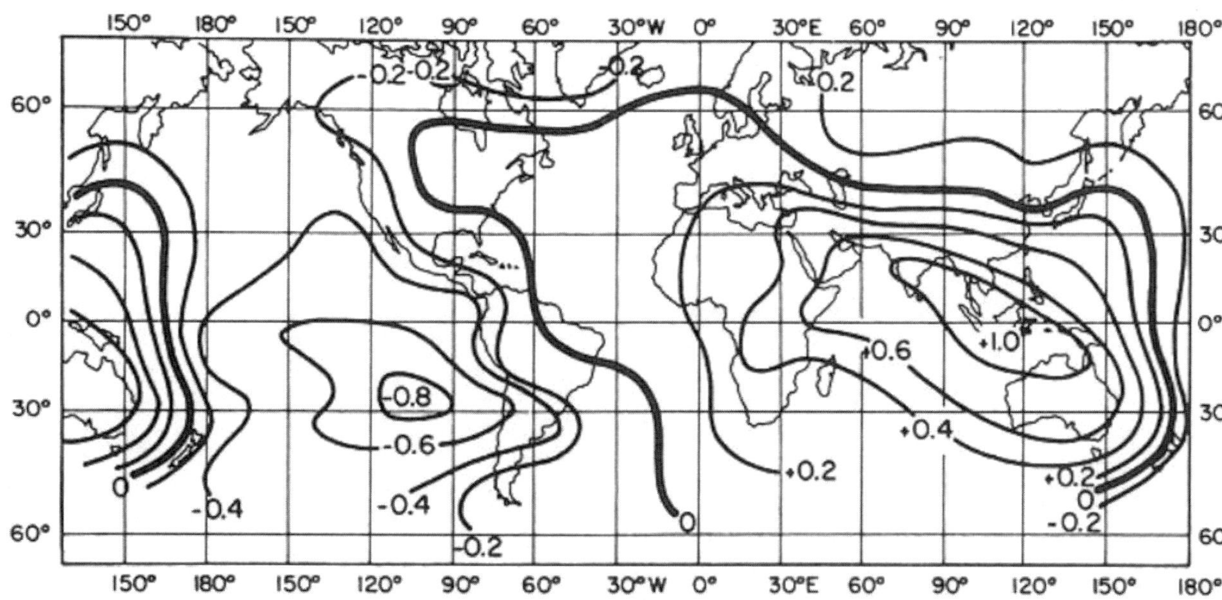

Figure 2. *Walker's Southern Oscillation, as shown in Berlage (1957). This map shows the worldwide distribution of correlations of annual pressure anomalies with simultaneous pressure anomalies in Djakarta, Indonesia. (Quinn, 1992).*

fall over the Ethiopian highlands, resulting in greater runoff and consequently a more abundant flow of the Nile. Contrariwise, a season of low or negative SOI is likely to provide relatively low summer rainfall and a consequently low flow of the Nile. The co-incidence between droughts in western India and low Nile floods, as well as between seasons of unusually heavy rains in western India and high Nile floods, was noticed as early as 1908 and reported in *The Imperial Gazetteer* of India (Vol. 1, p. 127). However, the association between those occurrences and the El Niño phenomenon, with its wide-ranging teleconnections, was realized much later. Figure 3 shows the simultaneity of droughts occurring in Western India and the Ethiopian highlands during the El Niño of 1982-1983 (Glantz, 1996). Figure 4 shows the annual discharge of the Main Nile and the combined discharge of the Blue Nile and the Atbara River in the twentieth century. (Note that it is primarily the Blue Nile, together with the Atbara and the other streams originating in Ethiopia, that supplies the waters of the annual flood, whereas the White Nile plays very little part in generating that flood [Shahin, 1985; Hillel, 1994]).

The correlation between Nile flow and the SOI led at least one researcher (Quinn, 1992) to suggest that the SOI climatic record itself may be extended back to the early 7th century by using the annual Nile River flood crest at Cairo as a sort of proxy for SOI. Looking toward the future, however, the question arises where the signs of a developing ENSO event can be detected first, whether on the western or the eastern side of the "seesaw," and how long before their impacts are manifested. If it can be shown that an advance warning system is practical, then it may be possible to forecast on a seasonal basis, some weeks or even months ahead of time, the likely occurrence of plentiful or deficient flows in the Nile. In any case, there is usually a time lag of some weeks between the onset of summer rains, the local generation of voluminous runoff, the routing of that runoff via the river, and the eventual appearance of the flood many hundreds of kilometers downstream in Egypt.

The possibility of forecasting the Nile flood, if indeed it can be established rigorously (by means of statistical correlation and quantitative analysis using validated regional climate models), may help to guide the practical management of water in the

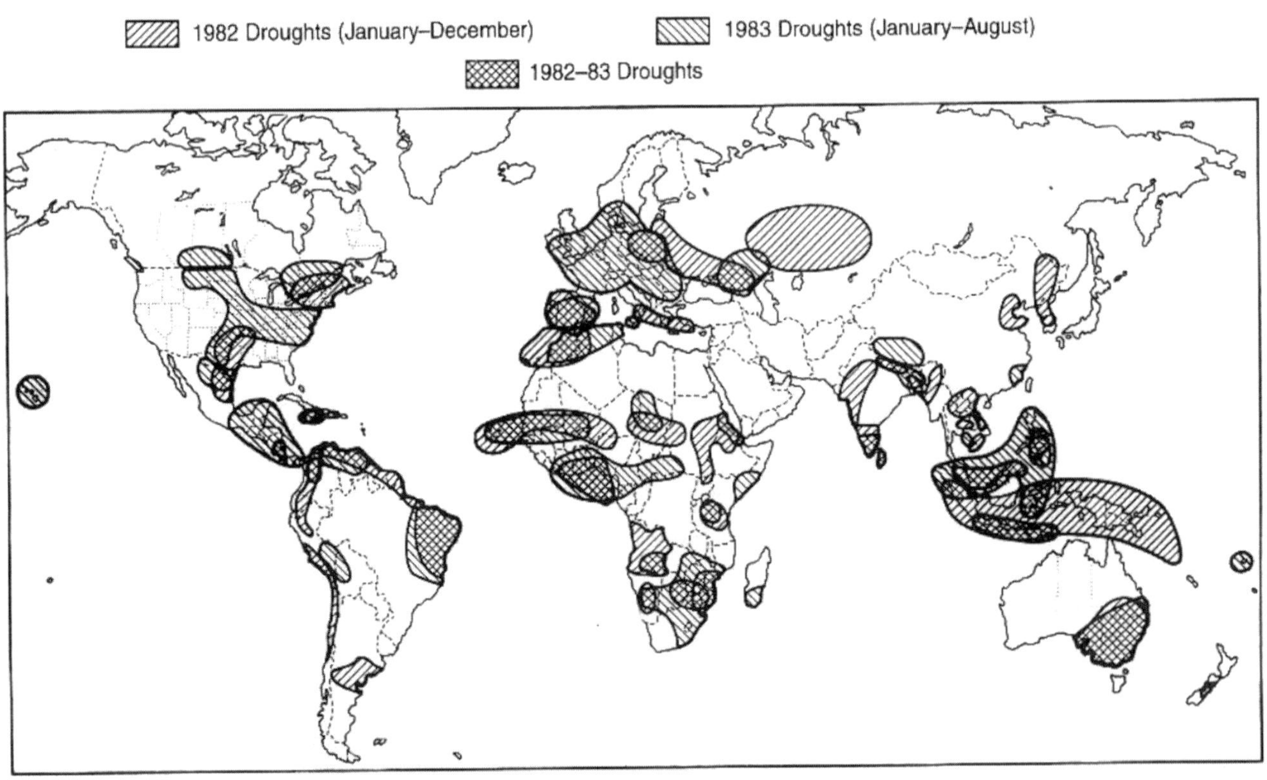

Figure 3. *Drought map of the 1982-1983 El Niño. (Glantz, 1996).*

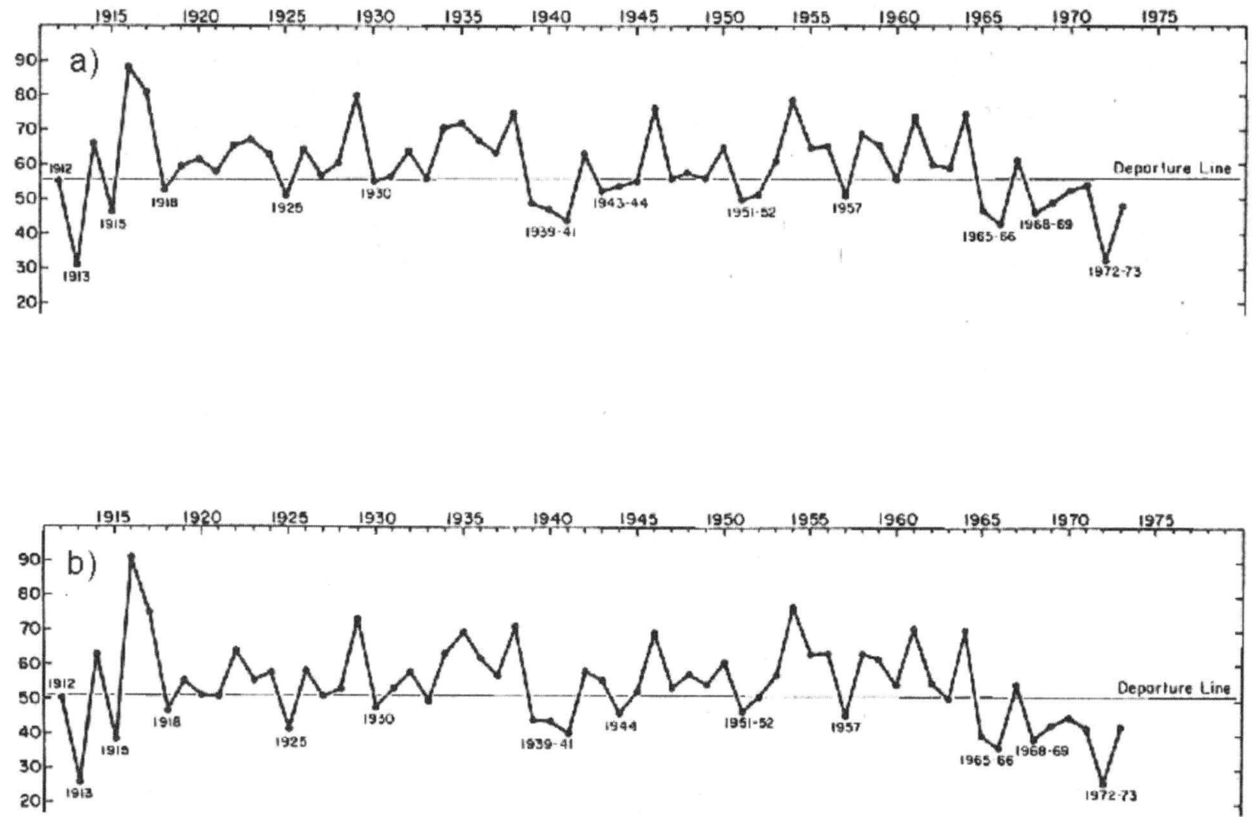

Figure 4. *(a) Annual discharge of the Main Nile River for July-October at Dongola (19°10′N, 30°29′) in $10^9m^3yr^{-1}$; the horizontal line at 55.92 x $10^9m^3yr^{-1}$ signifies the discharge level for the minimal degree 1 reduction from the long-term average discharge value. These years are identified below the corresponding points in the graph. (Quinn, 1992). (b) Combined annual discharge of the Blue Nile and Atbara Rivers for July-October at respectively 15°37′N, 32°E and 17°42′N, 33°58′E, in $10^9m^3yr^{-1}$, with the departure line at 51.11 x $10^9m^3yr^{-1}$ used to signify the combined minimal degree 1 reduction from the long-term average discharge value. (Quinn, 1992).*

existing facilities such as the Aswan High Dam and the smaller dams on the upper reaches of the Blue Nile. Moreover, it can help in the design of future facilities and in the efficient allocation of water over time and among actual or potential users upstream and downstream. Timely information on how much inflow can be expected along the river may facilitate optimizing the volume of water to be kept in storage or—alternatively—to be released downstream at a controlled rate. For example, if the expectation is of a diminished inflow over the coming season, the judicious strategy might be to release less water in the interim and to keep more in reserve so as to prevent the subsequent development of an acute shortage such that might cause total crop failure and paralysis of vital systems dependent on water supply (including the generation of hydropower).

Though Ethiopia, as mentioned, generates about 90% of the summer flood coursing down the Nile, that country utilizes hardly any of that river's water in its own territory. It is constrained from doing so owing to the lack of international agreement on what should be its rightful share, as well as the lack of investment and infrastructure. The country subsists mainly on rainfed rather than on irrigation agriculture (which is the major mode of production in the downstream countries, Sudan and Egypt). It is therefore highly vulnerable to recurrent droughts, and is in fact even now in the throes of a severe one. Following a prior occurrence of famine in the mid-1980s, during which hundreds of thousands of people perished (Glantz, 1996), Ethiopia's National Meteorological Services searched for early indicators of changes in rainfall that have been associated with crop failures. It issued charts showing the apparent linkages between El Niño occurrences and the paucity of rainfall in Ethiopia's principal rainy season from June to September (Tables 1 and 2). Note that a distinction is made between two distinct rainy seasons.

Quite another example where seasonal forecasting can greatly help in the management of water resources is the case of Israel, where surface water resources are regularly managed conjunctively with groundwater resources. How much water to withdraw early in the rainy season from Lake Kinneret (the country's major surface-water reservoir) in order to recharge the coastal aquifer (one of the two aquifers that together supply some 70% of the country's freshwater) obviously depends on the expectation of late-season (springtime) inflows to that lake, as well as on the expectation of natural recharge of the aquifer due to direct rainfall along the coast. Absent the ability to anticipate and moderate the fluctuations of demand and supply, the agricultural sector, which normally uses over 60% of the country's freshwater and which suffers an increase in crop-water requirements during drought seasons just when the supply is least, is ever vulnerable to sudden curtailments of water supplies.

In the case of rainfall over the catchments and aquifers of Israel, however, climatic teleconnections appear to be less clear than in the case of summer rainfall over the Blue Nile catchment. The summer season in Israel is totally dry. The amount

Table 1. *ENSO impacts on the rainfall regime in Ethiopia, as characterized by the National Meteorological Services Agency. (Quinn, 1992).*

ETHIOPIA Analogue Basis for Forecast		
ENSO Phase	Small Rains*	Big Rains**
I. ENSO	Heavy	Light
II. Post ENSO	Light	Heavy
III. Anti-ENSO	Light	Heavy
IV. Normal	Variable	Stable

*Mid-February to mid-May.
**June to September.

Table 2. *Recommendations issued in response to ENSO forecast of 1986 by the Ethiopia National Meteorological Services Agency. (Quinn, 1992).*

THE ETHIOPIAN EXPERIENCE January 1987	
(A) Small Rains*	
Prediction	Heavy long rains
Recommendation	Maximize land and sowed. Government increase issue of seeds/fertilizer
Climate	Heavy long rains
Results	Land use maximized. Bumper crops
(B) Big Rains**	
Prediction	Probably light; drought likely
Recommendation	Decrease land area sowed. Sow short-term crops. Conserve food and water. Government make early request to international economic assistance entities.
Climate	Light, short rains (2 weeks); drought
Results	Harvest severely impacted; minimal crops. Cattle weakened. International economic assistance entities granted Ethiopia early relief.

*Mid-February to mid-May.
**June to September.

of winter rain appears to be related positively to the occurrence of El Niño events, though the level of probability yet remains to be established.

At this point, a caveat is in order. Even where strong teleconnections can be established statistically and mechanistically, seasonal forecasts can never be certain, only probabilistic. Acting on the basis of uncertainty, however high the probability, entails risk. The risk is not only of economic loss during a given year in which the prediction does not pan out, but—beyond that—of loss of public confidence and a consequent reluctance to act on similar information in the future. Hence those who generate forecasts must attach warning to those forecasts and educate the public (and the institutional decision makers) to be prepared to bear the consequences of nature's penchant to deviate from seemingly established patterns.

REFERENCES

Berlage, H.P. 1957. Fluctuations of the general atmospheric circulation of more than one year, their nature and prognostic value. *Koninklijk Nederlands Meteorlogisch Institut, Mededelingen en Verhandelingen* 69. 152 pp.

Glantz, M.H. 1996. Currents of change: El Niño's impact on climate and society. Cambridge University Press. Cambridge, UK. 194 pp.

Hillel, D. Rivers of Eden: The Struggle for Water and the Quest for Peace in the Middle East. Oxford University Press. New York. 355 pp.

The Imperial Gazetteer. 1908. India. Vol. 1, p. 127.

Quinn, W.H. 1992. A study of Southern Oscillation-related climatic activity for A.D. 622-1900 incorporating Nile River flood data. In Diaz, H.F. and V. Markgraf. *El Niño: Historical and Paleoclimatic Aspects of the Southern Oscillation.* Cambridge University Press. Cambridge, UK. pp. 119-149.

Shahin, M. 1989. Review and assessment of water resources in the Arab region. *Water International* 14:206-219.

Climate and Food Security, Issues within a Matrix of Development

Graham Farmer
United Nations, Food and Agriculture Organization, Rome

SUMMARY

The relationship between climate prediction and food security is not as simple as it first appears. True, if we have reliable seasonal forecasts and optimal response mechanisms, we will be able to optimize our production possibilities and move towards improved food security. However, we live in an imperfect world and the optimal is rarely achieved. Are we, then, making any actual difference with our seasonal forecasting efforts? Alternatively, are we looking in the right places to see if we are making any difference?

The concept of food security provides the entry point to this discussion. If an individual or household does not have access to food for a prolonged period, even in a time of surrounding plenty, malnutrition and debilitating hunger will increase. If enough people are subjected to lack of access to food, for whatever reason, famine can occur. What we have to do is identify and separate those aspects, such as climate, that impact on food availability through production, from many other issues, primarily in the economic regime of poverty, that impinge on both food availability through trade and access to food through individual/household assets.

If we link food security to poverty, the whole development process is under investigation here. Larger scale economic improvements resulting from seasonal forecasts will improve food security, but only if the effects translate down to the local, household level. Similarly, the issue of increasing population size in low-income developing countries, those countries often of keen interest to climate forecasters, is presented as a threat to food security because of the increasing food need. The development process is crucial here also because improving standards of living moves the ownership of seemingly high birth rates to those directly involved. And from ownership of the problem, comes the solution.

Thus, if we are to maximize the impact of climate forecasting in the area of food security, particularly in developing economies, we must analyze it within the broader context of development issues. This paper highlights those issues and the need for an integrated, developmental approach to poverty reduction and food security.

Technology Transfer, Disasters, and Development

Raymond Meyer
Office of the U.S. Foreign Assistance, U.S. AID, Washington, USA

INTRODUCTION

There is no doubt that disasters[1] are increasing in frequency, and cost, all around the world. Developing nations are impacted disproportionately by natural and human-caused disasters and, as a result, increasing demands are being made on the disaster response resources of the international donor community. In order for the disastrous effects of future natural and human-caused hazards to be mitigated, or prevented, we need to insure that disaster mitigation and preparedness programs are incorporated into both emergency response and broad development programs. In order for these efforts to succeed, there needs to be clear acceptance on the part of policy-makers of the philosophical and fiscal priority that must be given to mitigation, preparedness, and prevention efforts. In addition, institutional structures that facilitate the integration of relief, mitigation, and development need to be developed.

Information and technology transfers are critical components in responding to—and mitigating and preventing—disasters. However, experience has shown that information and technology transfers, in and of themselves, are insufficient. The transfer and application of knowledge and understanding is what is needed, and this requires a better understanding of the needs and capacities of the end users and how the information and technology will be used. In addition, information and technology need to be delivered in an appropriate format to the right people at the right place and time. If these criteria are met, information and technology can effectively assist in disaster response, mitigation[2], preparedness, prevention, and more resilient development.

NATURAL DISASTER TRENDS

Over the years, we have been able to identify a number of alarming trends relating to natural disasters.

- Disasters are increasing in frequency. There were three times as many great natural disasters in the 1990s as in the 1960s.
- The economic and social costs have also increased. Between the 1960s and the 1990s, the economic costs increased more than ninefold.• Developing countries are hardest hit by disasters. Ninety percent of disaster victims worldwide reside in developing countries.
- The vulnerability of those living in risk-prone areas is perhaps the single most important cause of disaster casualties and damage—and this vulnerability is only increasing. For example, 40 of the 50 fastest growing cities in the world are in earthquake zones (Figure 1).

Hazards[3]—such as drought, earthquakes and flooding—do not have to become disasters. Disasters occur when hazards result in immense societal impacts affecting lives, livelihoods, social or political economies, and environments. Mitigation, preparedness, and prevention can reduce the destruction and suffering caused by natural hazards. For example, in the case of Hurricane Mitch—a severe storm that led to the largest disaster in the Western Hemisphere in over two centuries—proper watershed management, land use, and sustainable infrastructure development would have significantly decreased the substantial destruction and loss of human life.

Experiences like this demonstrate that future efforts should focus on three critical areas—better preparedness through application of information from early warning systems, more effective response policies, and reduced vulnerability of the population. Work in these areas must be incorporated into development programs and strategies. Without addressing these areas—and effectively preventing large-scale disasters—social and economic development will not be sufficiently resilient to withstand the large economic costs of

DECLARED NATURAL DISASTERS:
Type and Magnitude of OFDA Response
FISCAL YEARS 1977 - 1999

DISASTER TYPE	#	$ RESPONSE
Flood	291	52,361,956
Severe Storm*	158	109,600,216
Drought	125	301,873,959
Earthquake	94	60,768,785
Infestation	75	33,145,148
Volcanic eruption	22	5,551,947
Landslide	15	653,731
Avalanche	4	97,500
Extreme cold	3	75,000
Tsunami	2	1,084,668
TOTAL	**789**	**565,212,998**

• Cyclone, Hurricane, Typhoon, Other Storm
Source: USAID Office of Foreign Disaster Assistance (OFDA

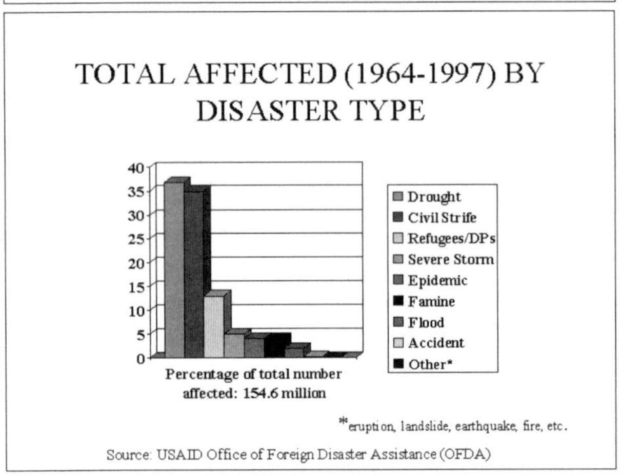

Figure 1. *Declared Natural disasters and total affected*

rehabilitation and reconstruction. A single natural disaster occurring in a country that has failed to take preventive measures, has failed to develop effective forecasting and response mechanisms, and where a large portion of the population remains vulnerable, places development investments at-risk and is likely to reverse years—even decades—of development.

THE ROLE OF INFORMATION AND TECHNOLOGY TRANSFER

From experience, we have found that information and technology transfer has a critical role to play in the three critical areas of preparedness, effective response, and reducing vulnerabilities. However, the simple dissemination of information and technology to vulnerable countries is not sufficient. The appropriate information and technology need to be provided to the right place, and the right people, at the right time. Once received, the information needs to be used by critical thinkers in the target, vulnerable population who understand the meaning of the information so that important choices can be made. If all of these conditions are met, the transfer of data and technology becomes the transfer of knowledge and understanding and can be used for effective disaster response, mitigation, and prevention.

Information and technology must be provided after answering, and taking into account, the following:

- Who are the end users?
- Who are the decision-makers?
- How will the information or technology be used?
- What information is needed/required by those on the ground?
- What are the local capacities to process and effectively use this information or technology? Is the information or technology appropriate for the region?

Without understanding and incorporating this knowledge into information and technology transfer, it is not likely to have its desired effect. The danger is that the information and technology provided would be entirely useless, a waste of financial resources, or—even worse—be misunderstood or misapplied in such a way that it causes more harm than good.

Examples of areas where information and technology can be used for preparedness, prevention, and mitigation include the following:

- Effective use of Early Warning Systems for natural phenomena
- Public awareness campaigns and emergency planning
- Development and implementation of improved (i.e. earthquake or cyclone resistant) building design
- Hazard mapping for earthquakes, volcanoes, landslides, etc.
- Weather monitoring stations, river stage monitoring, and satellite image downlinks and analysis

In each of these areas, the key is to build the local capacity to effectively use the information and technology so that future hazards can be managed by local institutions and resources. One challenge is to insure that there is political will in the country to use, maintain, and share information and technology appropriately and effectively.

SUSTAINABLE DEVELOPMENT: THE INTEGRATION OF DISASTER MITIGATION AND RESILIENT DEVELOPMENT

Effective information and technology transfer is a key element in the integration, and intersection, of relief and development. However, more needs to be done to insure that information and technology transfer, and disaster mitigation and preparedness as a whole, be incorporated into development. This is necessary because in order for development to be sustainable, it must be resilient to hazards.

Despite the common sense behind this sort of integrated approach, it is rarely implemented. Development and disaster response remain largely separate domains within the organization of international and national agencies that grapple with these issues and bureaucratic structures and regulations often discourage implementation of integrated approaches. Therefore, despite intellectual agreement that such integration is necessary, relatively little has been done in practical terms to achieve this.

More thought needs to be given to practical solutions for improved integration. These might include:

- Developing institutional structures that are given the mandate to improve integration of development and disaster mitigation, prevention, and preparedness;
- Providing forums where development and relief practitioners work together on these issues and create linkages between their programs;
- Increasing public awareness (including media and elected officials) of the need for prevention, preparedness, and mitigation. Do this in part by making known concrete examples where this type of work has had an impact.

Perhaps the greatest obstacle to the implementation of an integrated approach is the divergent mindsets between the development and disaster response camps. While development experts think in terms of long-term programs and results, disaster experts are conditioned to think in terms of shorter-term, emergency response. The media, public sentiment, and political pressure further the tendency to focus primarily on saving lives within the wake of a disaster rather than taking steps to ensure that future disasters are prevented or mitigated. Unless all the actors involved come to a clear and firm consensus that disaster mitigation and prevention need to be incorporated at all stages and levels—from emergency response to long-term development—then it is unlikely that the increasing trend in the number, scale, and costs of disasters will be reversed.

NOTES

1. Disaster is defined as the impact of a hazard upon a population or area that is vulnerable to such impacts causing substantial damage, disruption or casualties beyond the capacity of the affected population to cope with unaided.
2. The definition of mitigation used by OFDA is "any step taken to reduce the loss of life, livelihoods, or property at risk from natural or human–caused hazards."
3. A hazard is defined as phenomenon that is potentially damaging and implies a risk to a population because of the potential for its occurrence

Famine Prevention

Maxx Dilley
The World Bank, Washington D.C., USA

SUMMARY

Famine is a complex phenomenon and agricultural production is only one of many factors involved. Seasonal climate forecasts have the potential to provide early warning for food insecurity but prevention of famine necessarily involves many other strategic inputs.

Climate variability and shocks affect different household food security strategies differently, depending on the relative importance of different sources of food in the household (Figure 1). Impacts of climate shocks may extend beyond crop production to livestock health or holdings and access to off-farm income.

Food security is affected by numerous other factors such as war, government policy and prices (Figure 2), complicating interventions.

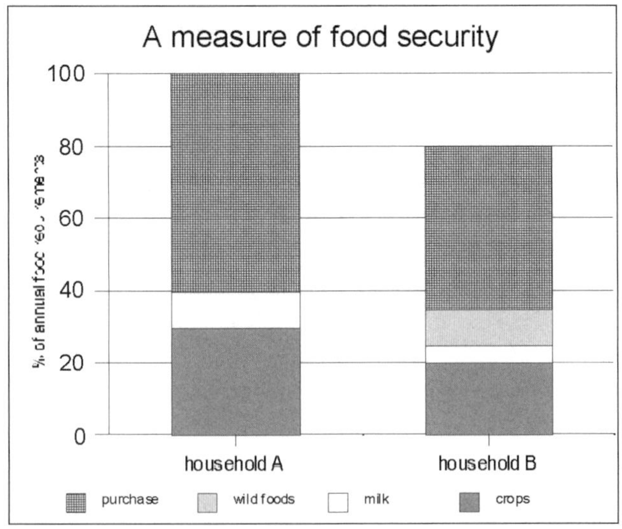

Figure 1. *Sources of food vary from household to household.*

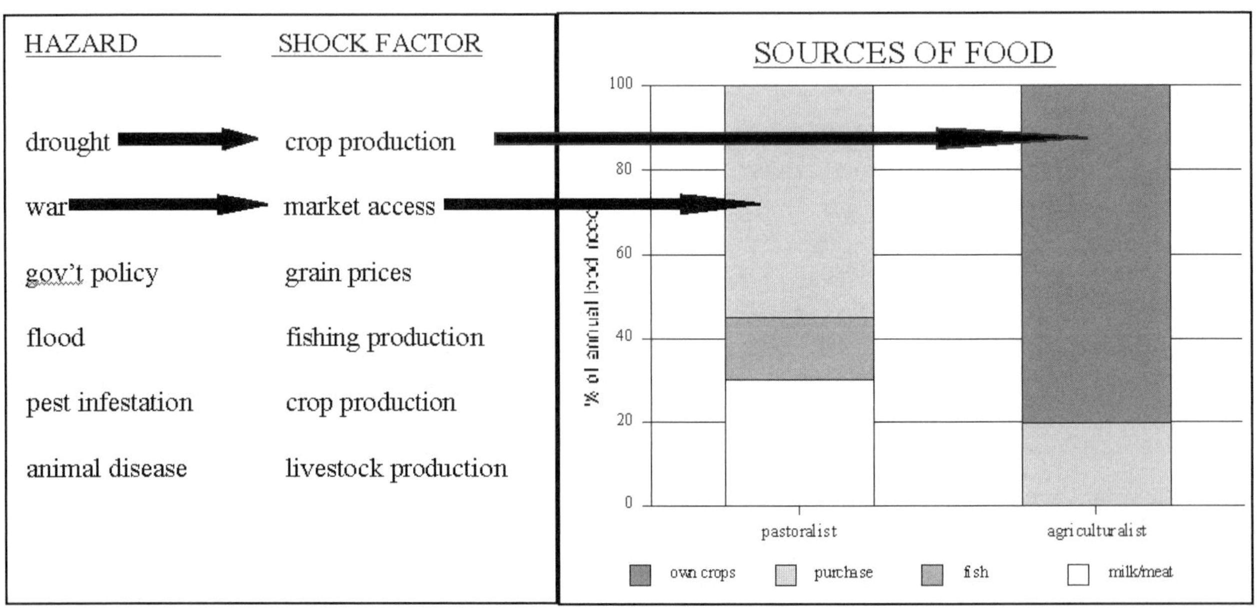

Figure 2. *Social, political, and environmental hazards send shocks to different components of the food security household.*

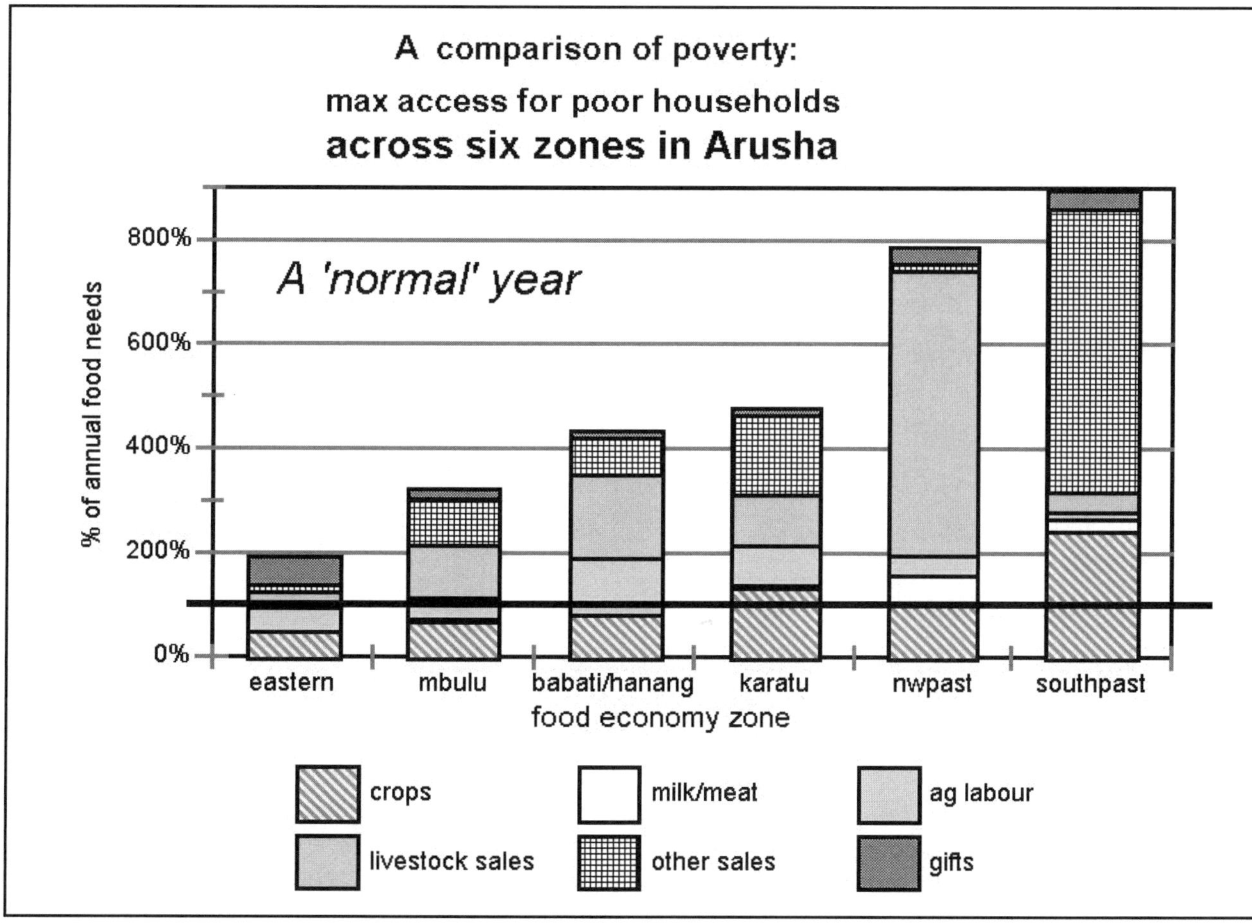

Figure 3. *Food access in a "normal" year in Arusha, Tanzania.*

A study of food access and vulnerability in Arusha, Tanzania, illustrates the variable impact of climate shocks on food access as a function of household poverty. In a "normal" year, annual household food needs were satisfied in all zones in Arusha though a combination of sources, including crop and livestock sales. Poorer households tended to satisfy a greater percentage of their food needs by growing crops than other households. These households were also operating dangerously close to the minimum of required food needs (Figure 3). In a year after total crop failure, food access for households relying largely on crops to meet household food needs fell by almost thirty percent (Figure 4). For household with greater reliance on other means of food access, climate shock had less of an impact on their food security.

Seasonal climate forecasts can play a role in mitigating food insecurity but forecast uncertainty makes interventions more difficult. Uncertainties are associated with:

- Temporal distribution for rainfall
- Spatial distribution
- Tercile range boundaries
- Interpretation of probabilities

The climate community needs to partner with institutions that are already working on food security problems in order to have maximum influence. The Famine Early Warning International Assistance system, aimed at responding to food insecurity after it has developed, is well established and can benefit from advance warning of climate shocks. More recently, efforts have been aimed at famine prevention, which requires involvement of intermediaries and identification of populations at risk. The range of institutional partners with whom the climate

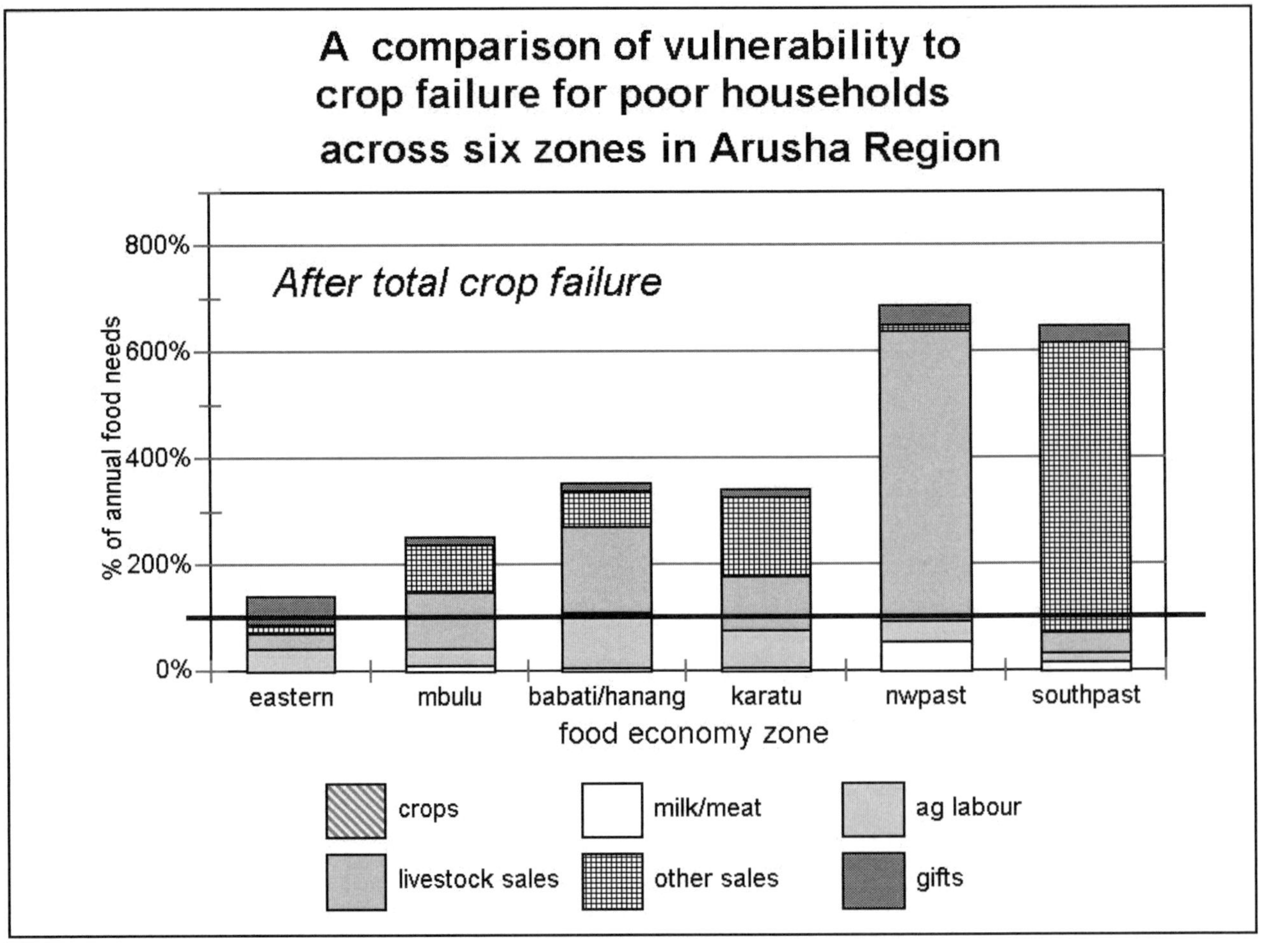

Figure 4. *Food access after total crop failure in Arusha, Tanzania.*

forecast community can work in utilizing forecasts in this regard is even greater. Through improvements in forecast skill, and experience in incorporating uncertain information into mitigative strategies, progress in utilizing seasonal climate information for improving food security can be made.

Notes

All figures are by Tanya E. Boudreau, of the Food Economy Group, and are from the paper "Vulnerable to Limping: the Achilles' Heel of 'Vulnerability to Food Insecurity' ", submitted March, 2000 to *Food Policy* by M. Dilley and T.E. Boudreau.

SUMMARY REPORT OF THE WORKING GROUP SESSIONS

Edited by Jennifer Phillips, James Hansen and Shardul Agrawala
International Research Institute for Climate Prediction, USA

Forum participants were randomly assigned to one of ten groups so that each group represented a range of experience. Groups were charged with addressing a series of questions with the objective of obtaining feedback from the research community in identifying common themes and goals as well as helping to guide the development of agricultural applications activities at the IRI. Each group was asked to consider and report back to the Plenary Session on the following:

1. What are the main common lessons we have learned about the application of climate prediction information to support agriculture and development from our experience to date in different parts of the world?
2. What are the main impediments to the effective use of climate prediction information, and what research and development work is needed to address these?
3. What key recommendations can be made to advance the field, both in general and with respect to the IRI's activities?

COMMON THEMES IDENTIFIED

The reports from each group all contain some common threads, indicating a degree of coherence among experiences and needs identified. The most common of these themes, understanding who the users are, communications issues, problems of scale, and the importance of partnering, were all articulated from the perspectives of lessons, impediments, and recommendations to address these problems.

Understanding User Groups

An important lesson learned from the past decade of activities in applications of climate forecasts is that users are diverse and cannot be lumped into a homogeneous set. Even within the agricultural sector, the needs of agribusinesses such as seed suppliers and grain traders vary distinctly from primary producers. Among producers, limitations in access to resources or risk exposure condition responses to new information. One characteristic of users is self-sufficiency or survival (subsistence producers) or profit maximization (commercial). This plays a critical role in what information they would be willing to use. Second, how rigid are their operating environments and decision rules? Family farms potentially have somewhat greater flexibility in moving dates for critical activities than those who have to lock in labor in advance or meet production deadlines. Early assumptions about the value of climate forecasts were often exaggerated due to a lack of understanding of the variety of user-decision making environments. Misperceptions on the part of the climate and applications community about user needs, currently held beliefs regarding climate, risk, and constraints on responses to information are likely to lead to poorly designed interventions.

A common recommendation from the breakout groups on this theme was that more attention needs to be paid to identifying users and user needs. Some forecast products and tools for interpretation may be flexible enough to be useful to a wide range of users. But it may also be the case that institutions like the IRI may have to clearly define targeted user groups, then make the necessary effort to understand those groups more thoroughly. This also implies that user groups need to be involved from the inception of product or intervention design so that feedback can be incorporated. Additionally, the intervention points may be intermediary groups, such as agricultural extension workers, rather than the farmers themselves. Understanding the needs of the end user as well as the institutional framework of the intermediaries is therefore required.

Communications

A number of lessons, limitations, and recommendations revolved around the theme of communications. The primary lesson learned from experience to date is that communicating uncertainty in climate forecasts is one of the major challenges we face. The tendency to translate probabilistic information into deterministic predictions appears to be universal (it was remarked that even the climate scientists often do this!). Yet appropriate management decisions based on strong comprehension of risk and uncertainty are often likely to differ from those based on deterministic information. Therefore, there is a very real demand for clear, honest communication of forecast uncertainty or level of confidence. We recognize the problem of communicating uncertainty even within the English-speaking community, but it is further complicated by regional languages often encountered in the developing world, many of which are limited in expression of abstract concepts. In addressing this problem, it was recognized that all players involved in the chain of "end-to-end" activities, from how the forecast is presented graphically by the climate community, to how the media interprets that information, to translation by agricultural advisors into impacts and expected outcomes, will need to be included in education and training for progress to be made.

A second aspect of communications centers on the tools necessary for interpretation of impacts and potential management responses. This aspect is interpreted more broadly than the specific problem of translating uncertainty, and extends to issues of climate information generally with respect to site specificity, differences in socio-cultural, agroecological and sectoral needs. The linkage from climate and forecast information to appropriate responses needs to be supported by the development of tools and methodologies. Tools should be interpreted in a variety of ways, such as models of interaction between weather and agroecological systems, demonstration projects, graphical displays, and workshop curricula designed to explore and assess impacts at the scale of decision making. Ideally the primary tools developed will be general enough to have applications in widely diverse settings. However, attention to the distinct needs of various social groups and regional settings, such as outlined in the section under *Understanding user groups,* will require adaptation of these tools to each setting. This implies the need for strong two-way communications, so that forecast providers build understanding of local notions of climate, uncertainty, and how these notions fit with current uses of climate information.

Spatial and temporal scale of forecasts and impacts

A commonly recognized problem in the application of climate forecasts in the agricultural community is that currently available seasonal forecasts are at too large a scale to be useful for site-level planning. Both spatial and temporal scales need to be refined for agricultural management applications. This point was reiterated by almost all the breakout groups. Although the applications community, through interaction with climate scientists in the context of activities over the past few years, now has more realistic expectations of skill levels to be expected from high resolution climate forecasts, there exists a suite of approaches to be developed for down-scaling forecast products. The recommendation to push the development of regional climate models was paired with the suggestion that there are additional techniques that can used to downscale forecasts, such as the use of weather generators or other statistical approaches, and the use of analogue years. The international climate and applications community has a role to play in developing these tools, but regional institutions should be assisted in their utilization and interpretation, given their familiarity with local climate.

Partnering

Partnering with other researchers and institutions was recommended by participants in two different contexts. The first is capacity building: given an underlying objective of improving regional ability to utilize climate information, partnering with regional and local institutions, particularly National Meteorological Services, serves a capacity-building function. Secondly, in order to extend the impact and effectiveness of the IRI's activities, it was recommended to partner with other institutions in our applications activities.

There are a number of different mechanisms for partnering to build capacity. Suggestions included "training the trainers" workshops, joint research activities, basic training in forecast development, and hosting research visits to the IRI. These activities do more than transfer capacity to regional institutions—they also provide an opportunity for us to learn more about regional political, social, and agroecological systems which are needed to shape regional activities. Additionally, as generic products and services are developed, application and testing in regional settings provides feedback on effectiveness.

Collaboration with other institutions provides a wealth of expertise beyond that which the IRI can sustain internally. Additionally, the model of the IRI functioning as a hub in support of a network of institutions reinforces the goal of developing blueprints for successful applications activities and the tools necessary for implementation which can then be tested and improved by a large number of groups on the ground performing data-gathering, adaptation of approaches to specific settings, and evaluation which feeds back into the "toolbox". The IRI can serve as a clearinghouse for information and the exchange of experiences among groups.

ADDITIONAL LESSONS, IMPEDIMENTS AND RECOMMENDATIONS

There were a number of ideas presented in the reports from the breakout groups that were unique to their groups, and although they were not cited by all groups, they should be given attention.

- Measures of skill and forecast confidence relevant to user's understanding need to be developed to facilitate the use of probabilistic forecasts. Transparency of method is important to user understanding as well as to building trust with the user community.
- Attention needs to be paid to improving our ability to do interdisciplinary research and applications, given the cross-cutting issues we are attempting to address.
- ENSO is often taken as synonymous with climate forecast—we need to broaden our perspective not only to include other drivers of interannual variability but to include general climate information as well. In many regions, improved use of climatology may be as valuable as an uncertain seasonal forecast.
- There is a need to de-emphasize the focus on forecasts of extreme events and emphasize use of climate information every year, even when climatology is expected. This may be more relevant to agricultural applications than to other sectors.

PARALLEL SESSION PRESENTATIONS

SESSION I. CLIMATE VARIABILITY AND PREDICTIONS

Seasonal Prediction and Climate Simulations Using CPTEC/COLA AGCM

Iracema F.A. Cavalcanti; José A. Marengo; Carlos A. Nobre; Christopher Castro; Marcos Sanches; Helio Camargo; Gilvan Sampaio
Instituto Nacional de Pesquisas Espaciais, INPE, Brazil

INTRODUCTION

The center for weather forecasting and climate studies, CPTEC, started the operational weather forecasting at the end of 1994, using CPTEC/COLA AGCM. In January 1995, seasonal predictions using the same model were initiated in an experimental way. CPTEC/COLA is a version of the Center of Ocean-Atmosphere-Land spectral model, and has a resolution of T62 L28 (equivalent to 1.890 grid point horizontal interval and 28 vertical layers). The main features of the model that has a sophisticated vegetation module, SiB (Xue, 1991) are described in Kinter et al (1997). From 1995 to 1997, four initial conditions were used to obtain an ensemble mean seasonal prediction. Persisted sea surface temperature (SST) anomalies were applied as boundary condition.

Currently, 25 initial conditions are used to form the ensemble mean and to analyze the members dispersion. The SSTs for the predicted months are predicted SST and/or persisted SST anomalies from the last month observed SST. In the tropical Pacific the predicted SST is given by NCEP coupled model forecast, and in the tropical Atlantic the SST is predicted using the statistical model SIMOC (Pezzi et al. 1998). Outside these regions persisted SST anomalies are used. From November to April, both SST sets are used in two sets of integrations (this is the time when SIMOC has a good performance), otherwise, the persisted SST anomalies are the best approach.

A ten-year climate ensemble simulation using the operational T62L28 version with nine initial conditions and applying monthly observed SST was performed recently at CPTEC. The climatology of several variables were compared with observational data sets and the interannual variability of precipitation in specific areas, as well as the behavior of the model related to ENSO were analyzed. In this paper, results are shown for the austral autumn (MAM) in South America, mainly for Northeast and Southern Brazil were seasonal predictions are important to guide agriculture procedures.

USE AND DISSEMINATION OF PRODUCTS

The numerical predictions are presented and discussed monthly in a "Climate assessment meeting" at CPTEC, attended by scientists, users and media.

The results are used together with other information of the atmosphere conditions considering also the model's behavior, to provide the prediction for the coming season. The climate forecasts are distributed in monthly bulletins issued by CPTEC: INFOCLIMA (Brazil and regional). Monthly analyses of weather and climate are published in CLIMANALISE journal, directed towards scientific and non-scientific users. The predictions are also distributed by mail, FAX, and are available on internet (http://www.cptec.inpe.br/products/clima). Precipitation predictions for DJF and MAM in South America have been published since 1998 in the Experimental Long-Lead Bulletin (e.g. Cavalcanti et al. 2000). The model results have been useful for government decisions and to guide the management of agriculture.

Other types of applications of the seasonal predictions are monitoring risk of fires in the entire country (PROARCO) and the monitoring of soil moisture conditions and climate conditions in Northeast Brazil (PROCLIMA). PROARCO is an effort between CPTEC and the Brazilian Institute for the Environment (IBAMA) and consists of monitoring and processing climate and weather conditions to elaborate charts of fire risk for the entire country, but with main focus on the regional called Deforestation Arc, that includes several Amazonian states and parts of the states in southern Amazonia where "cerrado" (savanna) vegetation type grows, and that are prone to fires. A model implemented by INPE's specialists is used in order to process local observations (conventional, satellite) and to feed a model that produces a map of fire risk. This model also feeds some other ecological models.

PROCLIMA program is an effort between CPTEC and the Agency for Development in Northeast Brazil (SUDENE), and consists of elaboration of maps of soil moisture produced by water balance models. Data from more than 800 rainfall stations, soil and vegetation conditions from RADAM BRASIL, and information from the automatic weather and hydrology stations, as well as from satellite, implemented on a GIS framework for Northeast Brazil are used in this process. The level of detail goes from the entire region to the local level. The maps of soil moisture plus the numerical seasonal predictions are taken into account in the monitoring of the Northeast Brazil rainy season. These two programs are in response to specific needs of the central Brazilian government, and the products are available via internet.

CLIMATE SIMULATION

Characteristics of Precipitation over South America in MAM

The model precipitation field of climatological MAM (Figure 1a), which is the rainy season of Northeast Brazil (NE) represents well the position of the ITCZ shifted southwards, when compared to other seasons. However the model shows a tilting of the ITCZ towards NE, that gives an overestimation of precipitation in parts of the semi-arid region and a sub-estimation at the Amazon mouth (Figure 1b-c). There is also a tendency of the model in linking the ITCZ and the convection over Amazonia region with a NW-SE band of precipitation similar to the South Atlantic Convergence Zone (SACZ) displaced northwards. In the observations, the band of precipitation NW-SE is located over Southern Brazil, and is associated with the presence of frontal systems in that region. The ITCZ over the East Pacific has also overestimated precipitation. The main characteristics of the atmospheric circulation are well represented, as the subtropical highs and the trade confluence, although the Atlantic trade winds are weaker in the model than in the observations (not shown).

Precipitation anomalies and influence of ENSO

Precipitation anomalies for MAM from the simulation of 1982 to 1991 are shown in Figure 2 for north of NE (3°S-9°S; 35°W-45°W); south of NE (9°S-15°S; 35°W-45°W) and Southern Brazil/ Uruguay (22.5°S-35°S; 58°W-48°W). These figures display the ensemble mean, CMAP and the minimum and maximum values of the member's ensemble. The interannual variability of the anomalies are very well detected and although there is more dispersion in the south region of NE, the ensemble mean has very close values to observations. The influence of El Niño and La Niña on the precipitation of South America is well simulated by the model. Drought periods in 1983 and 1990 are well simulated, as well as the wet periods of 1985 and 1989. In Southern Brazil/Uruguay, the

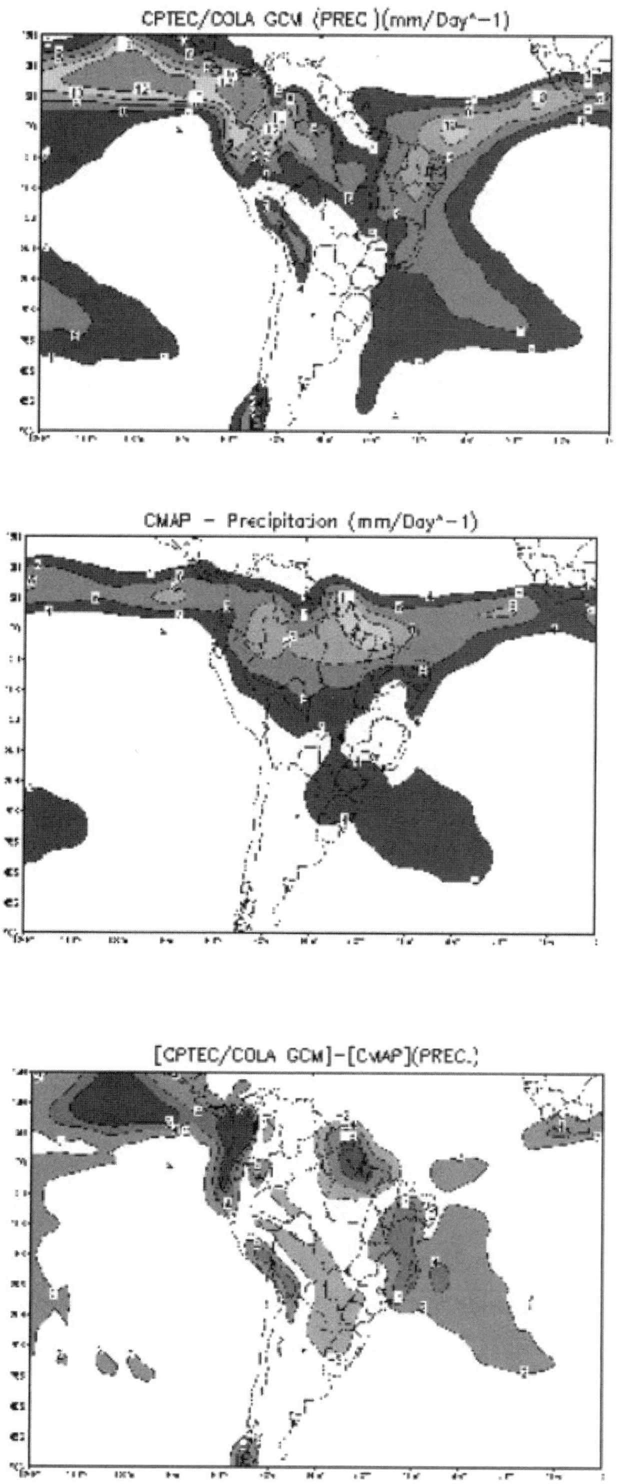

Figure 1. *(a) Precipitation simulated by the model (ensemble of nine integrations) in MAM (1982-1991); (b) Observed Precipitation in MAM (1982-1991) from CMAP (Xie and Arkin, 1998); (c) Differences between simulation and observation.*

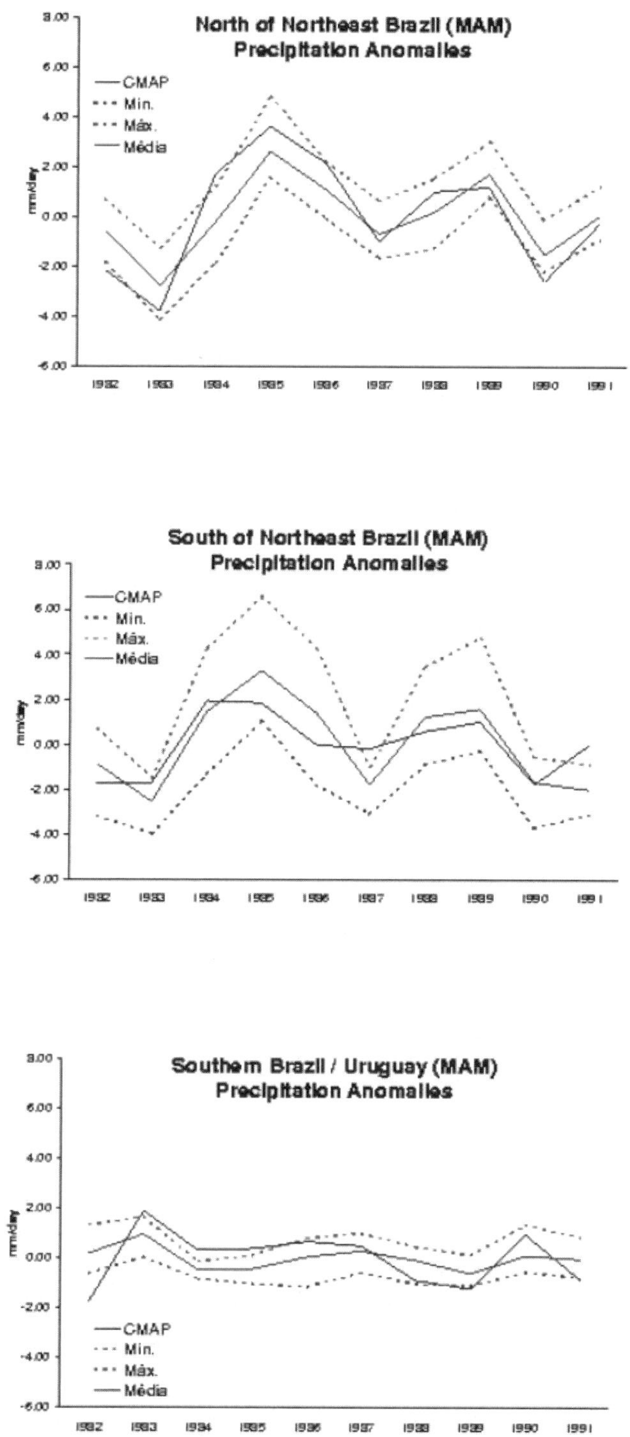

Figure 2. *Spatial average of precipitation anomalies (MAM, 1982-1991). (solid black line is the observational CMAP data; red line is the ensemble mean and dashed lines are the maximum and minimum values of the member's ensemble). (a) north of Northeast Brazil (3°S-9°S; 35°W-45°W); (b) south of Northeast Brazil (9°S-15°S; 35°W-45°W); (c) Southern Brazil/Uruguay (22.5°S-35°S; 58°W-48°W).*

opposite occurs, with wet periods in 1983 and 1990 and drought in 1989, which are well simulated although with less intensity.

VERIFICATION OF PRECIPITATION PREDICTION FROM 1995 TO 1999

Predictions for the rainy season of NE from 1995 to 1998 are discussed in Nobre and Cavalcanti (2000), comparing precipitation and atmospheric variables from the model results to observational fields. The intense drought in Northern and Eastern Northeast during the March-May 1998 rainy season was well forecasted by the model. The drought was also detected in northern Amazonia where El Niño conditions favored the development of forest fires. The dipole of rainfall anomalies northeast/south which usually occurs in El Niño years, was well reproduced by the model, showing also above normal precipitation in southern Brazil. The ability of the model in predicting extreme events in these two regions is worthwhile because of the impacts in agriculture of such regions and in hydroelectricity areas of southern Brazil where the major power sources are located. Predictions by the model can be applied to actions in these areas.

Verification for MAM 1999 is shown in Figure 3. Predicted SST in tropical Pacific and tropical Atlantic were used for the forcing boundary condition and the result is an ensemble mean of 25 members. The ITCZ position and its influence on the precipitation near the Amazon River mouth was well predicted, as well as the drought in the eastern Northeast. However the model predicted normal precipitation over Amazonia region and Southern Brazil, while the observations showed above normal precipitation over Amazonia and below normal over part of Southern Brazil.

Precipitation anomalies for MAM from the seasonal predictions of CPTEC/COLA AGCM from 1995 to 1999 are shown in Figure 4. It is noticeable the improvement of results from the ensemble of four members in 1995, 1996 and 1997, to the ensemble of 25 members in 1998 and 1999.

CONCLUSION

The simulation experiment showed that the model response to the observed SST is comparable to the observed fields, mainly considering the

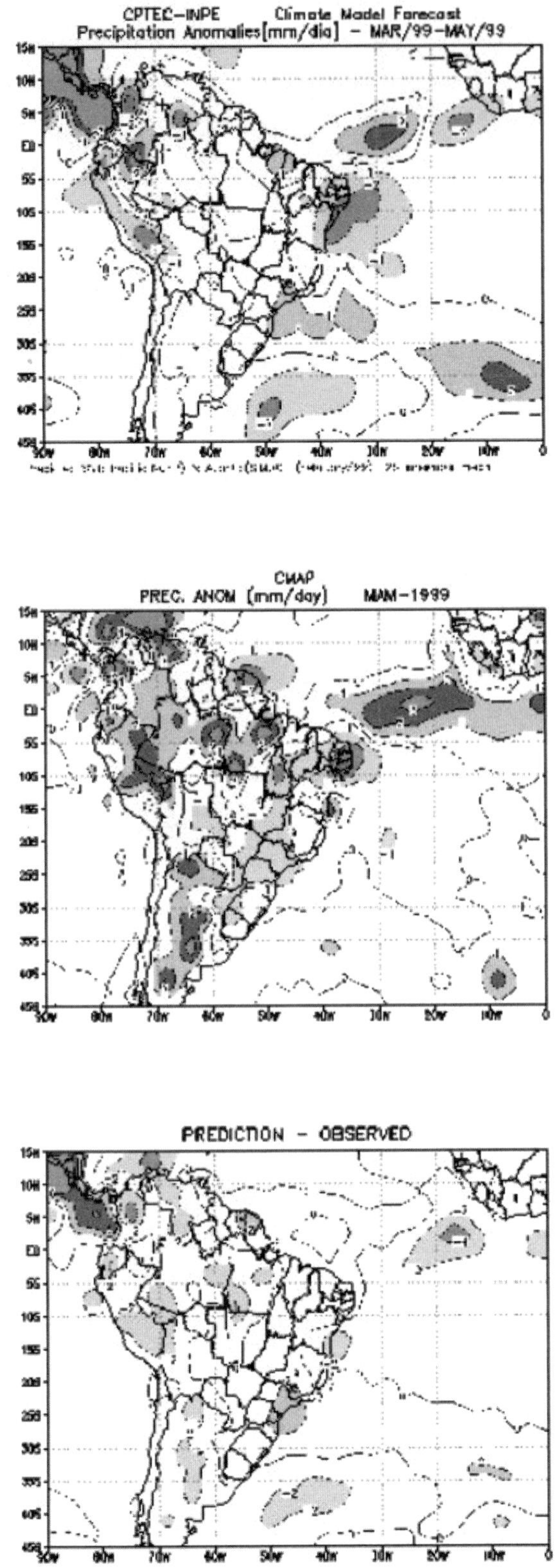

Figure 3. *(a) Predicted precipitation anomaly of MAM 1999 using predicted SST and 25 integrations; (b) Observed precipitation anomaly of MAM, from CMAP; (c) Difference between prediction and observation.*

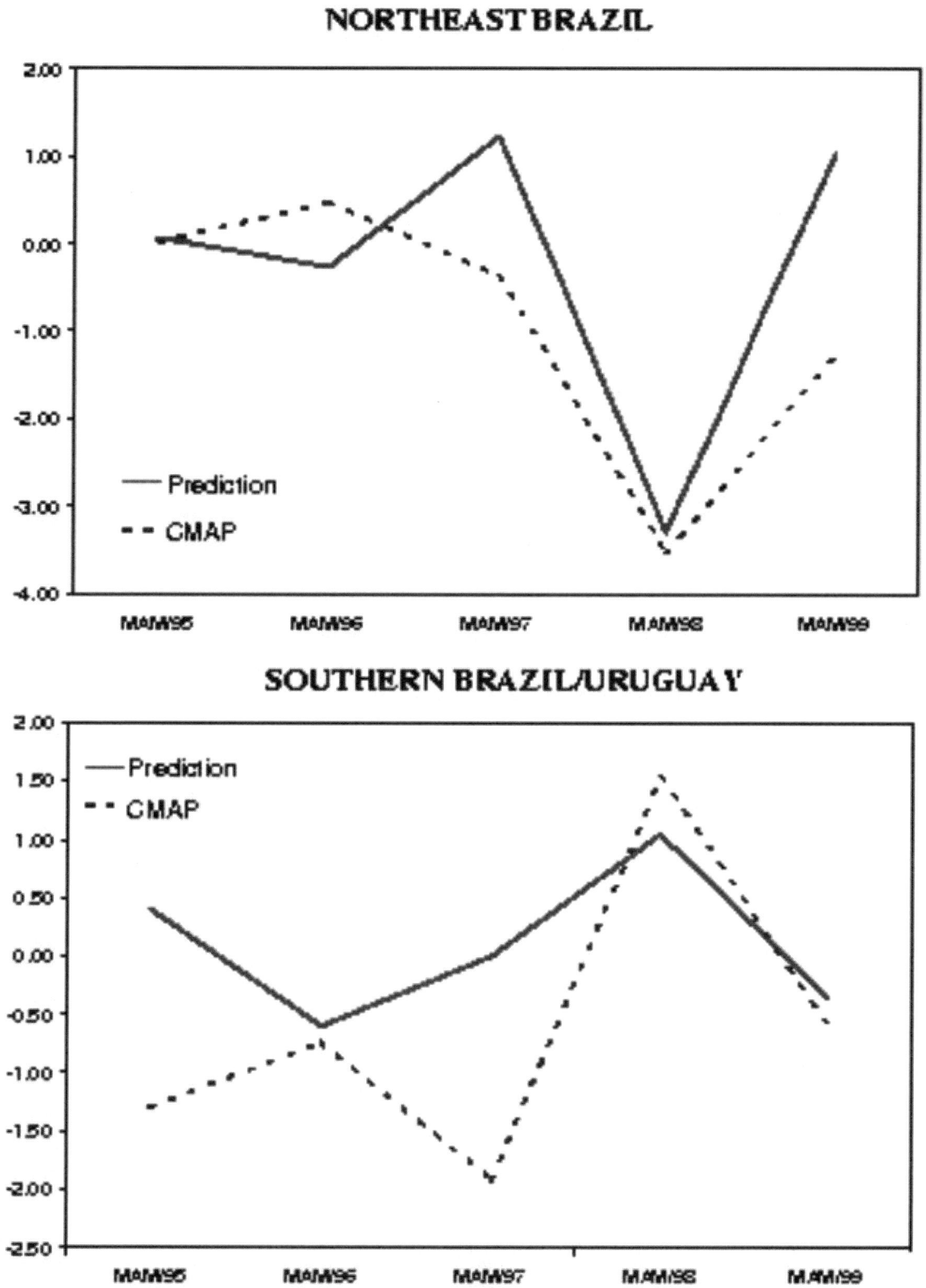

Figure 4. *Spatial average of precipitation anomalies from the predictions of MAM, (1995-1999). Solid black line is the observational CMAP data; and dashed line is the ensemble mean (4 members from 1995 to 1997 and 25 members to 1998 and 1999). (a) Northeast Brazil (20S-120S; 350W-450W (b) Southern Brazil/ Uruguay (22.50S-350S; 580W-480W).*

anomalies, which remove in part the systematic errors observed in the climatological fields. Results of the interannual precipitation variability of the Northeast and Southern Brazil /Uruguay shows the ability of the model in simulating extreme anomalies in these areas. There are still errors in the prediction of precipitation of some areas, when the tropical SST forcing is not strong. More studies and experiments are needed to improve the model's response.

Future developments at CPTEC comprises the use of a Coupled Ocean-Atmosphere GCM, increase of the model resolution and the continuous improvement of parameterizations processes. The limited area model ETA, which is used in weather forecasting at CPTEC is been tested as a downscaling tool for regional climate predictions. Monthly to seasonal predictions for a domain covering South America and part of the adjacent oceans will provide a detailed regional prediction for South America.

REFERENCES

Cavalcanti, I.F.A.; J.A. Marengo; H. Camargo; C.A.C. Castro;; M.B. Sanches, G.O. Sampaio. Climate prediction of precipitation for the Nordeste rainy season of MAM 2000. Experimental Long-Lead Forecast Bulletin, 9, No. 1.

Kinter III,J.L.; D. DeWitt; P.A. Dirmeyer; M.J. Fennessy; B.P. Kirtman; L. Marx; E.K. Schneider; J. Shukla; D. Straus, 1997. The COLA atmosphere-biosphere general circulation model, Vol.1: Formulation. Report nº.51, 46 pp, COLA, Maryland.

Nobre, P.; I.F.A. Cavalcanti, 2000. Seasonal climate prediction over Northeast Brazil using CPTEC/COLA Atmospheric General Circulation Model. (To appear)

Pezzi, L.P.; C.A. Repelli; P. Nobre; I.F.A. Cavalcanti; G. Sampaio, 1998. Forecasts of Tropical Atlantic SST anomalies using a statistical Ocean Model at CPTEC/INPE Brazil. Experimental Long-Lead Forecast Bulletin, 7, nº 1, 28-31.

Xue, Y.; P.J. Sellers; J.L. Kinter III; J. Shukla, 1991. A simplified biosphere model for global climate studies. J. Climate, 4:345-364.

Xie, P; P.A. Arkin, 1996. Analyses of global monthly mean precipitation using raingauge observations, satellite estimates, and numerical model prediction. J. Climate, 9:840-858.

Climate Downscaling Experiments at GISS/CCSR

Leonard M. Druyan and Matthew Fulakeza
NASA/Goddard Institute for Space Studies, USA

INTRODUCTION

The Regional Climate Model (RCM) at GISS/CCSR simulates the evolving climate over selected regions on a grid with 50-km spacing. This genre of model can be used to "downscale" coarser resolution observational analyses or climate predictions from a GCM by providing more spatial detail for the distribution of observed or projected temperatures, precipitation and circulation (Giorgi and Mearns, 1999). Since all the variables are better resolved, local gradients are likely to be sharper and the climatic effects of topography better modeled.

The challenge in downscaling is to simulate accurate representations of climate fields within a limited area domain even though information about the global climate system is provided only through the lateral boundaries at much lower temporal and spatial resolutions. Ultimately it is important to demonstrate that the high resolution results are more faithful to the real climate than those represented by coarse grid analyses or global GCMs. The best validation will be against individual station observations of surface temperature, wind and precipitation.

RCM simulations can be driven by GCM seasonal predictions or hindcasts whereby the seasonal specific information from the lower resolution global model is continuously communicated to the high resolution domain as synchronized lateral boundary conditions. The objective is to obtain regional climate details, such as time-space distributions of surface air temperature, precipitation and circulation, from the RCM that are implied by the GCM prediction. This has a high potential benefit for agricultural applications such as crop models which may require (or work better with) site specific climate data.

We have initiated and analyzed RCM simulation experiments over southern Africa, West Africa and northern South America.

SOUTHERN AFRICA

Our first example refers to simulations over Zimbabwe which demonstrate the RCM downscaling. In the standard GCM grid, the borders of Zimbabwe are included within parts of only four grid boxes. At this resolution, the GCM may be severely tested to represent local gradients within Zimbabwe, even though the country is really composed of distinctly different agricultural regions. For a given season, these different regions may not even experience the same climate anomalies. The RCM run on a 50-km spaced grid resolves Zimbabwe into 168 grid elements and therefore has the potential to represent climate heterogeneity within the country. Figure 1 shows the results of a pilot study which allow us to compare the observed precipitation distribution over Zimbabwe for December 1982 with the GCM hindcast and with the much more realistic RCM downscaled field. Note especially how the RCM, but not the GCM, resolves the dry area over Zimbabwe, even when the GCM simulation is used to supply the lateral boundary information.

Figure 2 compares the observed daily precipitation and maximum temperature at two Zimbabwe stations to RCM simulated (ECMWF initial and boundary conditions) values at co-located grid elements (average of four grid elements for each location). The major rain events were simulated by the model as were the major trends of temperature within the month. Although the simulation indicates too many rain events, the daily totals for actual rain days are reasonable. The amplitude of observed temperature oscillations is somewhat underestimated by the model, but the two major episodes of sudden cooling, on the 9-10th and 15-16th, were both captured by the model. Note that the model has successfully differentiated between the different timing of events at these two stations, *separated by only 2.7° of latitude* (300 km). In fact, both stations are located within the same grid ele-

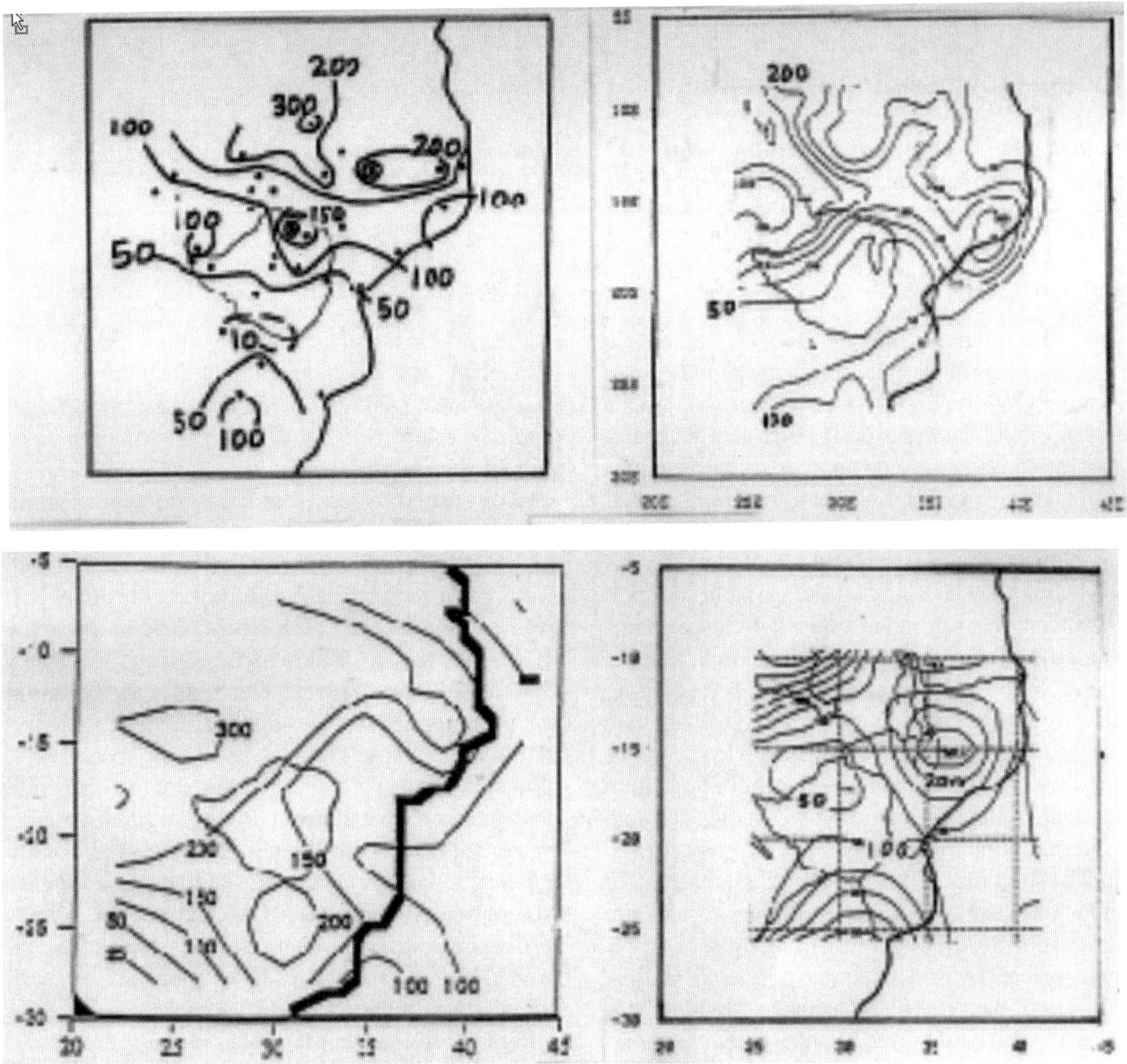

Figure 1a. *Observed Dec '82 precipitation (mm) over SE Africa. Except for the extreme north, Zimbabwe accumulations <100 mm.*

Figure 1b. *Regional Climate Model Dec '82 precipitation (mm) simulated with ECMWF forcing. Large deficits are correctly simulated over the southern half of Zimbabwe.*

Figure 1c. *GISS GCM 5-run ensemble Dec '82 precipitation (mm). Zimbabwe accumulations are too high.*

Figure 1d. *Regional Climate Model Dec '82 precipitation simulated with GISS GCM forcing. Zimbabwe accumulations are correctly <100 mm.*

ment of the standard GISS GCM. Degrading the resolution of the RCM results to that of the standard GISS GCM leads to an obvious and definite decrease in the correlation between modeled and observed precipitation and maximum temperature. Based on this preliminary evaluation we conclude that the spatial gradients of RCM simulated variables (50-km grid) have high potential as input for crop models. Note also that these results from a single run are likely to improve when ensemble averages can be constructed from multiple simulations.

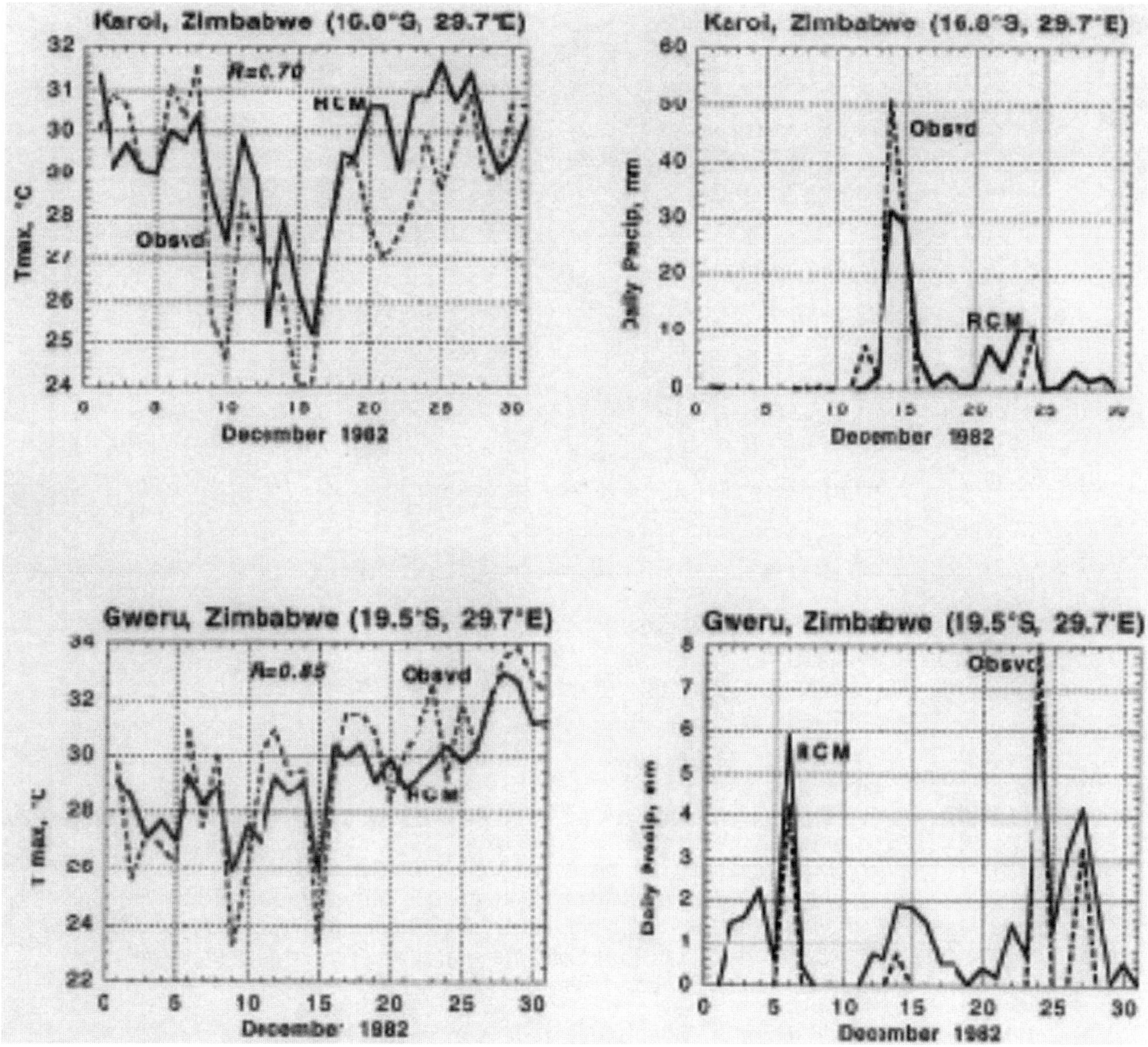

Figure 2. *Validation of downscaled daily maximum surface air temperature (left) and daily precipitation (right). RCM results are the average of the four grid points closest to each station. LBC are from twice daily ECMWF analyses.*

WEST AFRICA

Our second example relates to the rainy season in West Africa (June-September) when African wave disturbances (AWD) and squall lines are important precipitation triggers (Peters et al., 1989). Typical AWD wave lengths are about 2500 km (Burpee, 1972), but convergence zones, vertical motion extremes and squall lines are often confined to one wave sector (Reed et al., 1977). Recurring Sahel droughts with negative impacts on the agricultural economy reinforce the importance of improving our understanding of the intra-seasonal and interannual variability of this region's climate. In addition, improved daily weather forecasts offer obvious benefits to many sectors of the society.

The study of the local weather and climate of West Africa should appropriately relate to the analysis and simulation of AWD and their associated meteorological mesoscale structures. The RCM exploits the 50-km resolution of its grid in order to better analyze their detail.

Druyan et al. (2000a) showed reasonably skillful daily RCM forecasts of the mid-tropospheric cir-

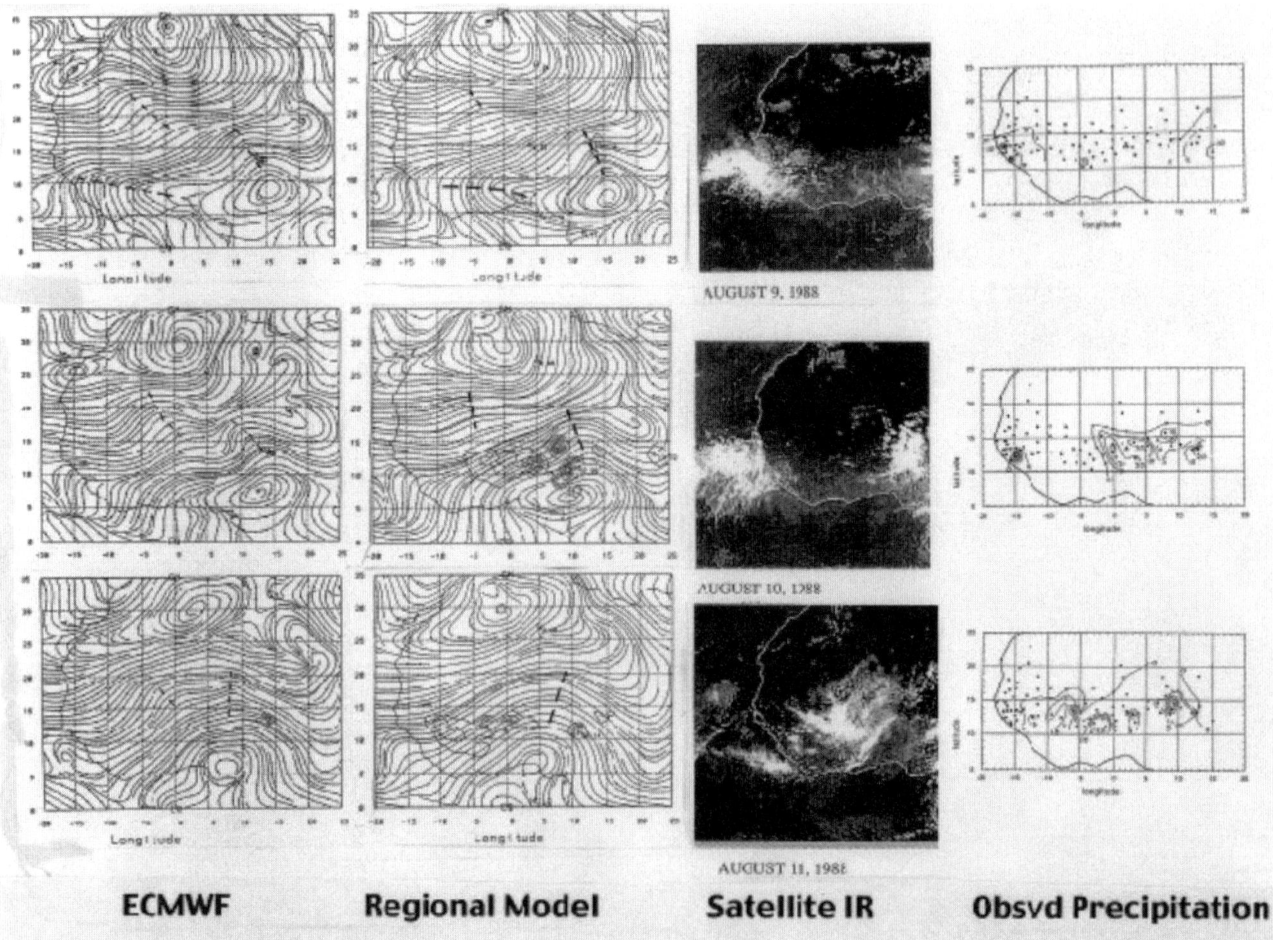

Figure 3. Regional model (50-km grid spacing) results for August 9, 10 and 11, 1988 (second column) are for daily precipitation (mm) and 700 mb circulation. Circulation is compared with ECMWF analyses based on observations and gridded at 2.5° horizontal resolution (left column). Modeled precipitation is compared with satellite IR imagery for 1500 UT and with the spatial analysis of daily precipitation accumulation based on about 85 station observations.

culation. Improvements to the model initialization procedure led to better circulation simulations and realistic daily precipitation distributions as described in Druyan et. al. (2000b). Figure 3 shows the modeled 700 mb circulation on August 9-11, 1988 (the first through third days of the simulation) compared to analyses based on observations gridded by ECMWF at 2.5° resolution. Figure 3 also shows RCM simulated precipitation distributions for the same days, which are compared with concurrent satellite IR imagery and spatial distributions of daily rainfall observed at about 85 stations. Note especially how rainfall is related to the wave-like structures whose troughs are marked, and how an area of heavy precipitation spreads westward as the wave propagates toward the Atlantic coast.

SOUTH AMERICA

Our main focus of interest is Brazil. Within Northeast Brazil, called Nordeste, some precipitation is recorded during every month of the year, but the largest totals occur during March-May. The large interannual variability of rainfall there has been marked by occasional droughts, and resulting famines have prompted large migrations of population (Chu, 1991). Nordeste experiences a smaller annual rainfall than the rest of Brazil, less than 800 mm over most of the area and less than 500 mm over certain interior portions (Kousky and Chu, 1978). Over northern Nordeste, the sharp March-April peak in rainfall corresponds to the southernmost excursion of the Inter-Tropical Convergence Zone (ITCZ) to about 5°S.

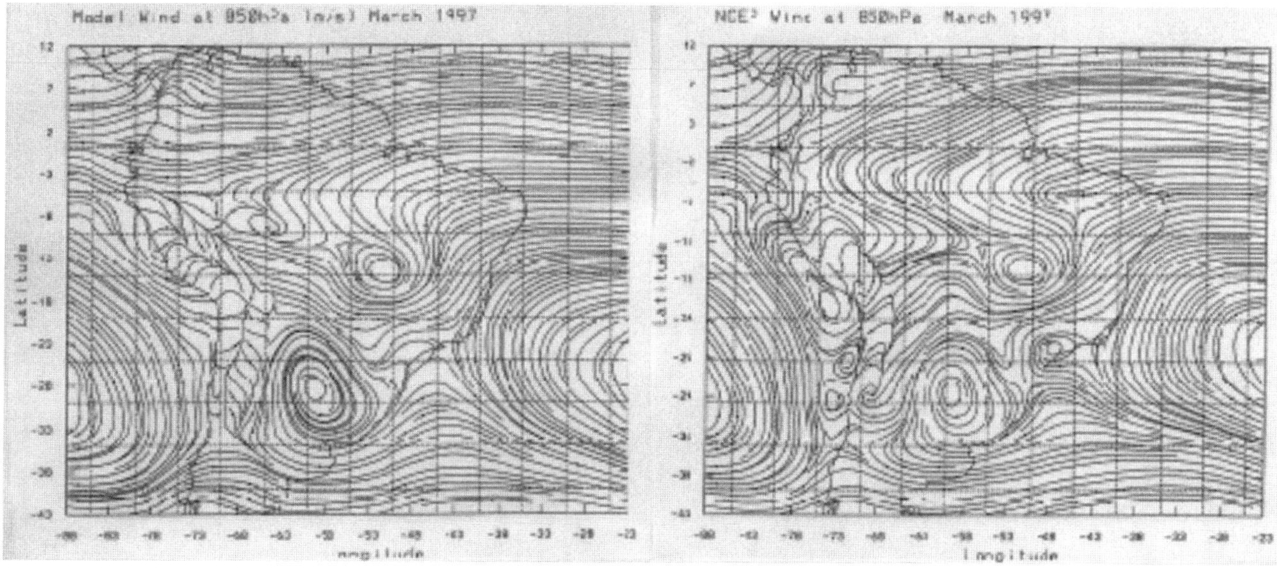

Figure 4. *Regional climate model 850 mb circulation for March 1997 (left) compared with NCEP reanalysis (right).*

Inter-hemispheric SST gradients modulate the north-south surface pressure gradient and consequently the near-surface meridional winds. Warm SSTA in the Atlantic north of the equator are associated with dry seasons in Nordeste. This situation results from enhanced pressure gradients south of the Equator that drive stronger southeasterly trade winds while weakening the northeasterly trades, thus stalling the southward migration of the ITCZ before it reaches Nordeste (Hastenrath and Druyan, 1993; Hastenrath and Greischar, 1993). Correlations show negative correlations between monthly Nordeste precipitation and North Atlantic SST anomalies during April and May, and negative

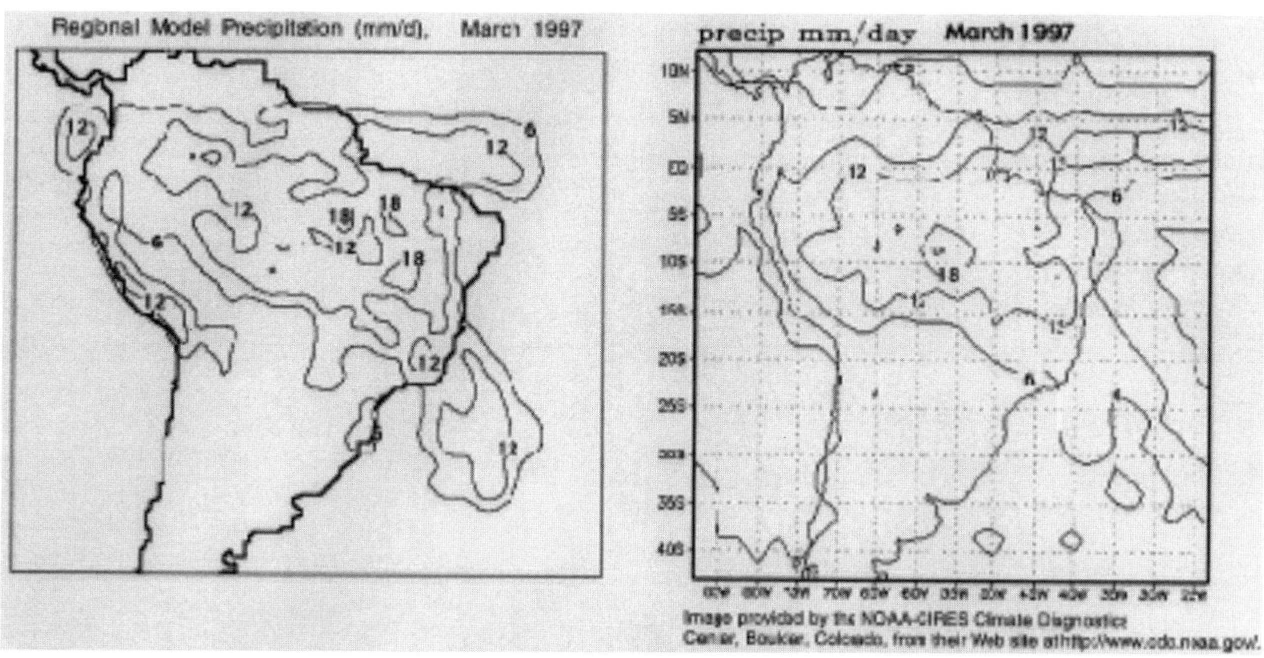

Figure 5. *Regional climate model precipitation distribution for March 1997 compared with estimated precipitation distribution based on satellite data.*

correlations with South Atlantic SSTA. Warm SSTA in the tropical South Atlantic during those months are associated with increased convection over Nordeste despite weakened southeast trades and therefore sluggish westward moisture flux from the South Atlantic.

Marengo's (1992) suggested that dry years in northern Amazonia correspond to weakened moisture advection from the northern tropical Atlantic, which often coincides with strong El Niño episodes. Wet seasons, on the other hand, which depend on strong northeasterly trades, are not always limited to the high phase of SOI. He also found strong negative correlations between SST over broad expanses of the tropical North Atlantic and Rio Negro water levels at Manaus during March-April. This indicates that wet seasons over the northern Amazonia coincide with cold tropical North Atlantic surface waters.

We began an RCM simulation which downscales the climate over South America, as given by NCEP reanalysis data, beginning February 20, 1997. Figure 4 shows the March 1997 mean 850 mb circulation for the RCM and for the NCEP 209-km resolution analysis. Figure 5 shows the RCM monthly mean precipitation distribution for the same month, compared to an estimate based on satellite observations. We note that the simulation correctly situates the ITCZ along 2°N, several degrees northward of its climatological position and consistent with an observed northward gradient of SSTA across the Equator in the Atlantic Ocean that was detected during February-April 1997. Precipitation maxima over Central Brazil correspond to the convergence of Atlantic moisture brought in by the northeasterly circulation which is clearly shown in the RCM results. Plans are to continue this simulation through June 1999 so that it captures the full ENSO cycle, El Niño to La Niña.

ACKNOWLEDGEMENTS

Research was supported by NSF Grant ATM-97-25142 , the NASA Climate and Earth Observing System Programs, the US National Research Council and IRI. Thanks to M. Saloum of Diréction de La Météorologie Nationale of Niger for precipitation station data used in the study.

REFERENCES

Burpee, R., 1972: The origin and structure of easterly waves in the lower troposphere of North Africa. *J. Atmos. Sci.,* 29, 77-90.

Chu, P-S., 1991: Brazil's climate anomalies and ENSO. Chapter 3 in *Teleconnections Linking Worldwide Climate Anomalies,* M. Glantz, R. Katz and N. Nicholls, eds. Cambridge University Press (NY).

Druyan, L., M. Fulakeza and W. Thiaw, 2000a: Regional model simulations of African wave disturbances. *J. Geophys. Res.,* 105, in press.

Druyan, L., M. Fulakeza, P. Lonergan, G. Jenkins and M. Saloum, 2000b: Regional climate model simulations over West Africa. *Mon. Wea. Rev.,* submitted.

Giorgi, F. and L. Mearns, 1999: Introduction to special section: Regional climate modeling revisited. *J. Geophys Res.,* 104, 6335-6352.

Hastenrath, S. and L. Druyan, 1993: Circulation anomaly mechanisms in the tropical Atlantic sector during the Northeast Brazil rainy season: Results from the GISS GCM. *J. Geophys. Res.,* 98 (D8), 14917-14923.

Hastenrath, S. and L. Greischar, 1993: Circulation mechanisms related to Northeast Brazil rainfall anomalies. *J. Geophys. Res.,* 98 (D8), 5093-5102.

Kousky, V. and P-S. Chu, 1978: Fluctuations in annual rainfall for Northeast Brazil. *J. Meteorol. Soc. Japan,* 57, 457-465.

Marengo, J. 1992: Interannual variability of surface climate in the Amazon basin, *Int. J. Climatology,* 12, 853-863.

Peters, M., G. Tetzlaff and W. Janssen, 1989: Rainfall intensity of West African squall lines. *Annales Geophysicae,* 7, 227-238.

Reed, R., D. Norquist and E. Recker, 1977: The structure and properties of African wave disturbances as observed during Phase III of GATE. *Mon. Weather Rev.,* 105, 317-333.

Predicting Shifts in Extreme Events Using a GCM

Lisa Goddard[1] and Simon Mason[2]
[1]International Research Institute for climate prediction, USA
[2]Scripps Institution of Oceanography, USA

INTRODUCTION

Climate forecasts provide potentially useful information of seasonally (3-month) averaged temperature and precipitation at lead times of 0 to 6 months (Mason, et al. (1999); Barnston, et al. (1999)). As the value and quality of seasonal climate forecasts increases, users of this information are already looking ahead and asking for more spatial and temporal detail from the forecasts. Dynamical and statistical downscaling techniques that spatially, and even temporally, refine the climate information from coarse resolution atmospheric general circulation models (AGCMs) are currently being assessed. However, there may be valuable information on the temporal (i.e. daily) characteristics of the seasonal rainfall, available directly from the AGCM predictions, that has been unutilized. Although it will never be possible to predict the weather for a particular day months in advance, it may be possible to predict the anomalous character of the weather within a season for an assumed shift in the seasonal climate.

The two characteristics of the daily rainfall contributing to the seasonal totals are the number of days it rains and the intensity of the rainfall on those days. Increases in either characteristic can substantially increase seasonal rainfall totals but with different consequences. The impacts of above-normal seasonal precipitation delivered by a few unusually intense storms are typically more devastating than those from precipitation that comes steadily but moderately through a given season.

Observed and simulated daily rainfall data is analyzed in relation to the total seasonal precipitation. The region over Queensland, Australia, for which risk of extreme above-normal seasonal precipitation was forecast (by the IRI— http://iri.ldgo.columbia.edu/climate/forecasts/net_asmt/apr1999/Wld_AMJ_99_prec.html) for April-May-June 1999, serves as the focus to assess whether it may have been possible to infer the nature of the daily rainfall that resulted in the high seasonal totals.

DATA

The observed rainfall data was provided by the Queensland Department of Natural Resources (QDNR). The data was collected by the Australian Bureau of Meteorology who maintains thousands of rain gauge stations across the country. QDNR then examined the station data for consistency and quality and placed the filtered collection of station data onto a 25km grid. Unlike some regions of Australia, a sufficient density of stations was retained in the filtered data set over Queensland, yielding a reliable gridded precipitation field. The daily rainfall data used in this study covers the period January 1960-June 1999.

The AGCM data comes from simulations using the ECHAM3 AGCM configured at T42 horizontal resolution (approximately 2.8x2.8 degrees) with 18 levels in the vertical. An ensemble of 10 members has been integrated for the period 1950-1999. Each ensemble member is forced with the same observed monthly-averaged sea surface temperatures from Reynolds reconstructed SST data through 1981, and Reynolds optimally interpolated SST data after 1981. These ensemble member runs are continuous runs that differ only in the atmospheric state assigned at the beginning of the runs. The daily data was saved for 8 of the 10 runs.

METHOD

The analysis was designed to calculate the anomalous frequency of extreme rainfall events within a season. First, base years are selected from which a base probability distribution function (PDF) is drawn. Since the focus is the anomalous frequency of extreme events during an extreme season, the base PDF is drawn from a set of near-normal years. Using the standard climatology period of 1961-1990, the base years are defined, at each grid point, as the near-normal one-third of the years (10 years) within that period. From this selection of base years at each point, a base PDF is constructed for the season by aggregating daily rainfall rates greater than 1.0mm for each day of the three month season for all 10 years. For the ensemble of AGCM runs, the 8 runs are aggregated also.

The base PDF quantifies the rainfall rate associated with the percentiles. For an event defined at the 75%-ile, the expected frequency, F_b, of a rainy days qualifying as an event in a normal year is 0.25:

$$F_b = 0.25 = ne_b/f_b,$$

where ne_b = number of events, on average, in base years; and
f_b = rainfall frequency in base years (= # rainy days in season divided by total days in season)

Similarly, in a particular year, the frequency of events, expressed as a percentage, at or above the pre-defined threshold is given by:

$$F_i = ne_i/f_i,$$

where ne_i = number of events in a specific year; and
f_i = rainfall frequency in specific year.

The anomalous frequency of extreme rainfall events during a particular season is then the fractional, or percentage, difference from the base rate.

$$F_{anom} = (ne_i/ne_b) - 1, \text{ or}$$

$$P_{anom} = 100*[(ne_i/ne_b) - 1].$$

RESULTS

The connection between seasonal rainfall variability and the two characteristics of the associated daily rainfall can differ. For example, for all of Australia, changes in the number of rain days in a season invariably correlates highly with changes in seasonal rainfall (not shown). However, the correlation between anomalous seasonal rainfall and anomalous average precipitation intensity exhibits a spatial pattern suggesting that changes in average intensity contributes little to seasonal variability during the rainy season, on average (not shown). Over Queensland, the April-May-June season marks a sharp transition season between very wet conditions, occurring at the beginning of the year, to very dry conditions, occurring near the middle of the year. During this season, changes in both the number of rain days and the average daily rainfall intensity are positively correlated with anomalous seasonal rainfall amounts (not shown). Thus, given a confident forecast for above-normal seasonal rainfall, it is not obvious how the characteristics of the daily precipitation are likely to change. This result highlights the potential benefits of using a dynamical model to predict the characteristics of the seasonal rainfall.

In many ways, the AGCM simulation over Australia agrees well with the observations for precipitation. The model can reproduce the evolution of the seasonal cycle with relative maxima in the appropriate locations (not shown). Quantitatively, however, significant biases do exist. The AGCM overestimates the seasonal rainfall frequency. Over Queensland during April-May-June, the model gives two to four times more rainy days than are observed; although, the AGCM does reproduce the spatial pattern of the rainfall frequency very well (not shown). Still, overestimating the rainfall frequency in regions where or when this characteristic contributes strongly to the seasonal variability will certainly degrade model skill. The average rainfall intensity is also too great by a factor of two throughout Queensland during April-May-June (not shown). These two characteristics combine to deliver a climatological rainfall that is too wet in this season. As a result, the quantitative thresholds that define the extreme events (e.g. 75%-ile, 90%-ile) will be higher in the AGCM than in the observations.

Despite the climatological biases, the AGCM simulates the interannual variability of rainfall with high skill over much of Queensland. Based on the spatial maps of the R.O.C. (relative operating characteristics, e.g. Mason and Graham, 1999) scores for extreme above-normal (85%-ile) seasonal variability, two sub-regions within Queensland are chosen for the extreme event analysis. Along the central eastern coast of Australia, near Brisbane, the AGCM has good skill in reproducing extreme-above normal, but far weaker skill for extreme below-normal conditions; over the area of central Queensland the model skillfully simulates both extreme above-normal and below-normal seasonal rainfall.

For the period 1961-1999, there are six years in the wettest 15% of the historical record. Table 1 below outlines the performance of the AGCM relative to what occurred in the observations. The categories "above" and "below" here refer to an increase or decrease in the number of extreme events relative to the base years for the predefined thresholds at the 75%-ile and 90%-ile. A hit is defined as the majority of the ensemble members agreeing with the observed shift in extreme event frequency; a miss is defined as the majority of the ensemble members not agreeing with the observed shift. An interesting point is that extreme rainfall events do not necessarily increase even when the seasonal totals are much above normal, as seen over the Brisbane region in April-May-June 1963.

For the extremely wet seasons of the 1960-1999 period the AGCM shows good agreement with the observations. For the events exceeding the 75%-ile, the AGCM correctly favors the observed outcome

Table 1 (a): *BRISBANE*

YEAR	OBS. (75%-ILE; 90%-ILE)	# RUNS ABOVE (BELOW) (75 %-ILE; 90%-ILE)	HIT/MISS (75%-ILE; 90%-ILE)
1963	Below ; Below	3 (5) ; 5 (3)	HIT ; MISS
1983	Above ; Above	6 (2) ; 4 (4)	HIT ; ----
1988	Above ; Above	3 (5) ; 5 (3)	MISS ; HIT
1989	Above ; Above	7 (1) ; 6 (2)	HIT ; HIT
1990	Above ; Above	3 (5) ; 3 (5)	MISS ; MISS
1992	Above ; Below	5 (3) ; 6 (2)	HIT ; MISS
1999	Below ; Below	3 (5) ; 3 (5)	HIT ; HIT

Table 1 (b): *CENTRAL QUEENSLAND*

YEAR	OBS. (75%-ILE; 90%-ILE)	# RUNS ABOVE (BELOW) (75 %-ILE; 90%-ILE)	HIT OR MISS (75%-ILE; 90%-ILE)
1963	Above ; Above	5 (3) ; 4 (4)	HIT ; ----
1968	Above ; Above	8 (0) ; 6 (2)	HIT ; HIT
1971	Above ; Above	3 (5) ; 2 (6)	MISS ; MISS
1983	Above ; Above	3 (5) ; 2 (6)	MISS ; MISS
1989	Above ; Above	7 (1) ; 5 (3)	HIT ; HIT
1990	Above ; Above	6 (2) ; 5 (3)	HIT ; HIT
1999	Above ; Above	6 (2) ; 4 (4)	HIT ; ----

in four out of six cases for both regions. For the events exceeding the 90%-ile, the AGCM seems to perform better over central Queensland where the majority of ensemble members simulated the observed outcome in three out of five cases, with an additional case in which the ensemble members were divided. Over the Brisbane region there are only two cases out of five considered "hits"; however, in two of the three "misses", the AGCM favored an increase in 90%-ile events when observations showed a small decrease.

For the recent case, April-May-June 1999, the Brisbane region experienced a decrease in extreme events, and this was the most likely scenario in the AGCM. For the central Queensland region, the observed frequency of extreme rainfall doubled for 75%-ile events and more than tripled for 90%-ile events even though the seasonal totals were not extremely above-normal. The AGCM did suggest that events exceeding the 75%-ile threshold would increase in this area, although the majority of the runs did not produce the above average amount of 90%-ile events.

Although this table presents the results crudely, evidence for potential predictability exists. More information can be gained from looking at the magnitude of the departures from normal in the individual model runs and in the observations (not shown). Using the above/below measure strongly separates the results without distinguishing a result that may be barely below or above average, from one that is far from average. Furthermore, the observations are not deterministic, they give only one case from a potential range of possibilities. Thus, a hit or miss in a particular year does not anoint or denounce the ability of the AGCM to give a fair estimate of increased or decreased likelihood of extreme events. Six or seven cases is clearly not enough to make rigorous statements on the potential of this tool, especially for very extreme events (e.g. 90%-ile) that may be inherently more unpredictable or may be due to processes acting on scales smaller than what this AGCM can resolve.

CONCLUSIONS

We have presented a preliminary examination of the ECHAM3 AGCM's ability to simulate shifts in seasonal frequency of extreme rainfall events. The focus of this study was Queensland, Australia for the April-May-June season, and in particular for the year 1999. Although many of the characteristics of the seasonal and interannually varying precipitation appear reasonable in the AGCM compared to the observed rainfall, significant biases exist. Most importantly, the daily rainfall in the AGCM is too intense, and it rains on too many days. This latter issue will clearly be problematic for predicting seasonal rainfall anomalies that are a result of increased number of rainy days. In spite of these noted shortcomings, the results were promising.

During April-May-June 1999, the observed frequency of extreme rainfall doubled for 75%-ile events and more than tripled for 90%-ile events in the central Queensland region. The majority of ensemble members from the AGCM simulation did suggest an increase in these extreme rainfall events for this region. During the same season, the small region near Brisbane experienced a marked decrease in extreme rainfall events which was also captured by the AGCM simulation.

Overall, our analysis involving 40 years of data shows some suggestion of potential predictability for the shifts in heavy rainfall frequencies. The locations that are potentially predictable are restricted to those areas where the model has demonstrated skill at simulating extreme above-normal seasonal rainfall (above 85%-ile of historical record). The years for which the shift in extreme events is potentially predictable requires more investigation, but for the years in which the observed seasonal rainfall was high, the AGCM appears to shift the frequency of extreme rainfall events appropriately during the April-May-June season.

NOTES

This work was originally presented at the Twenty Fourth Annual Climate Diagnostics and Prediction Workshop, and appeared in the proceedings as "Predicting Weather in Climate: An example from 1999".

REFERENCES

Barnston, A.G., A. Leetma, V.E. Kousky, R.E. Livezy, E. O'Lenic, H.Van den Dool, A.J. Wagner, and D.A. Unger, 1999: NCEP forecasts for the El Nino of 1997-98 and its U.S. impacts. Bull. Amer. Meteor. Soc., 80, 1829-1852.

Mason, S.J., L. Goddard, N.E. Graham, E. Yulaeva, L. Sun, and P.A. Arkin, 1999, The IRI seasonal climate prediction system and the 1997/98 El Nino event., Bull. Amer. Meteor. Soc., 80, 1853-1873.

Mason, S.J. and N.E. Graham, 1999, Conditional probabilities, relative operating characteristics, and relative operating levels. Weather and Forecasting, 14, 713-725.

Seasonal Forecasting Activities at the U.K. Met Office

Andrew Colman, Mike Davey, Richard Graham, and Sarah Ineson
The Met. Office, United Kingdom

INTRODUCTION

The Met. Office first started developing and issuing seasonal forecasts in the 1960s. The first prediction schemes were statistical, and due to the lack of accurate historical data to train the models the forecasts were not particularly skilful and were viewed with skepticism. Over the past decade, however, seasonal forecasting activity at The Met. Office and other meteorological and climate institutes has increased considerably. Computers have become powerful enough for the regular running of Atmospheric General Circulation Models (AGCMs) out to several months ahead. New high quality globally complete historical datasets such as the Global Ice and sea surface temperature (GISST) and the Comprehensive Ocean Atmosphere Dataset (COADS) have provided insights into seasonal climate and allowed the development of better statistical schemes.

SST information is particularly important in seasonal forecasting because anomalies in SST tend to be long-lasting, due to the large heat capacity of the ocean relative to the atmosphere, and can provide a systematic bias to the atmosphere on seasonal timescales. Statistical prediction schemes make use of links between SST and regional rainfall (for example) identified in the historical datasets. The AGCMs rely on predictions of SST, and for a range of a few months persistence of observed anomalies is one simple and effective strategy. Other ways of predicting SST are to use more advanced statistics, or to use coupled dynamical ocean-atmosphere models.

AGCM ENSEMBLE FORECASTS

For the past two years AGCM ensembles have been run weekly at The Met. Office to 4 month range. The AGCM is the standard Met. Office atmospheric climate model (known as HadAM3): it has 19 levels in the vertical and horizontal grid-points at 3.75o longitude and 2.5o latitude intervals. The 9 ensemble members are initialized with atmospheric analyses produced at 6 hourly intervals over a period of 48 hours immediately prior to the forecast start time. The SST is based on observations over the preceding 4 weeks, and SST anomalies are persisted after the start time.

The real-time basic rainfall and 850hPa temperature forecasts are made available to National Met. Services over the internet, on a password-protected site. (The 850hPa temperature can be used as a proxy for surface temperature.) These forecasts are presented as ensemble means and as probabilities of above or below average rainfall and temperature for months 1-3 (i.e. zero month lead) and months 2-4 (1 month lead) after the start time. The anomalies are relative to a model climatology. The basic probability forecasts are calculated from the distribution of the 9 ensemble members: thus if all 9 members predict above average rainfall for a specific grid box then the probability forecast for above average rainfall for the grid box is 100% and the probability of below average rainfall is 0%.

Examples of these forecast maps are shown in Figure 1. Figure 1a is a global rainfall map (expressed as percentages of normal) for the 3 month period starting 30 days after the forecast date (2nd September). Regional maps are also available on the website: the grid is the same as for the global maps but the map scale is enlarged for easier viewing; probabilities of above average temperature for European grid points are shown in fig. 1b (in the fig. 1b example, probabilities of above average rainfall exceed 0.6 in almost all European areas where there is skill). The global and regional maps are provided on the website, with and without skill masking. The skill masks simply blank out areas where the prediction skill is below some preset level (see below). The maps with and without masks enable users to see the large-scale pattern of anomalies and also to have guidance on model skill.

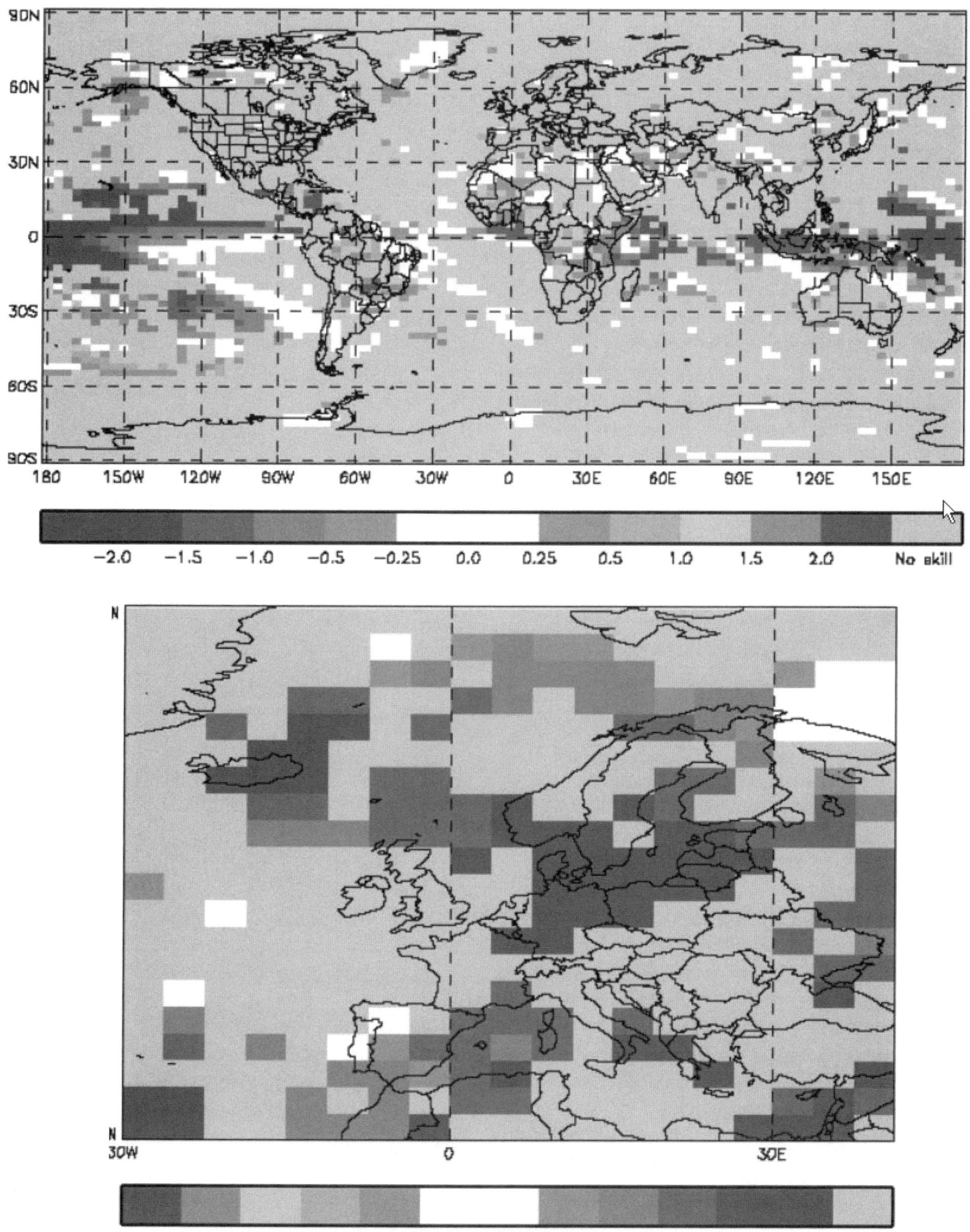

Figure 1. *Typical ensemble forecast maps available to NMSs. (a) Forecast rainfall anomaly in mm for gridpoints worldwide for the period September 3rd -December 2nd 1999 issued September 2nd 1999. (issued in color). (b) Forecast probabilities of above average temperature for European gridpoints for the period September 3rd -December 2nd 1999 issued September 2nd 1999.*

Forecast for selected areas are extracted and sent to regional Climate Outlook Fora when requested. Unless the area concerned is within one of the special areas (see section 3), the forecast format is the same as above. Examples of forecasts issued in this way include those contributed to the Southern Africa Climate Outlook forum.

The potential skill of the ensemble forecasts was first assessed by running trial AGCM simulations. The simulations were initialized with observed atmospheric initial conditions and forced with observed SST, as part of the PROVOST (PRediction Of climate Variations On Seasonal and interannual Timescales) project supported by the European Union. Quarterly sets of nine member ensembles were produced for 19 trial years (1979-1997). Starting dates were 1st March, 1st June, 1st August and 1st December and the simulations were all run for 4 months. Subsequently this set of simulations was repeated, using the same SST persistence strategy as for the real-time forecasts. The skill with persisted SST was slightly lower than that obtained with observed SST. The climatologies and skill assessments used for the forecasts are based on the model simulations over these 19 years.

The AGCM probability forecast skill has been assessed using various measures. The skill masking is based on Relative Operating Characteristic (ROC) scores (Stanski et al. 1989). ROC scores are based on predicted and observed events (where there are just two possibilities, for an event to happen or not to happen), and calculated using the Hit Rate (the number of times an event is predicted correctly divided by the number of times an event occurs) and the False Alarm Rate (the number of times an event is predicted but does not occur divided by the total number of times the event does not occur).

Probability forecasts can be assessed by ROC scores by deeming an event to be predicted when the probability is greater than a given probability threshold. By determining hit rates and false alarm rates for a number of probability thresholds a graph like that shown in Figure 2 can be produced. (Here threshold probabilities of 0.2, 0.4, 0.6, 0.8 are used for illustration.) As skill increases the hit rate exceeds the false alarm rate further and the curve is nearer to the top left hand corner than to the bottom right hand corner. The area under the curve is called the probability ROC score (PROC score. Monte-Carlo tests indicate that for random forecasts over the PROVOST assessment period, PROC scores of 0.6 are exceeded about 10% of the time (Graham et al, 2000). This PROC score threshold of 0.6 has been selected to mask the ensemble forecasts.

The deterministic forecasts are verified using anomaly correlations. The 90% significance level for anomaly correlations was determined by Monte-Carlo tests. Deterministic forecasts for areas where the anomaly correlation is below the 90% significance level are masked.

A detailed review of the AGCM forecasts and their performance is in Graham et al. (2000)

SEASONAL FORECASTS FOR SPECIFIC REGIONS

As well as producing gridded forecasts for the globe, more detailed forecasts are produced for several specified regions and seasons of special interest. These are regions where there is known to be strong predictability of seasonal climate anomalies or regions where there is a particular demand for forecasts. The regions for which we produce regular forecasts are shown in Figure 3. These forecasts are made using a combination of dynamical and statistical forecasts, except that for NW Europe which is purely statistical. Forecasts of wet season rainfall for the Sahel and neighboring regions of Africa and for NE Brazil have been issued for more than a decade following well publicized droughts and the discovery of predictability of seasonal rain-

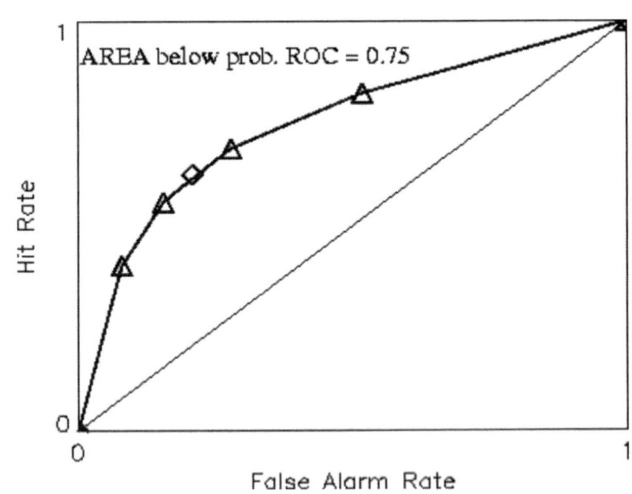

Figure 2. *Example of a ROC curve*

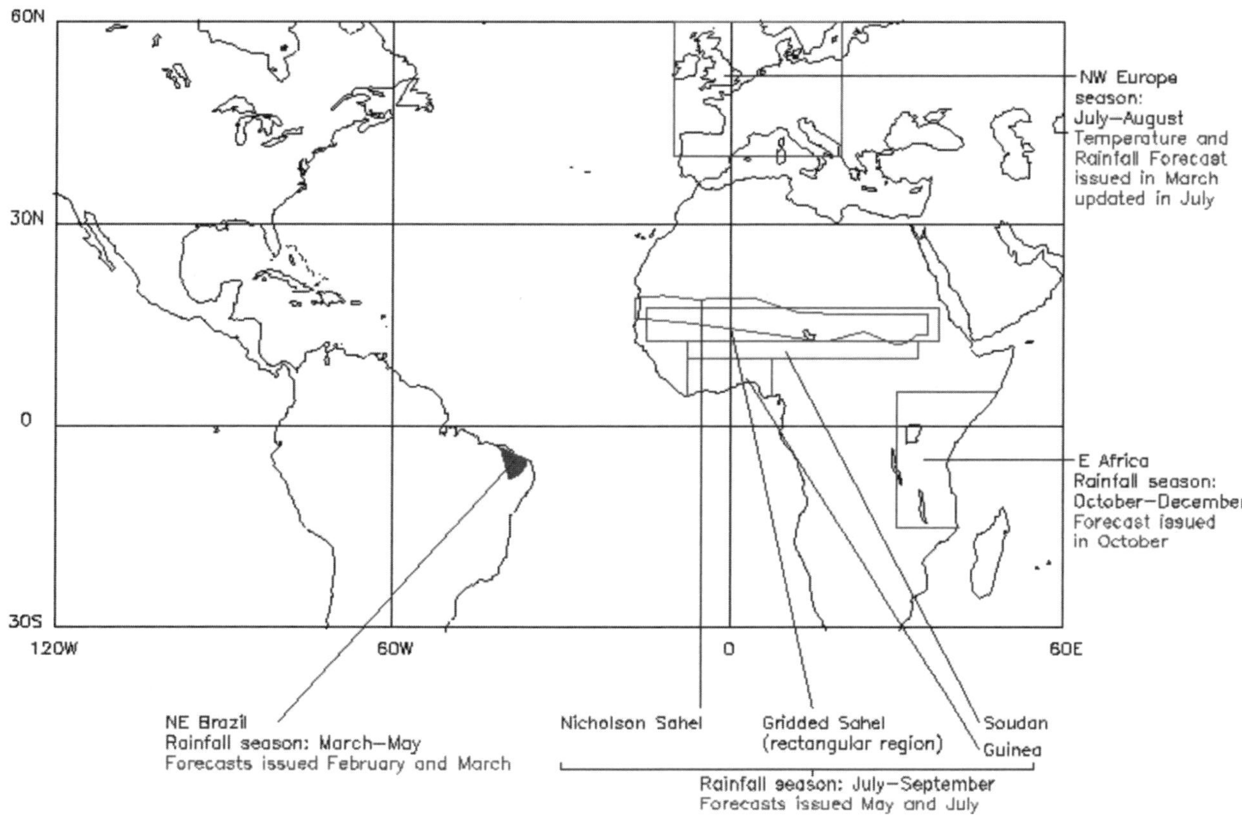

***Figure 3.** Areas of special interest for which probability seasonal forecasts are issued.*

fall anomalies from regional and global SST anomaly patterns (Folland et al., 1991, Ward and Folland, 1991). More recently, forecasts have been extended to East Africa (Mutai et al., 1999) and to NW Europe (Colman and Davey, 1999). These forecasts are distributed to institutes in the region concerned, and have been published regularly in the Experimental Long-Lead Forecast Bulletin, along with skill scores.

For the tropical seasonal rainfall forecasts, the historical record for 1961-1990 is categorized into 5 equi-probable quintiles. Best estimate and probability forecasts are made for the occurrence of these quintiles. Linear regression is used to make statistical best estimate rainfall forecasts and discriminant analysis is used to make statistical probability forecasts. The regression and discriminant equations are calculated using historical data back to 1901. The predictors are SST indices represented by the time series of eigenvector patterns. A set of statistical probability forecasts for north African rainfall is shown in Figure 4 as an example. Ensemble mean and probability forecasts are also made from the AGCM output.. These probability forecast take account of the skill of the model in predicting rainfall in trial simulations for 1961-1990 (Rowell 1998). Firstly, the model rainfall quintile predicted by the AGCM is identified according to the 1961-1990 model climatology. Probabilities of observed rainfall quintiles are then calculated to be directly proportional to the number of time the quintile is observed following a simulation of the forecast model quintile over the period 1961-1990. This method of probability forecasting has been used to provide forecasts for other regions. For example, a special long lead forecast of UK temperature and rainfall over the millennium period (December 15 1999- January 15 2000), included probabilities calculated in this way from AGCM forecast output.

The statistical and AGCM forecast information is combined subjectively to produce an overall seasonal prediction. For some areas, statistical and AGCM forecasts are made on the same grid to aid this process.

The only regular forecasts, which does not currently use the AGCM, are the predictions of NW Europe July-August temperature and rainfall from

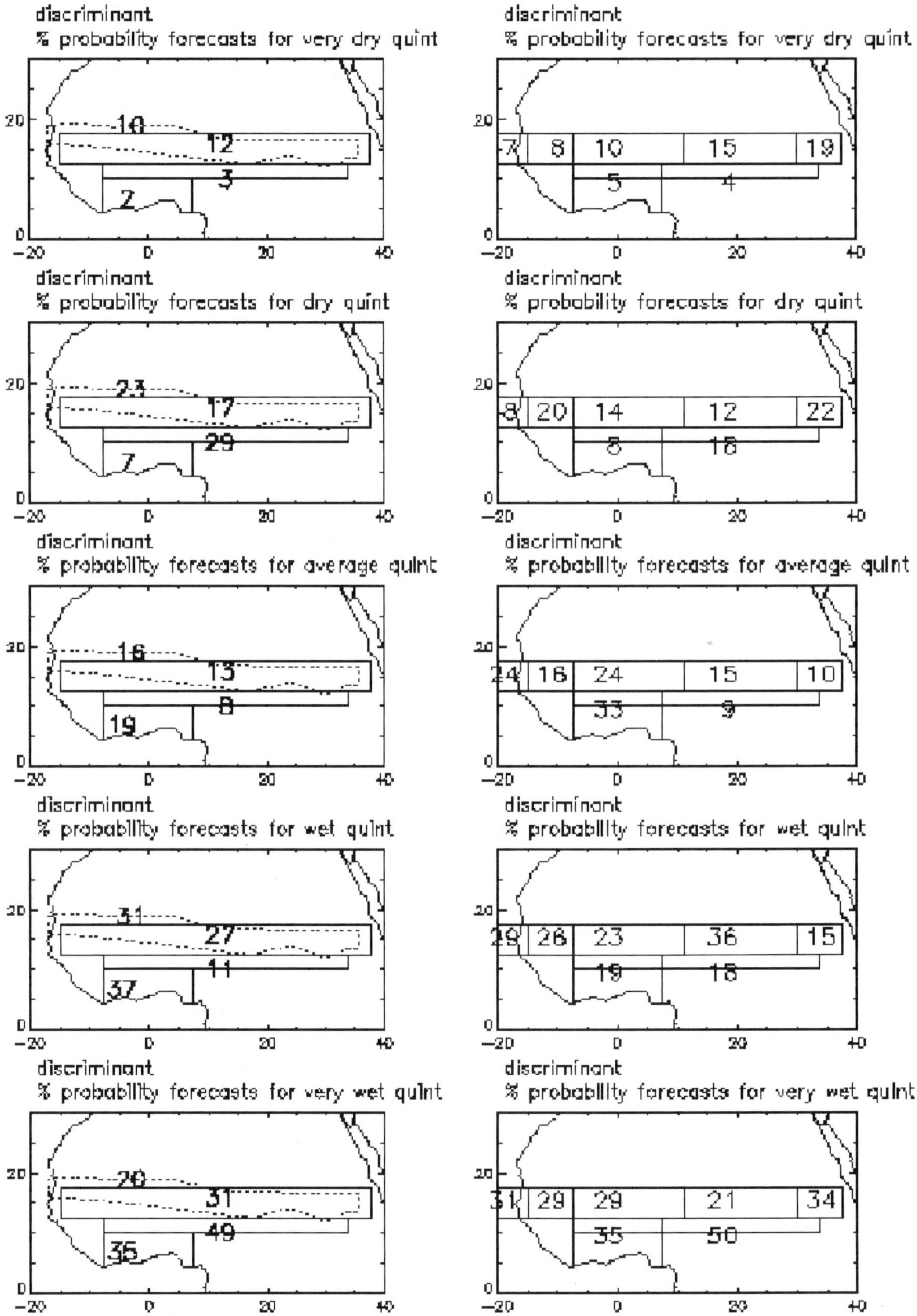

Figure 4. *Example of discriminant analysis probability forecast for July-September 1999 in North Africa*

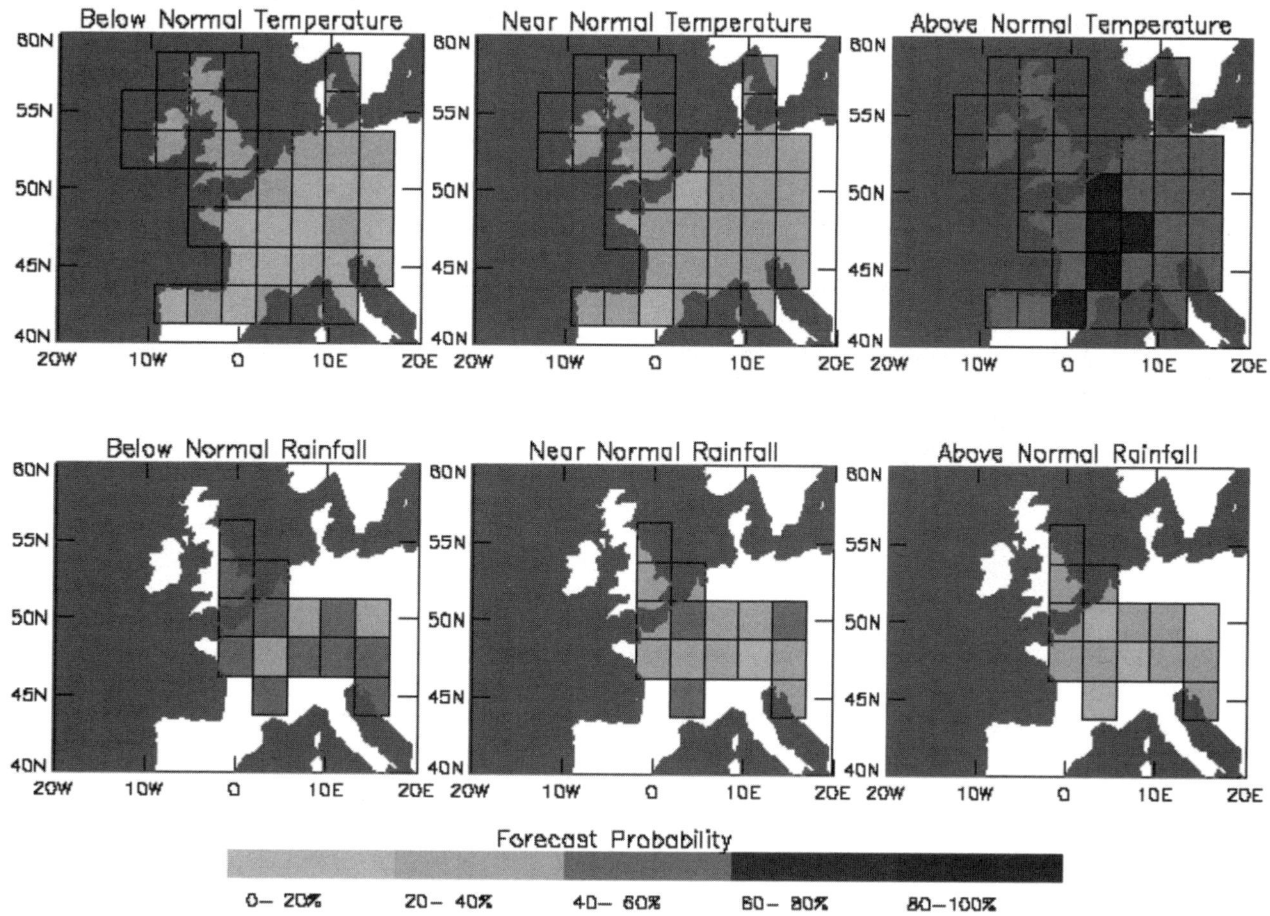

Figure 5. *The forecast for July-August 2000 published on the Met Office web site*

a preceding January-February north Atlantic SST pattern. Tercile forecasts of temperature and rainfall are made for 3.75o longitude and 2.5o latitude grid box means and temperature forecasts are also made for the UK subdivided into 10 districts. Figure 5 shows an example of the forecast format

FUTURE PLANS

DEMETER (Development of a European Multiple-model Ensemble system for seasonal to inTERannual climate prediction) is a major new EU-supported seasonal prediction project that started in April 2000. Several GCM-based seasonal prediction models (mostly coupled ocean-atmosphere) will be assessed over a multi-decadal period by making ensemble hindcasts to 6 months range. The consortium includes partners from the health and agricultural sectors, who will test the impact of the forecasts for their particular applications. To this end, the DEMETER model output will include a comprehensive set of variables relevant to crop modeling.

One outcome of the PROVOST project was the demonstration that skill could be improved by combining the output of several GCMs to form multi-model ensembles. This strategy will be included in the DEMETER assessments.

Another aspect being developed further is the use of statistics to correct model systematic errors.

For example, at the Met. Office relations between observed Sahel rainfall and AGCM simulations of global gridded winds, pressure and rainfall have been used to improve the skill of the model in that region substantially.

REFERENCES

Colman, A.W. and Davey, M. 1999: 'Prediction of summer temperature, rainfall and pressure in Europe from preceding winter north Atlantic ocean temperature.' Int. J. Climatol. 19 513-536

Folland, C.K., Owen, J., Ward, M.N and Colman, A.W. 1991: Prediction of seasonal rainfall in the Sahel region using empirical and dynamical methods. Journal of Forecasting, 10, 21-56.

Graham, R.J, Evans, A.D.L, Clark, R.T., and Robertson, K.B. 2000: The Met. Office's experimental seasonal prediction system. To appear in Quart. J. Roy. Met .Soc.

Mutai, C.C., Ward, M.N and Colman, A.W. 1998: Prediction of East Africa seasonal "short rainfall" rooted in evidence for widespread SST-forced variability during October-December. Int. J. Climatol.18 975-997.

Rowell, D.P., 1998: Assessing potential seasonal predictability with an ensemble of multi-decadal GCM simulations. J. Climate, 11, 109-120.

Stanski, H.R., L.J. Wilson and W.R. Burrows, 1989: Survey of common verification methods in Meteorology. World Weather Watch Technical Report. No. 8, WMO/TD 358, 114pp.

Ward, M.N. and C.K. Folland, 1991: Prediction of seasonal rainfall in the north Nordeste of Brazil using eigenvectors of sea-surface temperature. Int. J. Climatol., 11, 711-743.

Seasonal Changes in the Rainfall Pattern in São Paulo State, Brazil

Luci Hidalgo Nunes
The Secretariat for Environment of S. Paulo State, Brazil

INTRODUCTION

Agriculture is quite sensitive to the ordinary rhythm of the climate, being adjusted to make use of the seasonal changes. Failure of the expected seasonal rainfall or any other kind of disturbance of the rhythm leads to economic losses (Hare 1985). Fluctuations are internal to the climatic system. Rainfall in particular presents notorious variability, which is a necessary part of the climate. Nevertheless, such unpredictability poses a threat of major disruption to agriculture. Because of the broad sensitivity of most crops to climate fluctuations, the study of the regional effects of contemporarily climate variability is of great interest. This study focuses on São Paulo State, which concentrates a large proportion of population, farming, factories and hydroelectric power generation of Brazil. Furthermore, São Paulo City, the State's capital, is one of the largest in the world. For all these reasons, it is clear that São Paulo State is heavily dependent on adequate rainfall and is vulnerable to even moderate precipitation variability.

São Paulo State shows considerable climatic contrasts due to the complex interactions of physical and biological features, as well as atmospheric and ocean dynamics. It has a tropical regime and approximately 74% of the rainfall totals are concentrated in the summer-spring semester. São Paulo State is located in a transitional zone in terms of regional atmospheric circulation, being affected by tropical and extra-tropical air systems, as well as large-scale phenomena, like El Niño. Sugar cane, orange, soybean and coffee are examples of important crops cultivated in the state. Recent years have experienced increases of cultivated area and irrigation, mainly in the west and the north portions of the state.

DATA AND METHODS

Many authors have used the gamma distribution for fitting rainfall data (Juras, 1994; Wilks, 1995; Ropelewski and Halpert, 1996; Ben-Gai, 1998). The two parameters of this theoretical distribution were analyzed to evaluate the rainfall variability in time and space.

The stations and the period investigated were based on homogeneous precipitation records with good registration accuracy. The chosen stations had 10% or less of non consecutive gaps. Due to the great temporal variability of the precipitation in São Paulo State its autocorrelation is poor. Therefore, registers of stations with analogous behavior replaced the missing data (similar checked statistically).

The database consisted of 63 stations sorted from north to south, covering 42 years (1955-1996). The entire interval was divided into two consecutive periods (1955-1975—PI and 1976-1996—PII), which were seasonally compared in terms of the scale and shape parameters of the gamma distribution. Seasons are: summer (January-March), autumn (April-June), winter (July-September) and spring (October-December).

Changes in the gamma parameters between the first and the second periods were checked comparing their spatial correlation. The seasonal coefficient of variation (CV) was estimated using the first two moments of the gamma distribution: mean and variance.

Statistical comparison of the slopes of the seasonal rainfall mass-curve with the slopes of the correspondent regression line provided further information regarding the nature of the trends. Mass-curve analysis involves the accumulation of annual totals throughout the data period. If changes in the rainfall regime are present, an inflexion will appear on the graph near the point when the change occurred (Sumner, 1988).

Total and normalized rainfall averages were also evaluated to elucidate the character of the changes throughout the period.

The scale and shape parameters of the gamma distribution for the seasonal amounts at every station and the goodness of fit, checked by the K-S test, were calculated by the use of the Statistica Package Software, Statsoft Inc., version 5. Surfer for Windows, Golden Software Inc, version 6.01 created seasonal contour maps. All data search type and kriging method were chosen for girded interpolation.

RESULTS

For summer, PI and PII presented similar values of the scale parameter and were statistically correlated at the level of $p<0.01$. The shape parameter showed great spatial dissimilarity in both periods such that no statistical correlation was found. For spring, any spatial trend occurred for none of the parameters and no statistically significant correlation existed between the periods (figures not provided).

Autumn presented the most noticeable changes. The contour map for PI (Figure 1) shows that any significant spatial trend was verified for the scale parameter for this period. However, recent years (PII) experienced a remarkable northward decrease of this parameter (Figure 2). Regarding the shape parameter, the two periods presented an opposite spatial trend, with an appreciable northward increase in the second period, which also presented higher values in comparison to PI. The results evidenced the intra-annual variability in recent years in this part of the territory (Figure 3 and 4). The scale and the shape parameters did not show significant statistical correlation between PI and PII, emphasizing the opposite spatial trend of the periods and therefore the significant changes between PI and PII.

The spatial behavior of the scale and the shape parameters highlighted the existence of a north-south gradient in the state. Thus, further analyzes were provided for the two regions to identify the sort of the temporal changes between the two sectors. The north encompasses stations 1 to 22 and the south, 23 to 63. The normalized rainfall values for the two sectors demonstrated that the rainfall variability in the north was more pronounced and higher in PII (Table 1). However the south regis-

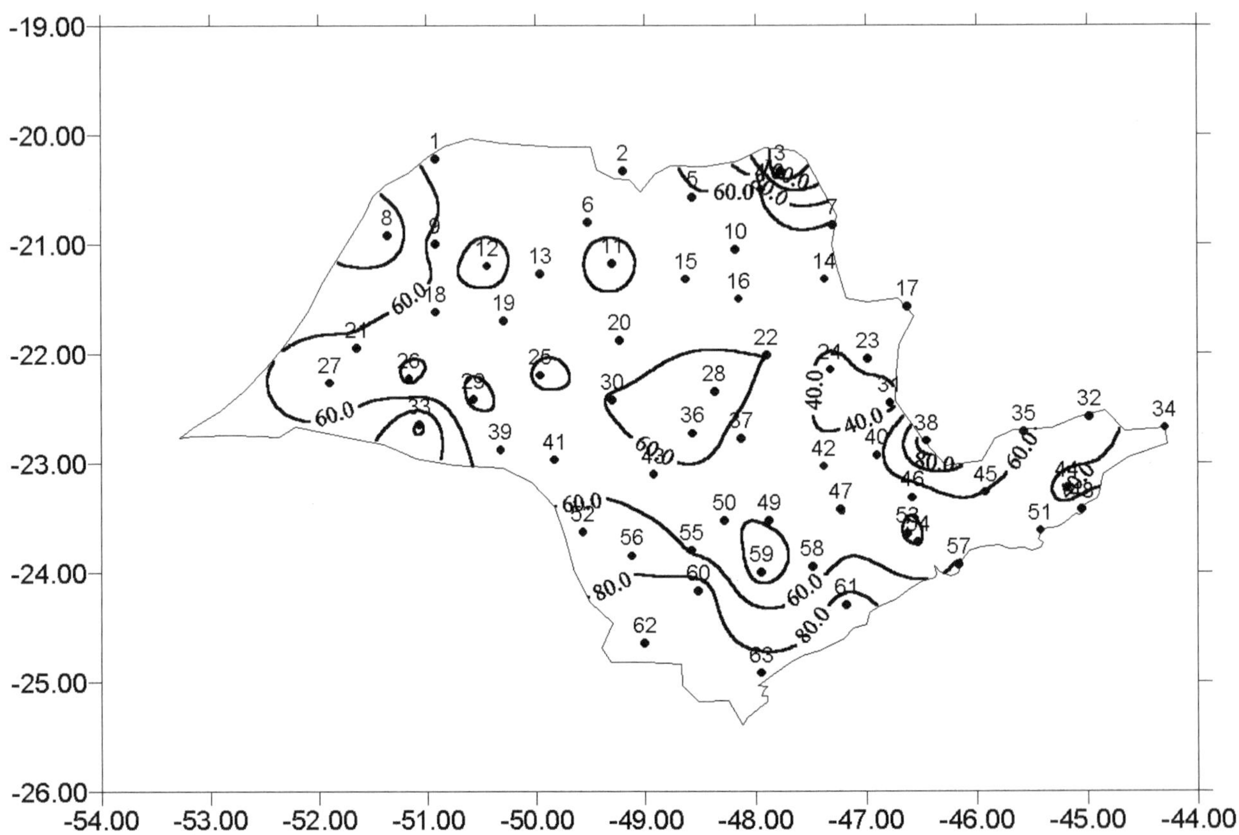

Figure 1. Spatial distribution of the scale parameter—Autumn, PI

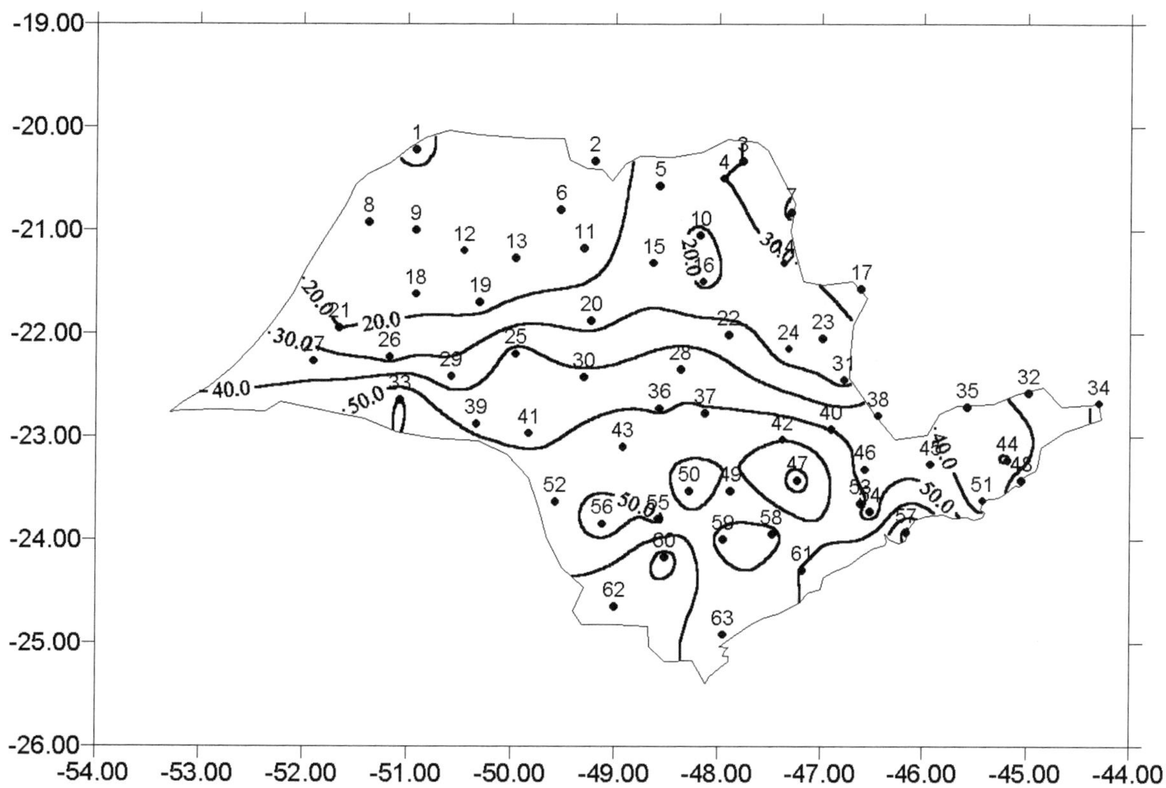

Figure 2. *Spatial distribution of the scale parameter—Autumn PII*

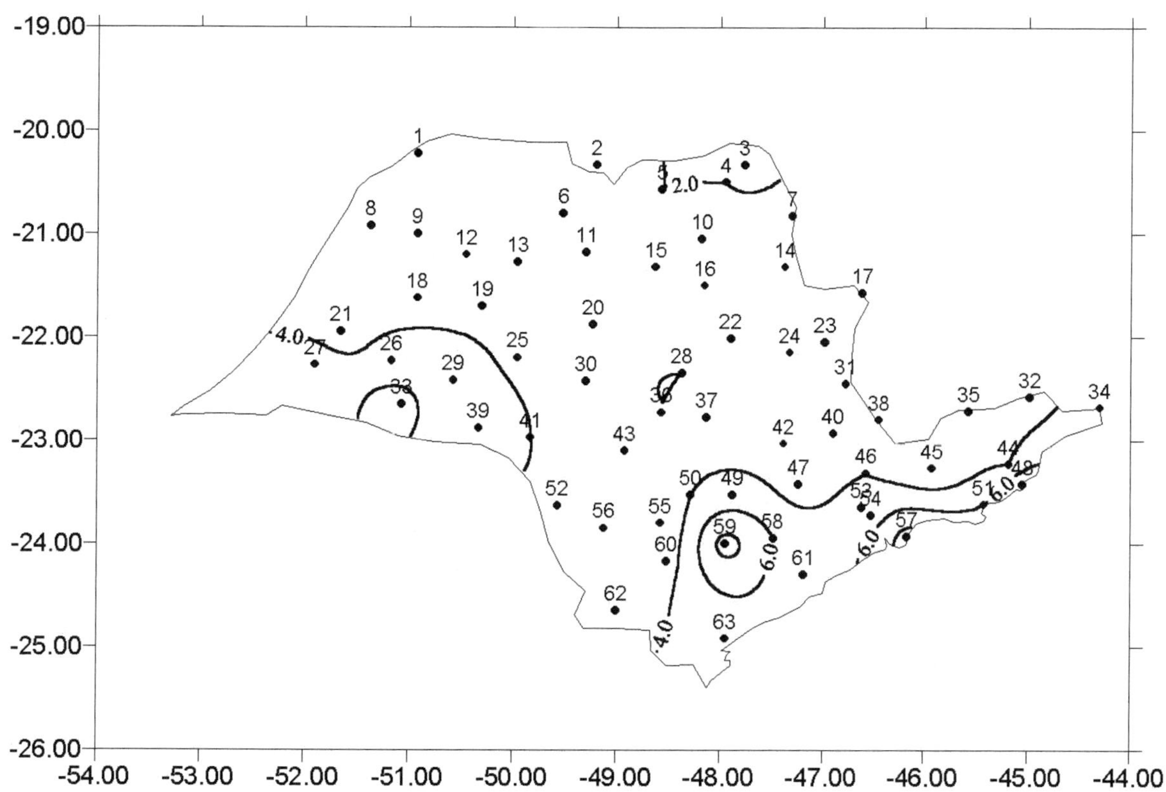

Figure 3. *Spatial distribution of the shape parameter—Autumn PI*

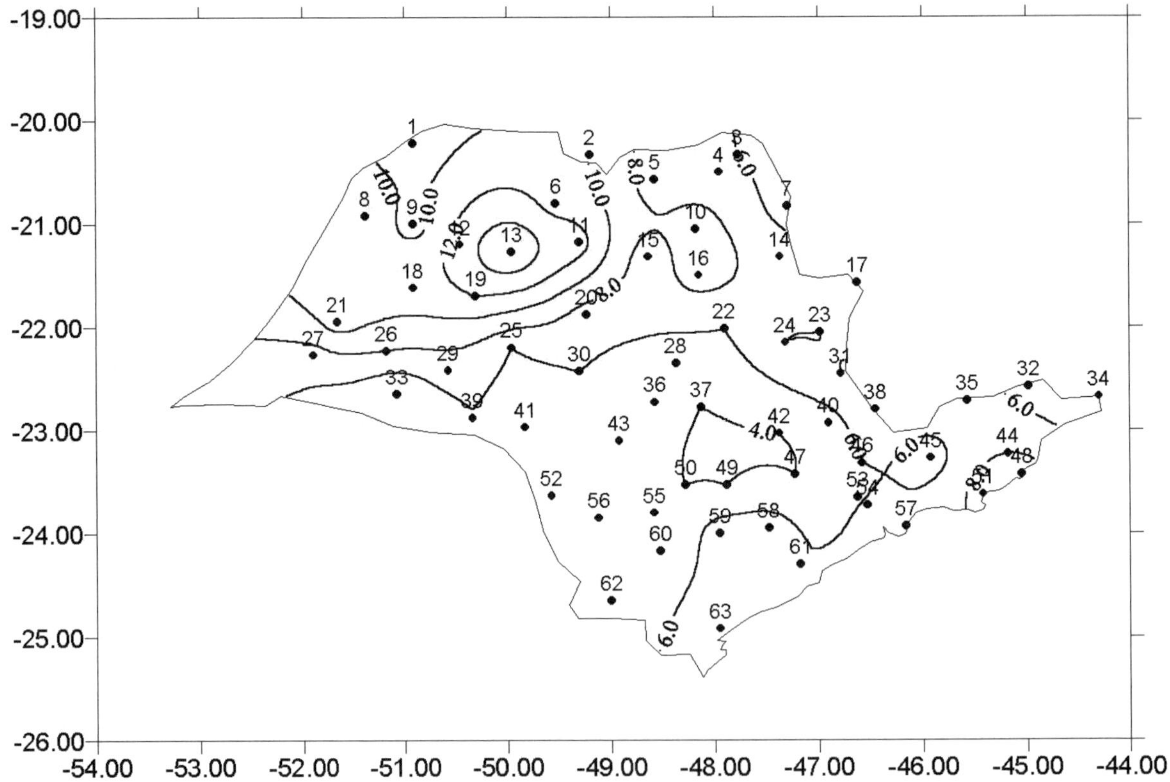

Figure 4. *Spatial distribution of the shape parameter—Autumn PII*

tered higher amounts of precipitation (figures not provided).

Some of the highest precipitation volumes were registered in El Niño years. The influence of this phenomenon in the rainfall totals in Southeastern Brazil is well known (Kayano and Moura, 1986; Sugahara, 1991; Xavier et al., 1995; Nunes, 1997). Notwithstanding, the low values of CV highlighted that the spatial variability for these years was not considerable (Figure 5), demonstrating that the rainfall increase in the autumn during El Niño years is a common feature for the whole state.

For winter, the scale parameter values presented close agreement between the periods, with lower values next to the coast. The recent period registered a slight decline of the values (Figures 6 and 7). For the shape parameter, the highest values were registered in the south for both periods, reinforcing the existence of a north-south gradient (Figures 8 and 9). The spatial similarity and the values between PI and PII for the two parameters were confirmed by significant statistical correlation at the level of $p<0.01$.

This season did not present any important alteration in the spatial trend but evidenced an increase in the intra-annual variably in PII.

Due to the north-south contrast, the state was divided into two homogeneous sectors. North encompasses stations 1 to 47 and south, the rest. The rainfall volumes were higher but less variable for the south, whereas the north had more pronounced variability (Table 1).

Also for winter, some of the El Niño years produced the highest amounts of rainfall but not particularly high variability (Figures 5 to 9), since the influence of this phenomenon affects the entire area.

It must be stressed that some El Niño occurrences impact the area more deeply in the autumn (1983, 1987 episodes), while some others increase the rainfall particularly during the winter (1957, 1972, 1976 events) (Nunes, 1997). Moreover, some of the strongest El Niño episodes occurred in the second period. Still, Table 1 highlights that autumn and winter presented more extreme situations than summer and spring, and that PI had a larger num-

Table 1. *Seasonal normalized rainfall totals (El Niño years in bold)*

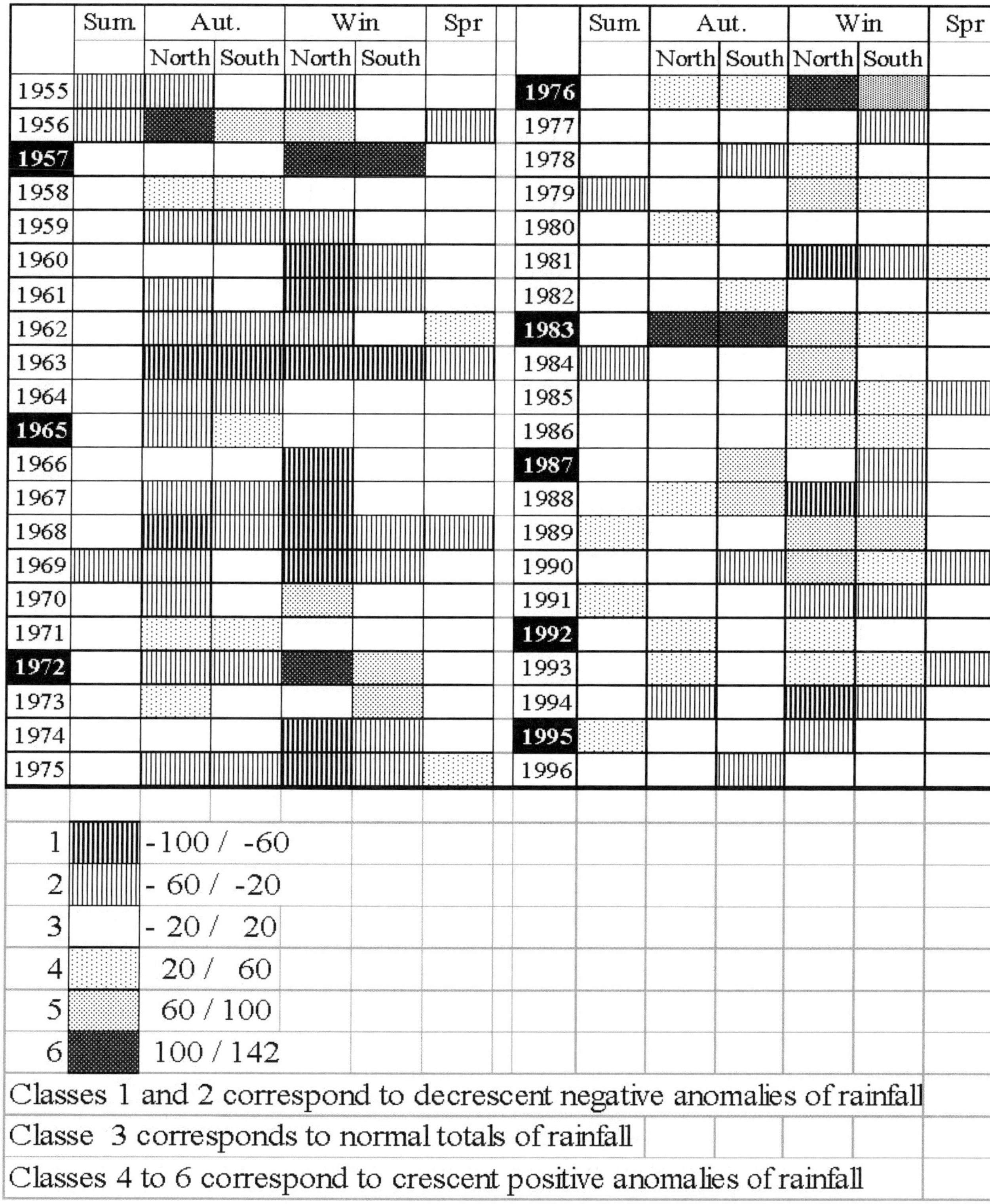

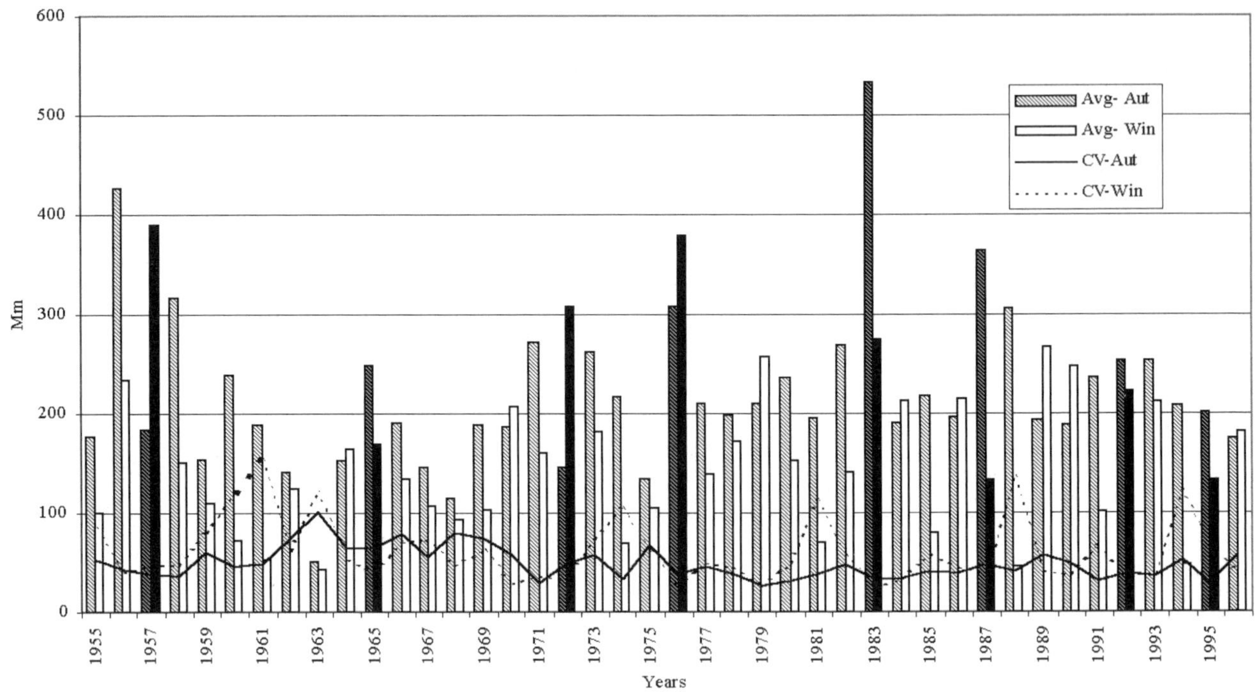

Figure 5. *Spatial rainfall average and coefficient of variation for the complete period (autumn and winter)*

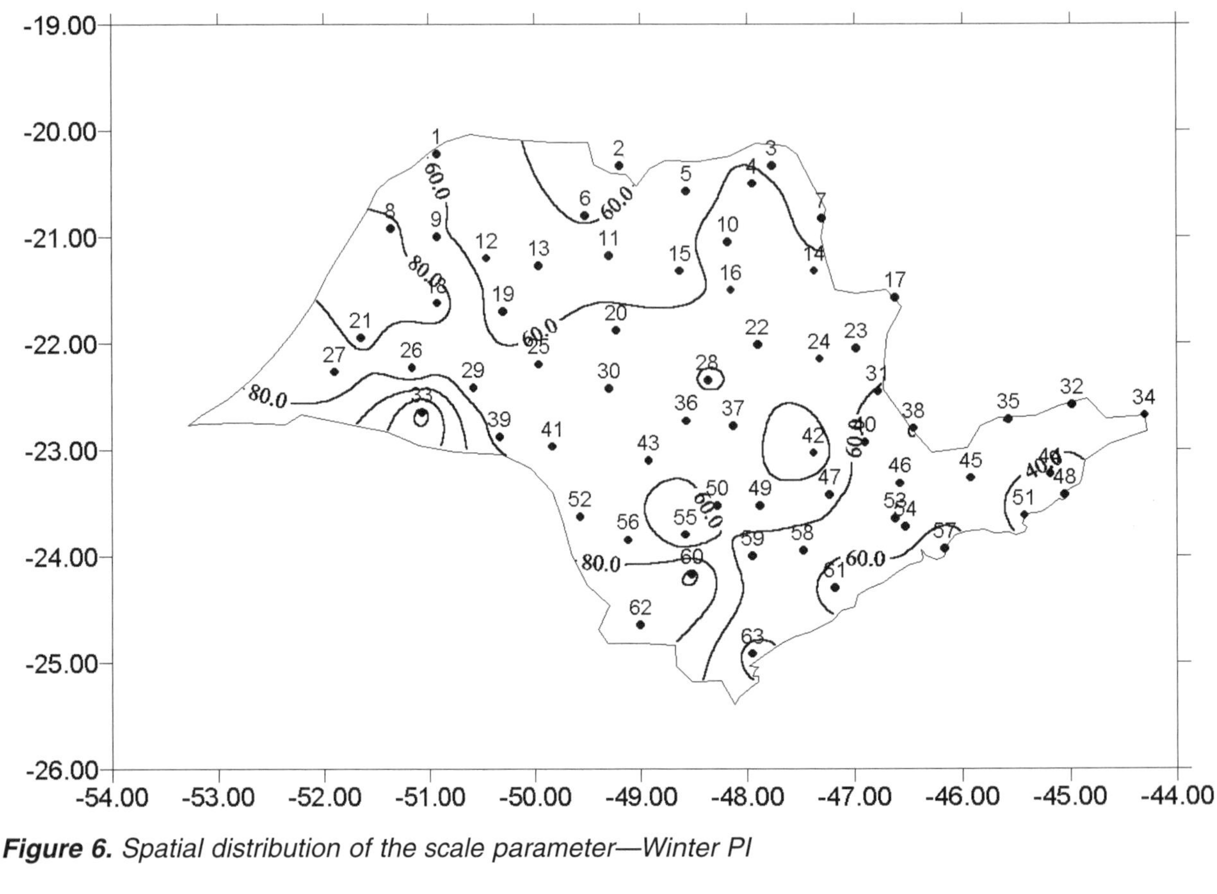

Figure 6. *Spatial distribution of the scale parameter—Winter PI*

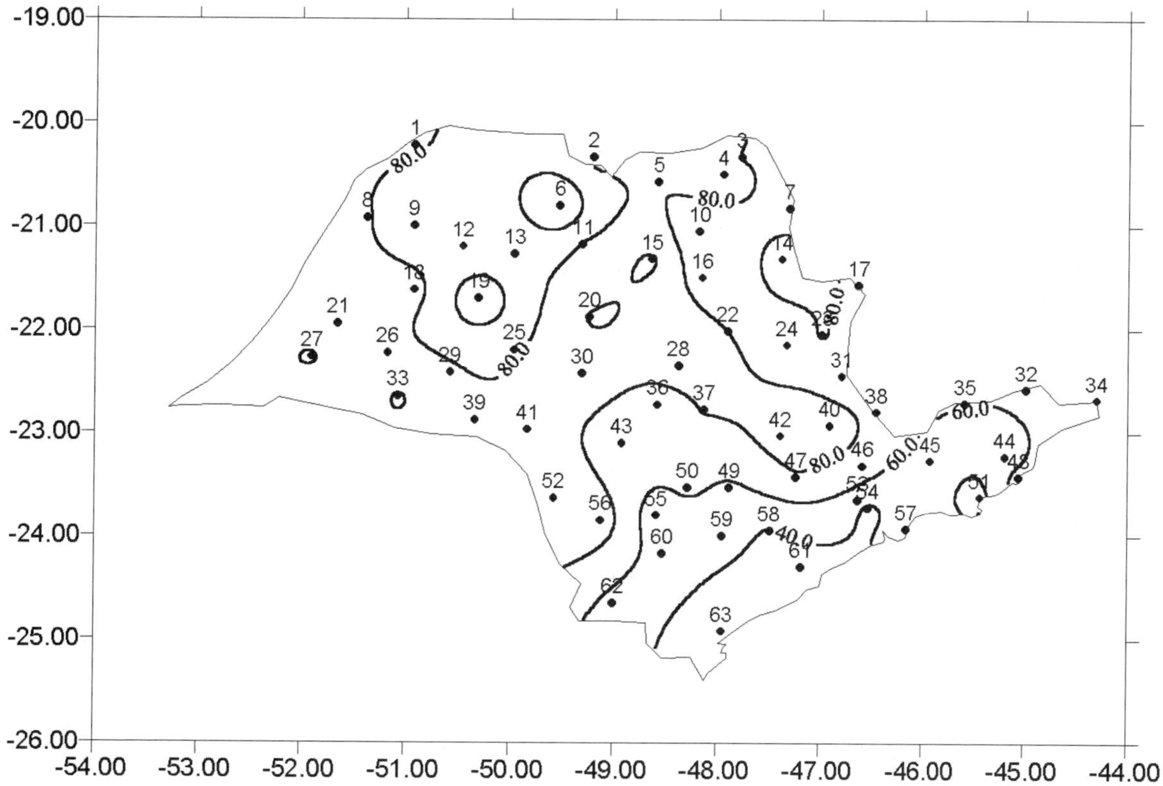

Figure 7. *Spatial distribution of the scale parameter—Winter PII*

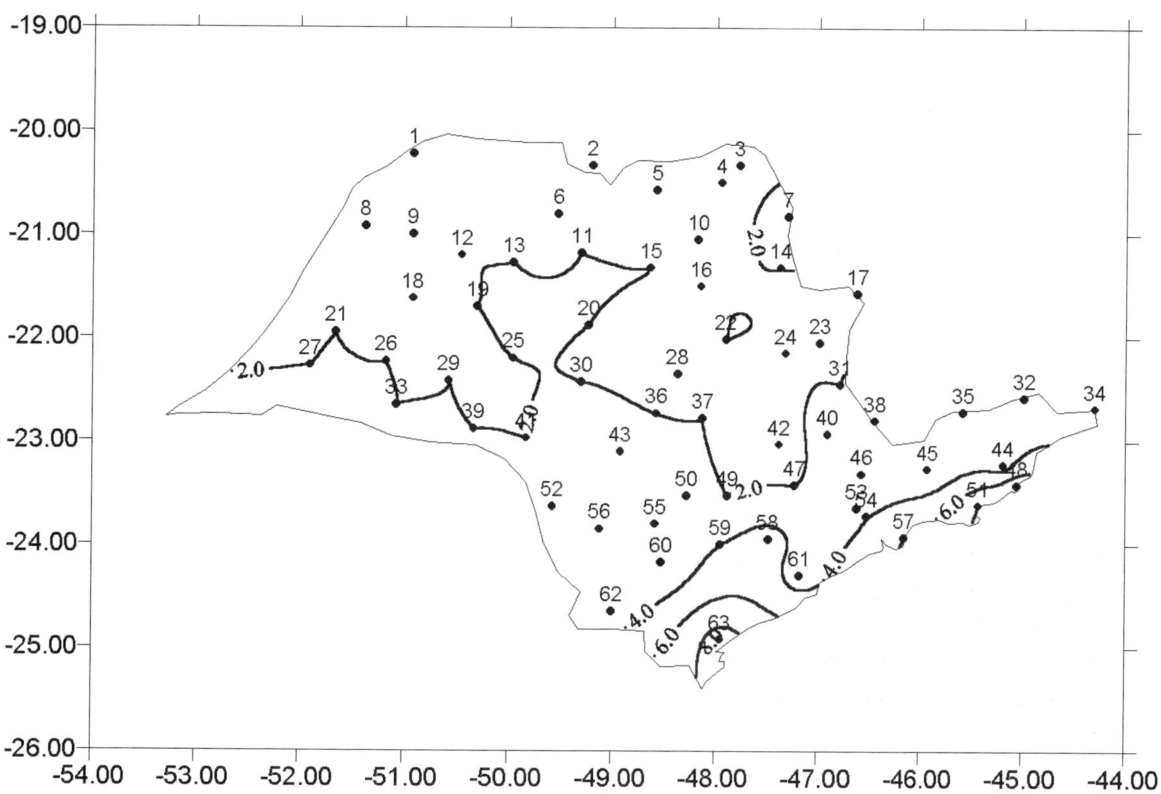

Figure 8. *Spatial distribution of the shape parameter—Winter PI*

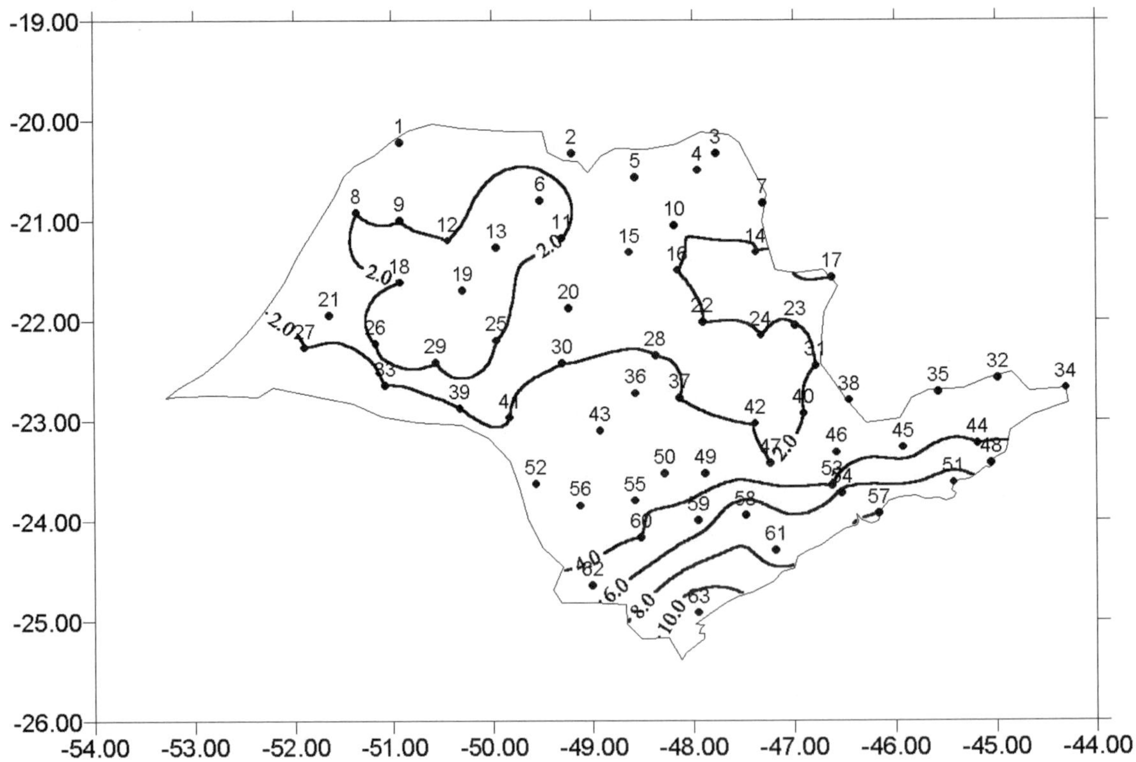

Figure 9. *Spatial distribution of the shape parameter—Winter PII*

ber of years with rainfall below average, in contrast to PII. Totals were higher for PII for all seasons except spring (data not shown).

Analyses of mass-curve did not reveal any consistent change in the precipitation regime for the seasons. Mass-curve graphs for spring and summer were very homogeneous, approaching a straight line. For autumn and winter they presented inflexions, which however did not continue throughout the time, being thus inconsistent (Figure 10).

Rainfall volumes were higher in the second period for all seasons except spring (summer: 5%; autumn: 24%; winter: 20%). Thus, the differences between the scale and the shape parameters for autumn and winter are related to changes in the totals as well as differences of temporal distribution of precipitation, revealed by the augmentation of variability in recent years. The seasonal values of the shape parameters were remarkably similar to the coefficient of variation, emphasizing that this parameter reflects the homogeneity of the values with time.

Precipitation over an area involves a variety of processes, including land use, occurring at different scales. This factor can influence rainfall conditions, causing variations in temperature, evapotranspiration, albedo, etc. During the last decades, the study area has undergone substantial changes in land use. The north in particular experienced a noticeable increase of agriculture. This area has water deficit during periods from 4 to 8 months, requiring irrigation. It is possible that these alterations during the long-term period might be one of the reasons for the enhanced rainfall variability observed in this study.

CONCLUSIONS

The present study revealed that recent years showed considerable intra-annual variation during winter and particularly autumn. These seasons are more affected by El Niño events. During the second period some strong episodes impacted the area considerably. Spatially, the most remarkable modifications were found in the northern part of the state, which has been experiencing an impressive increase in agriculture and irrigation, especially in the period, PII. Considering that any alteration of the precipitation pattern can seriously compromise the rhythm of agriculture, these recent trends can

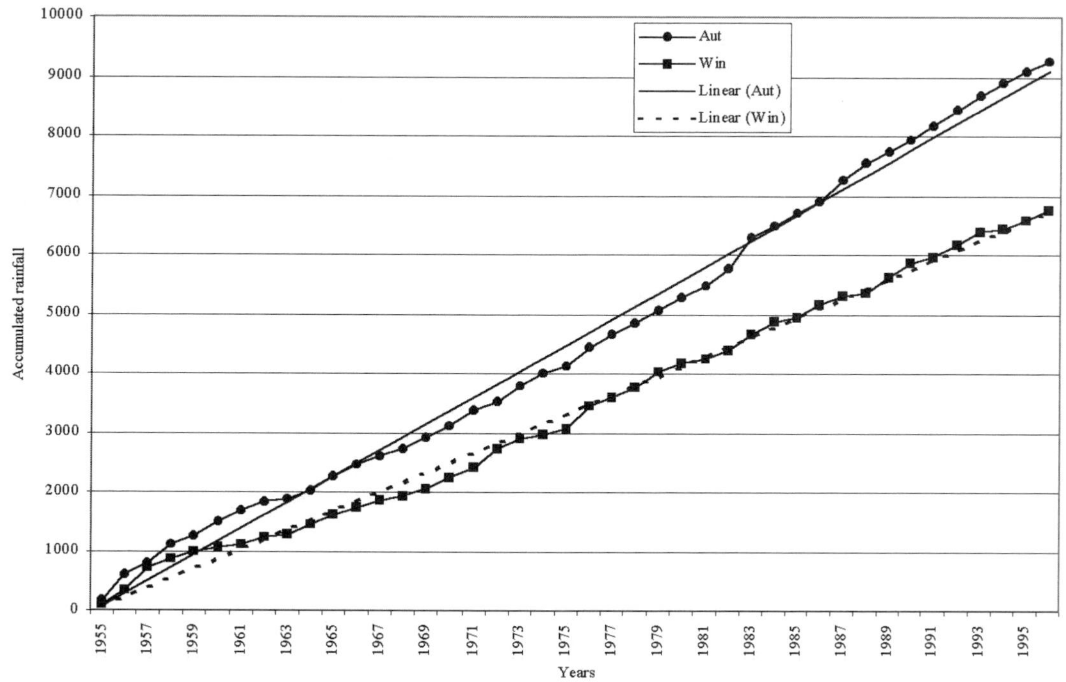

Figure 10. *Mass-curve*

lead to large economic losses, and promote environment disasters of great proportions in the state.

Climate change is one of the most important issues of environmental studies and the augmentation of the variability of climatic elements is one of its indicators. The present results, based upon the comparison between two short-period records of 21 years, limits more general conclusions about climate change and offer no scientific basis for assuming that the observed trends will continue indefinitely. However, it emphasizes that relevant variations are under way in the area, requiring further investigation.

In a second stage of this study, modifications in land use during the long-term period will be correlated with the results of rainfall variation.

REFERENCES

Ben-Gai, T., A. Bittan, A. Manes, A., P. Alpert, and S. Rubin. 1998. Spatial and temporal changes in rainfall frequency distribution patterns in Israel. Theor. Appl. Climatol. 61:177-190.

Hare, F.K. 1985. Climatic variability and change. p. 37-68. In: R.W. Kates, J. Ausubel and M. Berberian (ed.) Climate Impact Assessment: Studies of the interaction of Climate and society. John Wiley and Sons, New York.

Juras, J. 1994. Some common features of probability distributions for precipitation. Theor. Appl. Climatol. 49:69-76.

Kayano, M. and A.D. Moura. 1986. O El Niño de 1982-83 e a precipitação sobre a América do Sul. (In Portuguese, with English abstract) Rev. Brasil. de Geof. 4:201-214.

Nunes, L.H. 1997. Distribuição espaço-temporal da pluviosidade no Estado de São Paulo: variabilidades, tendências, processos intervenientes. (In Portuguese, with English abstract). Ph.D. thesis, Univ. of São Paulo, São Paulo.

Ropelewski, C.F. and M.S. Halpert. 1996. Quantifying Southern-Oscillation precipitation relationships. J. Climate 9:1043-1059.

Sugahara, S. 1991. Flutuações interannuais, sazonais e intrasazonais da precipitação no Estado de São Paulo (In Portuguese, with English abstract). Ph.D. thesis, Univ. of São Paulo, São Paulo.

Sumner, G. 1988. Precipitation analysis in time. p.365-366 In: Precipitation process and analysis. Bath, John Wiley.

Wilks, D.S. 1995. Theoretical Probability Distributions. P.86-93. In: Statistical methods in the atmospheric sciences. Academic Press, San Diego.

Xavier, T. de M.B.S., M.A.F.S. Dias and A.F.S. Xavier. 1995. Impact of ENSO episodes on the autumn rainfall patterns near São Paulo, Brazil. Intern. J. Climatol. 15:571-584.

SESSION II. PRODUCING USABLE CLIMATE INFORMATION

Climate Outlook Fora of Southeast South America. Performance of Two Years of Seasonal Forecasts.

Guillermo J. Berri
University of Buenos Aires and National Research Council of Argentina, Argentina

INTRODUCTION

A series of climate outlook Fora for Southeast South America began in Montevideo, Uruguay, in December 1997, in the middle of the last strong El Niño (1997/98). Participanta included specialists from the four countries of the Mercosur region, i.e. Argentina, Brazil, Paraguay and Uruguay. The motivation was concern about the ongoing El Niño in mid 1997, availability of experimental climate forecast, and knowledge about the impact of past El Niño events in the region. In little more than two years, additional seven meetings were conducted in the region: Foz do Iguacu (Brazil), June 1998; Buenos Aires (Argentina), August 1998; Salto Grande (Bi-National Technical Commission Argentina-Uruguay), December 1998; Asunción (Paraguay), March 1999; Buenos Aires (Argentina), September 1999; Montevideo (Uruguay), December 1999 and Cachoeira Paulista (Brazil), March 2000. Participants in the Fora included researchers from meteorology, agronomy, hydrology, oceanography, related environmental sciences, public health and social sciences, as well as users from agriculture, water resources management, hydroelectric generation, civil defense and emergencies. Sponsorship was received from different international organizations and research institutes in climate variability and short-term climate prediction, as well as universities, research institutes and research centers in the region, national weather services, water resources centers and hydropower plants, national and regional rural societies and other national and regional organizations. Each Fora produced seasonal forecasts of average precipitation and temperature anomalies expected for the upcoming 3-month period. This paper discusses results of the first six Climate Outlook Fora for precipitation. Validation of the last forum is still under way.

CLIMATE FORECAST

The backbone of the regional climate assessment is consensus agreement between coupled ocean-atmosphere model forecasts, physically based statistical models, results of diagnosis analysis and published research on climate variability over the region and expert interpretation of this information in the context of the current situation. The region considered for the forecast includes continental areas bounded between 20°S to 40°S and to the east of the Andes Mountains. Figure 1 displays the geographical location of the region. The inner rectangle is the sub-region where forecasts are validated. Seasonal climate forecasts of precipitation are expressed in probabilistic terms. For this purpose, three categories are identified: *above-normal, normal,* and *below-normal* coinciding with a ter-

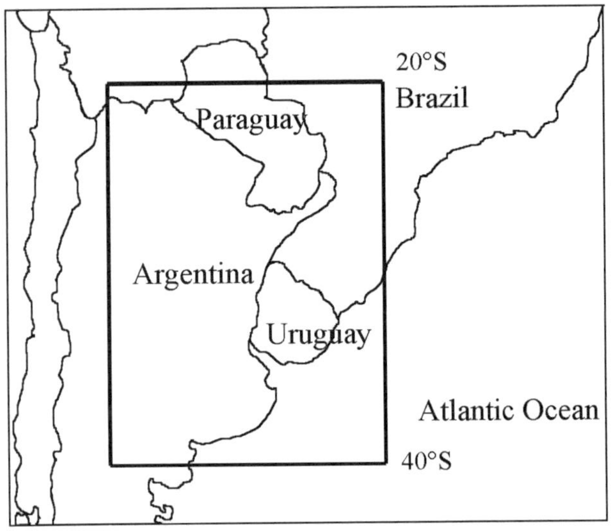

Figure 1. *Geographical location of the region. The inner rectangle is the region where the forecasts are validated*

cile of the precipitation distribution. Homogeneous regions are identified in each case and the seasonal climate forecast is presented as the probability of occurrence within each category during the upcoming 3-month period. Probabilities assigned to each category are determined by consensus agreement among specialists. Figure 2a presents the forecast produced in Montevideo (Uruguay) in December 1997 for the period January-March 1998. For each homogeneous region there is a set of three boxes, whose numbers represent the forecast expressed as the probability of occurrence of a 3-month precipitation anomaly within each tercile. For example, in the upper right corner of Figure 2a, the upper box indicates a 25% probability of *above-normal* precipitation, the center box a 35% probability of *normal* precipitation and the lower box a 40% probability of *below-normal* precipitation. Since there are only six cases to verify, this preliminary

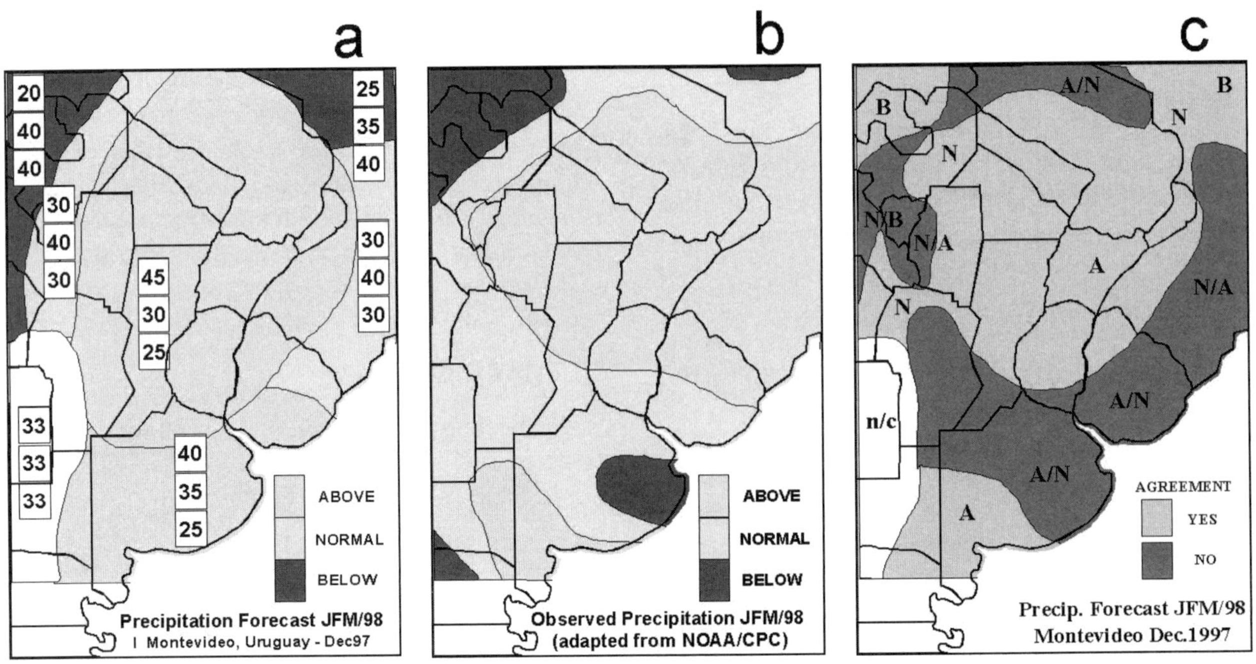

Figure 2. a) *4-month lead January-March 1998 forecast. Numbers in boxes represent the forecast for each tercile expressed as probability of occurrence (upper box is above-normal, center box is normal and lower box is below-normal, see text for details).* ***b)*** *observed precipitation during January-March 1998 adapted from NOAA/CPC maps (see text for details).* ***c)*** *forecast validation, light (dark) shading means categorical agreement (disagreement). Categorical agreement is expressed with one symbol (A for above-normal, N for normal B for below-normal). Disagreement is expressed with two symbols, for example A/N means that above-normal is predicted (first symbol) while normal is observed (second symbol). n/c means regions not considered for validation (33% probability forecast for every tercile). Regions bounded with a dashed line mean disagreement by two categories.*

evaluation considers that the forecast corresponds to the most likely category, i.e. the category having the highest probability. For example in Figure 2a, the precipitation forecast for the dark-shaded region in the upper-right corner is considered as *below-normal,* since the *below-normal* category has the highest probability. In order to facilitate the interpretation, regions in Figure 2a are shaded accordingly. There are cases in which forecasts assigned the highest probability to two categories simultaneously, for example *normal* and *above-normal* or *normal* and *below-normal.* In cases like these ones, the chosen forecast category is the extreme one, i.e. either *above-normal* or *below-normal* (there are no cases with highest probability assigned to *above-normal* and *below-normal,* simultaneously). Finally, there are cases in which the forecast assigns equal probability to each tercile, i.e. 33%. The Fora reports refer to such forecast as "climatology", meaning the lack of significant elements to favor the occurrence of a particular category. The interpretation of such forecast is that any category is equally probable to occur and therefore, any observed category would validate the forecast. Certainly, all situations like this are excluded from the analysis.

FORECAST VALIDATION

The only source of information on a regional scale available to the author to verify the forecast is the Climate Prediction Center (CPC/NOAA) 3-month precipitation maps. These maps present the observed 3-month precipitation totals as a percentage of normal precipitation (1961-1990), according to the following scale: <50%, 50-75%, 75-100%, 100-125%, 125-150%, 150-200% and >200%. Unfortunately, there is no coincidence between this scale and the one used for the predictions. The latter one uses 3 categories: *above normal,* defined as within the wettest third of observed precipitation in each region; *below normal,* defined as within the driest third and *normal* as the third centered around the median of observed precipitation. Therefore, the following criteria is adopted: i) observed precipitation below 75% of the normal is considered *below normal;* ii) observed precipitation between 75% and 125% of the normal is considered *normal;* iii) observed precipitation above 125% of the normal is considered *above normal.* Figure 2b shows the observed precipitation during the period January-March 1998 expressed in three categories according to the criteria discussed above.

DISCUSSION OF RESULTS

Figure 2c shows the result of the validation analysis of the 4-month lead forecast for the period January-March 1998, performed in December 1997, Figure 3 corresponds to the 7-month lead forecast for OND/98 performed in June 1998, Figure 4 corresponds to the 5-month lead forecast for OND/98 performed in August 1998, Figure 5 corresponds to the 4-month lead forecast for JFM/99 performed in December 1998, Figure 6 corresponds to the 3-month lead forecast for AMJ/99 performed in April 1999 and Figure 7 corresponds to the 4-month lead forecast for OND/99 performed in September 1999. The details of the symbols and shading are explained in the figure legend.

Several aspects conspire against the possibility of drawing definitive conclusions out of this analysis. They are basically the following: i) the limited number of forecast available to present, ii) the lack of coincidence between forecast categories (terciles) and observed precipitation anomalies (percentage of normal) and, iii) the fact that the maps employed to validate the forecasts are based on preliminary data and may not be strictly accurate.

Notwithstanding these limitations and based on the relative coverage of areas that present categorical agreement between forecast and observations, the following preliminary conclusions are drawn:

i) the forecast category "climatology", i.e. 33% probability for every tercile, can not be verified and seems to have no practical use;

ii) the OND/98 forecast (Figure 3c) is the worst one, although it must emphasized that is the only 7-month lead forecast;

iii) the 4-month lead JFM/98 forecast (Figure 2c), 4-month lead JFM/99 forecast (Figure 5c) and the 3-month lead AMJ/99 forecast (Figure 6c) show more coverage of areas with than without categorical agreement;

iv) the 5-month lead OND/98 forecast (Figure 4c), which could be verified over only a half of the region (the other half called for "climatology"), shows very little coverage of regions

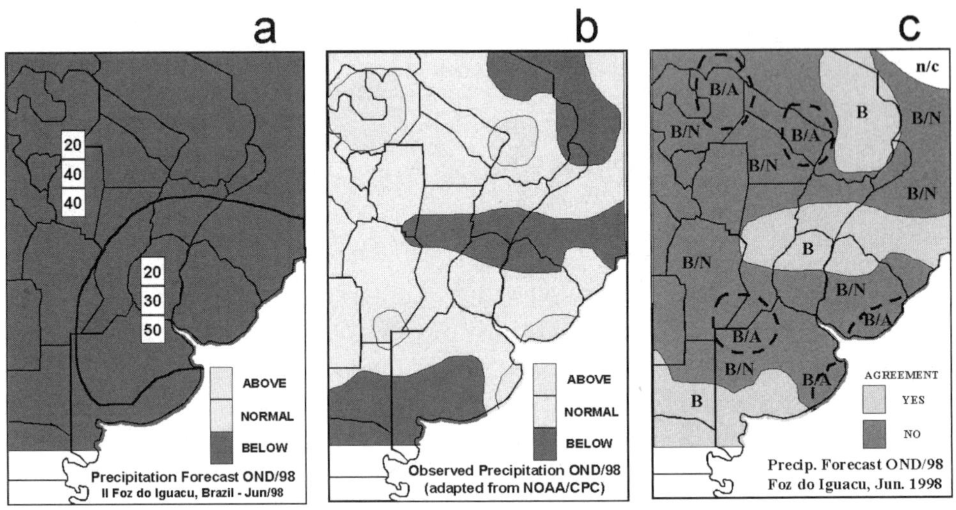

Figure 3. *Same as Figure 2 for the 7-month lead October-December/98 forecast.*

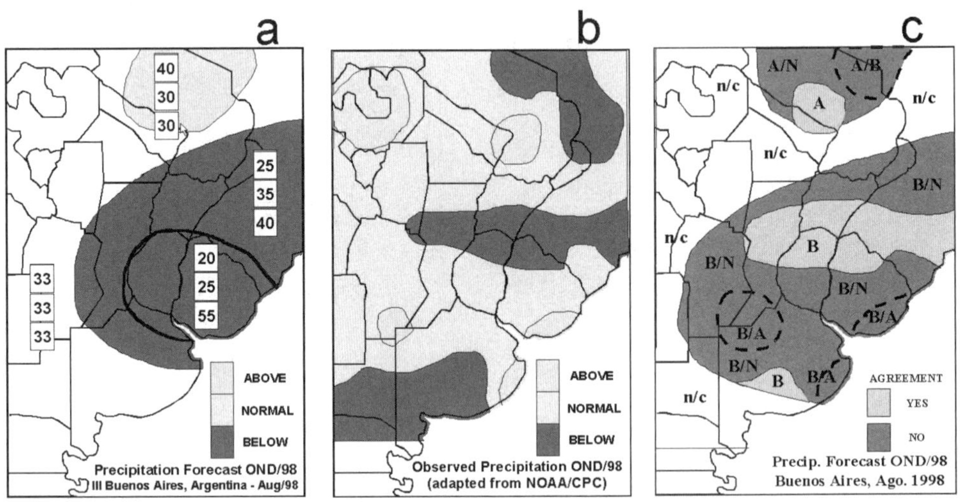

Figure 4. *Same as Figure 2 for the 5-month lead October-December/98 forecast.*

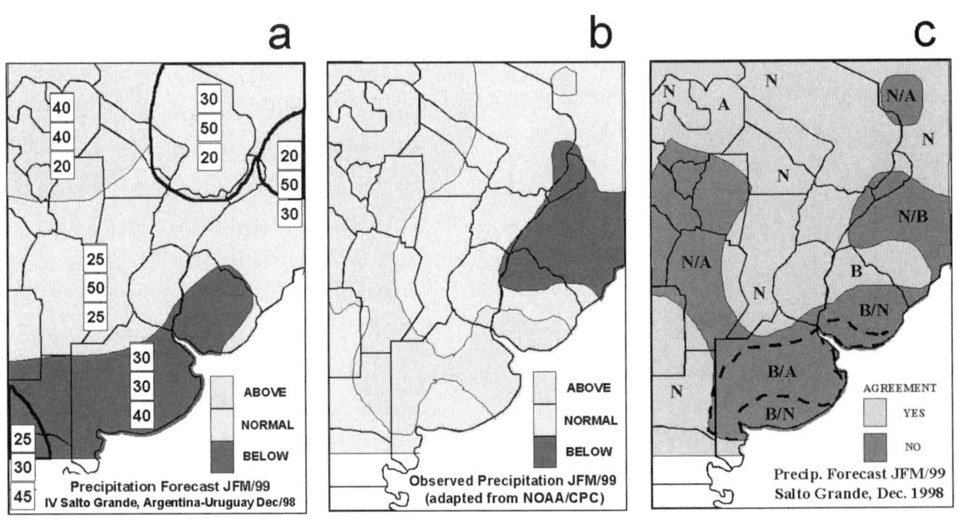

Figure 5. *Same as Figure 2 for the 4-month lead January-March/99 forecast.*

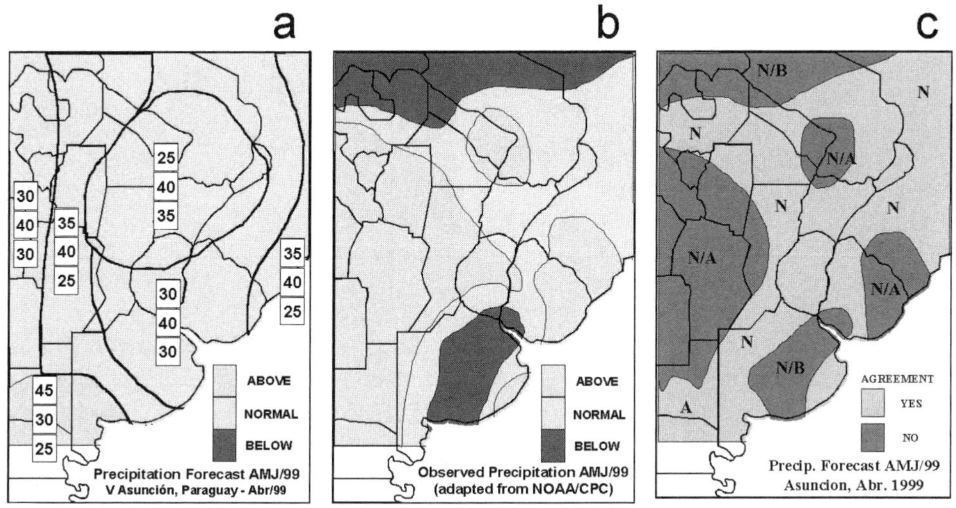

Figure 6. *Same as Figure 2 for the 3-month lead April-June/99 forecast.*

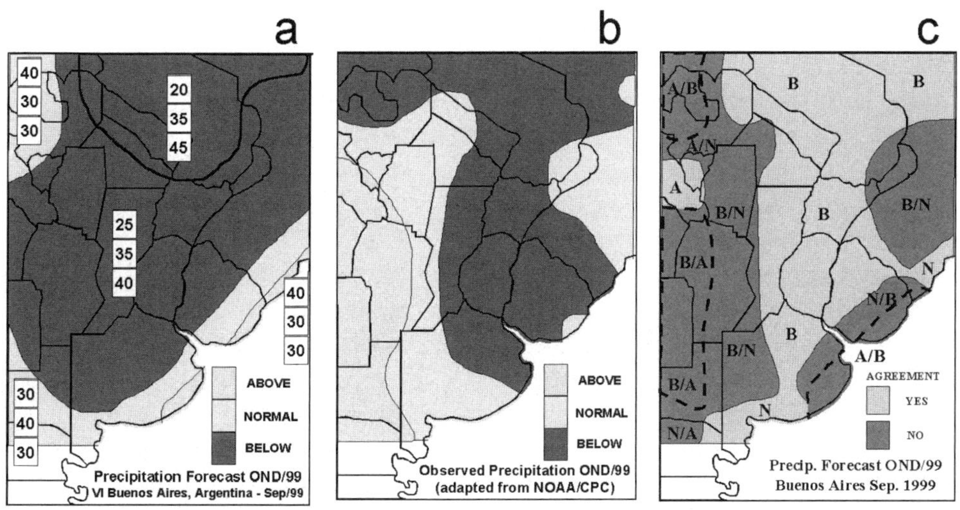

Figure 7. *Same as Figure 2 for the 4-month lead October-December/99 forecast.*

with categorical agreement, and it does not differ from the 7-month lead forecast performed in June 1998 (Figure 3c);

v) the 4-month lead JFM/98 forecast (Figure 2c) and the 3-month lead AMJ/99 forecast (Figure 6c) show disagreements of only one category, and

vi) the 7-month lead OND/98 forecast (Figure 3c), 5-month lead OND/98 forecast (Figure 4c), 4-month lead JFM/99 forecast (Figure 5c) and the 4-month lead OND/99 (Figure 7c) show some areas with disagreements of two categories.

ACKNOWLEDGEMENTS

This research was partially supported by Research Grant PIP-0389/98 of the National Research Council of Argentina.

Linking Climate Predictions with Crop Simulation Models

Mikhail A. Semenov[1] and Peter D. Jamieson[2]
[1]University of Bristol, United Kingdom
[2]New Zealand Institute for Crop & Food Research Ltd, New Zealand

INTRODUCTION

Crop simulation models have proved themselves to be powerful tools for analyzing the causes of production variations in many crops, and some have been incorporated into decision support systems. A common need of most crop simulation models is daily weather data as the principle driving variables. Such models are potentially useful, together with seasonal weather forecasts, for guiding management early in the season well before most growth has occurred.

Global climate models (GCMs), the tools most widely used for climate predictions, are very complex and their demand for computational power limits their output to spatial and temporal scales inappropriate for crop simulation. Output is on the scale of hundreds of kilometers and the associated time-scale is monthly, which means that sub-grid scale processes on a short time-scale, such as precipitation, are not adequately represented. Stochastic weather generators lack the predictive power of GCMs, but are able to reproduce site-specific climates on a daily time-scale quite well. The combination of these two methodologies should allow the production of climate scenarios for agricultural applications that: are site-specific with daily temporal resolution; include the full set of climate variables required by the crop model; contain an adequate number of years to permit risk analysis; and include changes in means and climate variability.

By combining the output from GCMs with a stochastic weather generator we are able to produce climate change scenarios with high spatial and temporal resolution. The importance of downscaling and the incorporation of climate variability into climate change scenarios are demonstrated in section 2 for two European sites.

An alternative to downscaling from GCMs is upscaling crop simulation models by simplifying them to meta-models with fewer parameters and equations, and that may run on inputs of coarser temporal resolution, e.g. monthly weather. An example of development of a meta-model from the Sirius wheat simulation model is presented in section 3, together with some performance data in comparison with its parent.

In an example of the potential use of forecasting, we show in section 4 that seasonal forecasts would allow management changes in sweet corn production for factory processing, especially where variations among seasons cause large changes in the date of maturity.

CLIMATE CHANGE AND CROP YIELD

In this section we demonstrate how climate change scenarios with high spatial and temporal resolution were constructed and used in agricultural impact assessments at two selected sites, Rothamsted (UK) and Seville (Spain). The climate change scenarios were produced in two steps. First, a stochastic weather generator (LARS-WG) was run for observed weather at each site, and parameter files characterizing typical weather at these sites were generated. Second, changes in climate characteristics, derived from GCMs (including changes in mean and variability), were applied to these parameters. New parameter files were used to generate site-specific climate change scenarios with a daily time-step. To construct scenarios, we used data from the UK Met. Office high-resolution GCM transient (UKTR; Murphy, 1995; Murphy and Mitchell, 1995) experiments. Instead of using climate change integration fields itself, the change fields were constructed by calculating the difference between a period in the climate change integration and the corresponding years of the control integration. We assumed that both the control and climate change integrations exhibit similar drift and long-term variability. This overcame an acknowledged problem of UKTR of climate drift in its control integration, that produces a noticeable deviation (approximately 1°C) from the initial ten-year average over the 75-year period of the simulation.

In order to produce scenarios of climate change at a site scale required by crop simulation models, it was necessary to downscale the coarse resolution GCM data to a site level. This procedure involved the development of relationships between the coarse- and local-scale data for the climate variables concerned. There are currently a number of downscaling methodologies in use, including circulation patterns (e.g., Bardossy and Plate, 1991; Matyasovszky *et al.,* 1993; Jones and Conway, 1995) and regression techniques (e.g., Kim *et al.,* 1984; Wigley *et al.,* 1990; Karl *et al.,* 1990; von Storch *et al.,* 1993). Both methods use existing instrumental databases to determine the relationships between large-scale and local climate (Wilby *et al.,* 1998). Regression techniques develop statistical relationships between local station data and grid-box scale, area-average values of say, temperature and precipitation and other meteorological variables. The circulation pattern approach classifies atmospheric circulation according to type and then determines links between the circulation type and precipitation. At the selected sites, regression relationships were calculated between local station data (temperature and precipitation; i.e., the predictands) and grid-box scale, monthly anomalies of mean sea level pressure, the north-south and east-west pressure gradients, temperature and precipitation (i.e., the predictors). The regression relationships were based on anomalies from the long-term mean in order to facilitate the use of the GCM-derived changes in the equations. To calculate changes in climatic variability, daily data for the appropriate grid boxes from the control and perturbed integrations of the UKTR experiment were used to calculate changes in precipitation intensity, duration of wet and dry spells and temperature variances. These changes were then applied to the LARS-WG parameters previously calculated from the observed daily data at each site.

Incorporation of changes in variability into climate change scenarios could have a significant effect of the agricultural impact assessment, although it does not effect monthly statistics such as, for example, monthly total precipitation or monthly mean temperature. These means were compared for the UKTR scenarios with and without variability for Seville. There is no significant difference between monthly mean temperatures for the scenarios with and without variability for all months. Results from a t-test indicate that precipitation totals were significantly different for four months out of seven during the vegetation period for winter wheat (January—July) (Semenov and Barrow, 1997). For three of these months (May, June and July) precipitation for both scenarios was so low that it did not make a big difference to total precipitation over the vegetation period, 184 mm and 210 mm with and without variability, respectively, compared to 496 mm for the base climate. The effect of climate change scenarios with and without changes in variability for simulated crop yield is significant. This is partly because crop simulation models have non-linear responses to environmental variables (Semenov and Porter, 1995). For the base climate the grain yield simulated by the Sirius wheat simulation model (Jamieson *et al.,* 1998c) was 5.6 t/ha and its coefficient of variation (CV) was 0.24 (Table 1). According to the UKTR scenario without variability, the grain yield does not change much (5.2 t/ha) and the CV remains about the same (0.23). If changes in climate variability are considered the results are very different. The grain yield drops to 3.9 t/ha and the CV almost doubles to 0.48. The reason for this is not the total

Table 1. *The effect of climate variability on crop yield and its coefficient of variation (CV), as simulated by SIRIUS Wheat, for UKTR scenario at Seville, Spain. Total precipitation and cumulative mean temperature were calculated for the winter wheat vegetation period from January to July.*

	Base	UKTR	UKTR with variability
Grain yield, t/ha	5.6	5.2	3.9
CV of yield	0.24	0.23	0.48
Total precipitation January-July, mm	296	210	184
Cum. Temperature January-July, °C	3630	4293	4323

amount of precipitation, but the change in precipitation distribution over the vegetation period and the prolonged dry spells. The probability of producing yields less than 3.5 t/ha is almost 50% for the UKTR scenario with variability and only about 10% for the UKTR scenario without variability or for the baseline climate. The high probability of obtaining low grain yields may make wheat an economically unsuitable crop in Spain under this climate change scenario.

The disadvantage of regression downscaling is that it is data intensive; observed data from several sites are required in order to calculate observed areal means and anomalies. Construction of site-specific scenarios of climate change may be aided by the current development of Regional Climate and High Resolution Limited Area Models (RCMs and LAMs, respectively). This methodology has been recently developed for climate change studies (Dickinson *et al.,* 1989; Giorgi, 1990) (Mearns *et al.,* 1999). The RCM is run with a high grid resolution (approximately 50 km) but only over a limited area of interest. The RCM is a physically-based model nested into the GCM and is able to reproduce regional climate in more detail than the GCM itself. However, recent work on the validation of a RCM has shown that there may be still large differences between model output and observed weather statistics, especially in the case of climate variability (Mearns *et al.,* 1995a, b). This means that the construction of local climate change scenarios from these models may be as problematic as from GCMs. Mearns et al. (1999) recently compared scenarios, which have been produced using a regional climate model, RegCM2, and a statistical downscaling method based on atmospheric circulation patterns. They demonstrated that substantial differences in the regional climate details of climate change are produced by two different means of downscaling from the same large-scale GCM experiments.

SIMPLIFYING A CROP SIMULATION MODEL TO A META-MODEL

An alternative approach to linking global climate models with crop simulation models is the development of a simplified crop simulation model that is able to use the output from GCMs directly at course spatial and temporal resolutions. Recently a sensitivity analysis and analysis of the structure of the Sirius wheat simulation model led to the development of a simpler meta-model that used relationships between simulated crop variables and aggregations of weather input variables (Brooks *et al.,* 2000). The meta-model produced very similar yield predictions to its parent, Sirius, for both potential and water-limited yields where fertility was not a limiting factor. The meta-model aggregates the three main Sirius components, the calculation of leaf area index, the soil water balance and the evapotranspiration calculations, into simpler equations. This reduces the requirements for calibration to fewer model parameters and allows weather variables to be provided on monthly rather than daily time-step, because the meta-model is able to use cumulative values of weather variables. This makes the meta-model a valuable tool for regional impact assessments with seasonal weather predictions, when detailed input data are not available. A brief description of the meta-model follows.

The meta-model calculates potential yield and then reduces it using a drought stress index. The essential elements of the meta-model are calculation of phenology to determine the duration of the main growth phases, calculation of the biomass accumulated during those phases, and calculation of a drought stress index.

The phenological submodel of Sirius is based primarily on the prediction of the rate of leaf production and the numbers of leaves produced on the main stem of wheat, in response to temperature and day length This part of the model has been retained intact in the meta-model, but there are prospects for simplifying it. It is important because it determines both the duration of growth and weather experienced by the crop during the critical grain growth period. The timing of anthesis is important within the meta-model because biomass accumulated by this time makes a contribution to grain yield.

Biomass accumulated from sowing to anthesis A is calculated from the ratio of mean daily solar radiation to mean temperature during this phase, and the phase duration. The simplified relationship encapsulates canopy response to temperature, affecting the amount of solar radiation captured by the crop, and the solar radiation itself. Similarly, the biomass accumulated after anthesis G, most of which is grain, is calculated from a single equation involving the ratio of mean solar radiation and temperature during the grain filling period. Potential grain yield is calculated as 0.25A + G. This is reduced by a drought stress factor based on a simplified calculation of the crop water balance, and the ratio between potential evapotranspiration and water supply, analogous to the Penman drought response model (Penman, 1971; Jamieson *et al.*, 1975).

A major difference of the meta-model from Sirius is that the growth of the plant is not simulated on a daily basis but, rather, biomass is related to the accumulated weather variables. Indeed, the only daily calculations are the adding up of the weather variables. An important characteristic is that the meta-model contains very little interaction between components. Once the anthesis date and leaf number are known, the anthesis biomass, the post-anthesis potential biomass and the water stress yield loss are all calculated separately.

The meta-model was run for Rothamsted 1961—1990 with 50% precipitation and for Edinburgh with a soil of low water holding capacity and the results compared with those from Sirius. Daily weather data was used in both cases. The scenarios were chosen to give a wide range of yields mainly due to variations in water stress. In both cases the meta-model performed well giving a root mean square error (RMSE) of 0.68 and 0.83 t ha^{-1} respectively on yields of 4-11 t ha^{-1} (cf. standard deviations in Sirius yields of 1.3 and 2.2 t ha^{-1}) and correlation coefficients of 0.92 and 0.95.

The good match of the meta-model yield values with those of Sirius indicates that the meta-model contains the most important aspects of Sirius and, in particular, that there are no other mechanisms within Sirius that substantially affect yield. The meta-model is based on analysis of the Sirius model, rather than just on its output, so it should be able to match the Sirius output well for most scenarios in Britain (e.g., different sowing dates or cultivars) and probably for many other climates without serious modifications.

The parameters and data used by the meta-model are the ones that need to be determined accurately for the meta-model or Sirius to perform well. As noted above, phenological development is crucial, so the parameters governing plant performance (responses to day-length and temperature) need to be well known, and temperature well specified. Solar radiation is important because it affects both potential yield and the water deficit. However, solar radiation is often not directly measured, but is estimated from sunshine hours, for example. To some extent the accuracy of the model will depend on the accuracy of those estimates.

Because the meta-model uses accumulated values of the weather variables, it is able to perform well with just monthly weather data. To demonstrate this, it was run at Rothamsted with 50% precipitation and at Edinburgh for the soil with low AWC (80 mm), using daily weather data and 30-days moving average weather data. The results of meta-model runs for daily and monthly weather were compared. RMSE for the anthesis day and water-limited yield are 1 day and 147 kg/ha, and 1.7 days and 470 kg/ha for Rothamsted and Edinburgh, respectively. Hence replacing daily data with disaggregated monthly data in the meta-model is unlikely to change the output significantly. This makes the concept of this type of model really useful in the forecasting arena where trends rather than daily weather data are the likely outputs of weather forecasting.

SWEET CORN

Sweet corn is produced for factory processing in the mid-region of the New Zealand's South Island in a climate that is close to marginal for the crop because of low temperatures and short season length. The crop cannot be sown much earlier than about mid-October (early summer in the Southern Hemisphere) because soil temperatures are too low before then. At the other end of the season, the risk of frost killing the crop becomes unacceptably large if sowings are delayed too much. There is, however, an operational requirement for a steady flow of product to the factory for processing. This is usually achieved by spreading plantings in time among the farms that supply the factory. The marginal nature of the temperature environment means that quite small variations in mean seasonal temperature can lead to substantial variations in both the time of harvest maturity, and the duration of the harvest from the earliest maturing to latest maturing crops. For instance, the maturity of a mid-October planting can change by about five weeks depending on whether the season is two degrees cooler than normal or two degrees warmer. Cool conditions increase the risk that late sowings will not mature before the crop is killed by frost. Warm conditions increase the risk that the harvest will be compressed sufficiently that some crops will have to be bypassed to maximize the success of the harvest when too many crops mature in a short time interval. Presently, harvest maturity predictions are made using scenarios of mean temperature, and of temperatures two degrees cooler or warmer than average (Jamieson & Gillespie, 1998). These forecasts are updated weekly based on current temperatures, with the remaining forecast periods based on the scenarios. As time progresses, the uncertainty in the forecasts diminishes. This system can forecast the consequences of seasonal variations, and is used as a guide by the processor, but is of limited help when planning a season's planting, because there is no probability attached to the temperature scenarios. If seasonal temperature forecasts could be produced with some reliability (probability) attached to them, then plantings could be compressed in predicted cool years and spread out in predicted warm years. For the decisions to be made there needs to be some numerical assessment of risk that can be interpreted using a simple thermal time model.

CONCLUSIONS

There is desire and perceived need for seasonal weather forecasts. An example of how useful it might be was given above for sweet corn. In this paper we have shown that methodology exists for linking climate predictions with crop simulation models either by downscaling GCM data, or by upscaling crop models.

ACKNOWLEDGEMENTS

Weather datasets were supplied by the U.K. Met. Office through the Climate Impacts LINK project. The work was supported by the European Commission's Environment Programme under Contract Number ENV4-CT95-0154 and ENV4-CT97-0496. IACR Long Ashton receives grant-aided support from the Biotechnology and Biological Sciences Research Council of the United Kingdom.

REFERENCES

Bardossy, A. and Plate, E.J.: 1991, 'Modeling daily rainfall using a semi-Markov representation of circulation pattern occurrence', *J. Hydrology* 122, 33-47.

Brooks, R.J., Semenov, M.A. and Jamieson, P.D.: 2000, "Simplifying Sirius: sensitivity analysis and development of a meta-model for wheat yield prediction", *European Journal of Agronomy,* (in press).

Dickinson, R.E., Errico, R.M., Giorgi, F. and Bates, G.T.: 1989, 'A regional climate model for the western United States', *Climatic Change,* 15, 383-422.

Giorgi, F.: 1990, 'Simulation of regional climate using a limited area model nested in a GCM', *J. Climate* 3, 941-963.

Giorgi, F. and Mearns, L.O.: 1991, 'Approaches to the simulation of regional climate change: a review', *Rev. Geophys.* 29, 191-216.

Jamieson, P.D., Martin, R.J. and Francis, G.S. 1995. Drought influences on grain yield of barley, wheat and maize. *N.Z.J. Crop Hort. Sci.,* 23, 55-66.

Jamieson, P.D., Brooking, I.R., Semenov, M.A. and Porter, J.R. 1998a. Making sense of wheat development: a critique of methodology. *Field Crops Res.* 55, 117-127.

Jamieson, P.D., Porter, J.R., Goudriaan, J., Ritchie, J.T., van Keulen, H. and Stol W. 1998b. A comparison of the models AFRCWHEAT2, CERES-Wheat, Sirius, SUCROS2 and SWHEAT with measurements from wheat grown under drought. *Field Crops Res.* 55, 23-44.

Jamieson, P.D., Semenov, M.A., Brooking, I.R., and Francis, G.S.: 1998c, 'Sirius: a mechanistic model of wheat response to environmental variation', *Field Crop Res.* 8, 161-179

Jamieson, P.D. and Gillespie, R.N.: 1998, "Predicting sweet corn maturity for factory processing in Canterbury, New Zealand." *Proc 9th Australian Agronomy conference,* 437-438.

Karl, T.R., Wang, W.-C., Schlesinger, M.E., Knight, R.W. and Portman, D.: 1990, 'A method of relating general circulation model simulated climate to the observed local climate. Part I: Seasonal statistics', *J. Climate* 3, 1063-1079.

Kim, J.-W., Chang, J.-T., Baker, N.L., Wilks, D.S. and Gates, W.L.: 1984, 'The statistical problem of climate inversion: Determination of the relationship between local and large-scale climate', *Monthly Weather Review* 112, 2069-2077.

Matyasovszky, I., Bogardi, I., Bardossy, A. and Duckstein, L.: 1993, 'Space-time precipitation reflecting climate change', *Hydrological Sciences* 38, 539-558.

Mearns, L.O., Giorgi, F., Mcdaniel, L., and Shields, C.: 1995a, 'Analysis of variability and diurnal range of daily temperature in a nested regional climate model—comparison with observations and doubled CO_2 results', *Climate Dynamics* 11, 193-209.

Mearns, L.O., Giorgi, F., Mcdaniel, L., and Shields, C.: 1995b 'Analysis of daily variability of precipitation in a nested regional climate model—comparison with observations and doubled CO_2 results', *Global and Planetary Change* 10, 55-78.

Mearns, L.O., Bogardi, I., Giorgi, F., Matyasovskey, I. and Paleski, M.: 1999, "Comparison of climate change scenarios generated from regional climate model experiments and statistical downscaling", *Journal of geophysical research,* 104(D6): 6603-6621.

Murphy, J.M.: 1995, 'Transient response of the Hadley Centre coupled ocean-atmosphere model to increasing carbon dioxide. Part I: Control climate and flux adjustment', *J. Climate* 8, 36-56.

Murphy, J.M. and Mitchell, J.F.B.: 1995, 'Transient response of the Hadley Centre coupled ocean-atmosphere model to increasing carbon dioxide. Part II: Spatial and temporal structure of the response', *J. Climate* 8, 57-80.

Penman, H.L. 1971. Irrigation at Woburn VII. Report Rothamsted Experimental Station for 1970, Part 2. 147-170

Semenov, M.A., and Porter, J.R.: 1995, 'Climatic variability and the modeling of crop yields', *Agric. Forest Meteor.,* 73, 265-283

Semenov, M.A., Wolf, J, Evans, L.G, Eckersten, H. and Iglesias, A.: 1996 'Comparison of wheat simulation models under climate change. II. Application of climate change scenarios.' *Climate Research,* 7: 271-281

Semenov, M.A. and Barrow, E.M.: 1997, "Use of a stochastic weather generator in the development of climate change scenarios", *Climatic Change,* 35: 397-414.

von Storch, H., Zorita, E. and Cubasch, U., 1993, 'Downscaling of global climate change estimates to regional scales: an application to Iberian rainfall in wintertime', *Journal of Climate* 6, 1161-1171.

Wigley, T.M.L., Jones, P.D., Briffa, K.R. and Smith, G.: 1990, 'Obtaining sub-grid-scale information from coarse-resolution general circulation model output', *Journal of Geophysical Research* 95, 1943-1953

Wilby, R.L. et al.: 1998, "Statistical downscaling of general circulation model output: A comparison of methods", *water resources research,* 34(11): 2995-3008.

Development of Combined Seasonal Weather and Crop Productivity Forecasting Systems

Andrew Challinor, David Grimes, Julia Slingo, Chris Thorncroft, Peter Craufurd, Tim Wheeler
University of Reading, United Kingdom

SUMMARY

The purpose of this three-year project is to combine meteorological and agricultural modeling research in order to predict seasonal crop yield. The project began four months ago, and this summary presents some results and an outline of future work. The preliminary focus of the project is groundnut in India. Later work will involve the study of other crops in Africa.

Food production in seasonally arid areas is inherently risky, and productivity shows variability on a number of time-scales. A simple analysis of this variability shows that improvements in technology and bulk levels of precipitation (e.g. all-India total) can be the dominant factors which determine large-scale yields. However, spatial variability is also important, and the issue of downscaling General Circulation Model outputs to the high levels of resolution required by crop models is an important one.

An important challenge will be to find the appropriate level of complexity for the crop model. Complicated models require many input parameters and therefore a degree of tuning. The crop model should be detailed enough to produce accurate results, and simple enough to remain as portable as possible.

Experimental evidence shows the importance of variability in temperature, independent of any substantial changes in mean seasonal temperature, for the yield of annual crops. Seed yields are particularly sensitive to brief episodes of hot temperatures if these coincide with critical stages of crop development. Three research needs have been identified in order to provide a framework for predicting the impact of episodes of hot temperatures on the yields of annual crops: reliable seasonal weather forecasts; robust predictions of crop development; and crop simulation models which are able to quantify the effects of brief episodes of hot temperatures on the seed yield.

On the seasonal time-scale there is a need to combine the largely functional models of crop adaptation and yield responses to drought and heat stress, to the probabilistic models used in seasonal forecasting. The skill of seasonal forecasts in providing the necessary crop sensitive parameters will be assessed, along with the identification of potential shortcomings in the meteorological and crop numerical models. Part of this will be the determination of appropriate outputs (and frequency of outputs) from GCMs.

The Challenge of Developing Agroclimatic Scenarios from General Circulation Models

S. Gameda, R. De Jong, and A. Bootsma
Land Resource Management Agriculture and Agri-Food Canada

INTRODUCTION

Projected changes in greenhouse gas (GHG) concentrations could have dramatic impacts on climate (IPCC, 1996a), with associated impacts on agricultural production that can vary widely among regions (IPCC, 1996b). Significant changes can take place in factors such as land suitability for production, growing season length, and crop yields. The capacity to evaluate the potential range of impacts of climate change would be useful for preparing the agricultural sector for adaptation and mitigation measures. General circulation models (GCM) can provide information for assessing the impacts of climate change on agriculture. However, there is a need to determine the ability of GCMs to simulate existing conditions in order to develop confidence in their predictions of future climates for input into crop growth models.

Several datasets from the Canadian first generation coupled general circulation model (CGCM) are available for long-term climate change studies specific to Canada (CCCma, 1999). The most recent set includes a 201-year "transient" simulation of the effects of atmospheric concentrations of GHG observed in 1900 to the present and increasing at an annual rate of 1% thereafter, with the direct effects of sulphate aerosols included (CGCMI+A). The CGCMI+A dataset consists of monthly climatic values on a 3.75° latitude/longitude grid. Daily data for three 21-year time windows (1975-1995, 2020-2040, 2080-2100) are readily available.

Long-term climate daily climate data have been collected at numerous locations in Canada. The objective of this study was to conduct comparisons of observed climate data from two locations with those simulated by the CGCMI+A model for the 1975-95 period and to evaluate the impacts of simulated versus observed daily values on estimates of crop yields using a crop growth model.

METHODOLOGY

Two locations were chosen for comparative purposes—one in Western Canada (Regina, 50°26' N, 104°40' W) representing a moisture deficit region, and one in Eastern Canada (Sarnia, 43°00' N, 82°18' W) representing a moisture abundant region. The GCM node nearest (within 50 km distance) to each location was used for simulated values. The daily variables considered were maximum (Tmax) and minimum (Tmin) temperatures, and precipitation. Comparisons of crop yields under observed and simulated climatic conditions were determined using the EPIC crop growth model (Williams, 1995), assuming near optimum fertility conditions.

RESULTS AND DISCUSSION

There were significant differences between observed and simulated Tmax, Tmin and precipitation values averaged over the 1975-1995 period for the two locations (Tables 1 and 2). Simulated values of precipitation significantly exceeded observed values (Figures 1 to 4). Simulated values of the number of days with precipitation were also significantly higher than observed.

Tmin was consistently overestimated for Regina, with simulated average monthly values ranging between 1.8 and 15.1°C higher than observed. During the growing season, simulated average monthly Tmin values were between 1.8 to 5.5°C higher than observed. Tmax values were significantly underestimated for four of the 12 months, with most of the underestimates occurring during the spring period. Tmax was overestimated during winter months, with the largest difference in monthly averages (8.1°C) occurring in December. Precipitation values were significantly overestimated for all but four months of the year. The number of days with precipitation was substantially

Table 1. *Comparisons between 1975-95 observed and CGCMI+A simulated weather characteristics for the Regina climate station.*

	T_{max} (°C)		T_{min} (°C)		Precipitation (mm)		Number of Days with precipitation	
Regina	Measured	Simulated	Measured	Simulated	Measured	Simulated	Measured	Simulated
Jan	-9.8	-3.2	-20.6	-10.7	13.1	48.4	8.3	20.6
Feb	-6.9	-4.4	-17.2	-12.9	9.6	41.4	6.5	18.5
Mar	0.8	-1.3	-9.6	-7.4	20.4	37.8	7.0	18.4
Apr	11.4	6.2	-1.8	0.1	20.5	39.2	6.9	17.0
May	19.2	15.2	5.0	6.8	52.3	80.0	9.6	20.3
Jun	23.4	20.7	9.7	12.1	74.1	74.0	11.8	17.5
Jul	25.9	25.6	12.1	16.6	70.5	55.2	10.0	13.3
Aug	24.8	25.9	10.6	16.1	44.1	44.7	9.0	15.4
Sep	18.8	20.9	4.9	9.4	36.0	34.0	7.8	14.0
Oct	11.4	13.2	-1.8	3.8	19.1	36.4	6.1	15.7
Nov	0.0	5.1	-10.7	1.3	12.3	32.6	6.6	16.6
Dec	-8.0	0.1	-18.6	-3.5	16.5	46.6	8.4	20.3
Annual	9.3	10.4	-3.1	2.7	388.5	570.4	98.1	207.7

greater than observed throughout the year. Simulated monthly values were between 33 and 184 % greater than observed.

Simulated Tmin values were significantly higher than observed at Sarnia for all but three months of the year. During the growing season (May-September), simulated average monthly Tmin values were between 2.9 and 5.7°C greater than observed. Simulated Tmax values for the growing season had generally good correspondence with

Table 2. *Comparisons between 1975-95 observed and CGCMI+A simulated weather characteristics for the Sarnia climate station.*

	T_{max} (°C)		T_{min} (°C)		Precipitation (mm)		Number of Days with precipitation	
Regina	Measure	Simulate	Measure	Simulate	Measure	Simulate	Measure	Simulate
Jan	-2.1	-0.8	-9.1	-6.1	46.8	65.6	13.0	20.5
Feb	-1.0	-2.8	-8.6	-11.4	44.4	62.5	11.0	18.6
Mar	4.8	0.2	-3.6	-5.1	61.5	84.4	11.8	21.1
Apr	11.5	9.6	1.7	3.2	78.5	124.9	13.1	21.7
May	18.5	17.8	7.1	10.0	65.5	155.0	9.8	23.4
Jun	23.4	22.4	12.2	15.4	83.1	132.7	9.4	21.9
Jul	26.4	26.5	15.6	20.8	74.2	128.3	9.8	23.6
Aug	25.2	26.7	14.9	20.6	83.7	108.8	9.9	21.3
Sep	21.1	22.6	10.6	15.2	96.6	77.8	11.0	17.2
Oct	14.3	13.8	4.9	5.6	71.0	57.2	11.0	14.6
Nov	7.7	6.4	0.2	1.6	82.6	53.1	12.3	17.0
Dec	0.9	1.2	-5.9	-1.2	61.1	65.0	12.3	21.5
Annual	12.6	12.0	3.4	5.8	849.0	1115.3	134.4	242.5

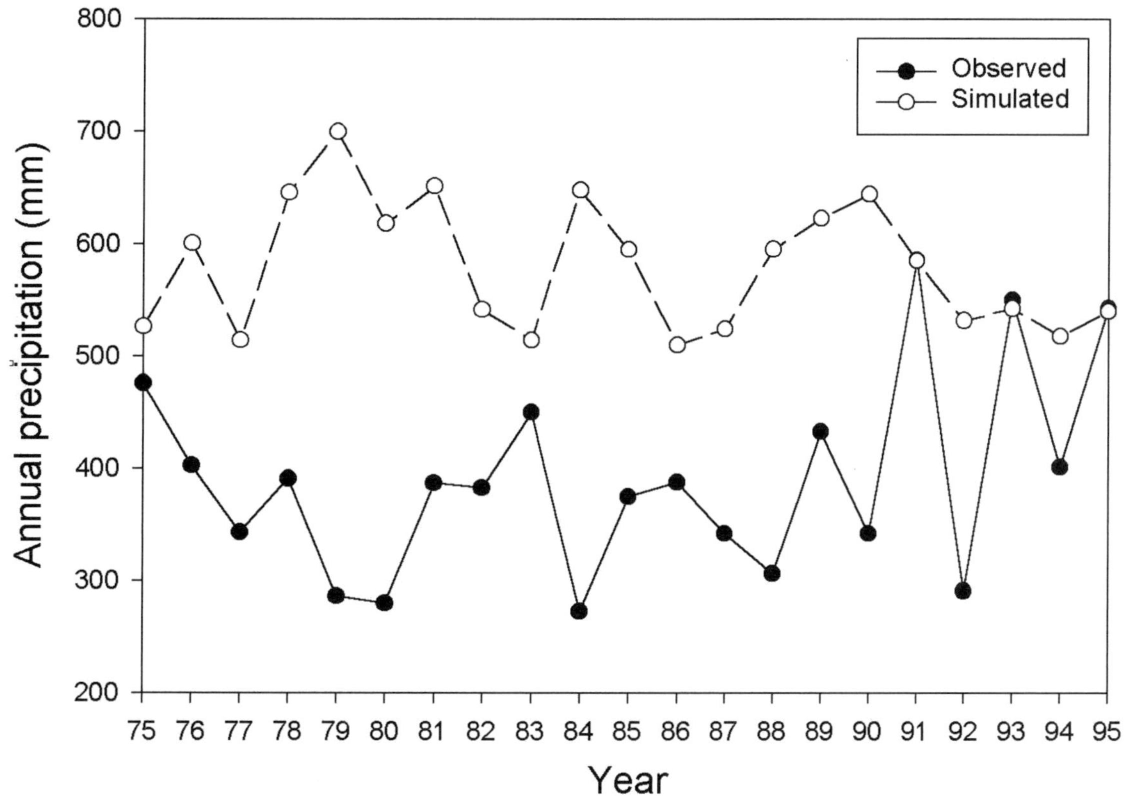

Figure 1. *Annual precipitation, Regina.*

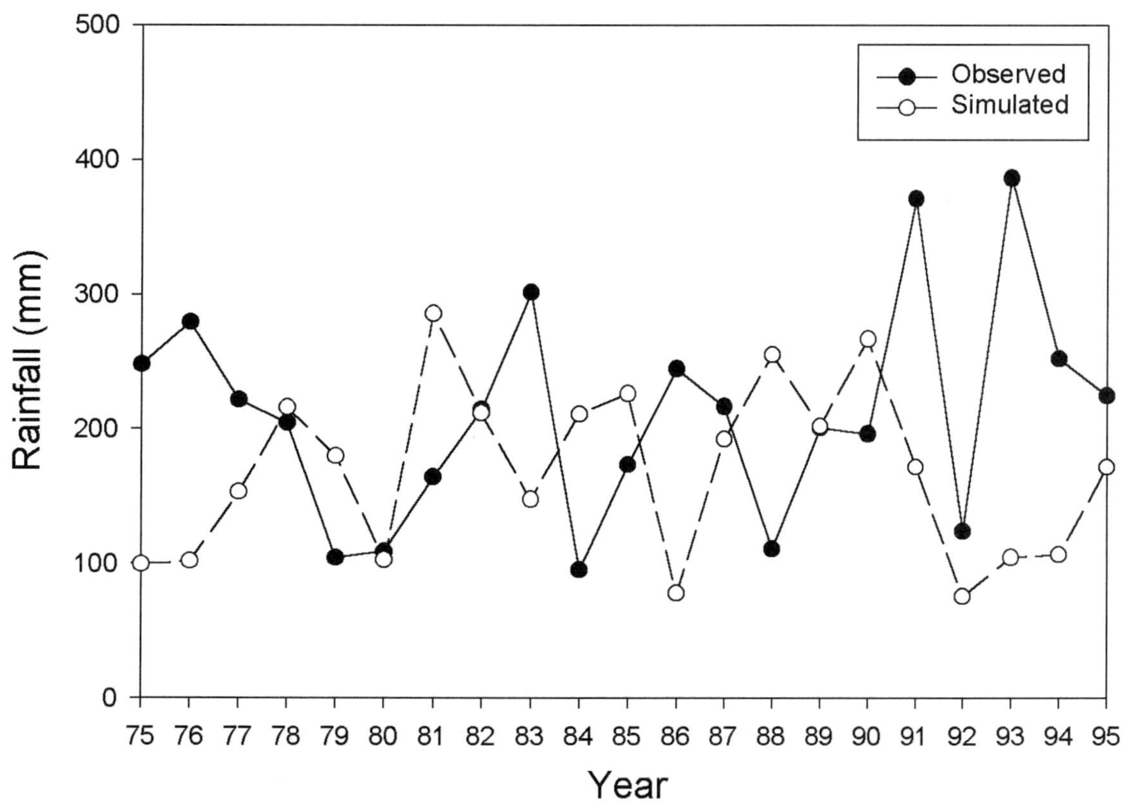

Figure 2. *Growing season rainfall, Regina.*

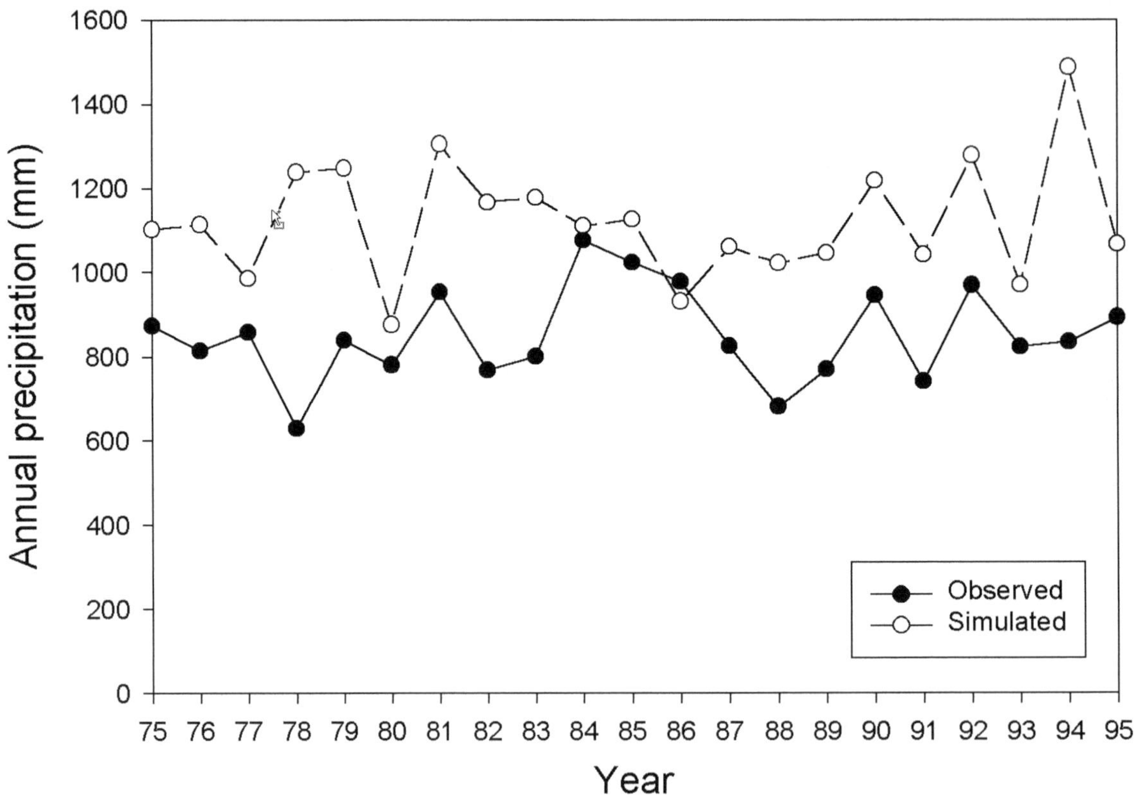

Figure 3. *Annual precipitation, Sarnia.*

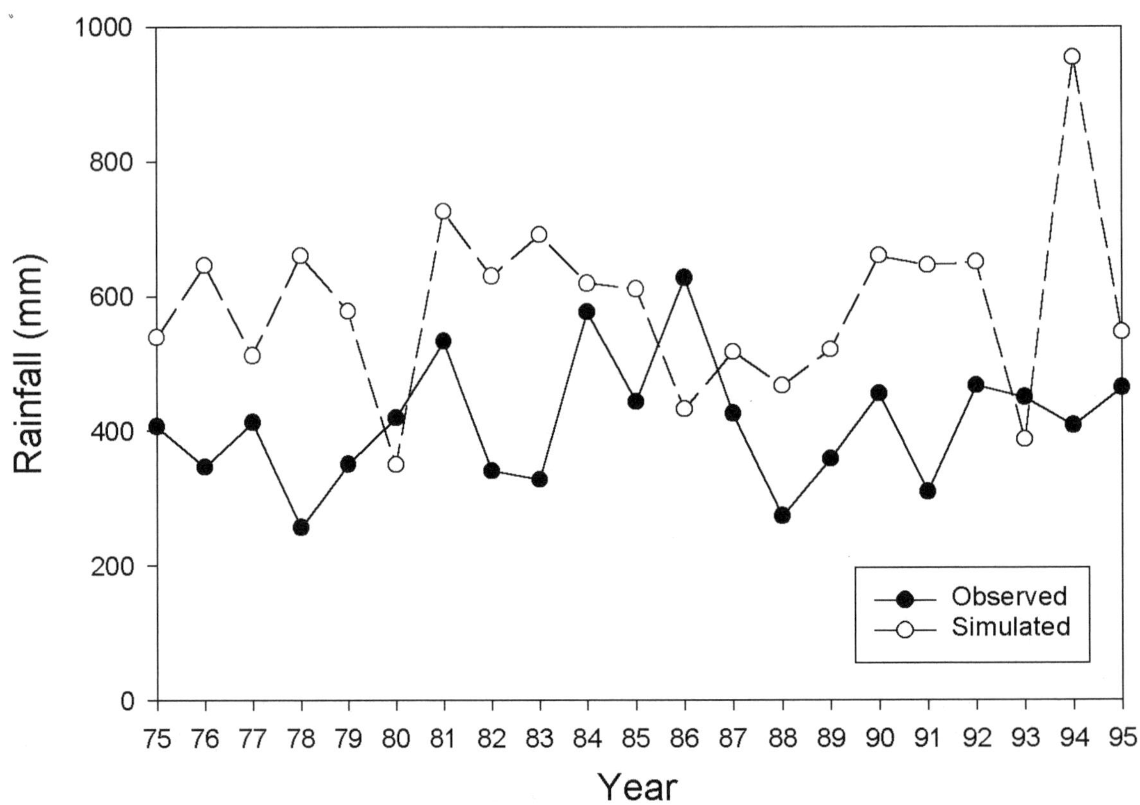

Figure 4. *Growing season rainfall, Sarnia.*

Table 3. *Statistics related to crop yield simulations for the Regina climate station.*

	Observed weather		Simulated weather	
	Average	CV (%)	Average	CV (%)
Yield (t/ha)	2.437	46	3.997	13.4
Annual precipitation (mm)	391.8	22.3	579.3	9.7
Seeding date	May 2		May 26	
Growing season length (days)	102.2	6.8	86.2	6.6
Spring soil water content (mm)	229.1	12.5	316.1	0.6
Growing season rain (mm)	211.5	37.2	169.5	36.6
Growing season PET (mm)	456.9	4.5	318.5	8.5
Water stress	37.5	33.8	7.8	59.6

observed data compared to Tmin, with monthly averages ranging between 1.0°C lower to 1.5°C higher than observed.

Modeled yields were significantly higher under simulated climate conditions as compared to observed ones for both locations (Tables 3 and 4; Figures 5 and 6). Yield variability, on the other hand, was significantly lower under GCM-simulated weather conditions.

The estimated average seeding date under simulated weather was 24 days later than that under observed weather (May 26 versus May 2), probably as a result of excessive soil moisture. The growing period from seeding to harvest was shorter (86 days versus 102 days) due to the elevated temperature regime under simulated weather conditions. Average wheat yields were 64% higher when the crop growth model used GCM-generated weather data. Yield variability, as expressed by the coefficient of variation decreased significantly from 46.0 to 13.4%.

Estimated seeding dates based on simulated weather were similar to those based on observed weather data. Unlike wheat seeding dates at Regina, which were based on both temperature and soil moisture, the seeding date for corn at Sarnia

Table 4. *Statistics related to crop yield simulations for the Sarnia climate station.*

	Observed weather		Simulated weather	
	Average	CV (%)	Average	CV (%)
Yield (t/ha)	5.635	30	9.532	9.3
Annual precipitation (mm)	851.2	12.7	1122.5	47.1
Seeding date	May 15	±5.4 days	May 10	±8.6 days
Growing season length (days)	148	3.6	155	5.4
Spring soil water content (mm)	312.9	1.2	317.7	0.7
Growing season rain (mm)	412	22.2	587.7	21.7
Growing season PET (mm)	545.3	6.9	479.8	8.4
Water stress	35.8	55.8	5.8	184.8
P stress	2.9	81.1	12.3	81.2

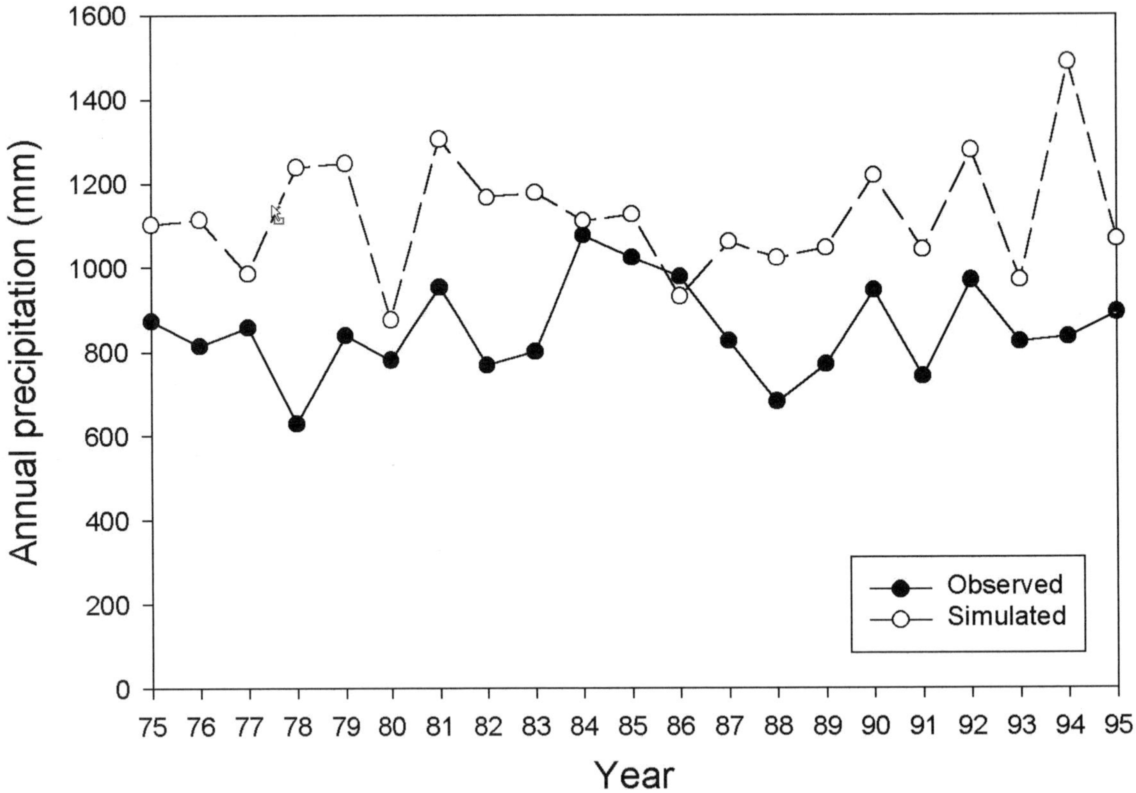

Figure 5. *Simulated wheat yields, Regina.*

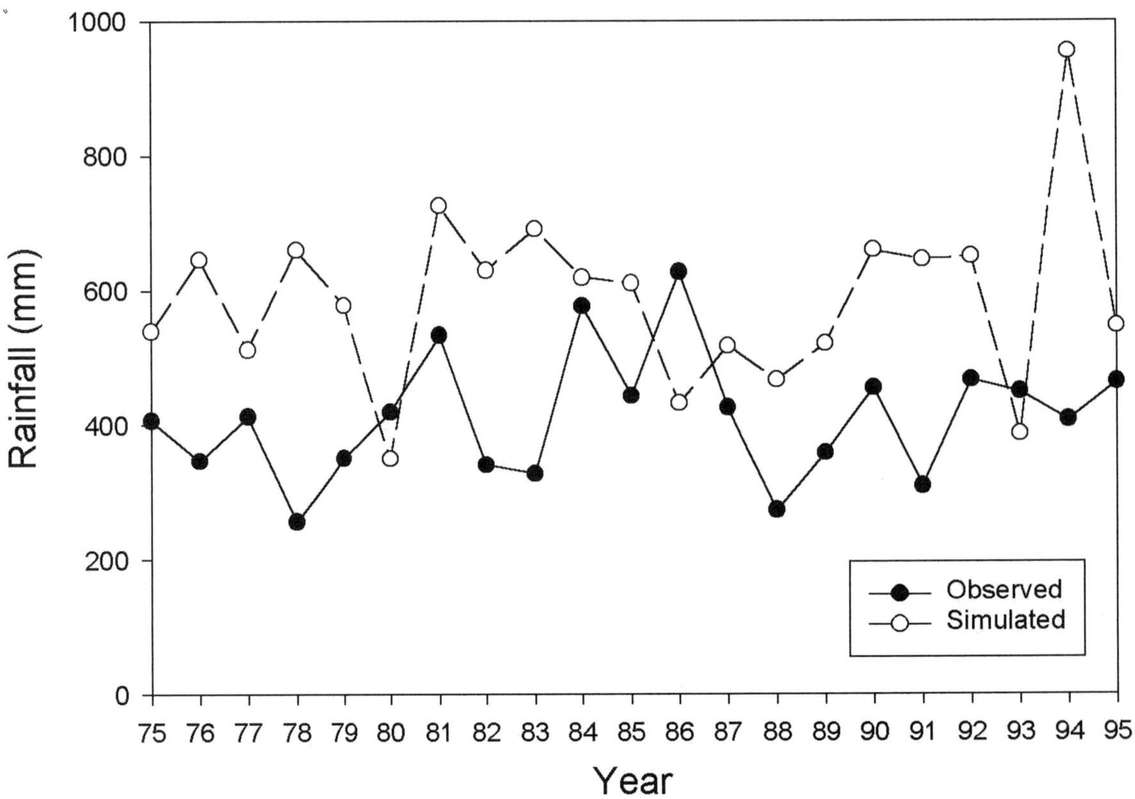

Figure 6. *Simulated corn yields, Sarnia.*

was determined by temperature criteria only. Growing season length was also similar under simulated and observed weather conditions. Average corn yields were 69% higher when the simulation used simulated weather data. Yield variability, as expressed by the coefficient of variation, decreased significantly from 30.0 to 9.3%.

CONCLUSIONS

There were significant differences between observed climate variables and those simulated by GCMs at the two locations studied. Crop yield and yield variability estimates were also significantly different between observed and simulated weather. Results indicate that direct use of generated daily weather data from existing GCMs in crop yield models is not appropriate if impacts assessment researchers and policy makers are to make use of model outputs. Until more accurate GCMs are constructed, current strategies for making use of GCM outputs using appropriate downscaling methodologies such as weather generators (Semenov and Barrow, 1997) and statistical downscaling (Wilby et al., 1998), which apply changes in predicted climate to historical weather data, will need to be used. The ability of GCMs to simulate variability in the parameters under consideration for existing climate data (e.g. 1975-95) would greatly enhance confidence in the use of climate change scenarios under increased CO2 concentrations.

REFERENCES

Canadian Centre for Climate Modeling and Analysis, CCCma. 1999. http://www.cccma.bc.ec.gc.ca/

IPCC, 1996a. The Science of Climate Change. Contribution of Working Group I to the Second Assessment of the Intergovernmental Panel on Climate Change. JT Houghton, LG Meira Filho, BA Callender, N Harris, A Kattenberg and K Maskell (Eds). Cambridge University Press, UK. pp 572

IPCC, 1996b. Impacts, Adaptations and Mitigation of Climate Change: Scientific-Technical Analyses. Contribution of Working Group II to the Second Assessment of the Intergovernmental Panel on Climate Change. R.T. Watson, M.C. Zinyowera, R.H. Moss (Eds). Cambridge University Press, UK. pp 878

Semenov, M.A. and Barrow, E.M. 1997. Use of a stochastic weather generator in the development of climate change scenarios. Climatic Change 35: 397-414.

Wilby, R.L., Wigley, T.M.L., Conway, D., Jones, P.D., Hewitson, B.C., Main, J. and Wilks, D.S. 1998. Statistical downscaling of general circulation model output: A comparison of methods. Water Resources Research 34: 2995-3008.

Williams, J.R. 1995. The EPIC model. Chapter 25 In: V.P. Singh (ed.), Computer models of watershed hydrology. Water Resources Publications, Littleton, Colorado.

Dissemination of Near Real-Time Weather Data for Agricultural Applications

Gerrit Hoogenboom, Goshko A. Georgiev, and D.D. Gresham
University of Georgia, USA

INTRODUCTION

Traditionally the U.S. National Weather Service (NWS) has been responsible for recording local weather conditions and providing weather forecasts. During the last few years, however, most weather data that are being reported by NWS are based on observations that are made at major airports. In many cases these conditions represent weather conditions for the cities where the airports are located, but do not necessarily represent the weather conditions in rural areas and farming communities. Cities show a heat island effect, especially at night, due to the large number of concrete buildings and roads. For instance, it is well known that the temperatures reported at the Hartsfield International Airport in Atlanta, Georgia, are several degrees higher than the surrounding area. The National Weather Service also maintains a Cooperative Network that includes more stations in remote locations. However, the variables collected by this network are limited and only include maximum and minimum temperature and daily precipitation. The data are normally collected at 8.00 am in the morning or 5.00 pm in the afternoon, and therefore also do not truly represent a daily observation. The objective of this paper is to describe an automated weather station that has been developed to support agricultural and environmental research, as well as agricultural, forestry, environmental, construction, and other industries that require detailed weather information on a regular basis.

NETWORK DEVELOPMENT AND SITING

In 1991, the University of Georgia (UGA) initiated the development of an automated weather station network similar to networks that had been developed in other states. One of the main goals of this weather station network is to collect weather and other environmental data in areas where agriculture is being conducted. Although the state of Georgia is well known for its peanuts, it has a diverse range of other crops. This not only includes traditional agriculture, such as row crops, but also fruits and vegetables, ornamentals, turf for golf courses and recreational activities, and other non-traditional agricultural operations. The Georgia Automated Environmental Monitoring Network (AEMN) has grown from four stations in 1991, to more than 40 stations in 2000. The current locations of the weather stations are depicted in Figure 1. It is expected that at least four more stations will be installed during the remainder of 2000. Stations are installed according to guidelines of the World Meteorological Organization (WMO) and in close collaboration with NWS.

Some of the overlap of the stations, such as in Griffin and Watkinsville, is due to installation of the weather stations on agricultural research stations. Besides research stations, automated weather

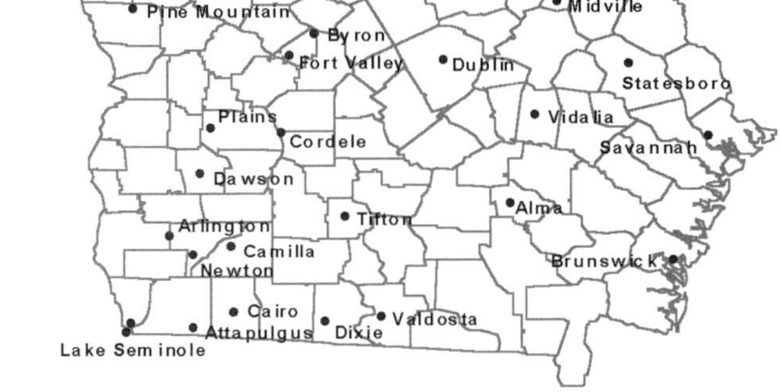

Figure 1. *Locations of the University of Georgia Automated Environmental Monitoring Stations in Georgia*

stations have been installed at local farms, nurseries, orchards, golf courses, high schools, universities, and recreational parks. The most recent installations were conducted in Lake Seminole, in the southwest corner of the state, to study the water balance of the lake. This study is part of litigation between the states of Alabama, Florida and Georgia with respect to water use and water availability.

OPERATION

Each station monitors air temperature, relative humidity, wind speed, wind direction, solar radiation, precipitation, and soil temperature at three different depths. Some stations also monitor photosynthetic active radiation (PAR), net radiation, barometric pressure, pan evaporation, and soil moisture. A listing of the standard and optional sensors is shown in Table 1. The sensors are connected to a Campbell Scientific CR10 or CR10X data logger. The CR10 data logger provides 2 pulse channels and 12 single-ended channels for voltage readers. It has temporary as well as permanent memory that can store the program which controls its operation as well as the data that are being collected. The data loggers can be programmed through a small keyboard or a program can be uploaded through a serial port connection from a personal computer. Programming of the data logger is rather flexible and provides a range of options with respect to data collection and simple processing.

Table 1. *Weather Variables Observed by the Georgia Automated Environmental Monitoring Network*

Variables	Standard	Optional
Air Temperature	x	
Relative Humidity	x	
Solar Radiation	x	
Wind Speed	x	
Wind Direction	x	
Precipitation (Rainfall)	x	
Soil Temperature—5 cm	x	
Soil Temperature—10 cm	x	
Soil Temperature—20 cm	x	
Barometric Pressure		x
Soil Moisture (Top 30 cm)		x
Open Pan Evaporation		x
Water Temperature (Pan)		x
Photosynthetic Active Radiation		x
Net Radiation		x
Leaf Wetness		x
Soil/Turf Surface Temperature		x

All sensors are scanned at a one-second frequency and data are summarized every 15 minutes. This provides a total or average value for each variable, as well as extremes for wind speed and rainfall. At midnight, daily extremes, including maximum, average and minimum, and daily totals are calculated. Eastern Standard Time (EST) is maintained by each data logger. Each unit is powered by a battery, which is recharged by a solar panel during daylight hours. This provides some protection from lightning which might come in through normal power lines.

COMMUNICATIONS

Communications are handled through a modem and either a dedicated telephone line or cell phone that are connected to each data logger. A dedicated computer at the UGA College of Agricultural and Environmental Sciences Campus in Griffin communicates with each weather station on a continuous basis. Ten weather stations in the greater Atlanta area are called every 15 minutes and 15 additional stations across the state are called every hour to download the most current weather information. The remaining stations are called once a day at midnight. After downloading the most recent data from the data logger to the personal computer, the data are processed immediately and spooled to a data server, as well as a web server. Users can view current weather conditions for 25 sites and retrieve daily data for all locations via the World Wide web (http://www.georgiaweather.net). A data flow diagram is presented in Figure 2.

APPLICATIONS

The most critical aspects of the weather data delivery, however, are the applications. Simple

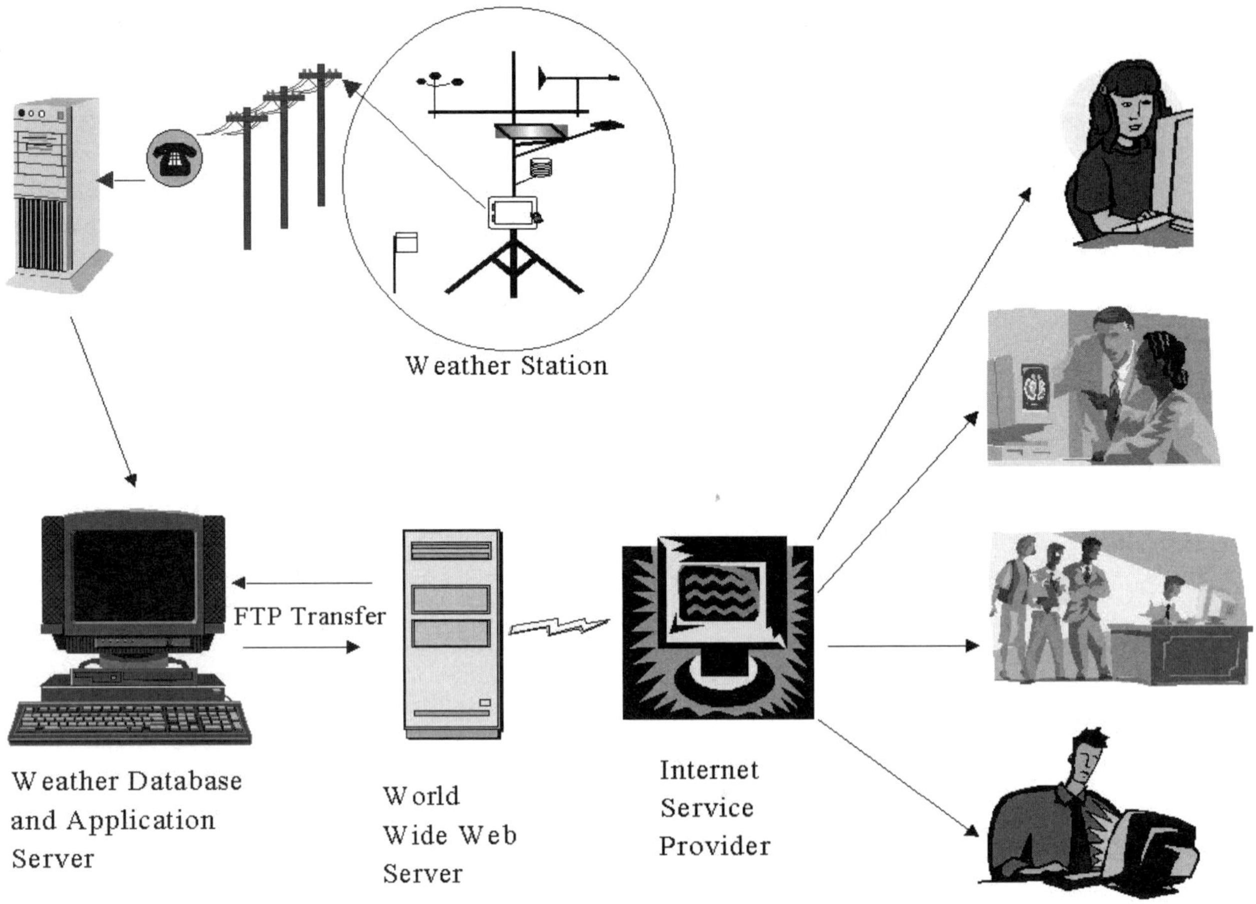

Figure 2. *Weather data and information transfer to the user.*

models have been implemented that provide a link between the user and the data (Georgiev and Hoogenboom, 1999). A user can specify both the starting date and ending date for these calculators, with the default values being set to January 1 and yesterday's date. Users have the option to calculate chilling hours during the winter, degree days during the summer, and a water balance that determines both cumulative precipitation as well as cumulative potential evapotranspiration. These calculators have become critical tools for farmers and others associated with agriculture. A prototype system has also been developed that links the crop simulation models with the current weather data (Georgiev and Hoogenboom, 1999). This provides more advanced capabilities for decision making (Hoogenboom, 2000; Tsuji et al., 1998).

There is scope to add short- and long-term climate forecasts to this system and to improve the prediction capabilities (Hoogenboom and Georgiev, 1999). First of all, the outputs of the regional climate models can be used as inputs for the crop simulation models to extend the simulations to the expected end of the growing season and the start of the next growing season. However, the weather data from the network can also be used as a database for ground truthing some of the data provided by the regional climate models. Because of the density of the mesonetwork as well as the large number of variables that are being measured or calculated, there is potential for evaluation of the climate models.

REFERENCES

Georgiev, G.A., and G. Hoogenboom. 1999. Near real-time agricultural simulations on the web. Simulation 73(1):22-28.

Hoogenboom, G.. 2000. Contribution of agrometeorology to the simulation of crop production and its applications. Agricultural and Forest Meteorology 103(1-2):137-157.

Hoogenboom, G., and G. Georgiev. 1999. The potential for applying seasonal climate models in agricultural decision support systems. In: Proceedings of the 1st International Regional Spectral Model (RSM) Conference. Maui High Performance Computing Center (MHPCC), Kihei, Hawaii.

Tsuji, G. Y., G. Hoogenboom, and P. K. Thornton [Editors]. 1998. Understanding Options for Agricultural Production. Systems Approaches for Sustainable Agricultural Development. Kluwer Academic Publishers, Dordrecht, the Netherlands.

Application and Dissemination of Climate Information and Prediction Services in the Greater Horn of Africa

Simon Gathara
Drought Monitoring Center, Kenya

ACTIVITIES OF THE DROUGHT MONITORING CENTRE, NAIROBI (DMCN)

The main aims of the DMCN project are:

- Monitoring of drought with respect to intensity, geographical extent and duration, in a timely manner.
- Regular provision of weather and climate advisories.
- Provision of early warnings on droughts, floods and other extreme weather events to governments so that appropriate coping mechanisms and strategies to combat the adverse effects can be formulated.
- Assessment of drought/floods impacts to various socio-economic activities including agriculture, health and water resources management.
- Generation of regional climate outlook products and early warning advisories and disseminating them to the National Meteorological and Hydrological Services (NMHSs) of the participating countries for downscaling to national level.

In order to achieve these aims, DMC is engaged in the following major activities:

- Acquisition of near real time climate data from the participating countries.
- Acquisition of Numerical Weather Prediction (NWP) products from the advanced global centers.
- Obtaining El Niño/Southern Oscillation (ENSO) conditions, sea surface temperatures (SST) and other anomaly patterns from the Climate Prediction Centre (CPC) in Washington and the National Climate Centre (NCC) in Melbourne.
- Performing monthly and dekadal (ten days) statistical analysis on rainfall and temperature data.
- Predicting rainfall patterns.
- Producing seasonal forecasts based on main indicators such as ENSO, SSTAs QBOs etc.
- Carrying out monthly and dekadal impact assessment of extreme climate anomalies on various socio-economic activities.
- Research to improve the prediction skills.

MISSION AND OBJECTIVES OF THE DMCN

The Drought Monitoring Centre, Nairobi (DMCN) and the DMC Harare were established in 1989 with funding provided by the United Nations Development Programme (UNDP) and the World Meteorological Organization (WMO) as the executing agency. The centers were established in response to the frequent and widespread devastating droughts and other weather/climate, related disasters that afflicted the eastern and southern African sub-regions in the late seventies and eighties. In addition, there was the realization that alleviating the effects of drought in the sub-region could not be effectively tackled in isolation, but through well coordinated, regional collaboration.

The main objective was to contribute to the mitigation of the adverse impacts of these events on agricultural production, the mainstay of the economic activity in the sub-regions.

The DMCN now provides advisories on weather and climate, advance warnings on droughts, floods and other extreme weather events on the basis of which Governments of participating countries take appropriate mitigation measures. This vital information is input into the decision-making processes in key economic sectors such as agriculture, water resources and health. In addition, it allows the formulation of longer-term preparedness plans to reduce the adverse effects of weather related disasters. DMCN is also responsible for generating sub-regional climate outlook products and early warning advisories which are disseminated to the National Meteorological and Hydrological Services (NMHSs) of the participating countries for down-scaling to national level. Out of the regional products the NMHSs are expected to further develop specific-user tailored products for use by local sectors. The NMHSs ensure that these products meet specific user needs and are applied effectively. This process involves interpreting the DMC regional forecasts and early warning advisories, adding value and specific country details, as well as developing suitable dissemination techniques that target specific user groups.

During the last two years, DMCN has established unique partnerships with research scientists, universities, major national, regional and international climate centers world-wide, users and other partners in the organization of Climate Outlook Fora (COFS). These Fora have provided an opportunity for researchers from various advanced climate prediction centers to work with scientists and experts from the region to develop consensus or consolidated climate outlook guidance products, together with their interpretation and dissemination to users. There is now, a general recognition that the warnings issued by the DMCN contributes to the reduction of damage to property, economic dislocation and loss of life in the sub-region. Consequently, there are efforts being put to find ways of ensuring the long-term sustainability of DMCN.

THE MAIN PRODUCTS AND SERVICES OF THE DMCN

The following products are generated by the DMC and then disseminated to the NMHSs of the member countries, who then distribute them to a wide variety of users, including; policy and decision makers, planners, agricultural and water resources managers, farmers, research institutions and relevant regional and international institutions. Currently, DMCN has over 300 subscribers for the monthly bulletin.

Ten day weather advisories for the sub-region. These include:

- Decadal (ten day) climatological summary.
- Decadal drought severity conditions.
- Decadal socio-economic conditions and impacts.
- Decadal synoptic review and weather outlook.

Monthly drought monitoring bulletin for the sub-region. This includes:

- Monthly and three monthly climatological summaries and drought severity conditions.
- Monthly dominant synoptic systems and three months weather outlook.
- Socio-economic conditions and their impacts.
- Seasonal Weather Outlook.

Additional products issued by the DMCN include special advisories when needed. These advisories are disseminated to the countries which are expected to be directly affected by drought and other extreme weather conditions.

THE MAJOR TARGET AND USE GROUPS OF THE DMCN PRODUCTS

The following is a list of the major user sectors of climate information and services from the DMCN including the potential groups:

- The National weather services within and outside the sub-region.
- Governments Agriculture and Livestock ministries or departments, Agricultural Research Institutes, Locust-pests control services, Range management, Horticulture and Individual Farmers.
- Water Resources management authorities such as Lake Victoria Development Authority, Lake Victoria Conservation, River and marine authorities.
- Health: Malaria, Cholera and Water-borne diseases.

- Energy: Hydroelectric, Solar, wind and Renewable energy resources.
- Environment and Natural Resources: Governments and Non Governmental Organisation.
- Tourism: Government ministries or departments, Travel agencies and Hotels.
- Wildlife Services: Parks, natural game reserves and Non-Governmental Organisations (NGO's).
- Forestry: Forest Government conservation authorities and Research institutions.
- Natural Disaster Management Services: Governments, NGOs and International Organisations.
- Media: Electronic and print.
- Relevant Regional and International Organisations.

The information equips the users, the decision-makers and planners in the various sectors, with the necessary tools for the formulation of appropriate strategies to combat the vagaries of drought.

DISSEMINATION OF THE DMCN PRODUCTS

The products are sent to all the National Meteorological and Hydrological services (NMHSs) of the member countries. The NMHSs then distribute them to a wide variety of users, which include; policy and decision makers, planners, agricultural and water resource experts, farmers, research institutions and relevant regional and international organizations. The demand for the monthly bulletin has grown from 50 copies in 1991 to over 300 copies at present. The products are disseminated through the following methods:

- Ordinary postal mail system.
- The Meteorological Data Distribution (MDD) System.
- The High Frequency Radio Fascimile Broadcasting system.
- Ordinary fax (dekadal).
- The Internet (e-mail and web site).°

There are also institutions to which the products are sent directly.

SOME OF THE DMCN'S ACHIEVEMENTS

The DMCs have achieved commendable results in a number of areas:

- Creation of data banks which archive useful information for drought monitoring and weather forecasting.
- Conduction of three data processing courses in Nairobi and one Workshop on the interpretation and use of DMC products in Harare.
- Strengthened co-operation between the meteorological services of the participating countries and DMCN especially in the area of data exchange.
- Establishment and sustenance of collaboration with the advanced Regional and Global Centers of excellence in matters of Climate prediction and services.
- Issuance of regular and timely forecasts of seasonal rainfall characteristics.
- Successful organization of five Climate Outlook Fora in the Greater Horn of Africa. These fora have helped DMCN establish unique partnerships with research scientists, universities, major national, regional and international climate centers world wide, users and other partners. The fora have provided an opportunity for researchers from various advanced climate prediction centers to work with scientists and experts from the region to develop consensus or consolidated climate outlook guidance products, together with their interpretation and dissemination to users. In addition, through these Fora, the DMCN managed to issue early warnings of the extreme climate events associated with the 1997/98 El Niño, the subsequent La Niña events and even the on-going climate conditions over the Greater Horn of Africa.
- Capacity building workshops for climate scientists in the sub-region.

MAJOR PROBLEMS FACING THE DMCN

- Lack of sufficient data processing equipment. Hence there is need for faster, and greater capacity handling computers to facilitate data processing and archival, and the production of timely, accurate and more reliable products.

- Persistently inadequate communications in the sub-region.
- Inadequate knowledge by climate scientists of the users' requirements while the latter lacks the capability to understand and interpret the products and are therefore unable to use them effectively.
- Climate prediction products currently available from global climate centers have relatively low skill at regional and local levels.
- How to continue with DMCN activities when the USAID funding comes to an end.

VISION FOR THE DMCN

The current DMC project intends to carry out the following activities:

- Improve the capabilities of the DMCN to enable it to provide long lead forecasts and early warnings of extreme climate events such as drought, floods etc.
- Go a step beyond the generation of better products and look into their effective dissemination and application by linking the producers of climate prediction products with various end-users, while at the same time developing feedback systems for improvement of products.
- To improve product generation and delivery, and ensure their effective application at regional, national and finally at local levels, the project will undertake the following:
  —Improve computing facilities;
  —Improve communications between the DMCN and the NMHSs of the sub-region for timely collection, processing and dissemination of data and products;
  —Integrate and apply DMC products in the development of appropriate policies for sustainable socio-economic development, particularly in early warning, disaster preparedness and prevention, the management and conservation of the environment, etc.;
  —Train local and regional scientists and users of climate information and prediction products through periodic attachment to the DMC, short courses in the generation and application of climate products, and climate outlook Fora for seasonal rainfall.

Modeling Inter-Annual Variation of a Local Rainfall Data Using a Fuzzy Logic Technique

Halmar Halide and Peter Ridd
James Cook University, Australia

INTRODUCTION

Agricultural production of the world shows inter-annual variability that is often the result of variations in rainfall. Clearly the ability to predict rainfall would be very useful in many agricultural industries (Solow *et al.,* 1998; Mjelde *et al.,* 2000).

There are various way of predicting rainfall being practiced today. Common variables used for long term prediction of rainfall are the sea surface temperature (Berri and Flamenco, 1999), the Southern Oscillation index (Ropelewski and Halpert, 1987 and 1996), and the atmospheric circulation patterns (Bardossy *et al.,* 1995). The methods of relating these variables to produce the rainfall prediction also vary. Recently, neural networks (Silverman and Dracup, 2000) and fuzzy logic (Ozelkan *et al.,* 1996) techniques have been introduced. The latter study shows that the fuzzy logic technique gives better prediction than that of linear regressions. To get higher prediction skill, the use of SST data was recommended.

In this study, the fuzzy logic technique is used to model and predict local rainfall data. Several predictors are used: local rain of the previous two months, the SSTA from Niño 3.4 region of the Pacific, and the remote rainfall data of India IMR (Indian Monsoon Rainfall). Predictions based on each and various combinations are compared based on their prediction errors, the RMSE (root-mean-squared-error). The specific aim of this work is to produce a rainfall prediction for the month of January at Hasanuddin airport. This station is close to the largest rice-producing area of Indonesia, Pinrang. January is the month when rice is usually planted.

MATERIALS AND METHODS

Input Data

In this study, there are three data sets used for modeling and predicting the annual January rainfall at Hasanuddin airport. Firstly, the monthly local rainfall data from WMO station # 97180 Hasanuddin airport at Makassar city, South Sulawesi province of Indonesia [119o 30' E and 5o 5' S]. The rainfall data was obtained from the NCDC-NOAA at http://www.ncdc.noaa.gov. Secondly, the monthly all-India rainfall data which is taken from Parthasarathy et al. (1994). Thirdly, the monthly sea surface temperature Anomaly (SSTA) data of the Niño 3.4 region in the Eastern Pacific which is available at the CPC NOAA web-site http://nic.fb4.noaa.gov:80/data/cddb/cddb/sstoi.indices. To model and predict an output, for example the local rainfall for the month of January in a particular year, the inputs to the model are: (a) the previous November and December of the local rainfall, (b) the total rainfall of the previous seasons in India, and (c) the January Niño 3.4 SSTA of the same year. Note that, the Indian rainfall data evaluated here are: the Southwest Monsoon (March—May), the Summer Monsoon (June-September), and annual (January-December). These inputs were fed into the fuzzy logic model to produce the local rainfall in January at Hasanuddin airport. This particular month is chosen because it is the beginning of rice planting in that province. Water availability in this stage is crucial for sustaining growth and production (Mears, 1981).

The Fuzzy Logic Model

Fuzzy logic can be viewed as a means for mapping/relating a set of input(s) into an output(s) (Bezdek and Pal, 1992; Pedrycz, 1993; Zadeh, 1996). In our case, the inputs are the various rainfall data and SSTA mentioned above and the output is the January rainfall at Hasanuddin airport. There are two procedures involved in this exercise. They are (a) data training (modeling or fitting in-sample data) and (b) data testing (predicting out-of-sample data). Both procedures use three subroutines, available in the computational package called MAT-

LAB® (Jang and Gulley, 1995), called GENFIS2, ANFIS and EVALFIS. These subroutines are to be successively invoked to these I/O (input/output) pairs.

During data training, the relationship between the inputs and output is determined. The inputs are: (a) the Nov. and Dec. local rainfall from 1950-1969, (b) the Indian rainfall (the two Monsoons and the annual) from 1950-1969, and (c) the Niño 3.4 SSTA data from 1951 to 1970. The output is the Hasanuddin rainfall in January from 1951-1970. To find the relationship, the following steps are taken. First, the subroutine GENFIS2 is used to generate the so-called FIS (fuzzy inference system) matrix based on these I/O pairs. The subroutine extracts a set of rules that models the I/O behavior. Secondly, the resulting FIS matrix with the minimum error is then fed into another subroutine named ANFIS. This is the optimization algorithm which is capable of further reducing the error, (i.e. difference between data and model output). Having tried many combinations of the model's parameter, the smallest error during data training after 100 iterations is obtained when using a cluster radius and an initial step size of 0.2 and 0.3, respectively. Finally, these input data created during training are used in the resulting fuzzy model through the subroutine EVALFIS to produce the model output. This output is compared with the actual data and its root-mean-squared-error (RMSE) are calculated.

For data testing, the above resulting model and steps are used to create an out-of-sample prediction. The inputs are the local and the Indian rainfall data from 1970 to 1975, the Niño 3.4 data from 1971 to 1976, and the output is the January rainfall from 1971-1976. The results are as follows.

RESULTS AND DISCUSSION

In this section, the results of modeling (data training) and out-of-sample prediction (data testing) using various inputs are presented. Results of using the local rainfall, the Niño 3.4 SSTA, and the combined local and Niño 3.4 SSTA data to the fuzzy model is plotted in Figure 1 (a)—(c). Figure 1 (a) shows that predictions using the local rainfall

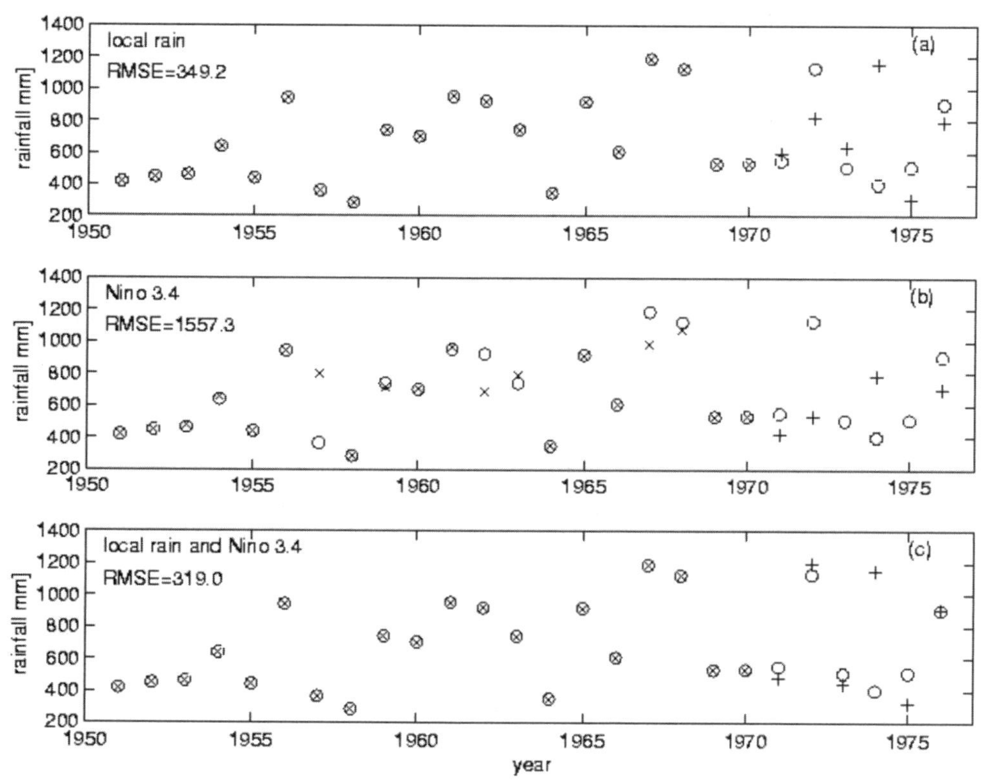

Figure 1. The annual in-sample and out-of-sample predictions of the January rainfall based on (a) local rain, (b) SSTA Niño 3.4, and (c) both local rain and Niño 3.4. Actual data, in-sample, and out-of-sample output are depicted using 'o', 'x', and '+', respectively.

is better then that using only the Niño 3.4 SSTA as an input. The former has an RMSE (out-of-sample) of 349.2 mm while the latter results in a much bigger RMSE of 1557.3 mm. Note that predictions in years 1973 and 1975 are off scale in Figure 1 (b). A similar poor prediction also found by Hunt (1997) and Nicholls et al. (1997) when predicting rainfall within Australia using either the SSTA or the Southern Oscillation Index (SOI). Even though, there has been considerable progress in predicting the ENSO events on a seasonal basis (Barnston *et al.,* 1999; Halide and Ridd, 2000), the use of this index alone for rainfall prediction is not always promising. Using local rainfall and SSTA data, the RMSE is found to be 319.0 mm (see Figure 1 (c)).

The results show that the prediction for the year 1974 is the worst in Figure 1 (a) and (c). The model overestimates the amount of rain in that year associated with the cold phase of the ENSO, La Niña 1974. If we remove the prediction in the year 1974 from predictions over the test data, the out-of-sample RMSE obtained using (a) the local rain, (b) Niño 3.4, and (c) both local rain and Niño 3.4 as inputs to the model is 184.3 mm, 1697.0 mm, and 98.4 mm, respectively. This is a substantial improvement in the prediction skill except for the model using only SSTA as an input.

Figure 2 (a)–(c) presents the effect of adding into the model the total Southwest Monsoon, Summer Monsoon, and the annual rainfall data, respectively. The RMSE over the data set is found to range from 258.1 mm to 342.1 mm. The best skill is shown when using the Summer Monsoon. Note that the total rainfall data of this season is used to derive the Indian Summer Monsoon Rainfall (ISMR) (Parthasarathy *et al.,* 1992), Even though the skill is much better when this particular Monsoon data is included in the model, the prediction in year 1974 is still poor. Note that during this particular La Niña year, the annual rainfall in India was also less than normal. It is possible that during El Niño, the IMSR got more rain (Shen and Kimoto, 1999) whereas during La Niña, the area had less rain.

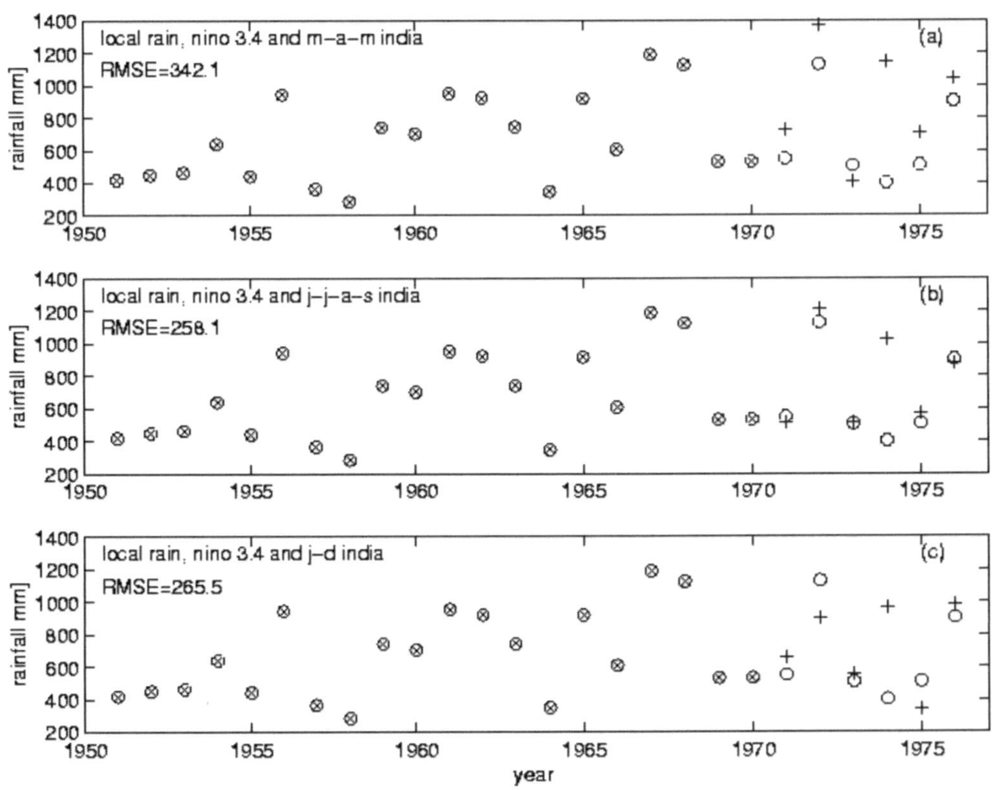

Figure 2. The annual in-sample and out-of-sample predictions of the January rainfall based on local rain, SSTA Niño 3.4, and various Indian Monsoon rainfalls. (a) For Southwest (Mar-Apr-May) Monsoon, (b) For Summer Monsoon (Jun-Jul-Aug-Sep), and (c) For annual Indian rain. The symbols are as above.

The low prediction skill in 1974 may be caused by the limited area covered by the ISMR. The use of the Extended Indian Monsoon rainfall and the large scale monsoon Hadley indexes proposed by Goswami et al. (1999) should be pursued in future work.

CONCLUSIONS

A fuzzy logic model for predicting the annual January rainfall from a local station in the tropics has been tested using various inputs as predictors. It is found that a better out-of-sample prediction skill is obtained when the model used both the local rainfall of the previous months of November and December, and the SSTA from the Niño 3.4 region of the eastern Pacific. The RMSE between data and model output is found to be 319.0 mm which is smaller than that by using either the local rain or the Niño 3.4 alone of 349.2 mm and 1557.3 mm, respectively. The skill was shown to be improved when the Indian Summer Monsoon rainfall was included as predictor. The RMSE drops to 258.1. This large RMSE is dominated by the large error on predicting the 1974 rainfall.

ACKNOWLEDGMENTS

We would like to express our gratitude to thank the NOAA Climate Prediction Center, cq. Dr. D. Garrett for letting us use the Niño 3.4 data set from the web site. We also would like to thank the National Climate Data Center NOAA for providing the rainfall data of the WMO station # 97180 of the Hasanuddin airport. HH would also like to express his gratitude to the Australian Government for the AUSAID PhD scholarship.

REFERENCES

Bardossy, A., L. Duckstein, and I. Bogardi. 1995. Fuzzy rule-based classification to atmospheric circulation patterns. Int. J. Climatol. 15: 1087-1097.

Barnston, A.G., M.H. Glantz, and Y. He. 1999. Predictive skill of statistical and dynamical climate models in SST forecasts during the 1997-98 El Niño episode and the 1998 La Niña onset. Bull. Am. Meteorol. Soc. 82: 217-243.

Berri, G.J., and E. Flamenco. 1999. Seasonal volume forecast of the Diamante river, Argentina, based on El Niño observations and predictions. Water Resour. Res. 35: 3803-3810.

Bezdek, J.C., and S.K. Pal. 1992. Fuzzy models for pattern recognition—Methods that search for structures in data. IEEE press, New York.

Goswami, B.N., V. Krishnamurthy, and H. Annamalai. 1999. A broad-scale circulation index for the interannual variability of the Indian summer monsoon. Q.J.R. Meteorol. Soc. 125: 611-633.

Halide, Halmar, and P. V. Ridd. 2000. Seasonal ENSO prediction using a linear neural network. Submitted to Bull. Am. Meteorol. Soc.

Hunt, B.G. 1997. Global climatic models: long-term predictions, annual to decadal. P. 31-44. *In* R.K. Munro, and L.M. Leslie (eds.) Climate prediction for agricultural and resource management. Australian Academy of Science Conf., Canberra, 6-8 May 1997.

Mears L.A. 1981. The new rice economy of Indonesia. Gadjah Mada Univ. press, Yogyakarta.

Mjelde, J.W., J.B. Penson Jr., and C.J. Nixon. 2000. Dynamic aspects of the impact of the use of perfect climate forecasts in the corn belt region. J. Appl. Meteorol. 39: 67-79.

Nicholls, N., W. Drosdowsky, C.S. Frederiksen, R. Kleeman, and N.R. Smith. 1997. Seasonal climate predictions. p. 45-59. *In* R.K. Munro, and L.M. Leslie (eds.) Climate prediction for agricultural and resource management. Australian Academy of Science Conf., Canberra, 6-8 May 1997.

Nicholson, S.E. 1980. The nature of rainfall fluctuations in subtropical West Africa. Mon. Wea. Rev. 108: 473-487.

Ozelkan, E.C., F. Ni., and L. Duckstein. 1996. Relationship between monthly circulation patterns and precipitation: fuzzy logic and regression approaches. Water Resour. Res. 32: 2097-2103.

Pearson, S., R. Naylor, and W. Falcon. 1991. Recent policy influences on rice production. p. 8-21. *In* S. Pearson, W. Falcon. P. Heytens, E. Monke, and R. Naylor. Rice policy in Indonesia. Cornell Univ. Press, Ithaca.

Parthasarathy, B., K.R. Kumar, and D.R. Kothawale. 1992. Indian summer monsoon rainfall indices. Meteorol. Mag. 12: 174-186.

Parthasarathy, B., A.A. Munot, and D.R. Kothawale. 1994. All-India monthly and seasonal rainfall series: 1871-1993. Theor. Appl. Climatol. 49: 217-224.

Pedrycz, W. 1993. Optimization of fuzzy models for system analysis, pattern recognition and knowledge engineering. p. 245-270. *In* P. Z. Wang and K. F. Loe (eds.) Between mind and computer—fuzzy science and engineering. World Scientific, Singapore.

Ropelewski, C.F., and M.S. Halpert. 1987. Global and regional scale precipitation patterns associated with the El Niño/Southern Oscillation. Mon. Wea. Rev. 115: 1606-1626.

Ropelewski, C.F., and M.S. Halpert. 1996. Quantifying southern Oscillation—precipitation relationship. J. Climate 9: 1043-1059.

Shen, X., and M. Kimoto. 1999. Influence of el Niño on the 1997 Indian summer monsoon. J. Meteor. Soc. Japan 77: 1023-1037.

Silverman, D., and J.A. Dracup. 2000. Artificial neural networks and long-range precipitation prediction in California. J. Appl. Meteorol. 39: 57-66.

Solow, A.R., R.F. Adams, K.J. Bryant, D.M. Legler, J.J O'Brien, B.A. McCarl, W. Nayda, and R. Weiher. 1998. Clim. Change. 39: 47-60.

Zadeh, L.A. 1996. Fuzzy logic = computing with words. IEEE Trans. Fuzzy Sys. 4: 103-111.

SESSION III. IMPACTS OF CLIMATE VARIABILITY ON CROP AND LIVESTOCK SYSTEMS

Predicting Vintage Quantity and Quality in Coastal California Using Pacific sea surface temperatures

Ramakrishna R. Nemani[1], *Michael A. White*[1], *Daniel R. Cayan*[2], *Gregory V. Jones*[3], *Steven W. Running*[1]
[1]University of Montana, USA
[2]United States Geological Survey, USA
[3]Southern Oregon University, USA

INTRODUCTION

California produces 90% of all wine within the U.S. and dominates the $33 billion/year domestic retail wine industry. Since the 1950s, wine grape growers in California have seen dramatic increases in premium wine quality, grape yield, and crop value. Advances in viticultural practices (irrigation, nutrition, pest/disease control, trellising etc.) and experience in wine making have certainly contributed to the success (Jackson and Lombard 1993). In spite of such advances, wine growers generally believe climate plays a significant role in determining the quantity and quality of a given vintage.

Widespread changes in climate have been reported globally during the last few decades, attributed mainly to the greenhouse effect of rising atmospheric CO_2 levels (Houghton et al. 1995). Depending on the magnitude and seasonality of climatic changes, their impacts on agriculture can be either positive or negative (Watson et al. 1998). For example, warmer winter/spring temperatures reduce frost damage and increase growing season length in northern latitudes. Given that high quality wines are generally associated with (Gladstones 1992), 1) low frost damage during mild winters (January, February, March), 2) early and even budburst, flowering and development during warm springs (April, May, June), and 3) low summer (July, August, September) temperature variability during maturation, the question arises: have regional climatic changes helped the California wine industry?

To answer this question, we analyzed daily climatic data (1951-1997, 47 years) from four places (Napa State Hospital, St. Helen, Healdsburg, Santa Rosa) in the premium California wine producing areas of Napa and Sonoma valleys. Here we report results of our analysis as: 1) observed changes in climate, 2) potential causes for the changes, 3) how observed climatic changes impact viticulture in coastal California, 4) predictability of quantity and quality of California vintages.

OBSERVED CHANGES IN NAPA/SONOMA CLIMATE

Consistent with reported global trends, annual average air temperature (Tave) over Napa/Sonoma valleys increased 1.13°C between 1951 and 1997. Nearly all the warming was caused by increases in night minimum temperature (Tmin, 2.06°C/47 yr), with very little change in daytime maximum temperatures (Tmax). As a consequence of the asymmetric warming, the diurnal temperature range (DTR, difference between daily maximum and minimum temperatures) declined by 1.87°C/47 years. Such asymmetric changes in temperature have been widely reported for various regions of the globe, and are presumed to be signatures of global warming. It is the asymmetric nature of climate warming, as will be discussed later, that has significant implications for agriculture in coastal California. Monthly analysis showed the warming trends to be highly seasonal (Figure 1). For example, average spring warming was nearly double that of rest of the year. Similarly, summer DTR showed the largest decline. Trends for Tave are significant at the 5% level for all months except December, while DTR trends are significant in March, May, July, August, September and October. Tmax increased during spring months and declined during summer, but changes are not significant in any month. There were no significant changes in monthly or annual precipitation.

PACIFIC OCEAN AND CLIMATE WARMING

While increased atmospheric CO_2 is considered to be the main reason for recent global warming, on a regional scale changes in atmospheric water vapor (another important greenhouse gas) also play a crucial role. Premium wine producing areas of California are strongly influenced by the maritime weather of the Pacific Ocean. Figure 2 shows the linkage between Pacific Ocean and coastal California climate, with the primary mechanism for the co-variation between ocean and land temperatures being the horizontal transport of water vapor. A strong relation was observed (Figure 3) between Pacific sea surface temperatures along coastal California and coastal dewpoint tempera-

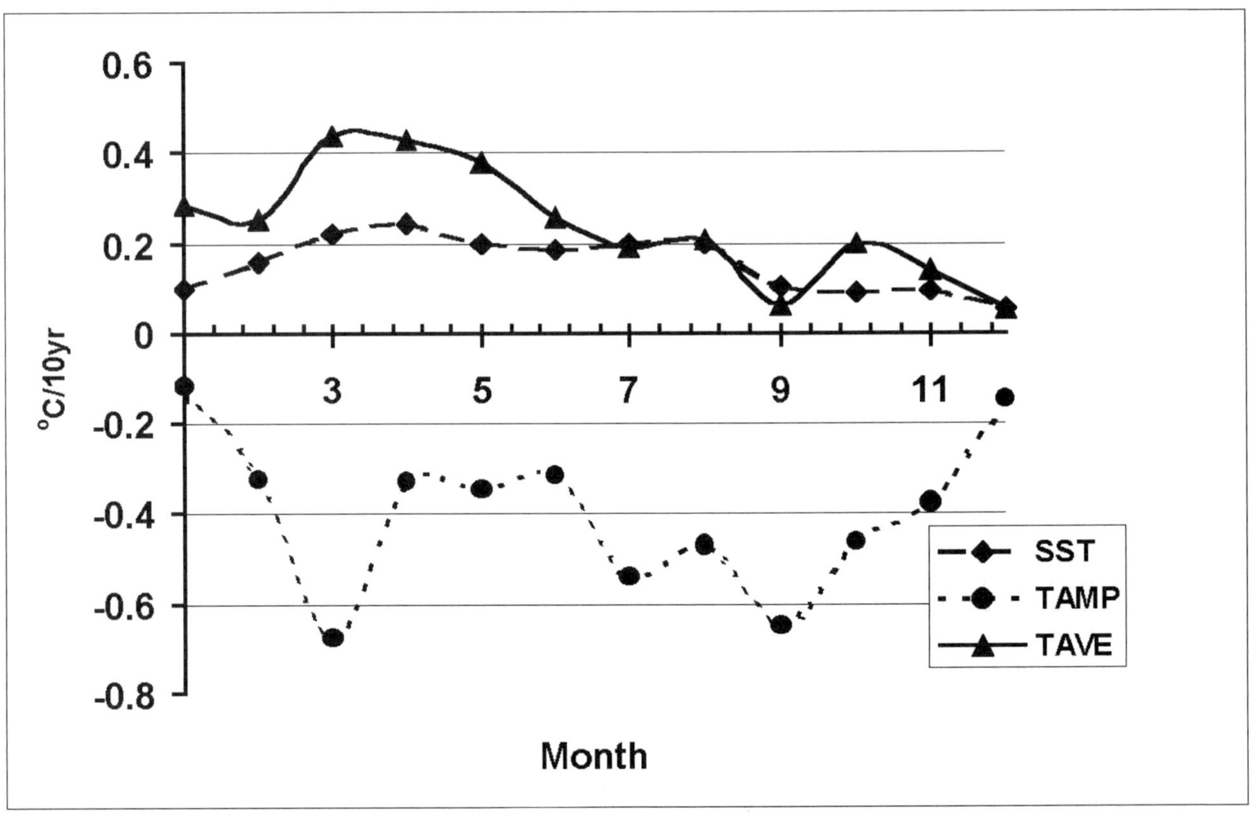

Figure 1. Monthly average temperature (TAVE) ,SST and temperature amplitude (TAMP) trends in Napa/Sonoma valleys, observed between 1950-1997. Higher spring temperatures and reduced DTR have been found to help improve the quality and quantity of vintages.

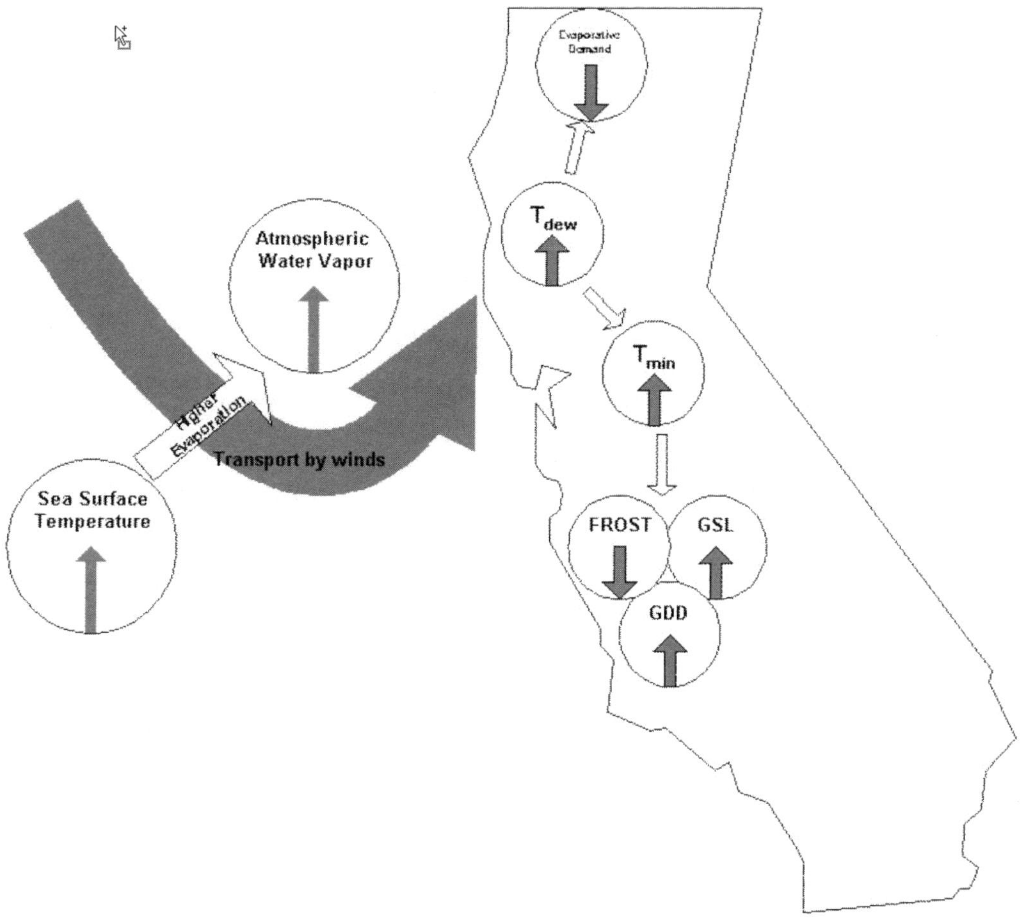

Figure 2. *Pacific climate influences coastal temperatures mainly through transport of water vapor. Changes in atmospheric water vapor, in turn, modify a number of biophysically important variables (frost frequency, evaporative demand, growing season length, GSL and growing degree days, GDD) for viticulture through changes in Tdew and Tmin*

ture (a measure of atmospheric water vapor, observed at San Diego and San Francisco), confirming the mechanisms shown in Figure 2. Pacific sea surface temperatures along the California coast increased by 0.7°C ($p = 0.0030$) between 1951 and 1997, with much of the warming occurring after the well documented shift in pacific climate during 1976-77 (Ebbesmeyer et al. 1990). Similarly, coastal dewpoint temperatures have also increased by 0.9°C/47 yr ($p < 0.001$). As a result of the proposed mechanism (Figure 2), there is a strong relation between SSTs and frost occurrence (Figure 4).

CLIMATIC CHANGES AND VITICULTURE

Reported as annual averages, the observed climatic changes in Napa/Sonoma are modest (1.13°C/47 yr for Tave), but biological consequences can be extensive. For example, the 2.06°C/47 yr increase in Tmin translated to a 71% decline in frost frequency (28 days/yr to 8 days/yr,) and a 25% increase in frost-free growing season length (GSL, 254 days/yr to 320 days/yr, $p<0.001$). Longer growing seasons allow vineyard managers greater flexibility in scheduling various viticultural operations (pruning, harvest, etc.). If the current trends in frost frequency continue, Napa/Sonoma will become a frost-free climate.

Enhanced water vapor shown by increases in Tdew, along with small changes in Tmax, resulted in an estimated 7%/47 yr reduction in growing season (March—October) vapor pressure deficit (VPD, $p=0.042$). We calculated VPD as the difference in vapor pressures at Tdew and Tmax. Based on a high correlation ($r^2=0.92$, Tdew=-0.35 + 0.98Tmin) between observed Tdew and Tmin

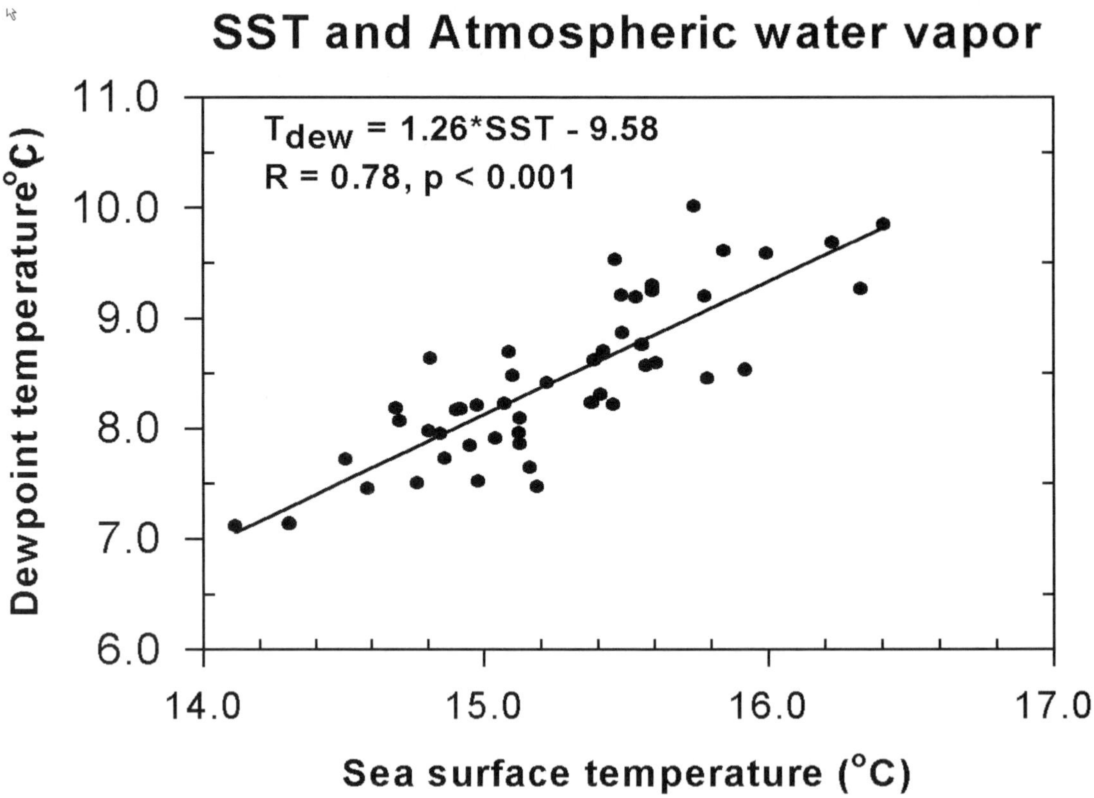

Figure 3. *Observed relation between Pacific sea surface temperatures and coastal dewpoint temperatures from 1951-1997. Eastern Pacific sea surface temperatures increased by 2.0oC following the 1976-77 shift in Pacific climate*

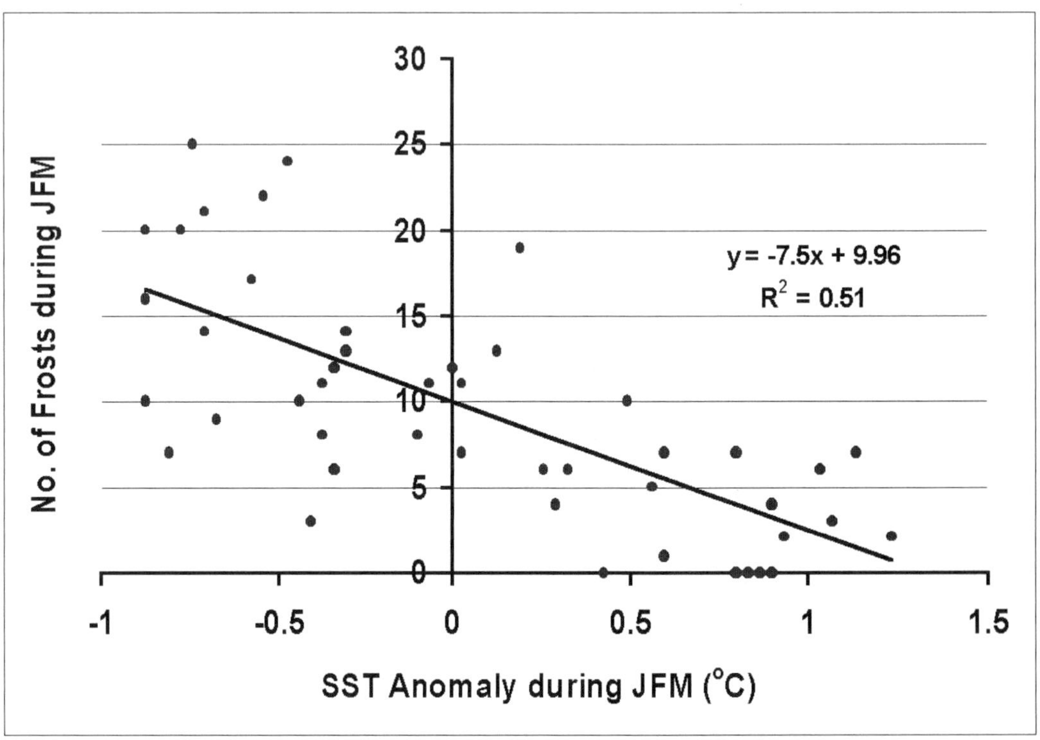

Figure 4. *Observed relation between JFM Sea surface temperature anomaly and the number of frosts*

along the west coast of U.S, we assumed that Tmin = Tdew at Napa/Sonoma (Gaffen and Ross 1999). Lower VPDs reduce evaporative demand and water stress and increase plant growth.

Use of growing degree days (GDD) is quite common in agriculture for predicting various phenological stages (budburst, flowering, crop maturity) and pest/disease outbreaks. Degree days are calculated as accumulated heat units above a base temperature generally taken to be 10°C for grapes. Observed warming trends increased GDD totals and accumulation rates. Between 1951 and 1997, GDD increased 14%, indicating higher sugar accumulation and improved quality (Gladstones 1992). GDD summations also showed that 1600 GDD, the amount required for harvesting grapes for wine making in Napa/Sonoma, were accumulated 20-25 days earlier in 1997 than in 1951. Faster accumulation allows vineyard managers to leave the grapes on the vines until the optimal balance of sugars and acids is achieved.

Temperature variability and temperature extremes, as measured by temperature variability index (TVI, Gladstones 1992), are related to wine quality. TVI = sum ((TDmax – TDmin) + (TMmax – TMmin)), where TD and TM represent daily and monthly values between March and October. Low TVIs favor high-quality wines. In Napa/Sonoma, the TVI declined from 36.1 in 1951 to 31.4 in 1997. TVI values under 30 indicate that any variety of table wine may be produced. Locations within a vineyard that maintain higher Tmin and low DTR as a result of soils or topography are regularly associated with high quality wines. Observed climatic changes (Figure 1) are likely to have similar positive influence on wine quality for entire vineyards.

CLIMATE CHANGE AND CALIFORNIA VINTAGES

Wine quality ratings by Sotheby (Stevenson 1997), available for California wines from 1963-1996 and dominated by north coast wines, increased by 0.22 points/yr (Figure 5, p=0.022). Wine ratings produced immediately after production, such as the Sotheby (Stevenson 1997) and Wine Spectator ratings (Laube 1996) are more indicative of climatic influences than are ratings updated on a yearly basis. However, annually updated ratings, such as the Wine Advocate vintage

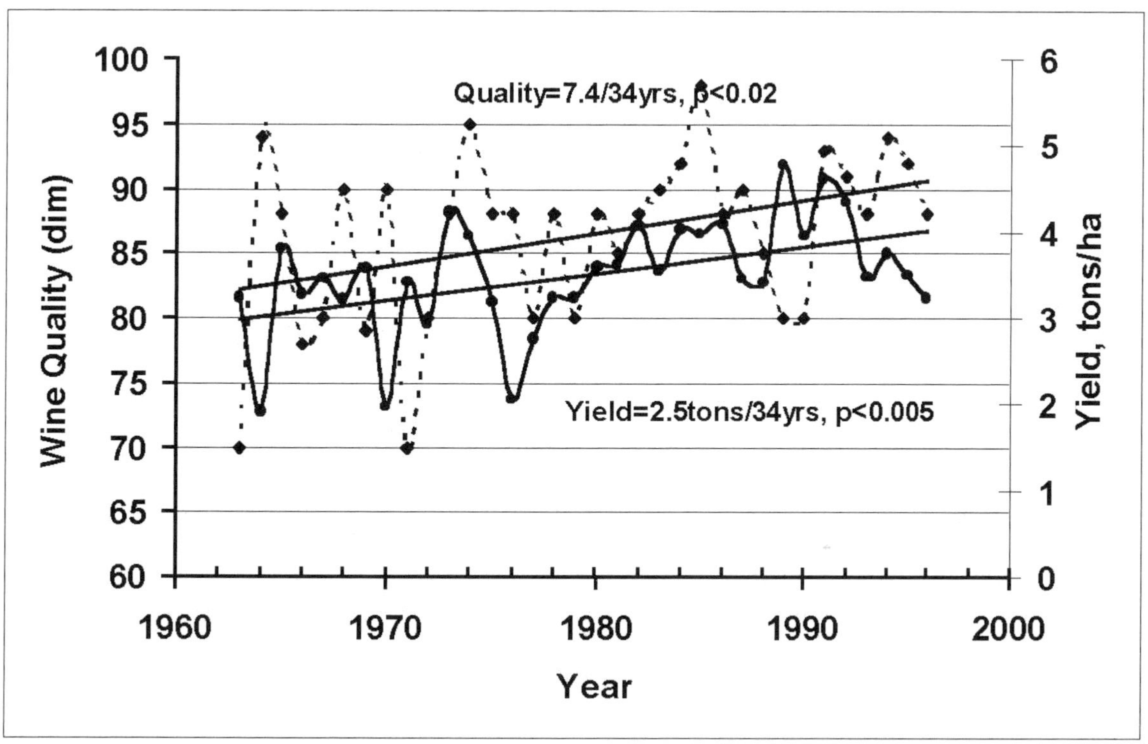

Figure 5. *Trends in Napa valley wine quality and yields. After the 1976-77 shift, consistently better vintages followed warmer temperatures during winter and spring month*

Table 1. *Climatic variables important for vintage quantity and quality (Gladstones 1992): 1951-1997 increases and mean (standard deviation, SD) before and after the 1976-1977 regional Pacific climate shift . All differences between periods were significant at the 1% level (t-test).*

Parameter	1951-1997	1951-1976	1977-1997
changes	mean (SD)	mean (SD)	
Winter Tmin (°C)	2.39	3.74 (0.88)	5.06 (1.16)
Spring Tmin (°C)	2.44	8.04 (0.63)	9.42 (0.77)
Summer DTR (°C)	–3.14	19.0 (0.85)	17.5 (0.98)
Number of Frosts/year	–20	22 (9)	10 (9)
Frost-free GSL (days)	65	272 (18)	311 (29)
Growing Degree Days	240	1714(102)	1830(112)
VPD (kPa)	–0.159	2.41 (0.12)	2.29 (0.15)

charts for north coast California wines (R. Parker, "The Wine Advocate's vintage guide, 1970-1997" (http://www.winetech.com/html/vintchrt.html, 1997)) show similar trends, as do all long-term datasets of California wine quality. Since long-term current year ratings were not available specifically for north coast wines and because the Laube ratings are on a limited 1-5 scale, we used the Sotheby system (1-100 scale) in this study. Among the variables listed in Table 1, the decline in frosts was significantly correlated with the increase in wine ratings (r^2=0.41, Figure 6). A possible explanation for such a relation could be that frosts damage buds on the vine, delaying subsequent phenological events lead-

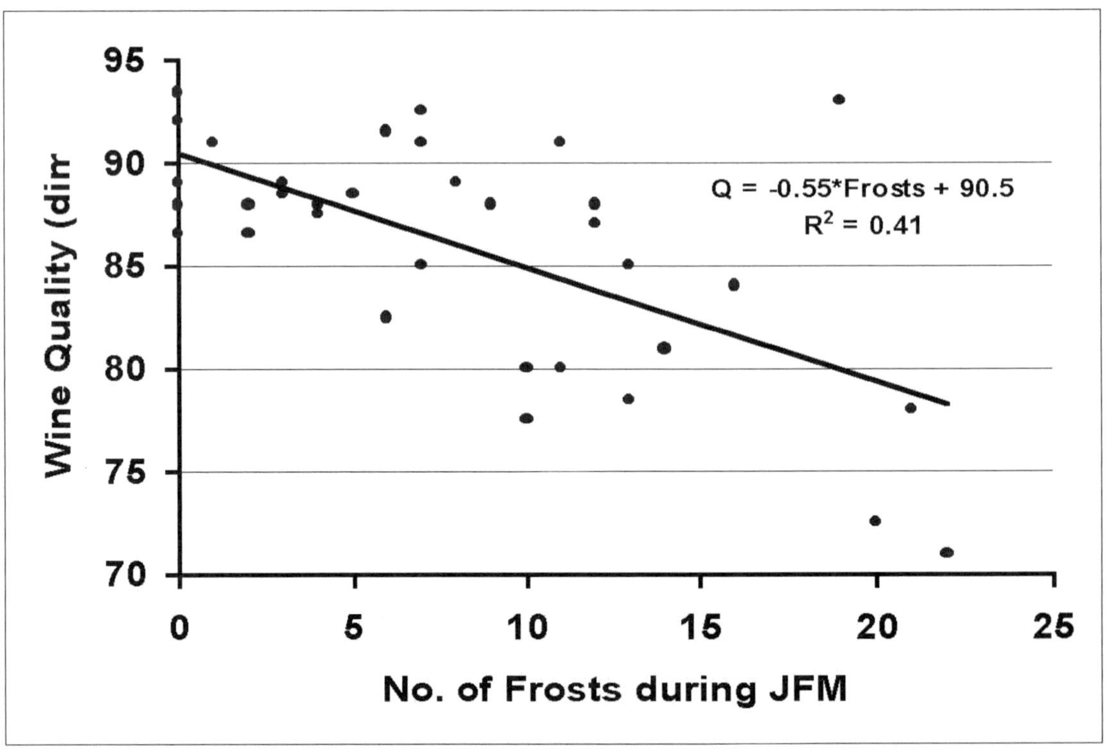

Figure 6. *Observed relation between wine quality (100 point scale) and the number of frosts during January, February and March. Since SSTs are persistent for 6-12 months, such a relation is useful for predicting vintage quality months in advance*

ing to uneven maturity and poor wine quality. Years with low frost also showed warm springs and low summer DTR, both of which promote wine quality (Coombe 1987). Wine ratings have a major impact on wine value. For example, analysis of Wine Spectator data showed that for 1995 Napa wines, a rating increase of 10 points translated to a 220% per bottle price increase. Grape yield grew (34%, Figure 5) from 7.3 ton/ha to 9.8 ton/ha from 1963-1996 (NAPA County crop reports), suggesting that high yield and high quality are not mutually exclusive. Increases in spring Tmin and decreases in summer VPD are able to explain more than 56% ($r2=0.56$, $p<0.001$) of the upward trend in Napa yields. In Napa valley, consequent with increasing quality and quantity, the value of the grape crop increased from $640/ha in 1963 to $19,600/ha in 1996 (NAPA County crop reports).

All parameters shown in Table 1 exhibit a pronounced increase over the 1951-1997 record, especially since the 1976-1977 shift in Pacific climate (Ebbesmeyer et al. 1990). Before 1976, a number of vintages had poor ratings associated with frequent winter frosts. However, after 1976, ratings as well as yields steadily improved with the near disappearance of frosts. Warmer SSTs after 1976, coupled with low sea level pressures enhancing the horizontal transport of water vapor (Figure 2) during 1977-1988 (Trenberth and Hurrell 1994), resulted in an unprecedented string of years with high quantity and quality. Similar warming trends accompanied advancement in phenological events and better sugar to acid ratios in Bordeaux, leading to higher wine quality over the last two decades (Jones 1999).

IMPLICATIONS FOR THE FUTURE

Unfortunately, along with the positive effects from recent climatic changes, there could be future negative impacts for the wine industry. Although Napa/Sonoma humidity levels are currently optimal (Gladstones 1992), trends toward increasing humidity and air temperature suggest that in the future, the risk of fungal and vector borne disease outbreaks may increase. Pierce's disease, a fatal bacterial (*Xylella fastidiosa*) disease transmitted by sharpshooter beetles (Cicadellidae family) and apparently limited by frost occurrence, is increasing in Napa/Sonoma. Climatic change may therefore require increased investment in pesticide application and disease-resistant rootstock.

Finally, the strong coupling between Pacific SSTs and coastal land temperatures suggests a possibility for predicting future vintages. For example, warmer winter SSTs, because of their persistence, also lead to warmer spring temperatures. This suggests that given winter SSTs, reasonable prediction of next-year wine quality may be possible. Warmer winter SSTs, on average, lead to higher quality wines from coastal California.

REFERENCES

Coombe BJ. 1987. Influence of temperature on composition and quality of grapes. Acta Hort., 206: 23-35.

Ebbesmeyer CC, Cayan DR, McLain DR, Nichols FH, Peterson DH, Redmond. KT.1976. Step in the Pacific Climate: Forty environmental changes between 1968-1975 and 1977-1984. Proc. Seventh Annual Pacific Climate (PACLIM) Workshop. Eds. JL Betancourt, Tharp VL, Interagency ecological studies program technical report, Dept. of Water Resources, Sacramento, CA, pp.115-126.

Gaffen DJ, Ross RJ. 1999. Climatology and trends of U.S. surface humidity and temperature. J. Climate, 12: 811-828.

Gladstones, J. 1992. Viticulture and Environment. Winetitles, Adelaide, Australia.

Houghton JT, Filho LGM, Callander BA, Harris N, Kattenberg A, Maskell K. 1995. Climate change 1995: The science of climate change. Intergovernmental Panel on Climate Change, Cambridge University Press, Cambridge, U.K.

Jackson DI, Lombard PB. 1993. Environmental and management practices affecting grape composition and wine quality: A review. Am. J. Enol. Vitic., 44: 409-430.

Jones, GV. 1999. Relationships between grapevine phenology, composition and quality for Bordeaux, France. Arboreta Pheanologica, 42: 3-7.

Laube, J. 1996. Wine Spectator's California wine. Wine spectator press, New York, NY.

Stevenson, T. 1997. The new Sotheby's wine encyclopedia. DK Publishing Inc., New York, NY.

Trenberth KE, Hurrell JW. 1994. Decadal atmosphere-ocean variations in the Pacific. Climate Dynamics, 9: 303-319.

Watson RJ, Zinyowera MC, Moss RH. 1998. The regional impacts of climate change: An assessment of vulnerability. Intergovernmental Panel on Climate Change, Cambridge University Press, Cambridge, U.K.

Impact of Regional and Global Climate Variability on the Production/Yield of Major Agricultural Crops in India

Krishna Kumar[1], *N.R.Deshpande*[1], *K. Rupa Kumar*[1] *and M.A. Cane*[2]
[1]Indian Institute of Tropical Meteorology, India
[2]Columbia University, USA

SUMMARY

Interannual variability of Indian summer monsoon rainfall is known to be strongly linked to the slowly varying boundary forcings such as the variations in snow cover, snow ice over the land and the sea surface temperatures over the global oceans in general and the Pacific Ocean in particular. Occurrence of large-scale droughts and floods over India generally coincide with the warm and cold phases of El Niño-Southern Oscillation (ENSO) respectively. Examination of summer total food grain production in India during the last five decades or so, shows significant relation with the total monsoonal rainfall in the country. In addition, many crops show significant relation with the local/regional rainfall and temperature variability at different crop stages. While technological innovations have dramatically improved the agricultural production, climate-related anomalies still continue to cause severe hardship to a large population in South Asia. Advance information on the possible climatic impacts on food production is crucial to ensure proper food security in the region. Keeping in view the strong links between food-grain production and the monsoon rainfall and also the monsoon rainfall and ENSO, an attempt is made here to examine in detail the data (1958-1997) on production/yield of major Indian crops such as, cereals, pulses and oil seeds both in terms of their variability and their linkages with regional and global climatic factors. Apart from these broad groups of food grains, analyses are also carried out separately for some important crops like rice, wheat, groundnuts, sorghum, etc. for the whole of India and on a smaller regional scale. Considerable efforts have been put in to generate long, homogeneous and well-representative data sets on food-grain production, reflecting its spatial and temporal variability. The analyses show a strong ENSO signal in most of the crop indices. However, the ENSO signal in winter wheat seems to be negligible. In order to get a clear climate signal in various crops, the production and yield data are worked out separately for the rainfed and irrigated areas of the country. Possibility of predicting the production/yield of some of the crops using regional climatic parameters as well as the ENSO forecasts from coupled models through various crop stages will also be explored.

ENSO Influences on Rice Production in Sri Lanka

Lareef Zubair[1] and Suganda Somasundea[2]
[1]International Research Institute for Climate Prediction, USA
[2]University of Peradeniya, Sri Lanka

INTRODUCTION

The climate in Sri Lanka is tropical with only a modest variation in temperature in different seasons. Rainfall is distributed bimodally with the main rainy season being in October-December and a second rainy season in April-May. Rice is the principal food crop in Sri Lanka. Its production has been increasing over the past century but is still below the island-wide demand. Rice is cultivated during two seasons principally in the Northern and Eastern plains of the country. The principal cultivation season, October to March, is known as "Maha". The second cultivation season, known as "Yala", is from April to September. There is some variability in the start and duration of cultivation within the island (Yoshino and Suppiah, 1983). The rice production during Yala is approximately half of that during Maha (Department of Census and Statistics, Sri Lanka, 1991).

From 1950 to 1990, there has been a steady increase in rice production in Sri Lanka. This rise has been attributed to increased area under cultivation, increased irrigation, improved seed varieties, increased fertilizer application and higher purchase prices for rice. Leaker (1984) reports that technological factors such as transplanting rice, fertilizer, and irrigation have had greater influence on rice yields rather than climatic variables. However, rainfall variations are a principal factor that affects the inter-annual variability of rice production (Yoshino et al., 1983).

The relationship between El Niño and La Niña events and the variability of rainfall in Sri Lanka has been recognized for over a decade (Rasmussen and Carpenter, 1983, Ropelewski and Halpert, 1989, Suppiah, 1997, Kane, 1998). This paper seeks to identify the rainfall deviations during the El Niño and La Niña events, to report on the variability of rice production, to interpret the trends and anomalies in agricultural production in the context of climatic variability and to assess the extent to which climate predictions based on ENSO could be useful for rice farming and agricultural management in Sri Lanka.

RAINFALL AND RICE PRODUCTION IN SRI LANKA

Sri Lanka can been divided into three principal agro-climatic zones (Figure 1). The wet, intermediate and dry zones have been demarcated based on hydrology, meteorology, soils and vegetation

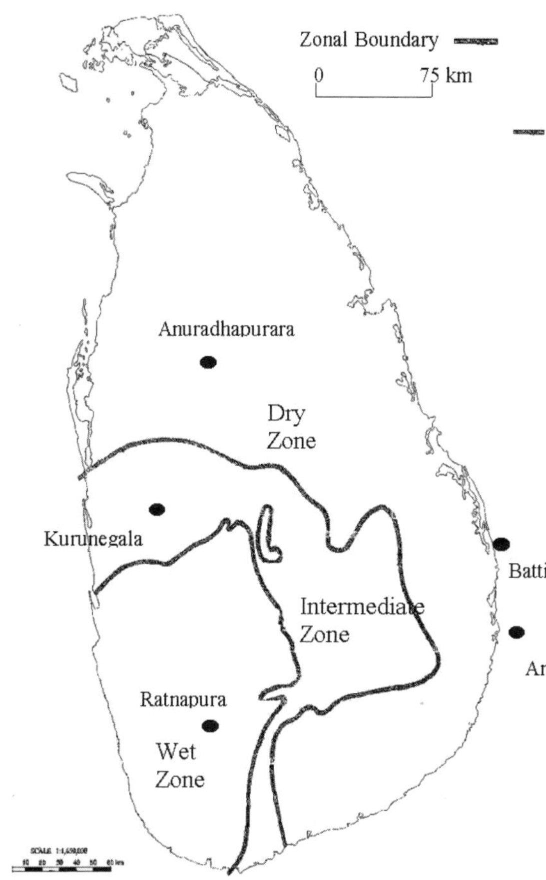

Figure 1. *The climatic zones of Sri Lanka and the rainfall stations used in this study.*

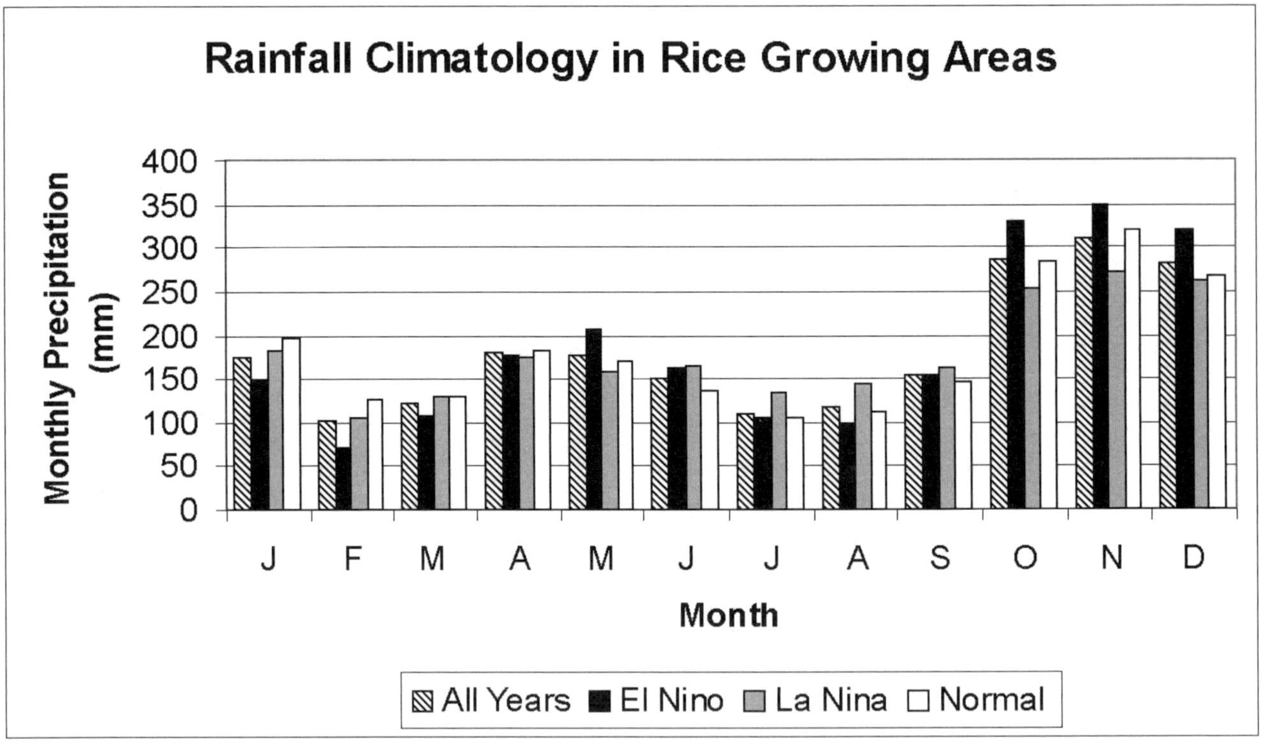

Figure 2. *The monthly climatology of rainfall in the rice producing areas based on data from 1869 to 1998. A climatology for El Niño, La Niña and normal years has been constructed by compositing the rainfall data for different months in which sea surface temperatures in the NINO3.4 area was greater than 0.4, less than -0.4, or otherwise respectively.*

(Somasekeram et al., 1988). The wet zone has an annual rainfall of more than 2500 mm; in the intermediate zone, the annual rainfall is between 1750 and 2500 mm; and the dry zone receives less than 1750 mm a year. An district-wise analysis of rice production (Department of Census and Statistics, Sri Lanka, 1993) shows that the contribution to the national rice production from the dry, intermediate and wet zone districts are in the ratio of 3:1:1. Three rainfall measuring stations of the Department of Meteorology of Sri Lanka were selected in the dry zone rice at Anurhadhapura, Batticaloa and Amparai; Kurunegala from the intermediate zone; and Ratnapura from the wet zone. The average of the rainfall in these five stations is used this paper as a representative index. The monthly climatology of rainfall from these stations is shown in Figure 2.

The trend of the rice production in Sri Lanka for both Maha and Yala from 1952 to 1997 (Department of Census and Statistics, Sri Lanka, 1991, 1998) was estimated from a five-year running mean. Obtaining the running mean approximately averages out the year-to-year fluctuations in production. The seasonal production anomaly is computed as the difference between the actual production and the trend.

The trend for both Maha and Yala rice production (Figures 3 and 4) shows a steady increase from 350,000 MT to 1,600,000 MT from 1952 to 1983. Cultivation was disturbed in the northern regions of the island after 1983 due to civil disturbances. A declining trend in Maha rainfall contributed to the declining production trends. The mean Maha rainfall from 1983-1997 was 1110 mm compared with 1233 mm from 1952-1982. However, the declining rainfall trends was not the only cause of declining production; a similar analysis for Yala rainfall shows that there was a decline in production from 1983-1997 although mean rainfall dropped by less than 2%.

The El Niño and La Niña years were identified following Trenberth (1997): El Niño and La Niña conditions persist when the average sea surface temperature in the NINO3.4 area of the equatorial Pacific Ocean (5°N–5°S, 120–170°W) exceeds a threshold of +/– 0.4°C. Positive anomalies are

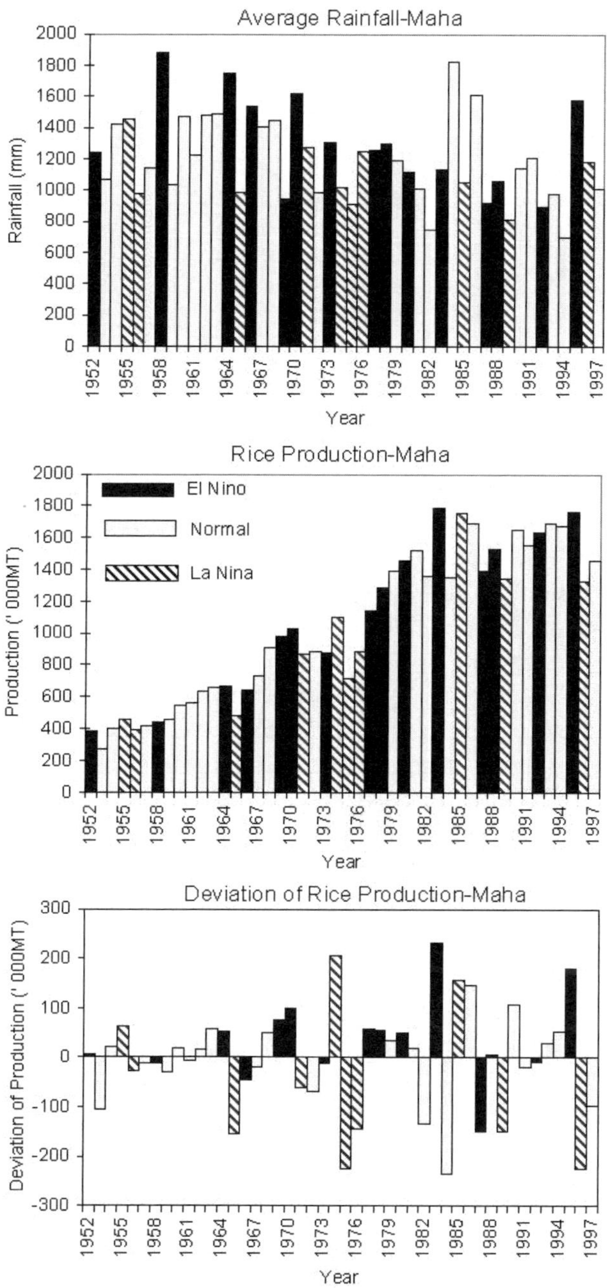

***Figure 3.** Top panel: the average rainfall in rice producing areas in Sri Lanka during the Maha season. The Maha in a particular year starts in the previous October and continues to the next March. The middle panel shows the variation of rice production during Maha. The bottom panel shows the rice production anomaly for each Maha computed as the difference between the production and the five-year running mean.*

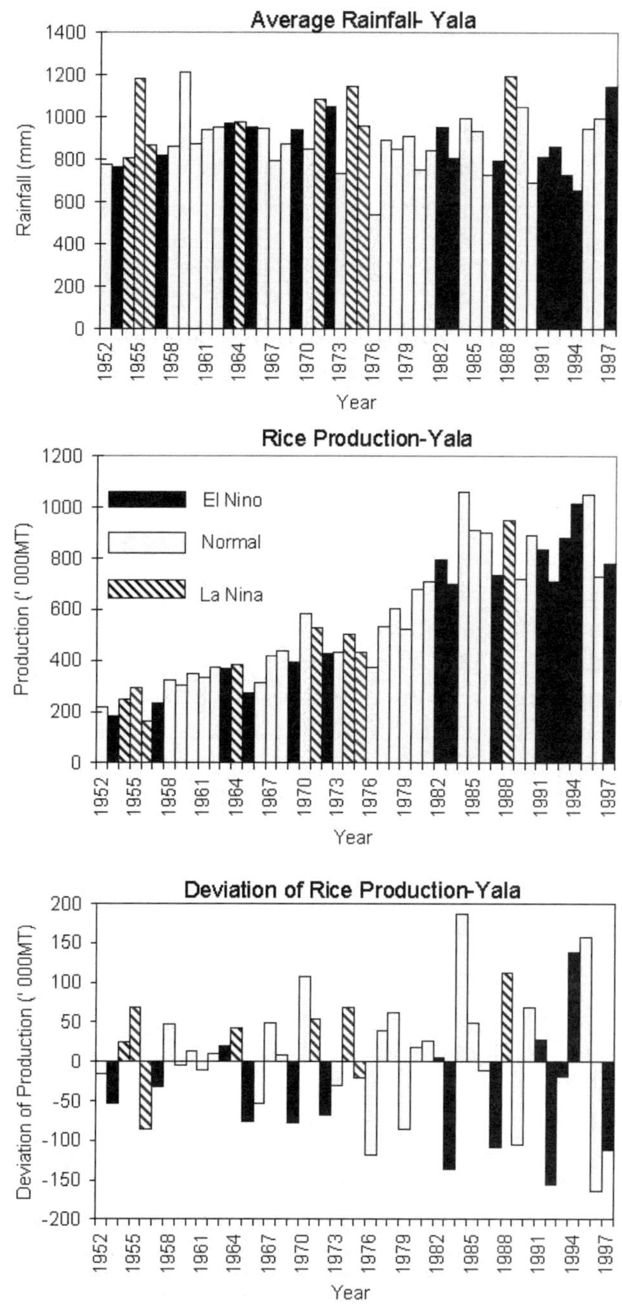

***Figure 4.** Top panel: the average variation of rainfall in Sri Lanka during the Yala season computed as the total rainfall from April to September. The middle panel shows the variation of rice production in Sri Lanka during Yala. The bottom panel shows the rice production anomaly for each Yala computed as the difference between the production and the five-year running mean.*

Table 1. *Maha and Yala seasons that are associated with El Niño and La Niña conditions in the NINO34 region of the tropical Pacific Ocean.*

Maha		Yala	
El Niño	La Niña	El Niño	La Niña
1951/52, 1957/58, 1963/64, 1965/66, 1968/69, 1969/70, 1972/73, 1976/77, 1977/78, 1979/80, 1982/83, 1986/87, 1987/88, 1991/92, 1994/95	1954/55, 1955/56, 1964/65, 1970/71, 1973/74, 1974/75, 1975/76, 1984/85, 1988/89, 1995/96	1953, 1957, 1963, 1965, 1969, 1972, 1982, 1983, 1987, 1991, 1992, 1993, 1994, 1997	1954, 1955, 1956, 1964, 1971, 1974, 1975, 1988

associated with El Niño and negative with La Niña. Maha and Yala seasons were identified with El Niño and La Niña when at least three months of the season had high SST anomalies (see table 1). A precipitation climatology for El Niño, La Niña and normal months has been constructed by composting rainfall in months chosen according to the above thresholds for NINO3.4 sea surface temperatures.

VARIABILITY IN MAHA

The Maha rainfall varied between a minimum of 702 mm in 1993/94 and a maximum of 1885 mm in 1957/58 with an average of 1220 mm. The Maha rainfall was 6% higher during El Niño years than during normal years. Both rainfall and rice production were higher in 10 out of the 15 El Niño years (Figure 3). The rainfall in the El Niño years of 1957/58, 1965/66 and 1972/73 was so high that floods were reported in some areas and production dropped (Central Bank Reports, 1958, 1966, 1973). The decreases in production in the El Niño Maha seasons of 1986/87 and 1991/92 are due to a decline in area under cultivation and a decline in rainfall.

During La Niña years, the Maha rainfall was 14% lower than average. There was a drop in rainfall in 7 out of 10 La Niña events. The exceptions were 1954/55, 1973/74 and 1984/85. In 1973/74 La Niña Maha season there was an increase in production notwithstanding the normally lower rainfall as the state reduced fertilizer prices and actively encouraged an increase in the area under cultivation. During 1984/85, farmers increased the acreage under cultivation due to poor harvests in the previous Maha (Central Bank, 1985).

In normal years, there were significant declines in Maha production in 1952/53, 1971/72, 1981/82, 1983/84 and 1996/97. The latter four Maha seasons had cold NINO3.4 conditions but the anomaly was not sufficiently high to be declared as a La Niña.

There were significant increases in production in the normal seasons of 1962/63, 1966/1967, 1985/86, 1991/1992 and 1992/93. Of these, the last two seasons were associated with moderately warm NINO3.4 temperatures. The remaining four declines were related to declines in rainfall unrelated to El Niño or La Niña conditions.

VARIABILITY IN YALA

The Yala rainfall had a maximum of 1212 mm in 1959 and a minimum of 541 mm in 1976 and an average of 901 mm. Rainfall dropped by approxi-

Table 2. *Average seasonal rainfall for Sri Lanka in the rice farming areas during normal, El Niño and La Niña conditions for the period from 1952 to 1997.*

Seasonal Rainfall (mm)	Normal	El Niño	La Niña
Maha (October to March)	1220	1290 (+6%)	1048 (-14%)
Yala (April to September)	862	832 (-4%)	992 (+15%)

Table 3. *Summary of variation in average rainfall in rice farming areas, rice production anomalies and yield anomalies for Maha and Yala during El Niño and La Niña events for the period from 1952 to 1997.*

Season	Anomaly	El Niño Seasons	La Niña Seasons
Maha	Rainfall	Rise in 10 out of 15	Drop in 7 out of 10
	Production	Rise in 10 out of 15	Drop in 7 out of 10
	Yield	Rise in 12 out of 15	Drop in 7 out of 10
Yala	Rainfall	Drop in 8 out of 14	Rise in 6 out 8
	Rice Production	Drop in 10 out of 14	Rise in 6 out of 8
	Yield	Drop in 7 out of 14	Rise in 5 out of 8

mately 4% on average El Niño years. Of the 14 El Niño events, the rainfall dropped in 8 of them. Of the exceptions, the average rainfall in the El Niño year of 1963, 1972 and 1982 is higher only on account of high rainfall at the wet and intermediate zone stations.

Although there was a decline in rainfall in 8 of the 14 El Niño events, there was a decline in production in 10 of 14 El Niño years (Figure 4). This may be due to 60% of the rice production being from the dry zone where there was a decline in rainfall in 10 of 14 El Niño years. The exceptions were 1963, 1982, 1991, and 1994.

The Yala rainfall during La Niña years increased by 15% compared to the overall average. In 6 out of 8 La Niña seasons, the Yala rainfall decreased with the exceptions of 1954 and 1956. In these exceptional years, the decline was modest and the NINO3.4 dropped below the threshold only for part of the season. There was a rise in Yala production in 6 out 8 La Niña years with 1956 and 1975 being the exceptions. Although, the decline in rainfall was modest in 1956, civil disturbances contributed to a drop in production. Although the rainfall was slightly above average in 1975, the application of fertilizer declined by 40% compared to the previous year subsequent to the rapid escalation of petroleum prices (Central Bank, 1975).

Of the normal years, there was a production anomaly greater than 5% of the average production during 1958, 1967, 1978, 1984, 1990, and 1995. Out of these years, the rainfall was significantly higher only in 1984 and 1995. There were cold NINO3.4 anomalies in 1967, 1984 and 1995, which was not sufficient to have been classified as a La Niña. The increase in Yala production in 1984 is partially due to farmers compensating for a poor rice harvest in the previous Maha by cultivating rice instead of other crops (Central Bank, 1984).

During 1966, 1979, 1983, 1989 and 1996, there were significant declines in production. Out of these years, civil disturbances affected cultivation in 1979, 1983 and 1989 and drought was reported in 1966 and 1989 in the dry zone although high rainfall in the wet and intermediate zones resulted in floods (Central Bank, 1966, 1989). 1966 and 1996 were also seasons, which had weak El Niño like conditions that may have been used to anticipate a deficit.

DISCUSSION

In El Niño years, the Maha rainfall in Sri Lanka is increased and Yala rainfall declines. In La Niña years, there is a rise in Maha rainfall and a decline in Yala rainfall. We may conclude that the overall impact of an El Niño event is to increase the Maha production and to decrease that of Yala. Similarly, the impact of a La Niña is to increase the Yala production and decrease that of Yala. The above prognosis is shown to hold in 33 out of 47 seasons with extreme ENSO conditions in the 92 seasons in the period under review. Of the 24 instances where this prognosis was incorrect, factors such as floods, drastic anomalies in previous seasons, acreage that is sowed, fertilizer and rice prices and subsidies could have been used to correct the prognosis in 9 seasons. Of the remaining, 15 seasons, allowance climate change and ENSO variations within seasons may correct for half of the erroneous interpretation.

Of the 47 seasons with ENSO extremes, production anomalies exceeded 5% of the average production in 31 occasions. In comparison, of the 45 seasons without extreme ENSO conditions, production anomalies exceeded 5% of the seasonal average production on 23 occasions. Of these 23 seasons, a lower threshold for declaring an onset could have corrected for the forecasts for 11 seasons.

ACKNOWLEDGEMENTS

Data used in this study was obtained from the Departments of Agriculture and Meteorology in Sri Lanka. Detailed comments by Jim Hansen are gratefully acknowledged. LZ thanks Chet Ropelewski, Gamini Seneviratne, Jennifer Phillips and Ranjith Punaywardhene for useful discussions. Part of the work reported here was carried out at the Institute of Fundamental Studies, Kandy, Sri Lanka. LZ was supported by a fellowship at the International Research Institute for Climate Prediction administered by the University Consortium for Atmospheric Research.

REFERENCES

Department of Census and Statistics (1991), Statistical Abstract of the Democratic Socialist Republic of Sri Lanka, Colombo.

Central Bank (1950-1997). Annual Reports, Central Bank of Sri Lanka, Colombo.

Leaker, Ann. (1984), An Investigation of the factors affecting Paddy yields from two districts, Journal of the National Science Council of Sri Lanka 12 (1): 71-92.

Kane, R.P. (1998), ENSO relationship to the rainfall of Sri Lanka, International Journal of Climatology, 18 (8): 859-872.

Rasmusson, E.M. and Carpenter, T.H. (1983), The relationship between eastern equatorial Pacific sea surface temperature and rainfall over India and Sri Lanka, Monthly Weather Review, 110, 354-383.

Ropelewski, C. F., and Halpert, M.S. (1989), Precipitation patterns associated with the high index phase of the Southern Oscillation, Journal of Climate, 2, (3), 268:284.

Somasekeram, T., Perera , L.A.G., Perera, M.P., De Silva, B.G., Karunanayake, M.M, and Epitawatta, D.S., (1988). *National Atlas,* Survey Department., Colombo, pp. 30-37.

Suppiah, R. (1997) Extremes of the southern oscillation and the rainfall of Sri Lanka, International Journal of Climatology, 17 (1):87-101.

Suppiah. R. and Yoshino, M.M. (1986), Some agroclimatological aspects of rice production in Sri Lanka, Geographical Review of Japan, 59 (Ser. B) (2), 137-153.

Trenberth, K.E, 1997, The definition of El Niño, Bulletin of the American Meteorological Society, 78 (12): 2771-2777.

Yoshino, M.M., Ichikawa, T., Urushibara, K., Nomoto, S., and Suppiah. R. (1983). Climate fluctuation and its effect on paddy production in Sri Lanka, Climatological Notes, Tsukuba, Japan, 33: 9-32

Sensitivity of Winter Wheat Yield in Denmark to Climate Variability and Climate Change

Jørgen E. Olesen
Danish Institute of Agricultural Sciences, Denmark

INTRODUCTION

Changes in the global climate will affect temporal patterns of temperature and rainfall at the regional level, and this will have important consequences for crop production (Parry, 1990). Several studies have examined the sensitivity of wheat yield potential to temperature changes using simulation models (e.g. Stockle et al., 1992). Comparisons of wheat models and of their sensitivity to climate change have also been performed (e.g. Wolf et al., 1996). Recently criticism has been raised on the use of such models for climate change impact assessment, because of the often poor results from comparison of simulated with observed yields from agronomic field experiments (Landau et al., 1998).

This study describes the use of a crop simulation model to evaluate the effect of temperature change on yield of winter wheat in Denmark. The model was evaluated by comparison with yields obtained at agricultural research stations.

MATERIALS AND METHODS

The CLIMCROP model describes the effects of climate on crop production and yield of winter wheat (Olesen et al., 2000). The model includes a crop model and a soil model that is updated in daily time steps.

The water balance model is based on a reservoir approach (Olesen and Plauborg, 1995) with three reservoirs: interception, root zone and sub-zone reservoirs. Both the maximum root depth and the capacity for plant available water are assumed to vary between soil types. Soil temperature is simulated using a simple resistance approach where air temperature and solar radiation drive the changes in soil temperature (Olesen et al., 2000).

The phenology submodel is identical to that of the SIRIUS wheat model (Jamieson et al., 1998) with a few exceptions (Olesen et al., 2000). The leaf area growth from emergence to 1st March is described as a linear function of soil temperature sum. The development of green area index during the growing season is described by a logistic equation in soil or air thermal time prior to start of senescence. From this point green area index declines linearly with thermal time.

Dry matter production is calculated using a radiation use efficiency which is restricted by low temperatures and reduced transpiration (Aslyng and Hansen, 1982). The crop is harvested at physiological maturity, and yield is based on 15% grain moisture.

Data on harvested grain yield in winter wheat were obtained during 1971 to 1997 from seven agricultural research stations in Denmark. The observed yield data represent yields from experimental plots, which were treated according to normal agricultural practice in the region (Olesen, 1991). The sites represent different soil types, and some of the observations represent irrigated treatments. The stations were grouped into four soil type classes, and the soil water capacities of the different soil types were estimated from measurements of soil texture and soil water retention curves (Hansen, 1976). The observed yields were detrended using the same linear time trend for all sites. These detrended yields were corrected to 1990 level. The simulation model was used to estimate grain yield of both irrigated and non-irrigated crops for all years and sites. A standard sowing date of 20 September was used in all cases due to incomplete information on crop management.

The effect of changes in mean temperature was analyzed using sets of 200 years of climate data generated by the LARS-WG stochastic weather generator (Semenov and Barrow, 1997). The LARS-WG was calibrated using daily observed climate data from Ødum (56°18′N, 10°08′E) for

1961 to 1997. The change in mean temperature was varied from –2°C to 5°C in steps of 0.5°C. The response of simulated yield was calculated for the four different soil types also used in the validation study. Both the mean yield and coefficient of variation of yield were calculated for each change in climatic parameters on the basis of 200 individual years of model runs.

The results of the change in temperature can be affected by the sowing date used. All scenarios were therefore run for sowing dates varying from 1 September to 30 October.

RESULTS AND DISCUSSION

Figure 1 shows the simulated versus detrended grain yields for the seven agricultural research stations. The results from the different stations are grouped according to the estimated soil water capacities of the different sites. Most of the simu-

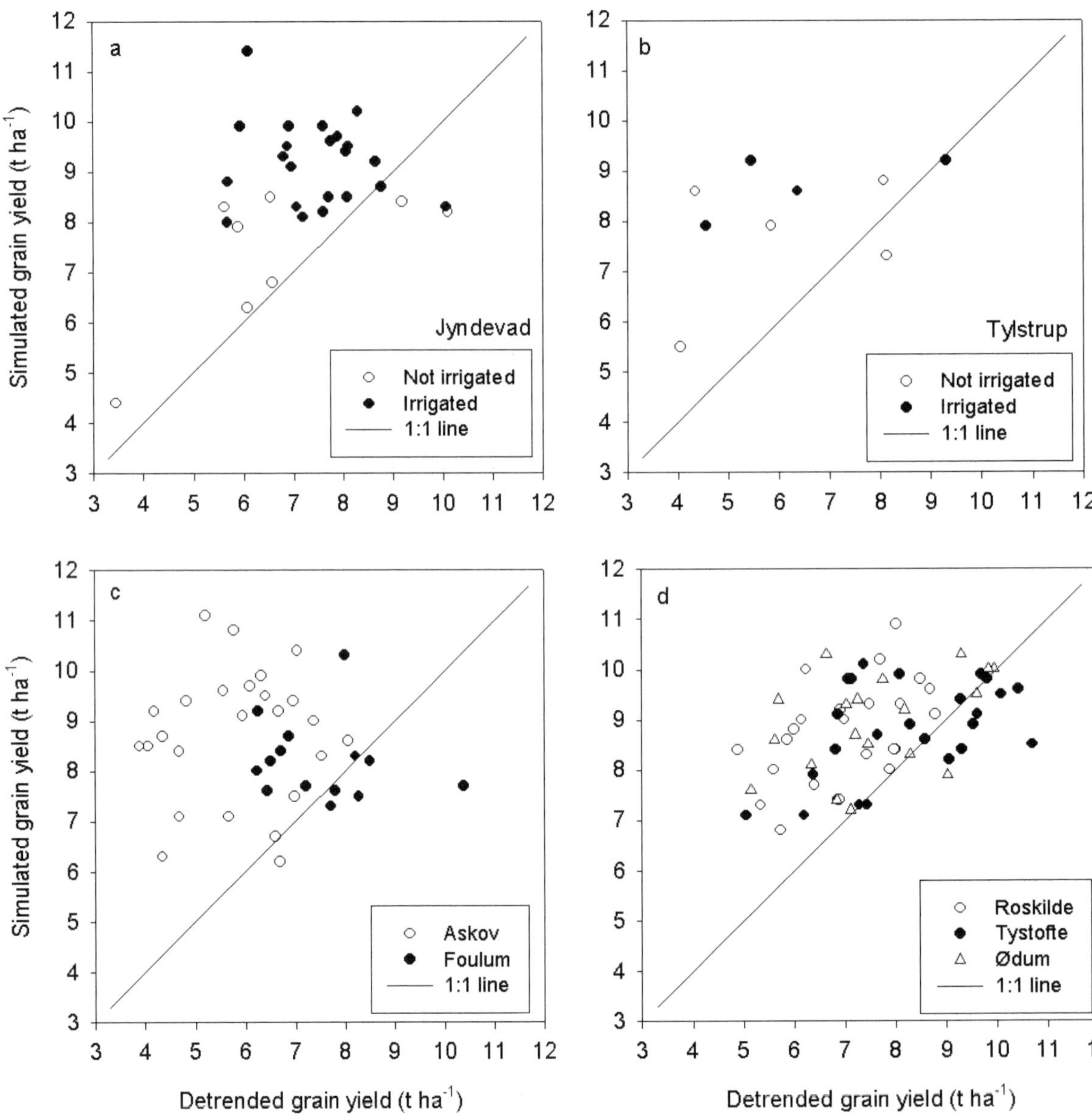

Figure 1. *Simulated versus measured grain yield detrended to 1990 level in winter wheat at seven Danish agricultural research stations for the period 1971 to 1997, grouped according to their estimated soil water-holding capacity; a: Jyndevad (coarse sand), b: Tylstrup (fine sand), c: Askov and Foulum (loamy sand), d: Roskilde, Tystofte and Ødum (sandy loam). Source: Olesen et al., 2000.*

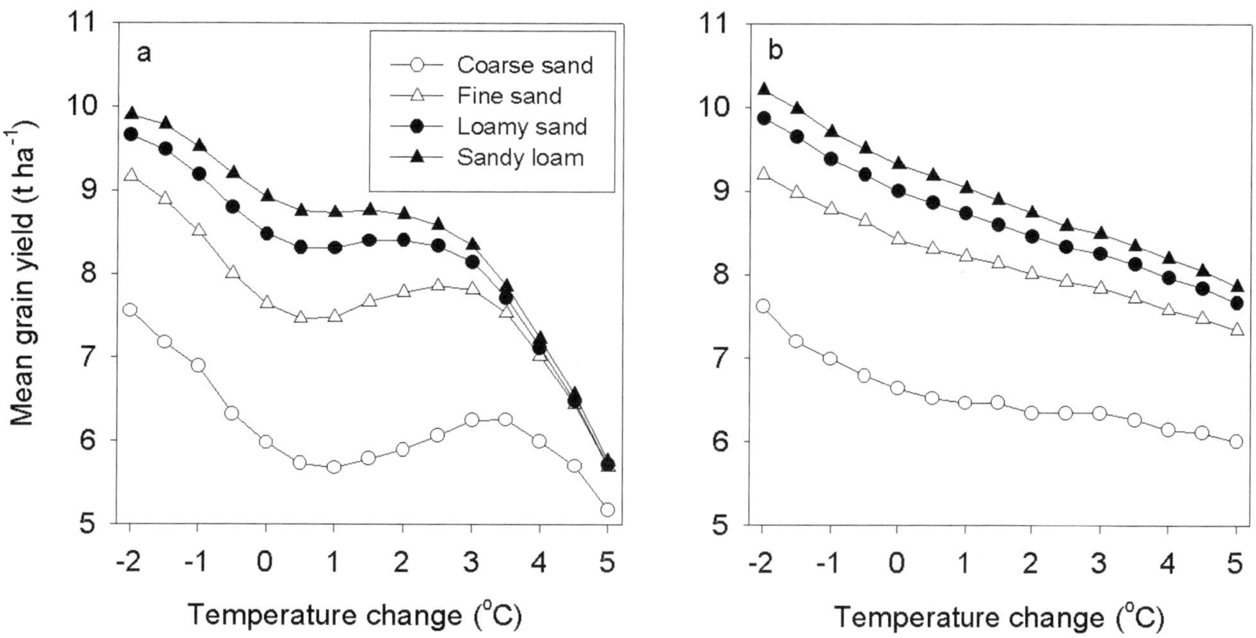

Figure 2. *Response of simulated mean rainfed winter wheat grain yield to changes in mean air temperature for four different soils. The results are shown both for a constant sowing date of 20 September (a) and for the sowing date giving the highest mean yield for each temperature scenario and soil type (b). Source: Olesen et al., 2000.*

lated grain yields were higher than the detrended yields, indicating that other factors may have been limiting yield. The correlations between simulated and detrended yields at agricultural research stations were very low with R2 in the range of 0.02 to 0.24. The highest correlation was seen for the sandy loam soils, which are the soil types where winter wheat is traditionally grown in Denmark. The larger variation and lack of correlation at Askov and Foulum may be due to a larger variation in soil texture and rooting depth at these locations (Hansen, 1976). The correlation between simulated and detrended yield in this study was higher than those obtained for winter wheat in UK using three different simulation models (Landau et al., 1998).

Figure 2 shows that the response of grain yield to temperature change depends on whether the evaluation is performed using a fixed sowing date or whether the optimal sowing date is used in each case. This is a result of the differential yield response to sowing time for the different temperature scenarios. There was a strongly non-linear response to a fixed sowing time of 20 September (Figure 2a). In this case yield may even increase with increasing temperature, especially on the sandy soils. For sowing at the optimal time there was a reduction in mean yield of about 0.29 t $ha^{-1}°C^{-1}$ for sandy loam and 0.13 t ha^{-1} $1°C^{-1}$ for coarse sandy soils.

Most other studies on the effect of temperature change on wheat yield have assumed a fixed sowing date, and studies have often been conducted for a single soil type only. Wolf et al. (1996) compared the response of four winter wheat models to temperature change at Rothamsted in the UK and found that the decrease in grain yield varied between 0.33 and 0.54 t $ha^{-1}°C^{-1}$ depending on the model used. Simulation results from the Great Plains in USA show yield decreases of about 7% for each °C increase in temperature (Stockle et al., 1992; Brown and Rosenberg, 1997). The estimated yield decrease in the current study was lower than the results reported in literature. This is mostly a result of using the optimal sowing date instead of fixed sowing date, and it shows the importance of considering both sowing time and soil type when evaluating effects of temperature on grain yield in winter wheat.

REFERENCES

Aslyng, H.C., and S. Hansen. 1982. Water balance and crop production simulation. The Royal Veterinary and Agricultural University, Copenhagen.

Brown, R.A., and N.J. Rosenberg, 1997. Sensitivity of crop yield and water use in a range of climatic factors and CO2 concentrations: a simulation study applying EPIC to the central USA. Agric. Forest Meteorol. 83:171-203.

Hansen, L.1976. Jordtyper ved statens forsøgsstationer. Tidsskr. Planteavl 80:742-758.

Jamieson, P.D., M.A. Semenov, I.R. Brooking, and G.S. Francis. 1998. Sirius: a mechanistic model of wheat response to environmental variation. Eur. J. Agron. 8:161-179.

Landau, S., R.A.C. Mitchell, V. Barnett, J.J. Colls, J. Craigon, K.L. Moore, and R.W. Payne, 1998. Testing winter wheat simulation models' predictions against observed UK grain yields. Agric. Forest. Meteorol. 89:85-99.

Olesen, J.E. 1991. Jordbrugsmeteorologisk årsoversigt 1990. Tidsskr. Planteavls Specialserie S2130.

Olesen, J.E., T. Jensen, and J. Petersen. 2000. Sensitivity of field-scale winter wheat production in Denmark to climate variability and climate change. Clim. Res. 15, 221-238.

Olesen, J.E., and F. Plauborg. 1995. MVTOOL version 1.10 for developing MARKVAND. SP report 27-1995, Research Centre Foulum, Denmark.

Semenov, M.A., and E.M. Barrow. 1997. Use of stochastic weather generator in the development of climate change scenarios. Clim. Change 35:397-414.

Stockle, C.O., P.T. Dyke, J.R. Williams, C.A. Jones, and N.J. Rosenberg, 1992. A method for estimating direct and climatic effects of rising atmospheric carbon dioxide on growth and yield of crops: Part II—Sensitivity analysis at three sites in the Midwestern USA. Agric. Syst. 38:239-256.

Wolf, J., L.G. Evans, M.A. Semenov, H. Eckersten, and A. Iglesias, 1996. Comparison of wheat simulation models under climate change. I. Model calibration and sensitivity analyses. Clim. Res. 7:253-270.

Forecasting the Impact of Climate Variability on Peanut Crop Production in Argentina

Marta Vinocur[1], Roberto A. Seiler[1], and Linda O. Mearns[2]
[1]Universidad Nacional de Río Cuarto, Argentina
[2]National Center for Atmospheric Research, USA

INTRODUCTION

Peanut (*Arachis hypogaea L.*) production in Argentina occurs in the central and southwest of the Córdoba province. In the area, the crop is an important source of economic welfare. In this mostly rainfed production systems, weather conditions cause interannual variability of yield mainly due to the shortage of water during critical periods of the crop. To determine the significance of impacts of climate variability on peanut crop growth, development and yield, the effect of changes in the mean value of temperature and rainfall and in the temperature variance were analyzed under different environmental conditions and management practices. A process-based model, CROPGRO V3.5—PEANUT, was calibrated and validated for the area, using a common runner-type cultivar under irrigated and rainfed conditions. Series of 99 years of simulated weather which include changes in the mean (no mean change, mean + 1.5°C and mean + 3.5°C) and changes in the variance of temperature from 0.5 to 2.0 in 0.25 increments were generated using an stochastic weather generator applied to 25–28 years of observed weather data. Changes in mean rainfall (+20% and –20%) were also included in the rainfed experiment. Results showed that mean temperature and rainfall changes and temperature variance changes caused bigger impact on yield in the rainfed experiment. The combination of increasing temperature variability and the different mean temperatures changes substantially decreased mean yields and increased their variability as a result of the effect of temperature and water stresses on crop growth and development.

Uncertainties in interannual weather variability create risky environments for agricultural production, affecting crop yield, the agricultural systems and food supply security. Climate variability is experienced by crops mainly through the occurrence of damaging extreme events such as droughts, freezes, and heat waves (Mearns et al., 1984). The probability of these occurrences is more sensitive to changes in variability than changes in the mean (Katz and Brown, 1992). Although, the evidence is for changes not only in climatic means but also in their higher-order moments under climate change conditions (e.g. Gordon et al., 1992; Mearns et al., 1995a, b), most research has analyzed the effects of changes in the mean value of the climate variables (mainly temperature and rainfall) on crop production (e.g., Smith and Tirpak, 1989; Rosenzweig and Parry, 1994; Magrin et al., 1997). Contrarily, few studies have considered the impact of changes in the climate variability (e.g., Wilks, 1992; Mearns et al., 1996, 1997; Barrow and Semenov, 1995), partly because of the uncertainty regarding how climate variability will change for any given region under perturbed climate conditions (Houghton et al., 1996).

Crop models have been used extensively to evaluate crop responses to changes in climatic variability in recent years (e.g., Mearns et al., 1996, 1997; Riha et al., 1996). The purpose of their application was to test the adequacy of crop model formulations to simulate crop responses to climate, and also as a tool for yield prediction and for assisting in management decisions to minimize crop production risk in different soil-climate conditions.

In Argentina, peanut is an important source of economic welfare with more than 60% of the production being exported. Ninety nine percent of the peanut production develops in the central and southwest of the Córdoba province. In this mostly rainfed production system, weather conditions cause interannual variability of yield mainly due to the shortage of water during critical periods of the crop. Using a crop model, the impact of the climate variability as changes in the mean value of temperature and rainfall and in the variance of temperature are analyzed for peanut crop growth, development and yield under different environmental conditions and management practices. Also, the sensitivity of the model is tested in its capability to predict crop responses and yield to climate change and different ecosystems.

MATERIALS AND METHODS

Crop Model

To simulate peanut crop growth, development and yield under different climate scenarios, CROPGRO V3.5 model is used. CROPGRO is a process-based model for grain legumes that simulates crop development, crop carbon balance, crop and soil nitrogen balance, and soil water balance. This model has a common set of FORTRAN code. All the species attributes associated with dry bean, peanut, and soybean are input from species files, plus the information contained in ecotype and cultivar files. Climate input variables are daily solar radiation (MJ m^{-2} day^{-1}), minimum and maximum temperatures (°C), and rainfall (mm day^{-1}). Other inputs include soil and cultivar specific parameters and management factors (e.g., plant population, row spacing, sowing depth). Description of the model structure, the processes, the interactions with climatic factors etc., may be found in Boote et al. (1997, 1998). Assumptions of the CROPGRO model used in this study are that nutrients are not limiting, and weeds, diseases, pests and poor management do not constrain yields.

Stochastic Weather Generator

WGEN is used to generate daily weather data (Richardson and Wright, 1984). WGEN provides daily values of maximum and minimum temperatures, incident solar radiation, and precipitation for an n-year period at a given location. The occurrence of rain on a given day has an influence on temperature and solar radiation for the day. The frequency of precipitation for a given day is generated independently and it is represented by a two-state, first-order Markov chain model. Precipitation amount on wet days is generated by the Gamma distribution. Maximum temperature, minimum temperature and solar radiation are then generated depending on whether a wet day or dry day was produced. The model is designed to preserve the dependence in time, the correlation between variables, and the seasonal patterns of the actual weather data for a selected location (Richardson, 1985). The relationship between the annual and the daily variability was described by Mearns et al., (1996). Approaches to changing the variability of stochastically generated climate time series were provided by Mearns (1989) and described in more detail by Wilks (1992).

Study Characteristics

CROPGRO V3.5 was calibrated and validated for peanut using data from eleven field experiments done with the same cultivar (Florman INTA) in three locations of the peanut productive region of Córdoba province (Manfredi, 31°48′S, 63°46′W; Río Cuarto, 33°07′S, 64°14′W; Carnerillo, 32°55′S, 64°02′W). For the climate variability study, the same runner type cultivar was assumed to be planted on a silty loam soil (Entic Haplustoll, Manfredi) and sandy loam soil (Typical Hapludoll, Río Cuarto) and at the same planting date every year (November 15). The plant density was 12 plants m-2 at 0.7 m distance between rows. Simulations were for rainfed and irrigated conditions (automatic irrigation, profile refilled when available soil water was less than 50% in the first 0.3 m of the soil), with no nitrogen stress and the soil water content was initialized to the same value every year.

The weather data sets used as the inputs for the simulation model were based on observed and generated weather data. Using the weather generator, series of 99 years of daily weather data including changes in the mean and in the variance of temperature were generated for each of the locations. The parameters for this model were estimated from a series of 28 years of observed weather data in Manfredi (1969-1996) and a series of 25 years in Río Cuarto (1974-1998). In the generated series the variance of the daily temperature was changed by factors from 0.5 to 2.0 in 0.25 increments and the mean of the temperature was changed as: T_{Mean} +

1.5°C and T_{Mean} + 3.5°C. Changes in the mean of precipitation (+20% and –20%) were also applied to the rainfed experiments, with no changes in the mean of the temperature. The same changes for the mean of the temperature and precipitation were applied to the series of observed data. Thus, a total of 24 cases (combinations) for the irrigated experiment and 40 cases for the rainfed experiment for each site were simulated as follows: the OBSER case (from the observed weather data set for the site plus changes), the baseline case (no mean temperature and no variance changes), the baseline case altered by two changes in the mean temperature, the variance changes with no mean change, the combined variance and mean changes and finally, the no mean temperature change and mean precipitation changes. Except for the first case, the rest were from the generated weather data set. The analysis of the results is based on the analysis of the mean, the standard deviation (SD) and the coefficient of variation (CV) of the crop yields, under the different scenarios. The date of harvest and the frequency of crop failure are analyzed using the same statistics.

RESULTS AND DISCUSSION

Simulated yields from the observed and generated climate and from the different cases are shown in Table 1 and 2 for Manfredi and Río Cuarto, respectively. Baseline case (in bold type) and OBSER case yields are very similar for the irrigated and rainfed conditions in both locations. SD and CV of the yields were much higher for the observed climate in Manfredi than in Río Cuarto. WGEN often underestimate the interannual variability of climate, and thus will likely result in lower variability of crop yields when generated time series are used in the model. The same pattern of variation is shown for the mean, SD and CV of the yields when all the cases for the observed climate simulations are compared to the generated climate simulation cases with no change in the variability of the temperature (variance change=1). Increasing the variability of the temperature (to variance change=2), causes mean yield reductions in each of the cases, either under irrigated or rainfed conditions (Table 1 and 2). Under irrigated conditions for the baseline and for T_{Mean} + 1.5°C cases, the SD and the CV of the simulated yields increase with the increase of temperature variability. Almost no effect in the yield variability is shown for the TMean + 3.5°C case in both locations. Under rainfed conditions, mean temperature increase and no change in the variance show in general peanut yield reductions. Strong yield reductions occur when the increase of the temperature variance is coupled to the increase in the mean of the temperature (e.g. in Table 1, 1416 kg ha-1 for the baseline case to 877 kg ha-1 for TMean + 3.5°C and variance change=2). The increase of the temperature variability exposes the crop to large fluctuations in temperatures and to higher frequency of extreme temperatures above the optimal level during the different crop stages causing crop yield reductions.

Moderate increase in the mean temperature (e.g., 1.5°C) may favor the crop dry matter cumulative processes and shorten the crop cycle while affecting more the period from flowering to maturity (data not shown). Under irrigated conditions, when no water stress is present, the interaction between the accumulation rate and the length of the crop cycle may cause the mean yield to increase. In contrast, the same scenario under rainfed conditions may cause yield reductions according to the water stress imposed on the crop by the lack of enough water. A more important increase in the mean temperature, such as 3.5°C, shortens the crop cycle significantly (33 and 26 days in Manfredi and Río Cuarto respectively) and also may exposed the crop to water and temperature stresses causing greater yield reductions in rainfed than in irrigated conditions (Table 1 and 2).The higher mean temperature increase (TMean + 3.5°C) may be related to temperatures outside the optimum ranges during the different crop processes. With a substantial reduction in the variance (to variance change=0.5), the yield reduction from increased mean temperature (TMean + 3.5°C) alone is greatly ameliorated as the crop is subject to lesser occurrences of extremes and temperatures outside the optimum ranges.

Increase in the mean of the precipitation, increased yields in both locations. Twenty percent increase in the mean of the precipitation with no change in the temperature variance, causes higher mean yield than the baseline case yield. The bigger

Table 1. *Simulated yields (kg ha-1) from observed (OBSER) and from generated weather data with changes in the variance of temperature and in the mean of temperature and precipitation, for the irrigated and rainfed experiments in Manfredi.*

Irrigated		Variance changes							
		0.5	0.75	1.0	1.25	1.50	1.75	2.0	OBSER
No Mean Change	Mean	5605	5527	5438	5317	5177	5039	4858	5364
	SD	239.7	249.1	284.5	344.8	445.2	505.4	649.3	539.5
	CV	4.3	4.5	5.2	6.5	8.6	10.0	13.4	10.1
TMean + 1.5°C	Mean	5681	5613	5547	5480	5416	5332	5234	5554
	SD	259.7	256.7	256.6	261.8	262.7	319.9	340.9	525
	CV	4.6	4.6	4.6	4.8	4.9	6.0	6.5	9.5
TMean + 3.5°C	Mean	5504	5447	5395	5332	5269	5214	5153	5389
	SD	311.4	304.6	298.9	290.1	284.7	278.4	288.5	572.6
	CV	5.7	5.6	5.5	5.4	5.4	5.3	5.6	10.6
Rainfed									
No Mean Change	Mean	1557	1486	1416	1345	1270	1196	1131	1319
	SD	838.3	817.8	800.6	775.2	747.4	723.2	687.7	1025.1
	CV	53.8	55.0	56.5	57.6	58.9	60.5	60.8	77.7
TMean + 1.5°C	Mean	1499	1428	1358	1286	1217	1148	1094	1317
	SD	834.4	816.8	792.4	765.5	736	706.4	683.3	1062.9
	CV	55.7	57.2	58.4	59.5	60.5	61.5	62.5	80.7
TMean + 3.5°C	Mean	1228	1153	1087	1031	974	920	877	1124
	SD	748.3	714.2	679.5	651.9	617	589.4	564.8	997.9
	CV	60.9	61.9	62.5	63.2	63.4	64.1	64.4	88.8
No Mean Change+20%	Mean	2118	2040	1960	1869	1773	1678	1568	1778
	SD	1041	1020	1009	989.8	964.9	939.3	899.3	1210
	CV	49.2	50.0	51.5	53.0	54.4	56.0	57.4	68.1
No Mean Change-20%	Mean	989.4	944	897	848	798	749	689	882
	SD	562.7	545.8	528.9	505.8	483.8	463.6	435.8	737.9
	CV	56.9	57.8	59.0	59.6	60.6	61.9	63.3	83.7

Variance=1: No change in variance; T_{Mean} + 1.5°C: maximum and minimum temperatures increased by 1.5°C; T_{Mean} + 3.5°C: maximum and minimum temperatures increased by 3.5°C; No Mean Change +20%: no change in mean temperature and 20% increase in the mean of rainfall; No Mean Change -20%: no change in mean temperature and 20% decrease in the mean of rainfall; Mean (kg ha^{-1}); SD (kg ha^{-1}) and CV (%) of the simulated yields.

increase is shown for Río Cuarto than for Manfredi. A 20% reduction in the mean of the precipitation reduces yields by 40% compared to the baseline case yield (Table 1 and 2). The biggest mean yield reductions (approximately 50%) and CV increments are observed for both places under rainfed conditions, when the precipitation mean is diminished and the variance of the temperature is twofold. Another description of the yields variability due to the different sceneries is shown in Figure

Table 2. *Simulated yields (kg ha-1) from observed (OBSER) and from generated weather data with changes in the variance of temperature and in the mean of temperature and precipitation, for the irrigated and rainfed experiments in Río Cuarto.*

Irrigated		Variance changes							
		0.5	0.75	1.0	1.25	1.50	1.75	2.0	OBSER
No Mean Change	Mean	6174	6101	6013	5920	5774	5613	5457	6011
	SD	253.6	310.9	346.9	405.5	474.3	564.2	631.8	355.9
	CV	4.1	5.1	5.8	6.8	8.2	10.1	11.6	5.9
TMean + 1.5°C	Mean	6293	6230	6179	6127	6061	5983	5893	6193
	SD	273.6	286.9	289.0	283.8	284.5	315.8	350.9	237.9
	CV	4.3	4.6	4.7	4.6	4.7	5.3	6.0	3.8
TMean + 3.5°C	Mean	6105	6067	6024	5968	5903	5851	5816	5947
	SD	380.4	358.8	350.1	346.2	332	323.2	320.5	292.4
	CV	6.2	5.9	5.8	5.8	5.6	5.5	5.5	4.9
Rainfed									
No Mean Change	Mean	3165	3084	2993	2895	2787	2674	2568	2907
	SD	1351	1330	1309	1274	1249	1231	1189	1622
	CV	42.7	43.1	43.7	44.0	44.8	46.0	46.3	55.8
TMean + 1.5°C	Mean	3174	3100	3024	2939	2852	2752	2666	2919
	SD	1397	1374	1357	1344	1326	1306	1289	1671
	CV	44.0	44.3	44.9	45.7	46.5	47.5	48.3	57.2
TMean + 3.5°C	Mean	2869	2757	2665	2581	2498	2412	2352	2658
	SD	1396	1360	1334	1305	1281	1258	1227	1627
	CV	48.7	49.3	50.0	50.6	51.3	52.2	52.1	61.2
No Mean Change+20%	Mean	4129	4042	3956	3854	3730	3592	3462	3677
	SD	1343	1326	1315	1298	1285	1281	1237	1652
	CV	32.5	32.8	33.2	33.7	34.5	35.7	35.8	44.9
No Mean Change-20%	Mean	2018	1948	1870	1790	1706	1627	1548	1824
	SD	1102	1074	1039	997.1	960.7	931.9	892.1	1190
	CV	54.6	55.1	55.6	55.7	56.3	57.3	57.6	65.2

Variance=1: No change in variance; T_{Mean} + 1.5°C: maximum and minimum temperatures increased by 1.5°C; T_{Mean} + 3.5°C: maximum and minimum temperatures increased by 3.5°C; No Mean Change +20%: no change in mean temperature and 20% increase in the mean of rainfall; No Mean Change -20%: no change in mean temperature and 20% decrease in the mean of rainfall; Mean (kg ha^{-1}); SD (kg ha^{-1}) and CV (%) of the simulated yields.

1 and 2, for Manfredi and Río Cuarto, respectively. Clearly, yields diminish with increasing mean temperature and variability for both places while water availability may offset the negative impact of the temperature changes.

CONCLUSIONS

Peanut yield in Argentina has shown to be sensitive to changes in the climate variability. Effects of changes in the mean of the temperature and precip-

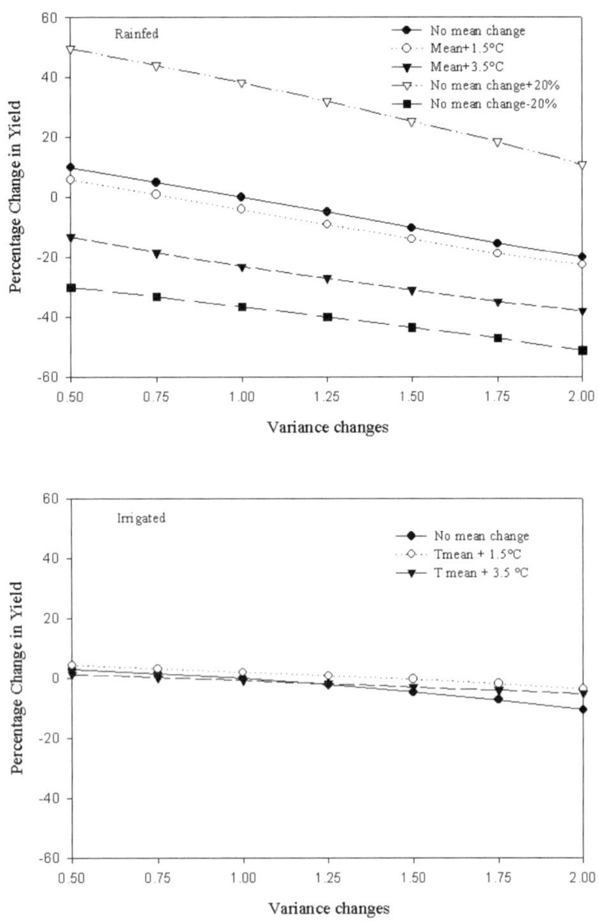

Figure 1. *Percent of change in the simulated yield of peanut with respect to the baseline case, for rainfed and irrigated conditions including changes in the variance of the temperature and in the mean of the temperature and rainfall, in Manfredi.*

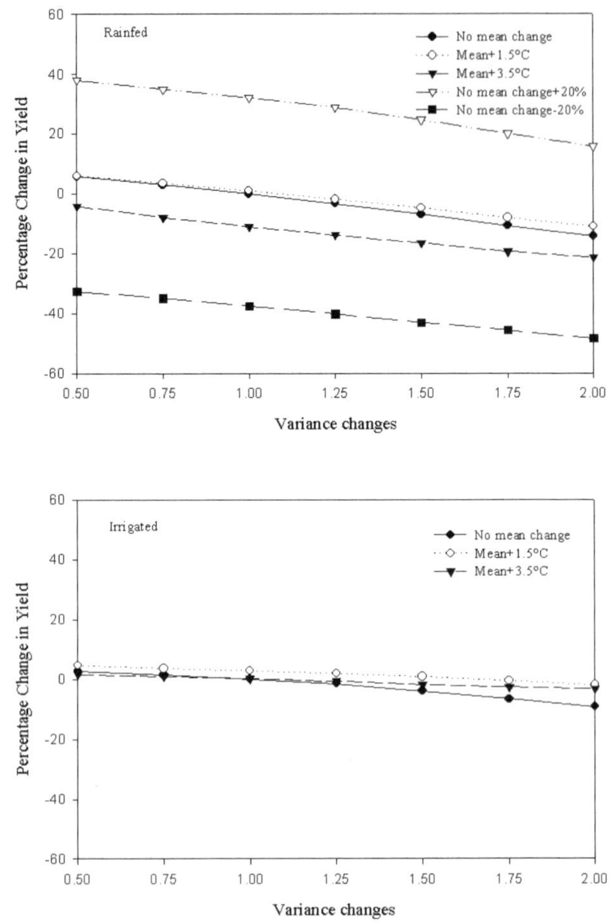

Figure 2. *Percent of change in the simulated yield of peanut with respect to the baseline case, for rainfed and irrigated conditions including changes in the variance of the temperature and in the mean of the temperature and rainfall, in Río Cuarto.*

itation and in the variance of the temperature on yields will be more important in rainfed than in irrigated conditions. Mean temperature increment of 1.5°C causes small changes in the mean of yields, but higher increments (e.g., 3.5°C) results in diminishing yields. Mean precipitation increments will favor increasing yields. The combination of increasing temperature variability and the different mean temperature changes reduce mean yields of peanut and increase yield variability as represented by the CV, in most cases. The peanut simulation model has shown to be sensitive in predicting yield variations under different weather conditions, management practices and ecosystems characteristics.

ACKNOWLEDGEMENTS

This work was partially funded by the National University of Río Cuarto, the National Scientific Research Council (CONICET) and the Córdoba Scientific Research Council (CONICOR). We thank Julio Dardanelli for providing Manfredi peanut field experimental data.

REFERENCES

Barrow, E., and M.A. Semenov. 1995. Climate change scenarios with high spatial and temporal resolution for agricultural applications. Forestry 68:349-360

Boote, K. J., J.W. Jones, and G. Hoogenboom. 1997. Simulation of Crop Growth: CROPGRO model. p.

651-692. In R.M. Peart and R.B. Curry (eds.) Agricultural Systems Modeling and Simulation. Marcel Dekker Publishers, New York.

Boote, K.J., J.W. Jones, G. Hoogenboom, and N.B. Pickering. 1998. The CROPGRO model for grain legumes.p.99-128. In G.Y. Tsuji et al., (eds.) Understanding options for agricultural production, Systems approaches for sustainable agricultural development Volume 7, Kluwer Academic Publishers, Great Britain.

Gordon, H.B., P.H. Whetton, A.B. Pittock, A.M. Fowler, and M.R. Haylock. 1992. Simulated changes in daily rainfall intensity due to the enhanced greenhouse effect: implications for extreme rainfall events. Clim. Dynamics 8:83-102.

Houghton, J.T., L.G.M. Filho, B.A. Callander, N. Harris, A. Kattenberg, and K. Maskell. (eds.) 1996. Climate Change 1995: The Science of Climate Change. Report Prepared for the IPCC Working Group I, Cambridge University Press, Cambridge, UK

Katz, R.W., and B.G. Brown. 1992. Extreme events in a changing climate: variability is more important than averages. Clim. Change 21:289-302

Magrin, G.O., M.I. Travasso, R. A. Díaz and R. O. Rodríguez. 1997. Vulnerability of the agricultural systems of Argentina to climate change. Clim. Res. 9:31-36.

Mearns, L.O., R.W. Katz, and S.H. Schneider.1984. Extreme high temperature events: changes in their probabilities with changes in mean temperature. J. Clim. Appl. Meteorol., 23:1601-1613.

Mearns, L.O. 1989. The simulation of meteorological time series for climate change scenario development. Paper presented at the USEPA Meeting on Climate Change Scenario Development at Boulder, 31 August, 1989. (Summarized in Scenarios Advisory Meeting Summary Report, December 1989, U.S. EPA, Washington, DC).

Mearns, L.O., F. Giorgi, C. Shields, and L. McDaniel. 1995a. Analysis of the variability of daily precipitation in a nested modeling experiment: comparison with observations and 2XCO2 results. Global and Planetary Change 10:55-78.

Mearns, L.O., F. Giorgi, C. Shields, and L. McDaniel. 1995b. Analysis of the diurnal range and variability of daily temperature in a nested modeling experiment: comparison with observations and 2XCO2 results. Clim. Dynamics 10:55-78.

Mearns, L.O., C. Rosenzweig, and R. Goldberg. 1996. The effect of changes in daily and inter- annual climatic variability on CERES-Wheat: a sensitivity study. Clim. Change 32: 257-292

Mearns, L.O., C. Rosenzweig, and R. Goldberg. 1997. Mean and variance change in climate scenarios: methods, agricultural applications, and measures of uncertainty. Clim. Change 35:367-396.

Richardson, C.W., and D.A. Wright. 1984. WGEN: a model for generating daily weather variables. USDA-ARS ARS-8. 80 pp.

Richardson, C.W. 1985. Weather simulation for crop management models. Trans. ASAE 28 (5): 1602-1606

Riha, S.J., W. Wilks, and P. Simoens. 1996. Impact of temperature and precipitation variability on crop model predictions. Clim. Change 32:293-311

Rosenzweig, C., and M.L. Parry. 1994. Potential impact of climate change on world food supply. Nature, 367: 133-137

Smith, J.B. and D.A. Tirpak.eds.1989. The potential effects of global climate change on the Unites States. USEPA Rep. 230-05-89-050. Washington D.C.,

Wilks, D. 1992. Adapting stochastic weather generation algorithms for climate change studies. Clim. Change 22:67-84

One Century of Frost Occurrences in Coffee Plantations of Southern Brazil

Laura R.B. Kiihl
Instituto Agronomico do Paraná , IAPAR, Brazil

INTRODUCTION

There are several publications about frost in Brazil. However, the characteristics of this phenomenon in the States of São Paulo and Paraná, where there was a high concentration of coffee plantations, had yet not been analyzed for a long period of historical data. As frost is a sporadic phenomenon at the Tropic of Capricorn, it was considered that a long period of survey would result in better identification of the phenomenon characteristics concerning frequency, time of occurrence, duration and extension. Therefore, the period between 1870 and 1975 was chosen to carry out this survey.

The objective of this work was to contribute with an alternative method to obtain series of climatic events of difficult observation, as it is the case of frosts, using non conventional sources. This approach is particularly relevant for countries that do not have large historical series of meteorological observations such as Brazil. The approach chosen was based on the relationship "coffee-frost". This relationship was particularly important because coffee was a permanent and important crop at those States. Frost occurrences were computed as the events that caused damages to coffee plantations of enough magnitude to be registered in the newspapers. The decision of using newspapers as a research source, despite its weakness as a source of scientific information, was made because it was the only way to make this work possible, since there were no historical records of minimum temperature available for all the period analyzed. However, the information from the newspapers was, whenever possible, compared with available meteorological data and bibliography. Major emphasis was given to the methodology used and the results obtained.

LITERATURE REVIEW

Frost is a widely-reported phenomenon, but there are only four publications investigating frost occurrence in Brazil, D'Utra (1919), Oliveira (1928), Salati (1957) and Camargo (1975). These authors provide useful information about the climatic phenomenon which occurs occasionally in the latitudes of São Paulo and Paraná states.

D'Utra (1919) used information from meteorological bulletins in his survey of a 48-year period (1870 to 1918), and provided dates of occurrence and locations affected by frost in the period. OLIVEIRA (1928) used the minimum temperature together with frost data collected from interviews with rural workers to elaborate a frost intensity graph, establishing coefficients which varied from 0 to 10. His graph showed the intensity of the frosts from 1884 to 1927. Salati (1957) and Camargo (1975) used the same type of data source, but adopted different observation criteria, periods and locations. Salati (1957) used daily meteorological data from the Luiz de Queiroz Agricultural College in Piracicaba, SP, for the period 1918 to 1956. The author drew two frost frequency graphs for publication in a newspaper article. The graphs show the number of days with frost in the months of May, June, July, August and September and the number of days with frost for the thirty nine years period. Camargo (1975) used a minimum temperature of less than 2.5o C in a first class meteorological station of the São Paulo State Agronomic Institute in Campinas, SP. The author set up a table with the collected information providing the dates of frost occurrence, the corresponding minimum temperature, wind direction and intensity for the periods 1890 to 1920 and 1929 to 1975.

The identification criteria and the type of data reported varied in each survey, making a uniform catalog of the frost characteristics impossible.

The present study resorted to unconventional data sources to correct this problem, maintaining within its scope finding the frost occurrence dates in the region. It intended to analyze frost frequency to identify the period of the year when they are most frequent and to establish a frost typology for the region, based on its extension. It is also within the scope of this study to assess frost duration and intensity and check the extension of the main frosts which affected coffee cultivation in São Paulo and Paraná states.

MATERIALS AND METHODS

The geographic area object of this research is situated between the 20° and 24° S and between 45° and 55° W, and belongs to the northern part of the Brazilian Meridional Plateau. The Tropic of Capricorn passes through a great part of the area that is a transition zone between climates controlled by tropical and polar masses and climates controlled by equatorial and tropical masses (Monteiro, 1973).

Research Sources

Newspapers and all documents that could refer to their collected data were considered a research source. The newspaper chosen was "Provincia de São Paulo", which, edited in 1875, circulated until 1889 when its name changed to "O Estado de São Paulo" and remained until today. The newspapers "Correio Paulistano" and "Gazeta de Campinas" were used for the five previous years. Meteorological bulletins edited by the National Department of Meteorology since 1889, temperature records from meteorological stations starting in 1944 and specific bibliographic sources such as periodicals, reports and magazines were also chosen as sources. They were used to validate the newspaper data.

Regarding the credibility of the main source, the history methodologists and critics nearly always point to the newspaper as an example of a suspect source when referring to editorials, which are the least credible part of a newspaper. Simple news reports without personal interest may contain errors due to bad observation, perception or representation. Advertisements, so useful to social and economic history because of the data they supply on market goods, should be used only after a critical examination. So much caution is required that a theoretician only accepts as true the disinterested facts, data from official sources, such as news on the weather, navigation, property transactions, markets, statistics, etc. In the case of news about frost occurrence, because of the importance of coffee for the Brazilian economy, the hypothesis assumed that the newspaper would identify the locations affected by the phenomenon every time it occurred. The long survey period would serve to ensure better identification of frost characteristics.

Data were collected from newspapers in the Public Archive of the Estado de São Paulo and from meteorological bulletins in the Geographic and Geological Institute. Temperature data was compiled from the archives of the 7th Meteorological District of the National Meteorological Department. The relevant literature was obtained in the Library of the Brazilian Coffee Institute and in the National Library in Rio de Janeiro. Index cards were used to store the information.

A period of the year most likely for frost occurrence, the months of May, June, July, August and September was detected, based on observations made while reading the first fifteen years newspapers and on observations by Serra and Ratisbonna (1945) and Titarelli (1972).

The systematized frost occurrence dates were then confronted with the meteorological information. All air temperatures equal to or less than 5°C registered in shelters of meteorological stations on the two horizontal trans sections localized on the 21°30' and 23°30' S were compiled. It was assumed that when temperatures were lower than this limit in the shelters of a station, the temperatures at soil or plant level were already close to 0°C, therefore indicating possible frosts (TARIFA et al, 1977).

The first trans section contains the locations of Araçatuba, Catanduva, Riberão Preto, São Simão and Mococa while Guaíra, Campo Morão, Apucarana, Londrina, Jacarezinho, Itapeva, Sorocaba, São Paulo (Horta Florestal) and Taubaté belong to the second. The data were ordered by occurrence of days with temperatures equal to or less than 5°C following the above sequence of meteorological stations, in each transection.

Frost Occurrence Selection Criteria

Frost extension and intensity were used as criteria to choose the events which deserved attention. Extension was taken to be frost occurrence in at least five locations. This criteria was used to eliminate the possibility of considering very frequent events due to the geographic situation of the locations. Intensity was considered as the lowest temperatures registered on the occurrence dates. This criteria was used to reduce doubts in cases of more than one occurrence with the same extension.

Three frost occurrence types were thus selected: those which affected less than five locations; those which affected five or more locations; and those which affected the greatest number of locations each year, within those frosts which affected five or more locations. Frosts within this last category were considered the "main frosts".

Data Mapping

The affected areas could not be mapped by county, because county contours had been modified with time. A special background map was prepared for the survey data to fix the locations affected by frosts along the time. A map of the County Division of Brazil was used, edited by the IBGE in 1968 and corrected in 1976, that contained the county division for the survey area. From this map a new one was drawn on transparent paper, which no longer contained the county divisions but a network of squares. Care was taken when determining the square size, so that each one contained at most three county headquarters. The main frost occurrence data was mapped from the two maps using the network of squares superimposed on the map of county divisions.

Statistical Analysis

The statistical treatment adopted was based on the Komolgorov-Smirnov non-parametric test at a significance level of 5%. In the cases where there was significance it was concluded that the frequency distribution of the N observations could not be considered a sample of the Fo frequency distribution. To compare the frost frequency distribution Fo was obtained from the accumulated frequencies of the total of frost occurrence data, while SN was the accumulated frequency of frosts which affected five or more locations. Thus it could be checked if the frosts which affected five or more locations had or did not have the same frequency distribution as the total of frosts occurred.

RESULTS

Frost Frequency

The graph in Figure 1 shows the negative asymmetric distribution of frost occurrences in the 106 years researched. The frequencies increased slowly during May until they reached a first peak on June 17, reaching another peak, the maximum period, on July 6 and declined sharply from mid-August. The shortest continuous period of the year which concentrated the greatest number of frosts was found by analysis of the frequency distribution of the number of days with frost and the number of frost occurrences. It can be observed that the tail of the graph is fairly clearly distinguished from the body, that is, from nine frost occurrences, the number of days of each class decreased compared to the number of days with eight or less occurrences. The criteria was stipulated that the period which encompassed nine or more frosts was the greatest concentration of frost occurrences. Figure 1 shows that such periods range from June 15 to August 6. In fact, the highest percentage of frost occurred (55.35%) in this period. Therefore, in only 53 days, from May 1st to September 30th there were 352

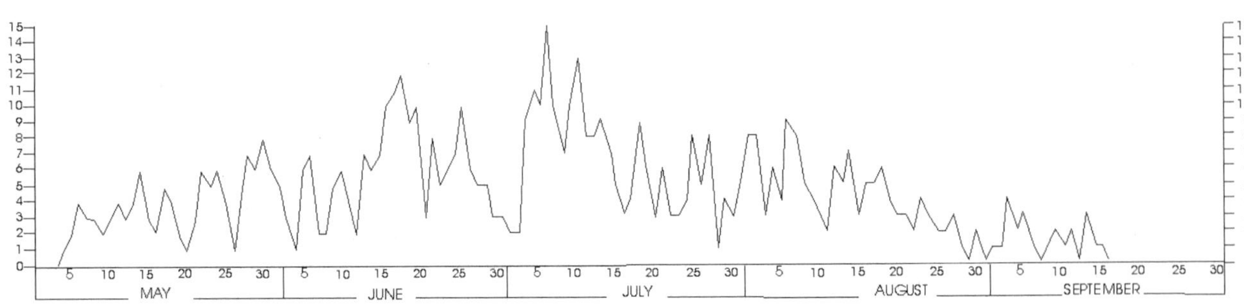

Figure1. *Frost frequency in the coffee regions of São Paulo Paraná states by day and month occurrence. Period 1870—1975.*

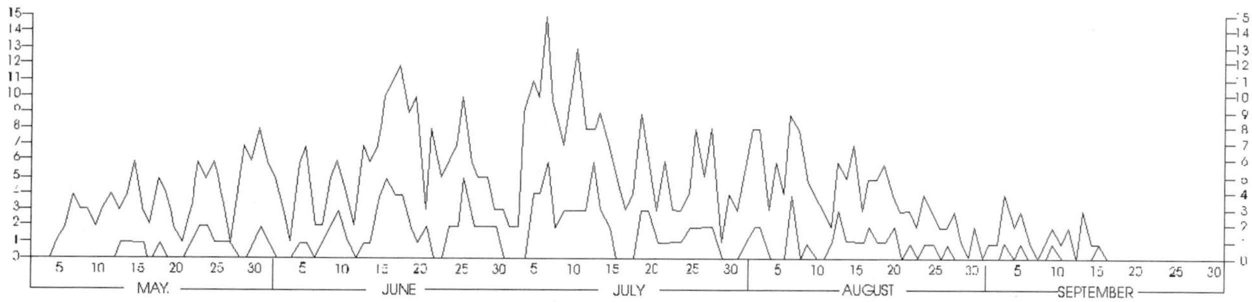

Figure2. *Frequency of frosts that reached five or more locations in the coffee regions in São Paulo and Paraná compared to all frosts that reached the region, by day and month occurrences. Period 1870-1975.*

frosts of a total of 636. From June 15 to July 10, 11 days with ten or more frosts were recorded. Also, in this 53 day interval there were 37 (59.6%) of the main occurrences of the 106 years period, in a total of 62 frosts.

There were 109 frosts in May (17%), 176 in June (28%), 201 in July (32%) and 127 (20%) in August, but only 23 (4%) in September. Thus the greatest frost concentration—60%—occurred in June and July, decreasing in August. Proceeding in the same way for the frosts which affected five or more locations, the greatest concentration of this type of frost—70%—also occurred in June and July. In the light of these results, the relationship between the distribution of the frequencies of all the frosts in the annual period, from May 10th to September 30th, and those which reached five or more locations in the 106 years was checked. By the Komolgorov-Smirnov test, a maximum deviation of Dmax = 0.09 was obtained, not significant at the level of 5% [P(Dmax £ 0.12)<5%]. This result shows that the frequency distributions are equal for the two variables, that is, the two curves had the same frequency distribution pattern (Figure 2). This finding enables frosts that affect five or more locations to be considered as a sample of the frost occurrences which affected the region in the 106 years researched.

Frost Duration and Intensity

Two frost damage variables should be considered for the coffee crop: the number of consecutive days with frost (duration) and the air temperature at each occurrence (intensity). Greater damage to the coffee tree is caused by longer duration and by lower air temperature. The greatest percentage of occurrences in the period were in the one-day frost class (58.59%). When each month is analyzed separately, the highest percent of one-day frosts happens in June (28.83%) and of two—(31.58%) and three-day frosts (37.14%) in July. The highest percentages of four- and five-day frosts occurred in June and July. Practically no frost periods longer than three days were found in August or September. The sequence of days with frost, as well as the day of the sequence when the frosts occurred in five or more locations, were also obtained. There were frosts in five or more locations in all the sequences observed (1, 2, 3, 4, 5, 6, 12), except for the six day sequence. The so-called main frosts occurred in all the classes, except for those of six and twelve days, and there were almost the same number of occurrences for classes of one or two consecutive days with frost.

The data supplied by the meteorological stations in the two trans-sections were used to check the frost intensity. It was found that on twenty-five dates of frost occurrences in less than five locations and when the meteorological stations of the two transepts recorded temperatures equal to or below 5.0°C, the temperatures remained above 0°C, except for the occurrences of 19 June 1952, when the Mococa station reported—2.0°C; 21 June 1963 when the Sorocaba station reported a temperature between –1.0°C and 0.0°C and 11 July 1969 when the stations at Campo Mourão and Sorocaba recorded temperatures between –1.0°C and 0.0°C.

On ten frost occurrence dates which affected five or more locations, the temperatures recorded by the meteorological stations were close to 0.0°C on only three occurrences. On the others, the temperatures were equivalent to those observed in the previous type.

There were stations which recorded negative temperatures in nine of a total of 14 so-called main frost occurrences, for which temperature data were available. The occurrences of 06 Aug. 1963, 09 July 1972 and 18 July 1975 had the greatest number of stations with negative temperatures, especially the last one.

Temporal-Spatial Distribution of the Main Frosts

There were sixty two "main frost" occurrences, according to the criteria adopted for selection which were distributed throughout the period covered by the research.

Some frosts stood out from the others because of the number of locations they affected. When twenty affected locations are considered, there are eighteen main frost occurrences that caused more extensive damage to the coffee plantations. They are the occurrences in the following years: 1886, 1902, 1904, 1910, 1918, 1919, 1923, 1931, 1942, 1953, 1955, 1957, 1962, 1963, 1966, 1969, 1972 and 1975. On the other hand, observation of the frost interval showed that until 1950 they were more spaced, but since then they became more frequent. This shows that prior to a large coffee region implantation in Paraná state, the frosts considered damaging to the plantations due to their damage extension were rare occurrences. But they became more frequent after 1950, as observed throughout the period, and became frequently damaging to coffee plantations because the crop had been implanted in a region unsuitable for coffee cultivation due to frost.

There were relatively few 'main frosts' in the period 1930-1950, when the crop was expanding from São Paulo state towards Paraná state. This must have encouraged the great expansion which occurred in the North of Paraná at that time.

All the frosts that caused large-scale damage to coffee plantations occurred in the months of June, July and August, except that of 1886. The main coffee-producing regions could be identified by checking the areas affected by the main frosts throughout the period covered by the survey.

Validity of Newspaper Data

In the 1944-1975 period, comparison of newspaper and meteorological station data showed a 64% correspondence between total frost data reported by the newspaper and dates of temperatures equal to or below 5°C in the trans-section meteorological stations. Taking each month separately, the correspondence was 58% in May, 59% in June, 72% in July, 58% in August and 100% in September. The fact that agreement was not 100% is attributed to the use of 5°C as the temperature indicator of the possibility of frost occurrence.

On the other hand, with the support of the specific literature, the newspaper data was confirmed for the main frosts in the following years: 1870, 1879, 1886, 1898, 1902, 1918, 1931, 1942, 1943, 1950, 1953, 1955, 1956, 1963, 1966, 1969, 1972 and 1975.

DISCUSSION AND CONCLUSIONS

Positive aspects of the methodology used were: a) the possibility of finding the date of frost occurrences that affected the coffee producing regions over a long period by using newspapers, in a continuous way; b) the opportunity to follow the evolution of the coffee crop from 1870 to 1975 in areas of the states of São Paulo and Paraná, by mapping the main frosts, and c) the establishment of a frost typology based on the extension of the phenomenon.

Perhaps the greatest limitation is that the reported data up to 1953 is only from the locations in the main coffee-producing regions of São Paulo state, with no information on Paraná state. Also, some data did not refer only to locations with coffee.

Sixty-two 'main frost' occurrences were found, of which eighteen affected twenty or more locations, and were therefore considered as those which caused extensive damage to coffee plantations during the survey period. They were: 12 May 1886, 19 Aug. 1902, 13 Aug. 1904, 19 July 910, 26 June 1918, 16 Aug. 1919, 13 July 1923, 29 June 1931, 06 July 1942, 05 July 1953, 01 Aug. 1955, 21 July 1957, 27 July 1962, 06 Aug. 1963, 06 Aug. 1966, 10 July 1969, 09 July 1972 and 18 July 1975.

The period 1930-1950 was identified as being relatively free of main frost occurrences, which must have contributed to the extension of coffee plantations in Paraná state. All the frosts that caused large-scale damage to coffee plantations happened in June, July and August, except for one occurrence. By mapping the area affected by the main frosts, the main coffee-producing regions could be identified throughout the 1870-1975 period.

Regarding the Validity of Newspaper Data

The 5°C temperature level cannot be taken as a limit indicative of possible frost occurrence, since there were many frost occurrences reported in the newspaper in temperatures above 5oC in the latitudes of São Paulo and Paraná states. Newspaper data were confirmed by the specific literature for eighteen of the main frosts.

In view of the observed results, the survey data have applications in the agronomic field, mainly for the coffee crop. It would be interesting in the field of history to check the relationship between the frequency of the main frosts in the 1930-1950 period and the expansion of coffee plantations towards the north of Paraná. The dates of the frosts for studies on the phenomenon genesis and later monitoring are of interest in the fields of climatology and meteorological research.

Sponsored by National Research Council—CNPq.

REFERENCES

Camanrgo, A. P. 1975. Geadas de um século. Boletim mensal do GERCA—IBC 5 .

D'Utra, G. R. P. 1919. As geadas e os meios possíveis de prevenir ou atenuar os seus efeitos sobre a vegetação., Secretaria da Agricultura Comércio e Obras Públicas, Serviço de Publicações, São Paulo, SP.

Monteiro, C. A. F. 1973. A dinâmica climática e as chuvas no Estado de São Paulo. Estudo geográfico sob a forma de atlas. Depto. de Geografia da Faculdade de Filosofia, Ciências e Letras de Rio Claro, Rio Claro, SP.

Oliveira, A. L. 1928. A meteorologia e a agricultura. Boletim Astronômico e Geofísico 1: 8-29.

Salati, E. 1957. Geadas. O Estado de São Paulo, Supl. Agric. 923.

Serra, A. and Ratisbonna, L. 1945. As ondas de frio da Bacia Amazônica. Boletim Geográfico 26 : 172-206.

Siegel, S. 1956. Nonparametric statistics for the behavioral sciences. McGraw-Hill, Tokyo.

Titarelli, A. V. 1972. A onda de frio de abril de 1971 e sua repercussão no espaço geográfico brasileiro. USP Instituto de Geografia, São Paulo, SP.

Tarifa, J. R. et al. 1977. A gênese dos episódios meteorológicos de julho de 1975 e a variação espacial dos danos causados pelas geadas a cafeicultura no Estado de São Paulo. Ciência e Cultura 29 : 1362-74.

Climatic Variability and Rice Production in Rainfed Rice Area in Northeast Thailand: Risk Analysis and Management Applications

Hatsachai Boonjung
Suranaree University of Technology, Thailand

INTRODUCTION

Rice production in Northeast Thailand relies mainly on rain because irrigation is limited to about 20% of the region. Rice yield during 1987-1993 have been clearly lower in the Northeast than in other regions of the kingdom even though yields have tended to increase as a result of adoption of new technology, such as new rice varieties, fertilizer application and weed management. This has been attributed to poor amount and distribution of rainfall, and to soil properties such as low water holding capacities and poor fertility (KKU-FORD, 1982; Pannangpetch, 1993; Jintrawet, 1995; Fukai et.al., 1997).

The complexity of the environment, including difference in soil type and weather condition, can be apprehended with the use of crop modeling. Simulation modeling is a powerful research tool for identifying drought development patterns and evaluating their impacts on crop growth and yield (Fukai, 1994). Coupled with a suitable geographic information system (GIS), crop modeling can be used as a tool for determining rainfed rice yield potential across different soil types. With the use of risk analysis, the best time for planting for each soil type and environment can be determined. This study aims to produce advice for farmers and agricultural managers on the most appropriate rice production schedule based on climate and soil characteristics using of both crop models and GIS. As part of the effort to help improve rice production of the region, this study is expected to contribute to improve rice yields.

METHODOLOGY

Site Selection

Nakhon Ratchasima was selected as a study area for demonstrating the effects of climate variability on rice production under rainfed condition. The province is administratively divided into 32 districts. It is located between 14°–16° N latitude and 101°–103°E longitude. The total area is approximately 2,049,396 ha, 28% of which is classified as paddy area (Nakhon Ratchasima Agricultural and Cooperative Office, 1998).

Climate

Nakhon Ratchasima has two pronounced seasons: wet during May to October and dry from November to April. Most precipitation is due to monsoon rain that reaches Nakhon Ratchasima from the southwest. Solar radiation is very high in April and decreased gradually during the wet season, which is also true for temperature and pan evaporation. The rainfall pattern is weakly bimodal with peaks in May and September.

Soil Parameters

By using overlay functions in the GIS (ARC/INFO from ESRI, 1991), the land use map (1:250,00; Figure 1) was digitized and overlaid on a soil series map to derive a paddy soil map of Nakhon Ratchasima. Ten of the 18 paddy soil series identified as paddy with the area greater than 2,000 ha were used as input to the crop model. Soil parameters used in the simulation were obtained from a DLD database, DLDSIS version 1.0.

Crop Model

CERES-RICE Model (Ritchie et al., 1986), a physiological based rice (Oryza sativa L.) model is selected as a tool in the project for simulating growth and yield of rice under different environment and management strategies. The model can estimate yield potential by combining the properties of crops, soil and weather. Its also helps to

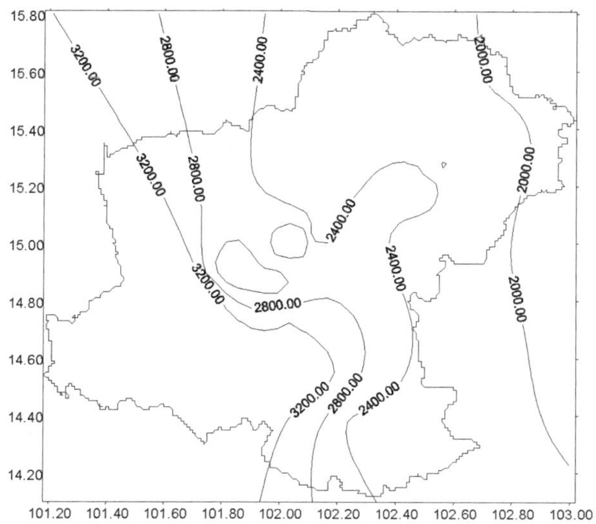

Figure 1. *Iso-yield map of rainfed rice of Nakhon Ratchasima interpolated form the 50th percentile of potential rainfed rice yield.*

assessment the variability and the risk of production and effects of planting dates on yields (Jones, 1986; Uehara, 1989). The model assumes complete control of growth limiting factors such as weeds, insects, diseases and other management variables (i.e., phosphorus, potassium, liming, etc.) (Ritchie et.al., 1986).

Crop Input Parameters

A photosensitive variety, 'KDML 105,' the most popular variety grown in Nakhon Ratchasima, was selected for this study. Genetic parameters derived from data from 12 consecutive monthly plantings at the Multiple Cropping Center, Chiang Mai University, Chiang Mai, Thailand (Jongkaewattana and Vejpas, 1998) were used as input to the model.

Meteorological Data

Twenty-three years (1975-1997) of daily precipitation in Nakhon Ratchasima were collected from Royal Irrigation Departments and the Meteorological Department. The combined map of 28 rainfall stations and 10 soils was obtained by overlaying the rainfall station map and paddy soil map. Simulation for each soil series used the weather data from the closest rainfall station. Daily solar radiation (MJ.m-2), maximum temperature (°C) minimum temperature (°C), and rainfall (mm) are required as the minimum weather data input for the CERES-RICE Model (Tsuji et.al., 1994). All the rainfall stations recorded daily precipitation, maximum temperature (°C) and minimum temperature (°C) but were equipped to record solar radiation. Therefore solar radiation in all weather data files were obtained from conversion of historic daily sunshine hours at the Muang Meteorological Station in Nakhon Ratchasima (Lat. 14.97° N and Long. 102.10°E).

Model Validation

A data set of experiments conducted under the ACIAR Rainfed Lowland Rice Project was obtained from the Rice Research Institute, Department of Agriculture. Conditions at seeding and transplanting in each experiment were used as inputs to simulation runs. The experiments were done during 1992 and 1993 at the Rice Research Station Surin, Khon Kaen, Chumphae, Phimai and the Rice Research Center at Sakon Nakhon and Ubon Ratchathani. Simulated and observed phenological events (panicle initiation, anthesis and maturity) and yields were compared.

Model Application

The simulation scenarios included 7 planting dates (every 2 weeks from 1 June to 1 September), transplanting of 25 days old seedling, rainfed condition, 23 years of weather data, and crop parameters for 'KDML 105.' The simulation was run for each of 10 dominant paddy soil series.

Risk Analysis

Using first-degree cumulative distribution function, the best planting date was identified from 7 planting dates in 23 simulation runs. Yield potential was based on 50th percentile simulated yields.

Output Maps

The GIS located the points of each soil series within 5 km of each rain station used in the 23 simulation runs. The Krigging interpolation function of the GIS was then used to interpolate rainfed rice yield potentials.

RESULTS AND DISCUSSION

Model Validation

Model validation was done using ACIAR's project data. Experimental and simulation results for crop phenology events were compared for all experiments at six locations in 1992 and 1993. The results suggested that the model overestimate days to panicle initiation and anthesis but are quite accurate in estimating maturity date. Grain yields were estimated well, with R^2=0.70**.

Model Application

In general, the farmers in Nakhon Ratchasima grow rice, especially 'KDML 105,' at different times starting in June each year, depending upon the onset of the rainy season. Base on analysis of Nakhon Ratchasima 1954–1993 rainfall data, Somchai (1996) reported that the onset of the rainy season is May 24th (R.M.S. = +25) and its termination date is September 28th (R.M.S. = +19). The average number of rainy season day is 155, long enough for growing rice. Farmer decisions of when to grow rice, based on their experience, are not always right.

The best time of planting was selected by stochastic dominance analysis of long-term simulation results. Then each best time of planting was determined based on 50th percentile potential yields. However the best time of planting for those 10 soil series tends to be either the 1st June or 15th June. This means that the longer the duration, the higher the rice yields. Longer duration also produces higher biomass. However, this may result in lodging with consequent yield loss.

Figure 1 shows the potential rainfed rice yield map for Nakhon Ratchasima. The map indicates that the rice growing areas in the Northeast of the province have lower potential yield (about 2,000 kg/ha) than in the Southwest of the province (about 3,200 kg/ha). It indicates that yield variability in the Northeast is higher than in the Southwest.

The combination of GIS and simulation modeling is a powerful tool for agro-ecological studies on regional scales (Aggarwal, 1993; Garcia, 1993 and Woperies et.al. 1993). Those studies have demonstrated that point simulation results can represent whole areas known as production zones. This study tried to demonstrate another way of interpolate point data to spatial data with the use Krigging interpolation.

ACKNOWLEDGEMENT

This research received equipment and financial support from IC-SEA, NOAA and START. Contribution of research space and time by Suranaree University of Technology is also acknowledged. I would like to thank Mr. Somchai Baimuang for providing meteological data, and Dr. Rajendra Shrestha, GIS specialist at A.I.T and the Royal Irrigation Department for providing rain data.

REFERENCES

Aggarwal, P.K. 1993. Agro-ecological zoning using crop growth simulation models: characterization of wheat environments of India. In Systems Approaches for Agricultural Development. Penning de Vries et al. (eds.) Kluwer Academic Publishers. 542p.

Baimuang, S. 1996. Agroclimatic Zone Classification for Rice and Sugarcane Using AVHRR Data in the Northeast Region, Thailand. Asian J. Sci. Technology Development 13(1)(1996) 37-77.

Fukai, S. 1994. Plant Improvement of Rainfed Lowland Rice in Drought prone Areas of Thailand and Laos. Summary of Research Achievement of ACIAR Project PN9045. Department of Agriculture, The University of Queensland. 31p.

Garcia, A.G. 1993. Crop simulation models as an analytical tool for delineation of production zones. Proceeding of the "International Workshop on Agro-Ecological Zonation of Rice" held at the Zhejiang Agricultural University, Hangzhou, P.R. of China, 14—17 April 1993./ B.A.M. Bouman, H, H.H. van Laar and Wang Zhaoqian (Eds.). Wageningen: CABO-DLO.-III. 201p.

Jintrawet, A. 1995. A Decision Support System for Rapid Assessment of Lowland Rice-based Cropping Alternatives in Thailand. Agricultural Systems 47: 245—258.

Jongkaewattana, S. and Vejpas, C. 1998. Validation of CERES-RICE Model. In the homepage name http://www.mcc.aggie.cmu.ac.th/research/DSSARM/ThaiRice/ricevalid.ht.

KKU-Ford System Project. 1982. An agroecosystem analysis of Northeast Thailand. Faculty of Agriculture, Khon Kaen University, Khon Kaen, Thailand.

National Statistics Office. 1997. The survey of migration in Bangkok. National Statistical Office, Office of the Prime Ministry, Bangkok.

Pannangpetch, K. 1993. Application of model simulation to evaluate rice production at the district level. In Agro-ecology of rice based cropping systems. B.A.M. Bouman, H.H. Van Laar and Wang Zhaoqian (ed.) Wageningen: CABO-DLO.- III. SARP Research Proceedings, October 1993. pp.16—26.

Ritchie, J.T., Alocilja, E.C., and Uehara, G. 1986. IBSNAT/CERES Rice Model. Agrotechnology Transfer. 3:1-5.

Tsuji, G.Y., Uehara, G., and Balas, S. 1994. DSSAT v3. University of Hawaii, Honolulu, Hawaii.

Wopereis, M.C.S., M.J. Kropff, J.H.M. Wosten & J. Bouma, 1993. Sampling strategies for measurement of soil hydraulic properties to predict rice yield using simulation models. Geoderma 59: 1 –2 0.

Impacts of ENSO on Rice Yields in Asia

Helen Grace S. Centeno[1], David D. Dawe[1], G. Hammer[2] and J. Sheehy[1]
[1]International Rice Research Institute, Philippines
[2]Agricultural Production Systems Research Unit, Australia

SUMMARY

Ninety percent of the world's rice is grown and is eaten in Asia. Its monsoon climate favors the growth of irrigated and rainfed rice. Seventy-five percent of the world's total rice production comes from the irrigated rice ecosystem in Asia.

Annual climate variability in tropical Asia is affected by the El Niño-Southern Oscillation (ENSO) phenomenon. Generally, the change in southern oscillation index (SOI) from April to May has been associated with 'likely' climate. El Niño is associated with low SOI while La Niña is related with high SOI.

With nonlimiting water and nutrient supply, radiation and temperature determine the potential production of rice. High radiation (20–25 MJ m^{-2} day^{-1}) during an El Niño event promotes high production through increased crop photosynthesis under irrigated rice farming. However, very high temperature (>35°C) causes spikelet sterility and consequently, low yield.

In the Asia-Pacific region, El Niño is associated with clear skies and droughts, while La Niña is related to overcast skies and floods. Filipino farmers respond to water availability by taking land out of production during drought years (Table 1). In the 1997 El Niño event, 40% rainfed rice lands were unproductive while only 13% irrigated lands were left as fallow land.

Table 1. % change in harvested area for total, irrigated and rainfed rice grown during the wet season in the Philippines.

Year	Total	Irrigated	Rainfed
1972	–17	–14	–24
1982	–9	3	–23
1986	–4	–5	–3
1991	–10	–8	–13
1997	–21	–13	–40

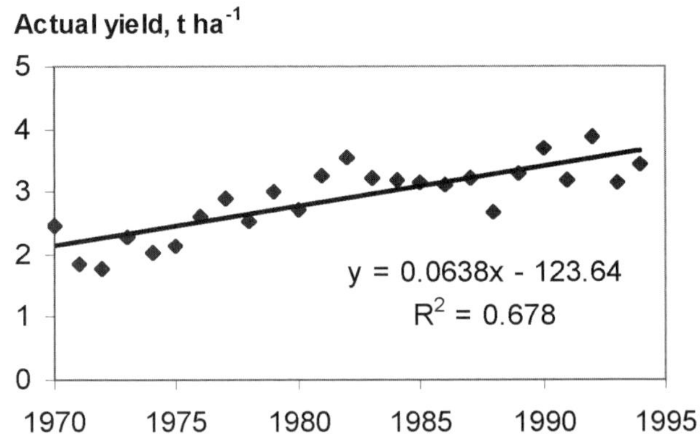

***Figure 1** Regression line for irrigated rice grown during the wet season in Central Luzon.*

Technology has improved rice yields linearly since the Green Revolution period in early 1970s. Use of modern rice varieties, improved irrigation systems, and proper timing of fertilizer applications are part of this improvement. When technology is accounted for, then the weather explains the variability in yield.

Figure 1 showed the regression line between irrigated rice and years in Central Luzon. Rice yield has been increasing by 64 kg ha-1 each year. 'Actual residuals' are calculated from the regression line to account for weather.

Assuming that biotic and abiotic stresses are not limiting, crop growth models give predictions of yield that solely depend on weather. One such model is based on a linear relationship between yield and intercepted solar radiation; the radiation coefficient factor (RCF) links the two. The simple RCF model has been compared with other more complex rice models (Mitchell et al, 1998) and was found to perform adequately for irrigated rice.

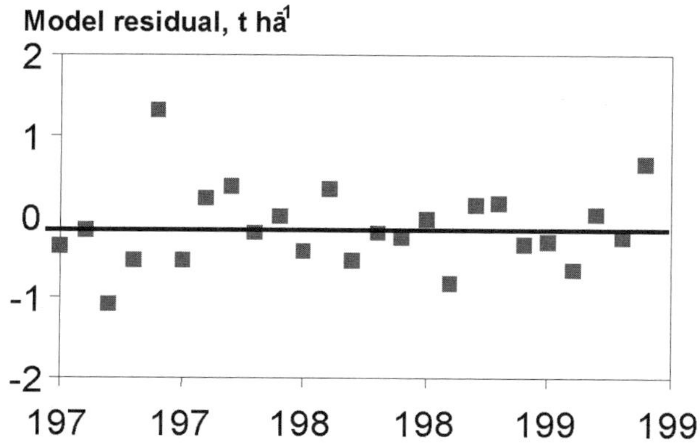

Figure 2. *Model residuals from predicted irrigated rice yields in Central Luzon.*

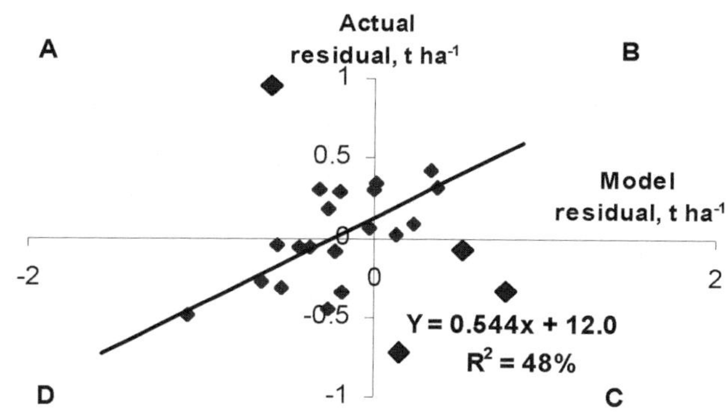

Figure 3. *Comparing actual yield residuals with predicted yield residuals.*

RCF model was used to predict yields for irrigated rice in each wet season for the years 1970-1995, and the average line was put through the predictions. Model residual is calculated as the difference between the predicted value and the average yield (Figure 2).

Actual and model residuals were compared in Figure 3. If solar radiation accounts for the variation in yield, we would expect a linear relationship from quadrant D to B. Data in quadrant C would be accounted for by the effects of water and water-related stresses. Data in quadrant A cannot be explained currently, and more research is required.

ACKNOWLEDGEMENT

This is a report from a pilot study that is jointly funded by SysTem for Analysis, Research and Training (START) and National Oceanic and Atmospheric Administration (NOAA).

REFERENCES

Cantrell, R.P. 1999. Abolishing poverty in Asia. In: IRRI Corporate report Rice: Hunger or hope? International Rice Research Institute, P.O. Box 3127 Makati City, Philippines. pp 2-4

Geng, Shu. 1999. Potential impact of global environmental change on world food production. In: Proceedings of the Int. Symp. on World Food Security, Kyoto, pp. 25-30.

Mitchell, P.L., J.E. Sheehy and F.I. Woodward. 1998. Potential yields and the efficiency of radiation use in rice. International Rice Research Institute Discussion Paper Series No. 32. Makati City, Philippines.

Oldeman, L.R., 1984: Climate of the Philippines. Paper presented at the Seventh International Forum on Soil Taxonomy and Agrotechnology Transfer, 12-23 March 1984, Los Baños, Philippines

Stone, R.C. and A. Aucliems, 1992: SOI phase relationships with rainfall in eastern Australia. Int. J. Climatol. 12, 625-636.

Wopereis, M.C.S. 1993: Quantifying the impact of soil and climate variability on rainfed rice production.

Vulnerability of the Agricultural Systems in Algeria to Climate Change: The Environmental and Socio-economic Incidences

Ahmed Adjez
National Climate Center, Algeria

SUMMARY

Algeria is probably the driest country in North Africa and around the Mediterranean basin. Since the mid 80's, Algeria has had a long run of below-normal rainfall with very few normal years: bad erratic rainfall and unfavorable rainfall distribution characterize the rainfall regime. In addition, the frequency and extent of drought conditions have tended to increase in recent years. Furthermore, the northern part of the country, where 70% of the population is concentrated in 10% of the territories, has experienced massive land-use changes in the last 30 years. Human population demographics, government policy in agriculture and transmigration and other socio-economic factors have all had an influence on the land-use patterns of the region. The combined consequences of these processes have been a noticeable impact on food security and have also decreased the livelihood options of households. If, for the developed countries, agricultural production is one of the pillars of their economy, Algeria is not sufficient, even in good years, and has to import large quantities of wheat and other agricultural products from abroad: Algeria spends around 5 billiard US$ to appropriate 5 to 10% of the world agricultural production put in the trade market (as much as China!)

Based on the study of the long-term temporal characteristics of climate variables, an overview of the impact of climate variability on crop productivity is given. In attempt to understand agricultural land use and its response to environmental change, data describing the spatial and temporal variability in crop yields are presented, as well as an assessment of yield changes in response to climate change, mainly rainfall variability. The agricultural sector has an incidence in the economical conditions of the country, and may hamper its development process in the future. Reliable predictions of rate and onset of climate variability, by permitting mitigation measures to reduce the variability of agricultural productivity and improve food security in the face of climate variability, could likely have had diminished the cost of our food bill.

Results of this study suggest that there is an urgent need to adapt seasonal forecasts for this sector, in order to adapt to climate change and that various adaptation strategies are needed, through new crops and crops varieties, improved water management, more efficient irrigation systems, etc.

Assessment of Irrigation Needs and Crop Yield for the U.S. under Potential Climate Changes

Kelly Brumbelow[1], Aris Georgakakos[1], and Kostantine Georgakakos[2]
[1]Georgia Tech University, USA
[2]Hydrologic Research Center, USA

SUMMARY

As a part of the Water Resources Sector Assessment of the U.S. National Climate Change and Variability Assessment, changes in irrigation demands and crop yields were assessed using the DSSAT suite of crop models (Tsuji et al., 1994). These models are physiologically-based simulation models with a proven record of technical development and field testing. The models were calibrated using historical values of crop yield obtained from the U.S. Department of Agriculture. The calibration rationale was to define the moisture stress threshold (MST) that matches the simulated and historical crop yields. These MST values were then used for all future climate scenario simulations, implying that farmers' drought tolerance and irrigation attitude will be in the future similar to what it has been in the past. The results are expressed as crop irrigation demands and associated potential yields. Soil moisture content in the root zone profile was initialized at the beginning of each growing season using the values calculated by the macro-scale soil moisture model of K. Georgakakos and Smith (2000). This is a distinguishing difference of this study compared to past and contemporaneous studies. Namely, the soil moisture sequence driving the crop models is hydro-climatically consistent with the atmospheric forcing generated by the Canadian Center for Climate Modeling and Analysis CGCM1. To match the spatial resolution of the soil moisture model, the crop assessments were conducted using the National Climate Data Center (NCDC) climate divisions as fundamental units. Because the crop simulation models work on a daily time step, the CGCM1 parameter values were downscaled from monthly to daily using a stochastic weather generator that preserved the monthly statistics of the climate scenario and the historical interrelationships of the atmospheric variables. Assessments were conducted for five crops: peanuts, durum wheat, soybeans, winter wheat, and corn. These crops were selected because they have a cumulative cultivation area that includes most of the conterminous U.S., their respective growing seasons extend throughout most of the year, and they represent a large share of the national agricultural economy (over $44 billion in 1998 farm-to-market value). Changes in irrigation demand and crop yield were assessed by comparing values from the 1994-2013 period of the CGCM1 climate scenario with values from the 2041-2060 period. These two periods were chosen to represent the current climatic conditions and the future conditions after significant warming and CO_2 accumulation takes place. The assessment was focused both on mean trends and extreme drought tendencies. All simulations accounted for the effect of atmospheric CO_2 on crop growth.

A summary of the some of the assessment results appears Figures 1 to 5. These figures present maps of changes in mean irrigation requirements for the five assessment crops. A more complete presentation of the assessment results is given in Brumbelow and A. Georgakakos (2000). From these results, several interesting conclusions were reached for regional agricultural changes. Peanut irrigation demands (growing season extending from mid-May to late summer) increased both in the Southeast (96% increase in mean demands) and Southern Great Plains (22% to 51% increase) with a significant increase in variability in the Southeast. Durum wheat irrigation needs decreased significantly in California (82% decrease; growing season extending from mid-December to early summer) and remained at near-zero levels in the Northern Great Plains (growing season extending

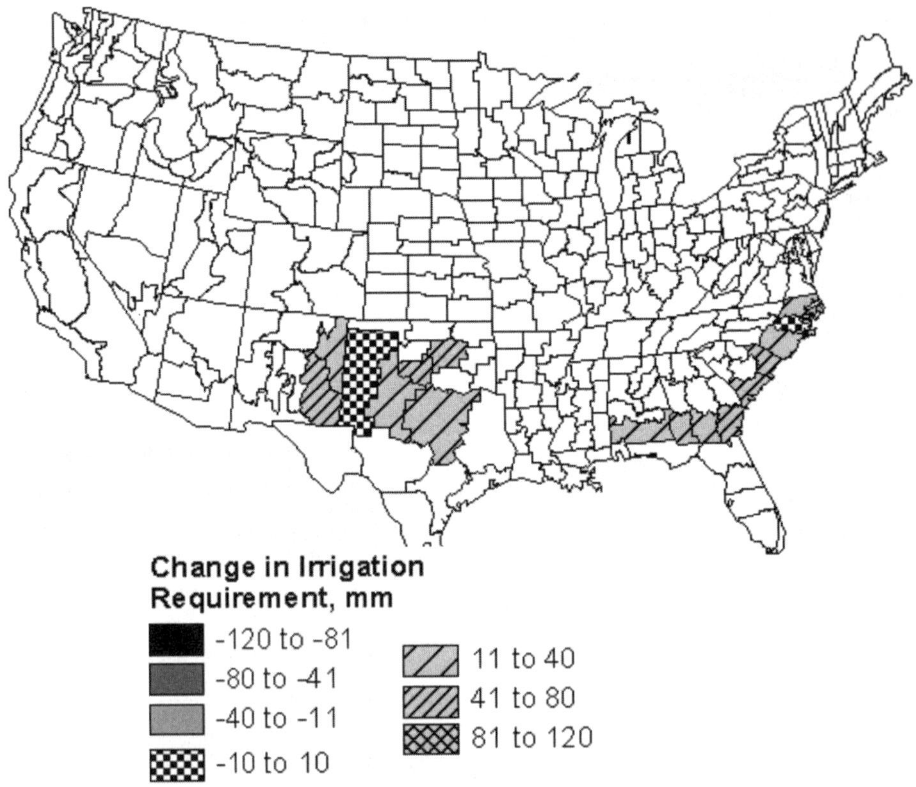

Figure 1. *Changes in Irrigation Requirements for Peanuts.*

from early Spring to mid-Summer). Soybean irrigation requirements (growing season extending from mid-May to late Summer) increased significantly in the southern portion of its cultivation range (86% to 158% increase) with an accompanying increase in variability, but northern areas had only slight changes in irrigation demands (6% to 31% increase). Winter wheat irrigation needs (growing season extending from mid-Autumn to late Spring) decreased in the northern and western

Figure 2. *Changes in Irrigation Requirements for Durum Wheat. (legend same as Figure 1)*

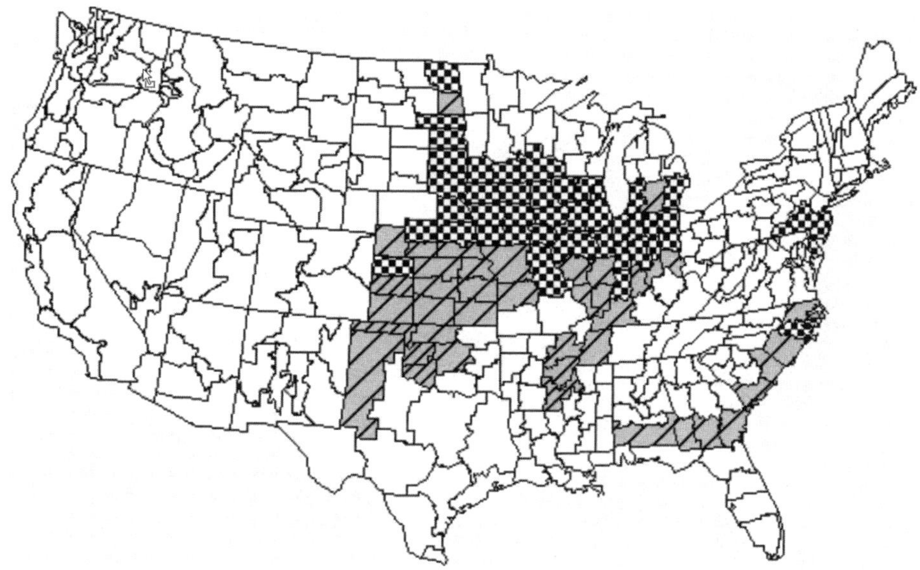

Figure 3. *Changes in Irrigation Requirements for Soybeans. (legend same as Figure 1)*

regions (27% to 74% decrease) and increased in the Southern Great Plains and Southeast (22% to 75% increase). Corn irrigation demands (growing season extending from mid-May to late Summer) strongly decreased west of the 104th Meridian (40% to 75% decrease) and were otherwise only slightly changed. While these trends are in general agreement with the Agricultural Sector assessments (of the U.S. National Climate Change and Variability Assessment), the variability estimates reported herein are consistently higher. The reason for this difference is attributed to the fact that the Agricultural Sector studies do not fully incorporate the soil moisture variability implied by the CGCM1 Climate Scenario. Soil moisture is both spatially and temporally variable (K. Georgakakos and Smith, 2000, and A. Georgakakos and Yao, 2000) and has a significant influence on irrigation

Figure 4. *Changes in Irrigation Requirements for Winter Wheat. (legend same as Figure 1)*

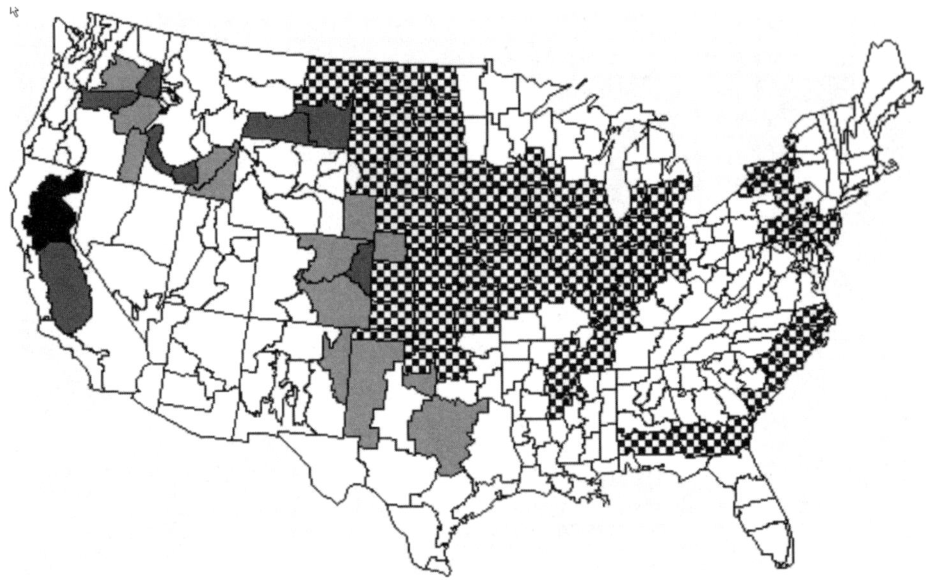

***Figure 5.** Changes in Irrigation Requirements for Corn. (legend same as Figure 1)*

demands and crop yields. In this regard, the need for reliable soil moisture measurements cannot be overstated.

Assessment results also indicated that crop yields mostly increased (assuming that the previous irrigation demands are met), with notable exceptions for winter wheat in the southern regions, and corn in all areas except the extreme northern and northwestern areas. Variability in crop yields increased for almost all areas for peanuts, durum wheat, soybeans, and winter wheat, and it increased for corn in the southern areas.

This study's assessment of the agricultural impacts in the US are in agreement with the CGCM1 climate scenario's trends of a wetter climate in the west, a dryer climate in the east, and warmer temperatures throughout the U.S. Depending on which particular factor is most limiting for crop growth over the growing season (i.e., water availability, temperature, or both) the US agricultural response exhibits marked regional changes in a west-to-east direction around the 104th meridian for corn, (b) a north-to-south direction around the 40th parallel for soybeans and durum wheat, and (c) a northwest-to-southeast direction for winter wheat.

REFERENCES

Brumbelow, K., and A. Georgakakos. 2000. Assessment of US Irrigation Needs and Crop Yields under Potential Climate Changes. J. Geophys. Res. (in review).

Georgakakos, A., and H. Yao. 2000. Climate Change Impacts on Southeastern U.S. Basins, 1, Hydrology. Water Resources Res. (in review).

Georgakakos, K.P., and D.E. Smith. 2000. Soil Moisture Tendencies into the Next Century for the Conterminous United States. J. Geophys. Res. (in review).

Tsuji, G.Y., G. Uehara, and G. Balas. (eds.). 1994. Decision Support System for Agrotechnology Transfer, Version 3. University of Hawaii, Honolulu.

SESSION IV. APPLYING FORECASTS IN MANAGING AGROECOSYSTEMS

Changes in Crop Management in Response to the Seasonal Climate Forecast in Zimbabwe during a La Niña Year

Jennifer Phillips[1], Leonard Unganai[2], and Ephias Makauzde[3]
[1]International Research Institute for Climate Prediction, USA
[2]Zimbabwe Department of Meteorology
[3]University of Zimbabwe

INTRODUCTION

Communal area farmers in Zimbabwe are targeted as a primary end-user of seasonal forecasts, with the expectation of reducing food-insecurity and improving livelihoods. Yet, a number of constraints in the production systems of resource-poor farmers potentially limit their ability of utilize newly-available climate forecast information (Phillips et al., 2000). We sought to track intended and actual usage of forecasts in a three-year study across a range of rainfall zones in Zimbabwe. Study objectives included assessing access to climate information in rural areas, use of traditional forecasting schemes, and identification of limitations and opportunities to improve farm management. This report focuses on management responses during the 1998/99 cropping season, a La Niña year in which above normal rainfall was predicted. Given that farming systems in this region are well-tuned to handle poor rainfall years (Scoones, 1996), we were particularly interested in seeing if farmers could take advantage of information to maximize production in a relatively good rainfall year.

Table 1. *Rainfall Zones in Zimbabwe*

Natural Region	Annual Rainfall	Characteristics
2	Well distributed	700-1500 mm
3	Seasonal drought	500-700 mm
4	Poor distribution	450-600 mm
5	Very poor distribution	< 500 mm

METHODS

Four hundred and fifty households were surveyed across four rainfall zones (Table 1). Interviews were conducted in the local languages of Shona (in Murehwa and Buhera Districts in the east) and Ndebele (in Inkosikazi and Tsholotsho Communal Areas in the west). The dominant crops in Natural Regions (NR) 2 and 3 are maize, millets, and groundnuts, primarily for home consumption but also for markets. In the drier zones 4 and 5 the millets and sorghum are predominant, with maize and groundnuts also grown. Cotton and sunflower are grown for cash in both areas. Field preparation is performed with ox-drawn ploughs when available, and hand labor. Access to credit is extremely limited, with family and neighbors being the primary source of assistance in purchasing inputs for the cropping season (Zinyama,1992).

Responses to questions posed before the season begins regarding what farmers intend to do might not represent actual practices because unforeseen limitations may alter decisions. Therefore, the same set of farmers was visited both before the beginning of the rainy season (October 1998) to assess intentions for the growing season, and after harvest (May 1999) to record actual outcome.

RESULTS

Climate Expectations and General Management

Both the Zimbabwe Met Service and traditional indicators supported the expectation of wetter than average conditions for the season. That year, the number of farmers who heard the "official" forecast was only about 36%, but most farmers had some expectation of "good" growing conditions based on traditional forecasting schemes. This expectation was associated with intended changes in crop management, which varied by rainfall zone.

The most prominent alteration in crop management identified by farmers surveyed in 98/99 (Figure 1) was altering the area planted. In some cases, farmers also mentioned changing planting date, and less frequently changing crop or variety. Farmers in this population are generally planting approximately 2/3 of their arable land in a season, but if draft power and labor can be secured, there is room to expand the area cultivated.

Regional Changes in Area Planted

In general, area planted per household is highest in the wetter zones and decreases with the decrease in average rainfall. This pattern was altered in the 1998/99 season as a result of expected excess moisture in the wetter regions, and hopes of good conditions in the drier zones. In October, before the rainy season began, farmers in the wettest area (NR 2) said they intended to decrease the area planted compared to what they indicated as "usual" practice (Figure 2). NR 3 showed similar but less extreme intentions. In contrast, moving into the drier zones 4 and 5, farmers planned to increase the area planted. In NR 5, the driest zone, farmers intended to more than double the size of the cropped area. In the end, these intentions were taken even further than originally planned, with more extreme changes made in response to observations of the rainfall during the season. In NR 2, some areas received record high rainfall and flooding was common. Many farmers complained of crop diseases associated with cool cloudy weather, and fields washing away or seeds rotting in the ground. In zone 5, however, rainfall was sufficient and well distributed warranting expanded cropping.

Intended Changes in Area Planted to Major Food Crops

Maize requires somewhat greater soil moisture levels than pearl millet, as is reflected in their dis-

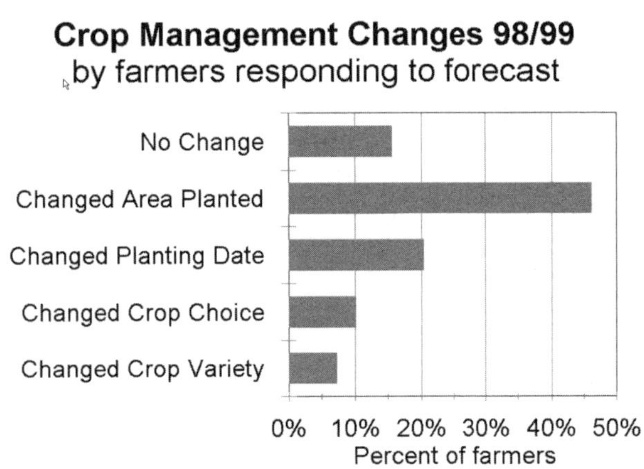

***Figure 1.** Percent of Communal Area farmers who had expectations of the seasonal rainfall in advance of the season in 1998/99 suggesting changes in crop management in Zimbabwe.*

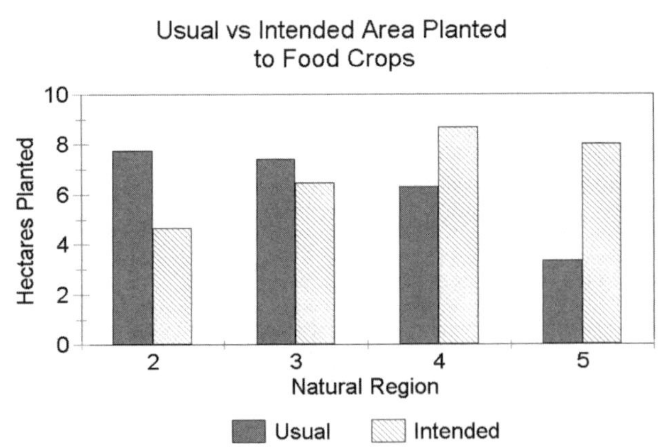

***Figure 2.** Farmers' responses to the question of what is the "usual" area they plant and intended area planted for the 1998/99 cropping season, combining maize, pearl millet, sorghum and groundnuts, shown by Natural Region.*

tribution across natural regions in Zimbabwe. Accordingly, the areas devoted to each was shifted in proportion to the expected rainfall. In the wettest zone, NR 2, farmers indicated intentions to decrease the area planted to both crops relative to their usual practice but proportionately favoring maize. In NR 5, intended area planted to both maize and pearl millet increased relative to usual. Farmers in NR 4 reported very little change in area for either maize or pearl millet. However, farmers in regions 4 and 5 hoped to take advantage of the expected rainy season to plant finger millet, which is important for beer brewing and generates income from sales but requires strong rainfall. That season, the farmers in our sample in NR 4 intended to increase the area planted to finger millet from 0.3 to approximately 2.5 ha, and in NR 5 from essentially none to about two hectares. Intended area planted to groundnuts, which are an important source of protein and also bring income from sales, also increased in the drier zones, with more than doubling of area in NR 5.

Final Changes in Area Planted to Maize and Millet

By the time of harvest, farmers in all four rainfall zones had been able to carry out their intentions with respect to changes in area planted (Figure 3). Acreage in both maize and pearl millet decreased in NR 2. Changes were minor in NR 3 and 4, but increases in area planted in NR 5 were greater than originally planned.

These changes are likely to reflect responses to the seasonal rainfall as it progressed, in combination with some expectation of rainfall for the period ahead. In NR 2, for example, many farmers attempted to plant, but had poor germination, or lost seedlings due to water logging. In NR 4 and 5, some farmers remarked that they continued planting quite late in the season when they saw that rainfall was good. Therefore the change in final area harvested cannot be interpreted solely as responses to a forecast, but was also shaped by "real-time" decision-making and adaptation.

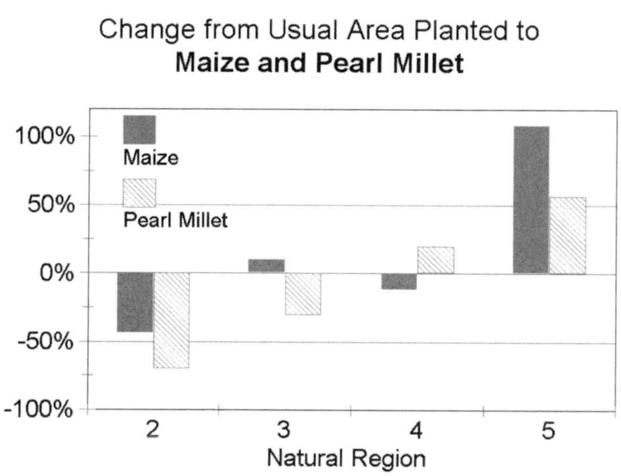

Figure 3. *Percent change in area planted to the two main food crops, maize and pearl millet, in the 1998/99 cropping season, compared to what farmers identified as "usual" area planted, by Natural Region.*

CONCLUSIONS

Although a number of factors limit decision-making for Communal Area farmers in Zimbabwe, there appears to be some flexibility in farm management that is taken advantage of in the context of seasonal climate forecasts. We found farmers in the high rainfall region (NR 2) cutting back area planted to all crops in the expectation of a very wet year in 1998/99, minimizing production losses. Farmers in the drier zones (NR 4 and 5) took advantage of the expected rainy season to substantially increase area planted, particularly in zone 5, thereby increasing total farm production for both food and cash-generating crops.

This is important in that it provides an example of farmers utilizing a forecast to improve production in a good year, as opposed to the more commonly assumed benefit of forecast as fore-warnings of drought. Appropriately communicated and supported, it is likely that for people living in low-rainfall zones, information regarding the rainfall conditions in a better than average year will prove to have higher value than warnings of a poor season (Phillips, 1998).

Nonetheless, some caveats apply: given that land shortages, labor and cash constraints as well as access to draft power for field preparation are limitations often cited by farmers in the Communal areas (Shumba, 1994; Zinyama, 1992; Phillips et al., 2000), one might ask if only a subset of farmers within the broad sample is able to take advantage of climate forecast information. Additional questions to explore might be: in what ways could strong fluctuations in area planted (and thus crop production) impact prices and food access for both producers and consumers? Another area of concern is that given that farmers appear to be responding to information, one needs to understand the degree of risk exposure this response creates, given probabilities for incorrect forecasts in the future. Finally, it will be important to assess the similarities and differences between this community of resource-limited farmers in Zimbabwe and other intended beneficiaries of season climate forecasts around the world.

REFERENC\ES

Phillips, J.G., L. Unganai, and E. Makaudze, 2000. Current and Potential Use of Seasonal Climate Forecasts for Resource-poor Farmers in Zimbabwe. American Society of Agronomy Special Publication "Impacts of El Niño and Climate Variability on Agriculture", in press.

Phillips, J.G., 1998. Comparing Optimal Use of ENSO Forecast Information in Two Socioeconomic Contexts. Invited talk at the American Association of Geographers Annual meeting, Boston, MA, March 1998.

Scoones, I., 1996. Hazards and Opportunities. Farming livelihoods in dryland Africa: lessons from Zimbabwe. Zed Books, London. 267 p.

Shumba, E.M, 1994. Constraints in the smallholder farming systems of Zimbabwe. In: E.T. Craswell and J. Simpson (Eds.) Soil Fertility and Climatic Constraints in Dryland Agriculture. Proceedings of the ACIAR/SACCAR workshop held at Harare, Zimbabwe, 30 August—1 September, 1993. ACIAR Proceedings No. 54. 137 p.

Zinyama, L.M., 1992. Local Farmer Organizations and Rural development in Zimbabwe. In: D.R.F. Taylor and F. Mackenzie (Eds) Development from Within: Survival in Rural Africa. Routledge, London. 284 p.

Experiences in Application of ENSO-related Climate Information in the Agricultural Sector of Argentina

Guillermo Podestá[1], David Letson[1], James Jomes[2], Carlos Mesian[2], Fred Royce[2], Andres Ferreyra[2], James O'Brien[3], David Legler[3], and James Hansen[4]
[1]University of Miami, USA
[2]University of Florida, USA
[3]Florida State University, USA
[4]International Research Institute for Climate Prediction, USA

INTRODUCTION

Recent scientific and technological advances have made it possible to predict, with usable skill, the occurrence of ENSO events with lead times of several months. Many assume that ENSO-related climate forecasts will benefit the agricultural sector by allowing farmers to mitigate potential negative consequences of climate variability or, alternatively, capitalize on potentially beneficial effects. However, we submit that the availability of climate forecasts is not sufficient to ensure that agricultural incomes will rise or that production costs will fall. We propose a conceptual framework for the successful application of ENSO-related climate forecasts. This framework, adapted from Pielke Jr. (1998), involves three components: (a) the generation of climate information, (b) the communication of that information, and (c) the use of climate information to modify decisions or policies. All three components have to be present for the climate information system to function effectively. A climate information system is defined broadly as the ensemble of functions and processes required for the production, dissemination, and use of climate information (including climate forecasts).

The proposed framework is illustrated with results from a pilot study focused on the agricultural sector of Argentina. The overall goal of this project was to develop a set of tools and methodologies for the effective use of climate forecasts in agriculture. Argentina is a major agricultural producer and the value of its agricultural exports was 50-60% of its overall exports and 5.5-5.9% of its GDP over 1989-1998 (Inter-American Development Bank, 1999). The central-eastern portion of Argentina, known as the Pampas, shows an ENSO signal (recent references include Grimm et al., 2000, and Montecinos et al., 2000; see references therein). Finally, interannual climate variability influences crop yields and returns in the Argentine Pampas (Podestá et al., 1999). Hall et al. (1992) describe the climate, soils, and crop production systems in the Pampas.

GENERATION OF CLIMATE INFORMATION

The first component of an effective climate information system involves the generation of climate information. Ideally, this information should not only be prognostic (e.g., a forecast of the occurrence of an ENSO event), but also diagnostic. Diagnostic information may provide relevant context, thus helping to interpret a climate forecast. For example, a forecast of enhanced likelihood of dry conditions may have different agricultural implications depending on current soil water availability. Our study did not aim to produce prognostic information; instead, we focused on historical diagnostics. A first necessary step toward the adoption and use of ENSO-related climate forecasts is the characterization of the ENSO impact on climate and agriculture in the Pampas. If there were no ENSO signal on crop yield or economic returns, it is unlikely that agricultural stakeholders would benefit from the forecasts (unless one considered impacts on other world producers of a commodity or on substitute commodities, which may influence output prices).

Statistical analyses of historical precipitation records confirmed the previously reported ENSO signature in the Pampas (Grimm et al., 2000; Montecinos et al, 2000). ENSO effects are most marked from November to February. Precipitation tends to be higher during warm ENSO events (or Niño years), whereas drier conditions are more likely during cold ENSO events (Niña years). When we computed the cross-validated percentage of variance explained by ENSO phase (Gershunov, 1998), we found that the variability of precipitation anomalies within a given ENSO phase (e.g., all Niño years) was comparable to the total variability (i.e., *all* years), except during a few well-defined periods. This means that although, on average, Niño summers tend to be wetter, a given Niño event may be dry, normal, or wet. The large intra-phase variability is typical of extra-tropical regions and has important implications for decision-making in agriculture and other sectors.

We explored ENSO impacts on agronomic (yield) and economic (net return) outcomes using two complementary approaches, statistical analyses of historical national-level crop records and modeling of current maize production systems. The historical analyses showed a positive effect of warm ENSO events on the yields of maize and soybeans. Above-normal maize yields (those in the upper third of the historical distribution detrended for technological changes) were twice as likely during warm ENSO events as by chance alone. A similar but opposite pattern was observed during cold events: low maize yields occurred twice as frequently as expected by chance alone. Sunflower, another summer crop, did not show much response to ENSO events. This finding provides a management alternative during cold event years, when low yields are more likely for other summer crops.

The historical analyses have some limitations: (a) crop records frequently encompass only a limited number of ENSO events, (b) it is difficult to establish the vulnerability of present agricultural production systems to climate variability, (c) it is not easy to separate effects of various sources of agricultural risk (e.g., climate vs. output prices), and (d) most statistical analyses are performed at national, state or county/district aggregation scales (due to data availability), but risk estimates from aggregated data may not be appropriate for decision-making at the farm or enterprise level. Modeling approaches can help overcome some of these difficulties, therefore we complemented our historical analyses with model-based analyses.

Our modeling effort addressed the question "what are the likely biological and economic consequences of ENSO events on current maize production systems in Argentina?" The approach involved a series of linked climatic, agronomic, and economic components. The climatic component simulated long synthetic daily weather series conditional on ENSO phase (Grondona et al., 2000). These series were used as input to the agronomic component, in which crop simulation models were used to produce distributions of maize yields by ENSO phase. In the economic component, simulated maize yields were combined with stochastically generated maize prices (assumed independent of ENSO phase) and information on production costs to obtain frequency distributions of economic net returns for each ENSO phase. Each outcome had an associated probability of occurrence. This is a result most relevant to policy/risk analysts, but which has not been reported in the literature very often (Schimmelpfennig 1996).

Extreme climatic events associated with ENSO are being increasingly discussed in the literature (e.g., Gershunov, 1998). However, much less attention has been focused so far on the influence of ENSO on extreme agricultural outcomes, probably because historical records frequently are short. In contrast, the numerous outcomes we simulated (990 crop cycles per ENSO phase) allowed exploration of extreme events (e.g., much-above normal yields). Focusing on central tendencies or extreme events may yield different perspectives of ENSO-related risk. For example, the central tendency (median) of simulated yields for cold events was about 71% of the median for neutral years in Pergamino (one of the locations examined). This indicated a relatively strong effect of cold ENSO events. On the other hand, when we considered extreme events (e.g., very high yields), differences between phases became more marked. We defined q90 as the yield that bounds the upper 10% of simulated yields for neutral years; this quantile serves as a reference value to explore ENSO effects. The chances of exceeding q90 were only 12% of the corresponding probability for neutral years, indicating a strong increase in yield risk from cold events. This increase in risk is proportionally much more pronounced than the differences in central tendency. The modeling approach also allowed us to explore the outcomes of various scenarios associated with initial availability of soil water or fluctuations in crop prices. Generally, high initial soil water reduced yield risks associated with ENSO events, whereas low initial water scenarios enhanced risks. An interesting result was observed in Pilar (a relatively marginal location for maize growing) for the scenario combining cold events and high initial soil water availability: the chances of exceeding any given yield were always higher (thus, yield risks were lower) than for a reference scenario (neutral years and medium soil water).

COMMUNICATION OF CLIMATE INFORMATION

This is the second required component of an effective climate information system, and the one that, until recently, received the least attention, possibly because it requires extensive interaction between natural and social scientists. Although this interaction has proven relatively difficult so far, there are encouraging signs that the situation is changing rapidly.

The routine availability of climate forecasts and the demonstrated impact of climate on agriculture do not by themselves argue persuasively for the economic value of climate information. Forecast dissemination and use are more complex than many researchers have assumed, and may be impeded by technical, financial and cultural barriers (Glantz, 1996). A necessary step in promoting agricultural use of climate information, or in assessing its value, is to gauge user perceptions concerning the use of that information (Stern and Easterling, 1999). As Morgan et al. (1992) advise "first, learn what people know and believe."

To assess perceptions of climate forecasting, we used personal interviews, focus groups, and a user survey to approach cereal and oilseed producers in Pergamino, a highly productive region in the Pampas (Letson et al., submitted manuscript). A recurring theme in our findings was that the 1997-98 Niño event was a key one in the development of awareness and concern about ENSO in our study region. Because of the magnitude of its climatic effects and the perceived success of climatologists in predicting both its occurrence and its effects on the regional climate, the 1997-98 Niño had an important influence on our respondents' attitudes towards climate forecasts. Nevertheless, most respondents (58%) did not change their decisions after receiving the forecast of a Niña event for 1998-99. This suggests stakeholders still are reluctant to use climate forecasts. The reasons respondents gave for their reluctance to change actions are consistent with our earlier findings of large inter-event variability and relate to the forecasts' spatial or temporal resolution. The combination of large spatial variability in the regional ENSO signal and known cognitive biases in decision makers' risk perceptions will influence their willingness to respond to ENSO forecasts. The spatial resolution of a forecast is important, as fewer than half of our respondents (43%) thought that their sub-region's climate was similar to that of others close-by.

What our respondents know about their regional climate and how they learned it help explain the patterns of forecast use we observed. Perhaps because of the strength of the 1997-98 event, most respondents (86%) knew that Niños tend to improve maize yields, but fewer than half (42%) were aware of La Niña's historically much stronger and detrimental effect on maize yields. When asked how they learned about their region's climate, most respondents said either from their own experience (61%) or from family elders (16%), rather than from sources such as extension or trade publications that might disseminate climate forecasts. Two positive findings are that increased exposure to climate forecasts and increased knowledge of regional ENSO effects each appeared to encourage forecast use. Parallel research on climate change has also found a correlation between the accuracy of users' understanding of climate and their acceptance of mitigation responses (Kempton et al., 1995; Bord et al., 1998). Promotion of climate forecasts should begin by educating users about the ENSO phenomenon and how it affects local climate, and those who deliver the message should learn how their audience thinks about climate.

USE OF CLIMATE INFORMATION

Simply documenting that ENSO affects crop yields or economic returns and having fluid access to a climate forecast does not imply that agricultural incomes will rise or that production costs will fall, regardless of how accurate and well-communicated a forecast is. For forecasts to have beneficial effects on this sector, and on society in general, they must induce changes in the decision-making process and in the actions taken by sector agents. The third required component of a climate information system, therefore, involves a choice process. To facilitate this process, a decision-support system is needed to evaluate the consequences of alternative management actions in response to an ENSO forecast.

To explore possible responses to an ENSO forecast, we have used historical weather series, biophysical crop models and optimization techniques to tailor land allocation among crops to ENSO phases for two representative farms in the Pampas (Messina et al, 1999). Differences in land allocation among ENSO phases were consistent with known climate response to ENSO and crop response to water availability. Crop mix and information value varied with crop prices, initial soil moisture and location, which again emphasizes the importance of context when interpreting climate information.

Differences between outcomes with and without climate information provide an estimate of the value of a forecast. This estimate, however, reflects the potential value over a long time horizon. That is, if farmers consistently react to the forecasts, in the long run they will profit, even though on a given year results may be unfavorable (e.g., if climate conditions did not turn out as expected). Agricultural decision-makers, however, are much more interested on the potential consequences over much shorter timeframes of acting upon wrong forecasts. For this reason, we are now simulating distributions of economic outcomes over decision horizons from 3 to 30 years to explore how information value changes with length of decision or planning horizon. Our simulations consider factors that may influence the value of climate information such as (a) the stochastic occurrence of ENSO warm and cold events (i.e., the temporal arrangement of ENSO events within the decision horizon), (b) the skill in forecasting an ENSO event, and (c) the large intra-phase variability in regional climate. The simulations also explore the economic context of decisions (e.g., various commodity price scenarios) and the decision-makers' attitude towards risk. These simulations will capture “the cost of being wrong” in a context more meaningful for decision-makers.

CONCLUSIONS

We have argued that an effective climate information system should involve three main components. The first component is linked with the generation of climate information (e.g., ENSO forecasts). Increased acceptance of climate information to support agricultural decision making will require going beyond a forecast of the occurrence of an ENSO event: regional quantitative forecasts (that incorporate local modulation of the tropical Pacific climate signal) are required. Furthermore, diagnostic information should be provided to assist in the interpretation of forecasts. The other two components of the proposed climate information framework should facilitate the dissemination and adequate comprehension of climate information, and the use of this information to modify actions of sector agents.

In more general terms, we submit that an effective climate information system will require close collaboration among scientists from various disciplines (both from the natural and social sciences) and institutions (governmental, academic, and non-governmental). Because the variety of talent required is beyond the capabilities of a single institution, innovative arrangements will be required to ensure fluid multi-institutional collaboration and information transfer. Most importantly, a climate information system ideally should involve stakeholders in some capacity. A “scientific-stakeholder collaborative” (e.g., as proposed by Cohen, 1997) can link scientific expertise with stakeholders' needs, expectations, and knowledge of factors beyond climate that influence agricultural decisions. In this context, the potential of ready-made intermediary institutions such as the agricultural extension system should be fully exploited. Translating imperfect ENSO-related climate forecasts into information useful for improved decision-making remains an interesting challenge, offering the opportunity to build bridges between the worlds of theory and practice.

ACKNOWLEDGEMENTS

This research has been supported by grants from the National Oceanic and Atmospheric Administration (Office of Global Programs) and the National Science Foundation (Methods and Models for Integrated Assessment Initiative) to a Consortium of Florida Universities (University of Miami, University of Florida, Florida State University). The Inter-American Institute for Global Change Research (IAI), Argentina's

Fondo para la Investigación Científica y Tecnológica (FONCyT), and the University of Buenos Aires provided additional funds. The National Meteorological Service of Argentina kindly provided climate data. The Pergamino office of the Cambio Rural program assisted in the field survey.

REFERENCES

Bord R.J., A. Fisher, and R.E. O'Connor. 1998. Public Perceptions of Global Warming: United States and International Perspectives. Clim. Res. 11: 75-84

Cohen, S.J. 1977. Scientists-stakeholder collaboration in integrated assessment of climate changes: Lessons from a case study of northwest Canada. Environ. Model. Assess. 2: 281–293.

Gershunov, A. 1998. ENSO influence on intraseasonal extreme rainfall and temperature frequencies in the contiguous United States: Implications for long-range predictability. J. Climate 11: 3192–3203.

Glantz, M.H. 1996. Currents of change: El Niño's impact on climate and society. Cambridge University Press, Cambridge.

Grimm, A.M, V.R. Barros, and M.E. Doyle, Climate variability in southern South America associated with El Niño and La Niña events. J. Climate 13: 35–58.

Grondona, M.O., G.P. Podestá, M. Bidegain, M. Marino, and H. Hordij. In press. A stochastic precipitation generator conditioned on ENSO phase: A case study in southeastern South America. J. Climate.

Hall, A.J., C.M. Rebella, C.M. Ghersa, and J.Ph. Culot. 1992. Field-crop systems of the Pampas. p. 413–449. In C.J. Pearson (ed.) Ecosystems of the World. Field crop ecosystems, Elsevier.

Inter-American Development Bank. 1999. Basic socio-economic data report: Argentina. Statistics and Quantitative Analysis Unit, Integration and regional Programs Department.

Kempton W, J.S. Boster, and J.A. Hartley. 1995. Environmental Values in American Culture. MIT Press, Cambridge, MA.

Letson, D., I. Llovet, G. Podestá, F. Royce, V. Brescia, D. Lema, and G. Parellada. User perspectives of climate forecasts: Crop producers in Pergamino, Argentina. Manuscript submitted to Climate Research.

Messina, C.M., J.W. Hansen, and A.J. Hall. 1999. Land allocation conditioned on ENSO phases in the Pampas of Argentina. Agric. Syst. 60: 197–212.

Montecinos, A., A. Díaz, and P. Aceituno, 2000: Seasonal diagnostics and predictability of rainfall in subtropical South America based on tropical Pacific SST. J. Climate, 13, 746–758.

Morgan, M.G., B. Fischhoff, A. Bostrom, L. Lave, and C.J. Atman. 1992. Communicating risk to the public: First, learn what people know and believe. Env. Sci. Technol. 26: 2048–2056.

Pielke Jr., R.A. 1998. What is a good forecast? WeatherZine February 1998. Available at www.dir.ucar.edu/esig/socasp/zine/.

Podestá, G.P., C.D. Messina, M.O. Grondona, and G.O. Magrin. 1999. Associations between grain crop yields in central-eastern Argentina and El Niño-Southern Oscillation. J. Appl. Meteorol. 38: 1488–1498.

Schimmelpfennig, D.M. 1996. Uncertainty in economic models of climate-change impacts. Climatic Change 33: 213–234.

Stern, P.C., and W.E. Easterling (Eds.). 1999. Making climate forecasts matter. National Academy Press, Washington, D.C.

On Applications of Regional Climate Predictions for Agriculture Activities in the State of Tlaxcala, Mexico: 97-98 El Niño and 99-00 La Niña

Tomas Morales-Acoltzi and Victor Magana Rueda
Center for Atmospheric Sciences, National Autonomous University of Mexico, Mexico

CLIMATIC VARIABILITY AT TLAXCALA STATE

In Tlaxcala State, as in other tropical and subtropical regions, the climate is characterized by a rainy period from May to September and dry period between November and April. Producers use this knowledge to select the time of sowing in the spring. One of the systems associated with climatic variability, is El Niño, which havs attracted increasing attention. It is probably the most important influence on interannual variability in many regions of Tlaxcala State.

RESULTS

El Niño impacts in Tlaxcala State are mainly in rainfall variability. For example, the 1982-83 event confirmed the relationship between SST anomalies in the NIÑO3 region and rainfall at Tlaxcala State (Fig. 1). The climate of Tlaxcala State has great spatial variability due to its complex orography and diversity of land use. However, the region's climate is modulated by large-scale, seasonal to interannual influences. The Tlaxcala climate depends strongly on the occurrence of El Niño or La Niña. Based on standard deviation criteria for SST and El

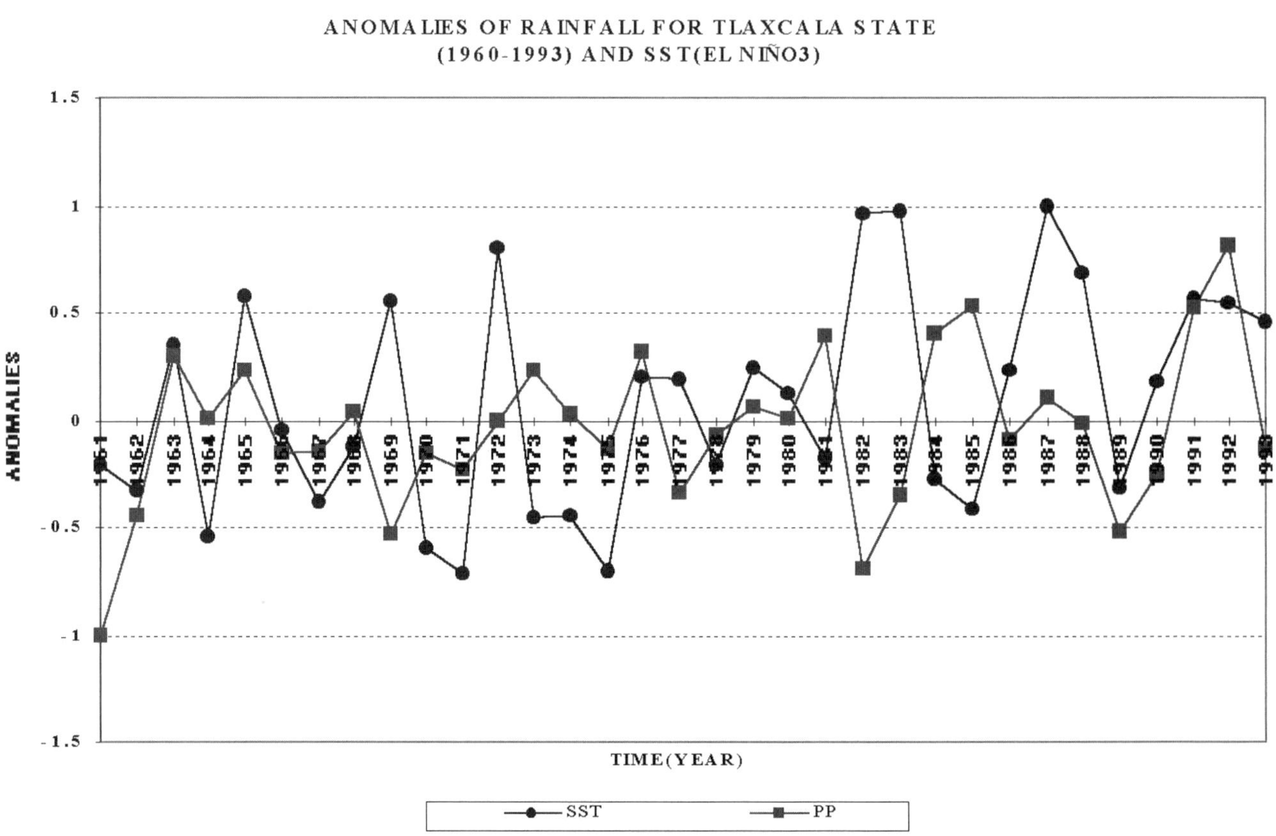

Figure 1. *El Niño impact of 82-83 on rainfall at Tlaxcala State*

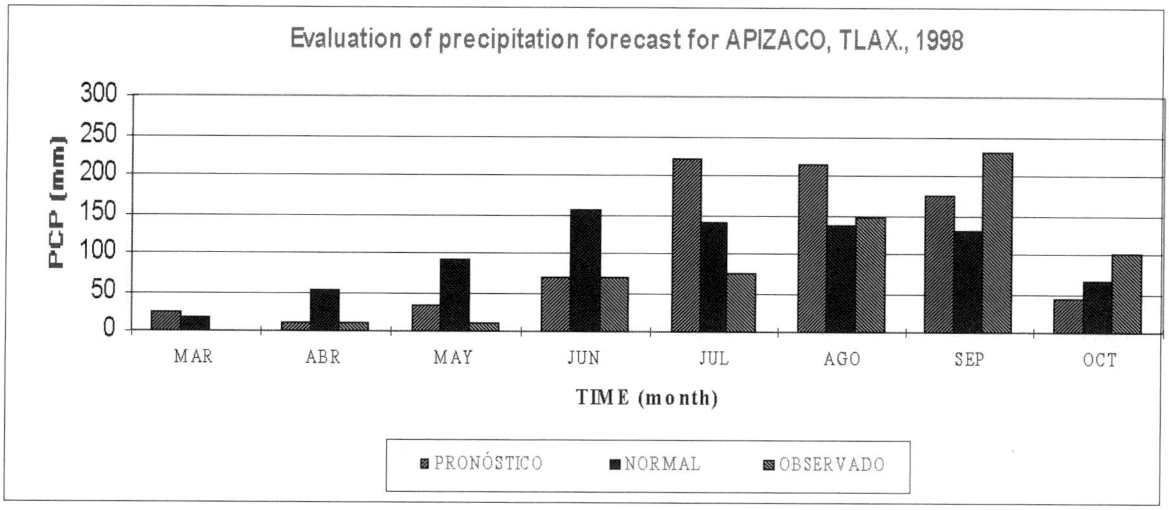

***Figure 2** Evaluation of precipitation forecast for APIZACO, TLAX. 1998*

Niño indices: summer in Apizaco in 1961, 1969, 1982, 1986, and 1993 may be classified as dry years, while . Wet years were 1958, 1987 and 1992, 1976 and 1991. Not all of them necessarily corresponded to La Niña years. 1980 and 1984 may be classified as normal conditions (Morales and Magaña, 1999).

THE FORECAST

The forecasts made in this study were verified against climatological medians and observations. The results presented here clearly indicate that there is a significant change in potential predictability from the warm to the cool phase of the ENSO cycle on a regional scale (Fig. 2). Although there is a strong relationship between ENSO and rainfall at Tlaxcala State, of a total of 9 ENSO years, only 4 were associates with deficient Tlaxcala rainfall. Tlaxcala State rainfall variability in the last 9 years of the present decade (1991-1999) demonstrates the complexity of the relationship.

DISCUSSION

The interaction between users and scientists involved in the Tlaxcala Project permitted the development of a "watchful forecast" in such way that the forecast system received feedback from the knowledge and observations of producers. Although is clear that climate forecasts have uncertainty, is not necessary to obtain "perfect forecasts" before producers begin use climate information.

ACKNOWLEDGMENTS

This work was financially supported by PRODUCE Foundation in Tlaxcala. Additional support was received from CONACyT SIZA-960603003-Autonomous University of Tlaxcala.

REFERENCES

Conde C., Ferrer R. M., Araujo R., Gay C., Magaña R. V. O., Pérez J. L., Morales-Acoltzi T. y Orozco S. 1999. EL NIÑO y la Agricultura: Cap. IV en Libro "Los Impactos de EL NIÑO en México" Editor Víctor O. Magaña R. Dirección General de Protección Civil, Secretaría de Gobernación. Págs. 229, (in Spanish).

Magaña R. V. O., Morales-Acoltzi T., 1998. Variabilidad Climática y Agricultura en México. CLARIDADES AGROPECUARIAS: Apoyos y Servicios a la Comercialización Agropecuaria-ASERCA, SAGAR, No. 57, Págs. 32-39, (in Spanish)..

Magaña R. V. O., Morales-Acoltzi T., Pérez J. L., Conde. C., Orozco S., Lezama J., Vázquez J. A. y Hernández M. 1999. Utilización de Pronósticos Climáticos en Actividades Agrícolas para el Estado de Tlaxcala. NOTAS:Revista de Información y Análisis, INEGI No. 7, Págs. 30-38, (in Spanish).

Morales-Acoltzi T. and Magaña R. V. O. 1999. Unexpected Frosts in Central Mexico During Summer. Annual Meeting 10-15 January, Dallas, Texas, USA pp. 262-263.

Orozco F. S. 2000. Cambio Climático Regional en Tlaxcala. Colegio de Geografía, Facultad de Filosofía y Letras, División de Estudios de Posgrado, UNAM. Tesis de Maestría (copias disponibles). Págs. 112, (in Spanish).

Regional Climate Forecast for the Summer of 2000 and it's Application in the Agricultural Activities of Tlaxcala, Mexico

Cecilia Conde, Rosa Ma. Ferrer, Víctor Magaña, Raquel Araujo, and Carlos Gay
National Autonomous University, UNAM, México

SUMMARY

Since 1998, producers, governmental decision—makers and researchers of the University of Tlaxcala and of the Center of Atmospheric Sciences of the University of Mexico, UNAM, have developed a strong interaction with the main objective of discussing regional forecasts and to design the consequent agricultural measures. Interviews and polls will be carried out during the months of March and April to collect information related to seed varieties and crop management that producers have decided to apply without using a climate forecast. For the spring-summer season of this year, posters and flyers describing the forecast will be distributed among the producers; some of them have accepted to use part of their fields as experimental, planting those seeds in the dates suggested by the forecast. Finally, the results of these three years of forecasts—decision making will be compared and discussed.

Rain in Mexico is highly dependant on conditions in the Pacific and Atlantic Oceans. The rain from May to October is fundamental for rainfed agriculture. We have learned that, in general, the occurrence of a strong El Niño leads to more winter precipitation in the northern part of the country, and less summer precipitation over most of Mexico (Figure 1).

In contrast, La Niña conditions lead to normal or even enhanced summer precipitation, but possible lower than normal precipitation in winter. Since 1997, Mexico has experienced extreme events such as droughts (1997, beginnings of 1998, 1999, 2000) and floods (September and October of 1998 and of 1999). In March and April of 2000, ten Mexican states were declared *Disaster Areas* because of a prolonged drought.

Since 1999, La Niña conditions have prevailed and might persist during all of 2000. Some forecasts indicate a possible transition to near normal summer conditions or even possibly El Niño conditions by the end of the year (Figure 2).

Given the weakening of the ENSO event, regional precipitation forecasts for Mexico for 2000 must include other factors such as the projected conditions (Penland, et al, 1998) for the North Tropical Atlantic (NTA). This is achieved using multilinear regression expressions that include the forecasted sea surface temperatures for El Niño 3 region and NTA (*http://www.cdc.noaa.gov/~lem/Atlantic.forecast.html*).

Our forecasts show that normal or above normal precipitation might appear in June for most of the country. The Mexican monsoon that begins in July might deliver rains slightly over normal to the northern States. During this month, the mid summer drought (MSD) in the center and south of Mexico (Magaña *et al,* 1999) is expected more acute than usual. For August, lower than normal rainy conditions are projected for the northwest region and above normal for the center and south of

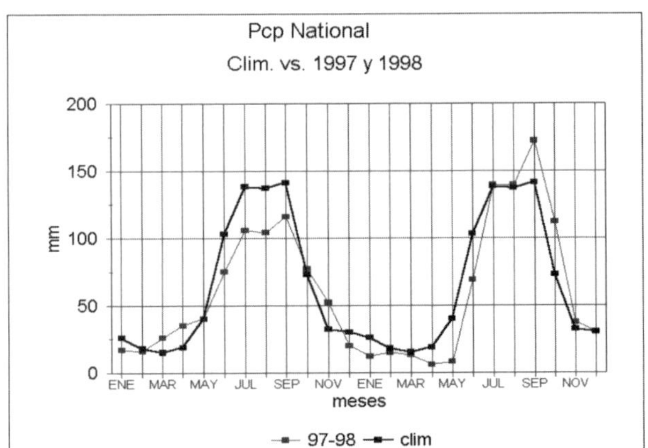

Figure 1. *Climatological Annual precipitation (1941—1998). Lower than normal rain appeared in the summer of 1997; a delay in the beginning of the rainy season in 1983 and upper than normal precipitation in September—October, 1983.*

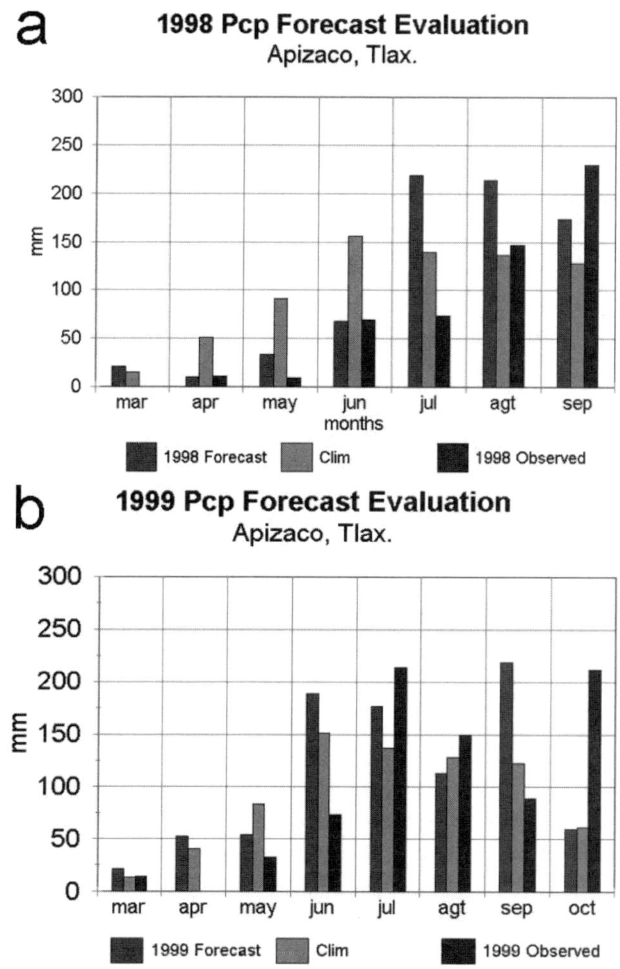

Figure 2. *(a) Forecasted 1998 precipitation compared with Clim (averaged 1961—1990) and with the observed data. (b) Forecasted 1999 precipitation compared with Clim (averaged 1961—1990) and with the observed data.*

Mexico. Finally, September might present below normal precipitation in most of the country.

If the easterly waves over the Atlantic are weakened, the precipitation in Mexico during the end of the summer season might be inhibited, particularly in the Gulf of Mexico. On the other hand, if La Niña conditions persist, summer precipitation might be normal or enhanced. These "opposed" tendencies impose the constant observation of both ocean conditions in the next months, and might determine the actualization of the presented forecast.

Since 1997, private, governmental and academic institutions are involved in the development of the project: "On the use of a climate forecast in the planning of agricultural activities in the State of Tlaxcala".

Seasonal forecasts are produced using statistical methods as described above. In the State of Tlaxcala and in some counties—like Apizaco—the summer rain is below normal during strong El Niño events, and normal or above normal during la Niña.

For monthly forecasts, an analogue method is used, comparing the current year with years with similar El Niño/La Niña conditions. Thus, 1983 was used as the analog year for 1998; the averaged of 1974 and 1984, as the analog for 1999 conditions. These results were presented on the month of March and discussed with farmers that gave us information about crop management and possible seed varieties that could be use.

In 1998, a severe drought affected most of the country; a preliminary cost analysis showed that the agriculture sector had losses near $1.4 billion USD. In Tlaxcala, the delivered forecast and the suggested measures by the Minister of Agriculture were used to decide the change of crops. By the month of May, the farmer's organization that collaborate with us (PRODUCE), decided to buy approximately 100 Ton of "non certified" seeds of oat (forage) at a lower cost; 900 hectares were planted and the oat harvested was use to fed their own cattle, was sold in the local market or kept to plant it in 1999.

Thus, the 1998 process of forecast—decision making was quite successful. The evaluation of these forecasts was made later on comparing with the observed data, and with the opinion of the producers, who found them very acceptable.

The floods in 1999 were called the "tragedy of the decade" by the President of Mexico. In Tlaxcala, 7,000 Ha were highly affected. The forecast was also satisfactory, even though the extreme event did not appeared in September (as forecasted), but at the beginning of October. Late June) and early September) frosts and the October floods destroyed many crops.

For the summer of 2000, producers expect a good year, based on their traditional forecast methods. Applying the analogue method, we compare this year with 1965 or 1976, two years that began with La Niña conditions and developed to neutral or El Niño conditions by the end of the year. The rain was above normal for both years. The 1965

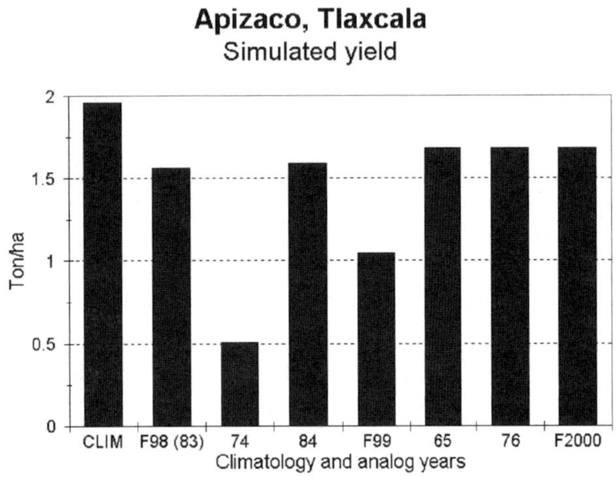

Figure 3. *Simulated yields for maize production in Apizaco, Tlaxcala. Clim represents the averaged yields (1961–1990). F98(83) is the yield for 1998 obtained using 1983 climatic conditions. F99 and F2000 are the yields using averaged climatic conditions for 1974 and 1984 or 1965 and 1976, respectively.*

conditions are more consistent with the current forecast for Mexico discussed above.

Given the forecasts, we applied the Ceres model to simulate the agriculture response in a specific site of study, Apizaco (Conde et al, 199). The yields for the forecasted 1998, 1999 and 2000 are shown in the Figure 3, the expected average yield (*Clim*) is approximately 2 ton/Ha.

The yields obtained for 1998 and 1999 are less than normal. The forecast for 1999 was done considering the average behavior of 1974 and 1984, but the Ceres model results are very different for those years; 1974 yield is less than a half of 1984, mainly because of severe frost events at the beginning of September, 1974. The results for 2000 using the year of 1965 as analog show a similar yield as the one obtained for 1976. Thus, the results shown can be considered the "forecasted" yield for years when the transition from La Niña to Neutral or El Niño event occur.

In conclusion, analogue methods used for forecasts are a powerful tool if they are combined with other statistical analysis. A great advantage is that daily temperature and precipitation data and other historic variables like radiation can be used in simulations with the Ceres model. Also, our experience in this project indicates that the participation of producers and decision—makers improve the usefulness of the forecast processes.

REFERENCES

Conde, C., M. Magaña, R.M. Ferrer. 1999. On the Use of a Climate Forecast in the Planning of Agricultural Activities in the State of Tlaxcala, Mexico. 11th Conference on Applied Meteorology . American Meteorological Society. 10–15 Jan. 1999. Dallas, Texas. Preprints: 101-102.

Magaña, V., Amador, J., Medina, S. 1999. The Midsummer Drought over Mexico and Central America. Am. Met. Soc. 12(6): 1577–1588.

Penland, C., and Ludmila Matrosova, 1998: Prediction of tropical Atlantic sea surface temperatures using linear inverse modeling. J. Climate. 11: 483-496.

Climate Variability, a Producer Typology and the Use of Forecasts: Experience from Andean Semiarid Small Holder Producers

Corine Valdivia, Jere L. Gilles, and Susan Materer
University of Missouri-Columbia, USA

INTRODUCTION

Farmers in highland semiarid production systems of the Andes face climatic variability, characterized by periodic droughts and El Niño Southern Oscillation (ENSO) events. These events, along with economic, social, cultural and biotic factors, affect the choice and mix of crops and livestock (Valdivia and Jetted, 1997; Valdivia et al. 1996), their productivity, and the food security of rural households. Since General Circulation Models (GCMs) are improving in accuracy (Blench, 1999), many argue that forecasts, such as ENSO's, could be more effective at reducing the negative impacts of events by improving disaster preparedness. Forecasts could also increase information provided to small holder farmers and therefore reduce vulnerability. In order for forecast information to be useful to farmers, it must address current needs and problems, be expressed in the language of the users, and provide the communities with appropriate alternatives to current production methods (Price, 1995; Stricherz, 1999; Blench, 1999; Stern and Easterling, 1999). It also requires that trust and communication exist between users and providers of climate forecast (Finan, 1999). Blench (1999) and Finan (1999) argue that these forecasts will probably be useful only to certain types of producers, as not all farmers can equally access nor use the information. An approach to evaluate rural livelihood strategies is developed to determine the conditions that may induce households to make use of information. Current sources of climate information for potato production decisions are analyzed, as this is a vulnerable crop grown in the highlands of Bolivia.

OBJECTIVES

Current research in this region aims to answer the following questions: 1) What are successful strategies developed by farmers to cope with climatic variation in the Andean region; 2) How do farmers currently use information from forecasts and local sources to make production and consumption decisions; and 3) Are there differences among these farmers in the actual or potential use of climate forecasts.

Forecasts are only useful if they are skillful, timely and relevant to actions, which potential users can incorporate into production decisions to improve potential outcomes (Stern and Easterling, 1999). Even when these requirements are met, Blench (1999) argues that forecasts are only relevant to producers that conform to the following profile: large and specialized operations, high in resources like education, safety nets and information, and dependent on rainfall. According to him only these type will gain by accessing forecast information, because they can bear the risks of the outcome, having more cash to protect themselves from inaccurate forecasts.

However, other studies show (Ellis, 1998; Cotlear, 1989) that there may be also a positive correlation between wealth and diversification in rural areas, which implies that not only specialized rich farmers may benefit from forecasts. This study presents an approach to identify groups of producer/rural livelihood strategies in the Andean region, through a peasant household approach that analyzes interactions between production and consumption, and agriculture and non agricultural activities. The level of diversification of farm households and their use of climate forecasts (local or modern) can be evaluated in this framework. The study proposes that diversification and use of forecasts may also go hand in hand.

RURAL LIVELIHOOD STRATEGIES, CLIMATE, AND FORECASTS

Rural livelihood strategies are shaped by several factors. In the Andean region climate plays an important role in production and consumption decisions. Other factors affecting what rural households can do, are access and control of resources, human, cultural and social capital (Bebbington, 1999), markets, institutions, and the political environment.

Interactions Between Climate, the Production Systems, and Potatoes

Periodic droughts, frosts, and ENSO events make climate an especially important factor in the Bolivian Altiplano. The impact of El Niño in 1997-98 was significant for Bolivia, with the drought contributing 53% of the total $527 million damage suffered by Bolivia, according to a study by the Coorporacion Andina de Fomento-Bolivia CAF (Jovel, 1998). The Ministry of Agriculture's evaluation of the agriculture in 1998-1999 found a decrease in the production of food crops. Among these were potato (*Solanum tuberosum ssp. Andigena*) and quinoa (*Chenopodium quinoa*), two important traditional food crops grown in the Andes. The decrease in potato production especially affected the poor and seed production (Jovel, 1998).

In the region of study, Aroma Province in the Department of La Paz, average rainfall is 404mm. During the 1983 El Niño total rainfall was 197.6 mm, while it was 231 mm during the 1989 event. The differences in total rainfall during El Niño years indicates a measure of unpredictability, even if producers know it will be an El Niño year. Rainfall is also erratic during non EL Niño years. In 1992-1993 the annual rainfall was 388.5mm, and fell to 241.9mm in 1994-1995. Besides the variation from year to year, farmers have to deal with the uncertainty of when the rains will start, and the variation in rainfall throughout the growing season. Between July and April the uncertainty is especially important, since this time period is crucial for crops. For example in 1998-99 precipitation was 422mm with 78 days of rainfall, while during the Niño event of the previous year (97-98) was 359 mm in 57 days. Major rainfall deficit months were December and January of 60% and 30% respectively, while October November and March had a 50% surplus in 1998-99 (Bolivia-MAGDR, 1999).

Coupled with droughts are frost events, which represent another risk to many crops. Potato and quinoa are vulnerable to both (Le Tacon et al., 1991). The probability that potato growth will be affected by frost is 30%. The probability of 90 days of potato production without frost in the Patacamaya region is 62%, while for 120 days is 45% (Le Tacon et al., 1991).

Two strategies to deal with production risk are spatial variation, with several plots of different soil types geographically dispersed, and the use of several potato varieties in staggered planting through the onset of the rainy period. Lighter soils are less subject to frost, so spatial diversification is important in dealing with frost (Le Tacon et al., 1991).

Livelihood Strategies, Sources of Income, and Vulnerable Incomes

Rural livelihoods are shaped by the ability of households and individuals to access resources, and the investments decisions they make to deal with climatic variability, economic events, policies and institutional change. These are diverse (Ellis, 1998), a function of the level and type of assets and resources families can access, both in and out of agriculture. Age, education and stage in the life cycle also affect these strategies (Valdivia and Jetté, 1997).

Life cycle stage, household's access to resources, and risk and loss management strategies to cope with a shock, determine the degree of diversification. In areas of greater risk household strategies are expected to be more diversified as a mechanism to minimize the effects of possible shocks from negative climate events, especially when loss management strategies are limited. Households with portfolios of economic activities that are diversified and have less covariant activities, will be able to cope better with climatic risk (Reardon et al., 1992). As income grows moving families away from food insecurity, some expect that households will specialize, and use insurance markets instead of diversification to deal with risk (Dunn et al., 1996).

Life cycle stage affects the degree of diversification of livelihood strategies. Households in their initial stages are starting to accumulate assets and therefore their ability to alter their portfolio is limited. As assets accumulate, diversification in agriculture and in non agricultural activities take place. As families become old and resources are bequeathed households become less diversified (Kusterer, 1989).

Diversification is also a strategy to maximize resource use (Ellis, 1993;1998). Therefore, diversification will persist with greater levels of wealth in environments such as the Andean region (Cotlear, 1989). The type of resources, natural, human and productive accessed, determines how households diversify. In agriculture households engage in several activities through the year to ensure a constant flow of income (Ellis, 1993).

When climatic perturbations or any idiosyncratic risk take place, households may access resources through networks of families and friends, defined as social capital. These are ex-post, consumption smoothing management strategies. Besides accessing networks, other strategies may include liquidation of assets, and temporal migration (Dunn et al., 1996; Valdivia et al., 1996 ;Ellis, 1998; Bebbington, 1999).

Access to Forecasts, use of Forecasts, and Local Predictions

Unlike previous ENSO events, information relating to the 1997 El Niño was available months in advance. Although this information was widely published and became common knowledge in developed countries, studies have shown that in Southern Africa and North East Brasil farmers did not respond to the information (Blench, 1999; Finan, 1999). The lack of utilization of the forecast information in regions that were vulnerable to there events suggests that: a) a gap exists between the information needed and delivered (Blench, 1999); b) there is a lack of trust or miscommunication between users and providers (Finan, 1999); and c) that even if the information is available the ability to respond by changing practices is limited or non existent (Finan, 1999; Broad, 1999).

What are current sources of information in the community? Are there differences according to the identified groups? Can the information provided by the forecast agencies replace the information generated from the local knowledge systems? Do these probabilistic models convey more information than the local knowledge systems convey?

In peasant communities the relationship between agriculture and climate is much more intricate than in western cultures, and farmers are able to identify, with the aid of many different indicators, specific and important weather patterns. Farmers base their crop and other production decisions on local knowledge systems, developed from years of observations, experiences, and experiments (Hatch, 1985; Bharara and Seeland, 1994; Osunade, 1994). Users' forecasts provide more than just information about the forecast. The forecasts provide a set of behavioral rules that households and communities follow when certain indicators are or are not observed, predicting a specific weather pattern or event. Predicting weather is an important cultural component for farmers, as it is common to discuss indicators on the street, markets and with family members (Hatch, 1985). Bharara and Seeland (1994) conclude that local indicators and local knowledge systems will not be replaced with scientific knowledge, because these are holistic, providing farmers with the ability to make informed production decisions and prepare themselves psychologically for the coming agricultural year. On the other hand Bebbington (1991) defines local knowledge as a changing system, where western knowledge has a place. Farmers access knowledge and incorporate it only if it is useful. In some areas, even with skillful forecasts of climate, producers may not be able to modify their current production strategies to reduce vulnerability due to economic or other structural constraints (Muchna and Iglesias, 1992; Finan, 1999). In other cases communication is a hindrance to incorporation of forecast information and steps should be taken to encourage communication and dialogue. Local forecasts are embedded, containing the culture and belief systems, and are fluid, adapting to change (Markowitz and Valdivia) as their livelihood strategies do.

The case study of San José Llanga illustrates two issues. The first is that rural livelihood strategies in the Andes are diverse, and decisions on when and how much to produce depend on both local knowledge systems and resources constraints. The next section provides information about the setting, followed by an analysis of the livelihood strategies using only local knowledge systems, during an average rainfall year and a drought. The following section looks at the effect of El Niño and La Niña on potatoes, and discusses the factors affecting planting in 1998. A final section looks at the actions taken by groups aware of El Niño, and its impacts on potato production, and effects La Niña. the following year.

LOCATION AND METHODS

San José Llanga is an agropastoral community, 116 Km south of La Paz, Bolivia's capital. Located in the central Altiplano, it is 3,786mts above sea level. Mean annual precipitation between 1943 and 1990 was 402 mm, with a coefficient of variation of 31%. Periodic droughts and ENSO events affect this area (Washington-Allen, 1993). This community exhibits diverse production strategies, growing traditional food crops such as potatoes and quinoa, as well as raising animals, cattle and sheep. The community manages 7,200 has of land, with six settlements or neighborhoods. (Valdivia and Jetté, 1997). Fallow agricultural land and crop residues are important in the integration of cropping and livestock production.

In order to unveil distinct household strategies, a household/peasant economic portfolio approach is used to develop a typology of producers. The need for climatic information may vary according to the diverse strategies identified. A survey was applied to 45 families (50% of the population) in San José Llanga, in 1993 and 1995, to identify the types of production strategies pursued by households, as well as understanding their distinct nature. Nine variables were used to identify groups of producers with similar strategies through cluster analysis (Valdivia and Jetté, 1996). The variables were chosen to reflect stage in the life cycle, types of technologies used, the incorporation of commercial activities, the ability to invest in technologies less vulnerable to drought and frost, the level of investment, and the capacity to generate income outside of agriculture. These set of variables correspond to a household peasant economics framework. The operational variables chosen were the following: labor available to the household measured in adult equivalents (Valdivia and Jetté, 1997), age of the head of the household, number of criollo/local sheep, number of improved sheep, criollo cattle, improved cattle, forage area, assets for investment (cattle numbers that can be liquidated), wages received, and consumption (estimated from in-kind production and cash expenditures).

Through cluster analysis with the variables diverse strategies are captured. The variables reflect assets or set activities that are distinct (Bebbington, 1999). Age and access to labor capture life cycle effects on rural livelihood strategies. Irrigated land, which is represented by forage area, represents resources less vulnerable to droughts. Criollo sheep and criollo cattle represent access to livestock that is less sensitive to drought and feed availability. Improved sheep measures availability of feed resources, while improved cattle reflect market integration, in this case commercial dairy production. Wages/off-farm income aims to capture a non agricultural activity to diversify income sources, and consumption (in-kind and cash production for consumption) measures the ability of the household to satisfy a safety first condition, food security. Net income from cattle measures the ability to capitalize and invest in new opportunities, as well as migrate through pull effects to other areas.

A diversity index measures the level of diversification of each group representing distinct strategies. The inverse Simpson (Valdivia et al, 1996) is used to calculate the number of activities and their share on the income being generated. Kusterer (1989) indicates that as families mature strategies of accumulation lead to diversification. Later in life assets are bequeathed to their children, and their level of diversification drops, as the children become their main source of support when old.

By analyzing the household strategies a pattern is revealed as to how household families produce in a climatic variable zone. The meaning of these activities pursued in 1993 and 1995 are explored to understand how households respond to two different climate events, 1993 a 'normal' rainfall year, and 1995 a year of 'delays and lower rainfall'. Changes in production strategies are consistent and reveal important insights into adaptation and security. In September of 1999 a household survey was applied to gather information about El Niño and La Niña years.

RESULTS

What Are Successful Strategies Developed to Cope With Climatic Variation

The analysis of the surveys from 1993 and 1995 provide information about the livelihood strategies pursued by households, with decisions informed by local knowledge systems. Table 1 presents the characteristics of each distinct strategy for both years. The first year rainfall patterns are near the average for 44 years, while 1995 was characterized by delayed rains and low rainfall levels. Forty-five families were interviewed in 1993 and 1995. The sample in 1995 consists of 39 families from 1993 and 6 new families to replace those that migrated. The purpose of comparing both years is to understand the change in the livelihood strategies, and the nature of the change.

Livelihood Strategies in 1993

The results of cluster analysis indicate that two distinct rural livelihood strategies exist, shaped family life cycle. Age and access to labor are the two variables that capture this. As a result two groups are identified, the productive and the elderly. The former has younger families that are engaged in several activities. The elderly have less access to all the resources, and other studies show that this group relies mostly on potato production and remittances from family members (Valdivia et al., 1996; Valdivia and Jetté, 1997).

The productive group is further subdivided in two groups, also pursuing different strategies. The first group is identified as the innovator group. This group adopted both improved sheep and cattle, and engaged in commercial production of milk, due to their access to forages. The households in this group were wealthier, consumption and income

Table 1. *Identified groups of household and their characteristics in 1993 and 1995 in San José Llanga, La Paz, Bolivia.*

	1993			1995			
	Productive Innovators	Productive Extensive	Elderly	Productive Innovator	Productive 2a	Productive 2b	Elderly
Age (years)	45.7	41.6	65	41.9	49	47.7	67
Labor (adult equ.)	3.4	3.3	1.5	2.9	2.8	3.9	1.4
Criollo Sheep (head)	4.2	20.2	3	4.7	16	19.1	1.2
Criollo Cattle (head)	0.3	2.5	0.7	1.3	4.4	0.4	0.6
Improved Sheep (head)	27.2	6.9	1	42.3	12.4	10.1	6.2
Improved Cattle (Head)	4.8	1.3	0.2	5.8	5.3	5.2	0.6
Forages (has)	4.9	1.6	0.6	4.1	4.2	7.3	1.6
Consumption (Dollars)	1,477	677	330	2,123	930	1,277	425
Off-farm Income (Dollars)	247	74	30	1,052	73	208	66
Diversity Index*	3.57	3.24	2.26	3.3	2.95	2.97	2.64

Sources: Surveys of rural households in San José Llanga in 1993 and 1995
*Valdivia et al.,1996.

from off-farm activities double that of the other productive group. The second subgroup, was defined as the extensive group, because criollo animals, sheep and cattle, were more important and raised on grazing areas.

Livelihood Strategies in 1995

Livelihood strategies, as in 1993, are defined by the stage in the life cycle. Two major groups reflecting distinct strategies are identified, again defined as the productive and the elderly. The average amount of improved livestock assets by the elderly increases, along with forage when compared to 1993, while the criollo animal numbers fall, though all the changes are small. The productive group is divided in two subgroups, one is further subdivided in two more groups. Table 1 presents these as three, productive innovator, and two productive, 2a and 2b. The productive innovator is younger, and depends on off-farm employment to complement farming activities. Their farming operation is stable, it has not changed when compared to 1993, with off-farm income being very important. The productive 2 a and b subgroups indicate different strategies. The 2a subgroup has less off farm income than innovators and group 2b, it relies basically on farming, with less that 10 percent of income from off farm sources. The productive 2b relies heavily on agriculture but has more labor and forage resources. The amount of off farm income they obtain is three times larger than group 2a.

In terms of shifting from one group to another between 1993 and 1995, five families moved from the extensive productive group to the innovators, while six families from the innovator group moved to the productive subgroups, three to each (the same or more forages). Two risk reducing activities are being incorporated to manage risk, off farm employment and increase in livestock/forage production. The former is reflected in the growth of off-farm income (Table 1). Those with significant off farm income increase in the productive innovator group, and those with more dependence on forage/livestock in 2 a and 2b. Forage production also grows in the elderly group.

Diversification

Diversification, as mentioned before, allows families to reduce the vulnerability caused by climatic variability. An Inverse Sympson Index is calculated for each household in 1993 and 1995. The greater the index, the larger the number of activities and the evenness in the share of income from each activity. Table 1 shows that in 1993 there were differences between the elderly and the productive in the degree of diversification. As expected (Kusterer, 1989) older households are less diversified than younger households. There were also differences between the innovators and extensive, with the former having a larger diversity index.

In 1995 the diversity index decreases for the productive groups, as the income from crop production falls. Those that maintain a greater index are the Productive Innovator that are incorporation off farm as crops fall. The productive 2b and 2a experience a loss in the diversity of income, as their strategy for that year relies mostly on the farm activities. Finally the diversity income for the elderly grows, as livestock and off farm income grow as shares of the total income.

The productive innovators perform better in terms of income growth and diversification. The correlation between diversity and income is positive, indicating that diversity and wealth grow together in this community (Ellis, 1998; Cotlear, 1989). Diversification is not only a risk reducing strategy, but also maximizes use of resources. The innovators improve their livelihoods in 1995 through off-farm income, suggesting that their linkages to markets allow them access to information and other income improving opportunities.

How do Farmers Currently Use Information from Forecasts and Local Sources to Make Production and Consumption Decisions

Regardless of the degree of wealth and the stage in the life cycle, all families in San José Llanga grow potatoes as their basic staple food. Several varieties are planted, some are native sweet varieties, others are introduced. Native bitter varieties, tolerant to frosts and droughts, mostly are no longer being farmed in this community. The main reason for potato production is consumption, and native varieties are grown primarily for this purpose. Introduced varieties are mostly produced for the market. Table 2 shows the mean area planted to potatoes in 1993, 1997-8 and 1998-1999. The area planted to potatoes has grown, especially for the group of extensive producers. While in 1993 post-

***Table 2.** Analysis of variance. Differences between groups in area (Has) planted to potatoes in 1993, 1997 and 1998, and area (Has) and plots (#) of food in 1993.*

	Potatoes		Food crops*		
Cluster	Area 97	Area 98	Area 97	Area 98	Plots 93
Innovators	1.49a	1.72a	1.1a	3.3a	6.3a
Extensive	1.39a	1.75a	0.7b	2.6ab	5.2a
Elderly couples	0.59b	0.61b	0.6b	1.5b	4.5b
Elderly single	0.85b	0.90b	0.6b	1.1b	2.8b

Post-hoc analysis: Least Significant Differences. Significant at 0.05. Differences between groups when letters are different in a given year.
*Data for 1993 from C. Valdivia and C. Jetté, 1996. Peasant Households in Semi-arid San José: Confronting Risk Through Diversification Strategies. IBTA 181 Technical Report 49/ SR-CRSP 47, and Valdivia, C. Household socioeconomic diversity and coping response to a drought year in San José. Chapter 6 in Coppock, L. and C. Valdivia (eds). Sustaining Agropastoralism on the Bolivian Altiplano: The Case of San José Llanga. (forthcoming) Published by the Global Livestock CRSP and International Livestock Research Institute.

hoc analysis of the area planted shows that there are only differences between the innovators and all other groups, in 1997-8 the extensive productive households have a similar area to the productive innovators, and both are different from the elderly. In 1998-9 similar results are obtained from the comparisons among groups. Mean area planted grows for all groups during La Niña.

Vulnerability to droughts: Potatoes

The change in potato production between 97-98 and 98-99 is significant. Table 3 shows the differences between area planted to potatoes, the number of plots used (indications spatial diversification), and total production of potatoes between 1997 and 1998. Area planted and plots are normally distributed, while quantity harvested is not, as there are some producers that are able to obtain very high yields. In the later case two approaches are used to test differences in total potato production between years, a non parametric test (Wilcoxon Signed Ranks Test), and a paired t- test that excludes outliers (2 in 1997 and 3 in 1998). In both tests there are significant differences between years in the amount of potato produced. The latter is presented in Table 3 along with the other analysis. The differences between means in area planted and quantity harvested are significant. There are no differences among groups on the plots of land managed each year, indicating that spatial diversification is similar. Staggered planting is a mechanisms to deal with uncertainty of rainfall distribution.

Table 4 shows that the only significant differences in area planted and quantity harvested are for the productive group. The non parametric test for area harvested is significant at 10 percent for the

***Table 3.** Mean differences in area planted (Has), plots and quantity (quintals*) harvested between 1997 and 1998. (Paired T -Tests)*

	Mean	t	df	Sig
Difference in area planted 97-98	–0.2249	–2.319	44	*
Differences total area potato 97-98	–0.2219	–1.419	44	ns
Differences in number of plots 97-98	–0.666	–1.427	44	ns
Differences in quantity harvested 97-98	–34.65	–2.357	44	*

*Significant at 0.05 probability level.
Source: Survey of 1999.
*Quintals: one unit equals 100 pounds, for potatoes one quintal equals 50Kg.

Table 4. *Paired T-Tests by clusters between 1997 and 1998 in area planted (Has) and quantity harvested (quintals) to potatoes.*

	Productive	Innovators	Extensive	Elderly
Difference Means				
Area 97-98	–0.2955	–0.2353	–0.3594	–0.0308
Paired Ts	–2.300	–1.255	–2.000	–0.471
Df	32	16	15	11
Significance	0.028	ns	0.064	Ns
Quantity 97-98	–37.7667	–28.733	–46.800	–26.1591
Paired Ts	–1.980	–1.147	–1.588	–1.428
Df	29	14	14	10
Significance	0.057	ns	Ns*	ns

*Non parametric S test -1.92, Asymp. Sig. 0.054 (Wilcoxon Signed Ranks Test).
Source: Survey of 1999.

extensive group. Table 5 shows the mean quantity harvested in both years. The groups experiencing the greater percentage changes are the extensive producers and the elderly, indicating these were more vulnerable to El Niño. These groups proportionally were less aware of El Niño. However, even those who were aware of el Niño were unlikely to change their production strategies (Table 6). Nonetheless, in absolute terms, the innovators achieved higher production levels in both 1997 and 1998.

Reasons why producers did not increase production in 1998

Reasons why producers did not increase potato production in 1998 are presented in Table 7. The lack of seed resulting from the El Niño year had equal impact across groups in planting for 1998-99. Proportionally more households in the extensive group had monetary constraints and could not purchase more seed and fertilizer, compared to other groups. The proportion of households where access to more land for planting was a constraint was similar for both extensive and elderly groups. Overall economic constraints were more important for the extensive and elderly groups. Very few households believed that the 1998-99 year would be bad for potato production. Households that did believe the year would not be favorable, did not rely on neighbors nor outside sources for their beliefs about climate and the year. These households were concentrated in the productive group.

Use of Information about El Niño and Impact on Decisions: Are There Differences Among These Farmers in the Actual or Potential Use of Climate Forecasts

For the most part, farmers knowledge of climate came from natural indicators and traditional knowledge (Table 8), which can be defined as local knowledge systems. Nearly all farmers (98%) used natural indicators such as the wind direction in

Table 5. *Mean Amount of Potato Harvested (quintals) by Group and Growth between 1997 and 1998*

Cluster	Harvest 97	Harvest 98	Growth %
Innovators (17)	107.882	179.706	67
Extensive (16)	98.375	179.125	82
Elderly (12)	44.271	115.667	161

Source: Survey of 1999 with 45 observations.

Table 6. *Knowledge of El Niño event, and change in planting area by group in 1997.*

Group	Knowledge	Change area	Did not change	Total
Innovator	No Yes	0 3	3 7	13 13
Extensive	No Yes	0 2	1 13	1 15
Elderly	No Yes	0 1	7 4	7 5
Total	No Yes	0 9	11 24	11 33

Source: Survey of 1999.

early August and biological indicators such as the flowering of T'ola (Parastrephia lepidophylla) and the behavior of birds and other animals. Traditional knowledge of weather patterns obtained from elders was also important for the productive households. Only 4% directly used mass media to obtain information, and 16% relied on neighbors. None used the technical knowledge produced by govern-

Table 7. *Percent of households and reasons for not planting more introduced potato varieties in 1998-1999 by groups.*

Reasons for not planting more	Groups (%)		
	1	2	3
R1: Did not have more seed	65	69	58
R2: Did not have more land available	18	25	42
R3: No more money to purchase seed	47	63	42
R4: There was no seed available in market	0	13	17
R5: Could not purchase more fertilizer	6	31	8
R6: Did not have cash to purchase pesticide	12	0	9
R7: Did not want to grow more of the type	0	0	0
R8: Feared the climate would not be good	6	6	0
R9: Feared the year would not be good	12	6	0
R10: Did not have more prepared land	29	38	50
R11: The variety is not marketed	6	0	0

Source: Survey 1999. Group 1: innovator. Group 2: extensive production. Group 3: the elderly.

Table 8. *How do you know if the climate will be good or bad: Sources of information in San José Llanga (percent responding they use the source).*

Factors that contribute to knowledge about climate	Respondents Yes %
1. Knowledge from our grandparents	56
2. Information from radio or TV	4
3. Neighbors in the community	16
4. Neighbors from other communities	0
5. Technical personnel from Organizations or Institutions	0
6. The Bristol Calendar (Farmers Almanac Bristol)	4
7. Natural Indicators (Three days in August, winds from north or south)	98
8. Biological Indicators (flowering of bushes, birds, other animals)	98
9. Other: Dreams, God's disposition	29

Survey of 45 households in 1999.

mental or non-governmental organizations and institutions. While farmers still relied on local and traditional knowledge to understand climate, there was a growing concern about these indicators. Many felt that weather patterns in the region had changed and traditional indicators were not as reliable as they were in the past. However they have not been replaced or modified to date.

While traditional indicators and local knowledge were the sources of virtually all of the farmer's ideas about climate, 75% were aware of the 199701998 El Niño event. Only the elderly appeared ignorant of the phenomenon (86% of the others had knowledge). This information raises two interesting possibilities: 1) that awareness of the El Niño phenomena is not connected to the farmer's understanding of climate and 2) that the knowledge of El Niño came indirectly through personal networks and not from forecasts or official sources.

On the individual farm level, only the farmers who reacted to climate forecasts were the most prosperous farmers who had significant off-farm sources of income. These farmers can include the probability of a wetter or drier year into their production decisions even though the information by itself has limited predictive value for producers. This limitation stems from the fact that timing of rainfall and frosts is more critical for crop yields than seasonal rainfall and temperature averages.

The users and potential beneficiaries of climate forecasting seem to be the richer farmers (innovators) who have off farm investments and sources of income. Knowing that there is an increased likelihood of drought or floods may help them decide how to allocate labor during the year. Larger farmers who do not have large sources of off-farm income do not have alternative uses for their labor. These farmers were more vulnerable to El Niño because it has increased it reliance on potato production, but their ability to adjust production in light of forecasts has not. As compared to the innovators, these groups did not react nor change practices with knowledge of El Niño. Furthermore, their ability to expand production after a dry year based upon forecasts is limited by their access to cash. Their ability to buy seed and fertilizer and to rent tractors is influenced by the conditions of the previous year. In other words, these farmers are unable to expand production rapidly after a drought because their cash reserves are depleted. Expansion will not occur even with information that the next year will be better. The elderly also are unlikely to benefit from forecasts. They too are limited by their income and perhaps their available labor. They plant as much as possible every year but are limited by their ability to access seed and land.

Interestingly enough, the group of farmers who are most likely to change farming practices in response to weather forecasts are those farmers who had significant amounts of off farm income. To some extent, this finding is contrary to what we would expect. We hypothesized following Blench's arguments, that specialized high risk farmers would find forecast information more useful than more diversified farmers. Since diversification is a strategy to reduce risk, diversified farmers are less vulnerable to climatic conditions and consequently would not value forecasts as much as more specialized producers whose livelihoods were more exposed to the elements. Our findings, belie these assumptions, those who changed their planting practices had significant off farm sources of income that made them less vulnerable to the vicissitudes of weather. It may be that these groups use forecasts to maximize income rather than to reduce risk. They can shift labor to non-farm activities if they feel that yields will be lower. Farmers who are most dependent upon agricultural production lack the means to change production strategies significantly. They lack the outside resources needed to shift strategies drastically in order to respond to possible weather shifts.

Basically under present conditions, the portion of the population whose well-being is most affected by weather conditions is least likely to alter behavior based on forecast information. This situation will not change unless there are institutional changes that allow the specialized agricultural producers to develop strategies to respond to forecasts.

CONCLUSIONS

The approach proposed unveiled household strategies and the ability of different households to access and take advantage of forecast information. This approach showed that wealthy diversified households have a use for improved forecasts and the flexibility to change activities in their economic portfolios. Overall in comparing strategies in 1993 and 1995 there is a move towards livestock production strategies, which is an activity less vulnerable to weather fluctuations.

The study showed that local knowledge systems are the basis for the forecasts in all groups of households. The few that accessed information from outside the community were in the innovator group, while those who accessed information from community networks were in the productive group. The greater proportion of households who knew of El Niño were in the wealthier group and shifted their production activities. Access to off-farm income provides flexibility and the ability to take advantage of improved forecasts.

Although focus groups indicated that farmers are observing their indicators to be less reliable than in the past, these are still essential sources of information for households, explaining the evolution of complex risk reduction strategies.

Two points should be kept in mind in observing the use of forecast information within the community of San José Llanga. The first is that although the El Niño event of 1997-1998 was stronger than the 1983 event, it did not impact the Altiplano nor San José as dramatically. This may have been due to the time of onset, if this is true research relating to the timing of El Niño may be very useful. The second is that periodic droughts are common during non El Niño years, and the ability to predict these events is more difficult. For example in 1995 San José and other areas of the Altiplano experienced a severe drought, affecting crops and livestock production. Modern forecast systems were unable to forecast this event. Therefore total reliance on modern forecast information should not be the only strategy employed by farmers in the Altiplano. If the forecasts are to be incorporated, it will be done in combination with local knowledge systems, providing farmers with a back up strategy in decision making processes. In Bolivia, information about El Niño in 1997 and La Niña in 1998 was provided through the National Early Warning and Food Security System (SINSAAT Sistema Nacional de Seguimiento de la Seguridad Alimentaria y Alerta Temprana), and was used mostly to assess the impact at the regional level. Given the variability observed this seems to be the most appropriate use of this information. For this information to be used at the local level, farmers need assistance in linking local situations to the regional forecasts produced.

It should be pointed out that in order to fully assess the benefits of the forecast to producers in this region we need a larger sample size of farmers that used or would have used the forecast information.

ACKNOWLEDGMENTS

We like to acknowledge the SR-CRSP team in developing the farmer typologies: Christian Jetté and Joao de Queiroz. We also want to thank the research team that developed the questionnaire with us as part of the NOAA funded grant, Christian Jetté of UNDP, and

Roberto Quiroz of the International Potato Center, and the field researchers that applied and processed that questionnaires. These are Edgar Cala, Justina Condori, Ramiro Carrillo, Federico Mamani, and Mariana Cruz of the International Potato Center who developed the data base. PROINPA's team members, Javier Aguilera, Bruno Condori, and Enrique Carrasco provided support and comments to the design of the survey. We want to especially thank the families of San José Llanga for their patience and assistance with the field surveys.

Funding for this research was provided by NOAA, Office of Global Programs, Human and Economic Dimensions Grant No. NA96GP0239, and the 1993 and 1995 data used here was funded under USAID Grant No. 138 G 00 0046 00.

REFERENCES

Agrawal, A. 1995. Dismantling the Divide Between Indigenous and Scientific Knowledge. Development and Change 26: 413-439.

Bharara, L.P. and Klaus Seeland. 1994. Indigenous Knowledge and Drought in the Arid Zone of Rajasthan: Weather prediction as a means to cope with a hazardous climate. Internationales Asienforum. 25:53-71.

Bebbington, A. 1991. Indigenous Agricultural Knowledge Systems, Human Interests, and Critical Analysis: Reflections on Farmer Organizations in Ecuador. Agriculture and Human Values. 8: 14-24.

Bebbington, A. 1999. Capitals and Capabilities: A Framework for Analyzing Peasant Viability, Rural Livelihoods and Poverty. World Development. 27(12):2021-2044.

Blench, R. 1999. Seasonal Climate Forecasting: Who can use it and how it should be disseminated? Natural Resource Perspectives. No 47, November 25. Overseas Development Institute. UK. 4Pp. Available: http://www.oneworld.org/odi/nrp/47.html

Bolivia- MAGDR. 1999. Bolivia Ministry of Agriculture, Livestock, and Rural Development. Assessment of Agricultural Production 1998-1999 (Evaluación de la Producción Agrícola 1998-1999). La Paz, Bolivia: May.

Broad, K. 1999. Climate Information and Conflicting Goals: El Niño 1997-98 and the Peruvian Fishery. Workshop Proceedings from Public Philosophy, Environment, and Social Justice. October 21-22. Carnegie Council on Ethics and International Affairs. New York, USA.

Céspedes, J.and L. Rodriguez. Assessment and Forecast of Production by Peasant Producers' Perspective. (Evaluación y Pronóstico de la Producción de Vista del Productor Campesino.) Not published. Mimmeograph Report.

Coppock, L. and C. Valdivia (eds). Forthcoming. Sustaining Agropastoralism on the Bolivian Altiplano: The Case of San José Llanga. Published by the Global Livestock CRSP and International Livestock Research Institute.

Cotlear, D. 1989. Peasant Development in the Andes. (Desarrollo campesino en los Andes.) IEP Instituto de Estudios Peruanos. Lima.

Dunn, E., N. Kalaitzandonakes, and C. Valdivia. 1996. Risk and the impacts of micro-enterprise services. Assessing the Impacts of Microenterprise Services (AIMS). MSI. Washington DC.

Ellis, F. 1993. Peasant Economics: Farm households and agrarian development. Second Edition. 1993. Cambridge University Press.

Ellis, F. 1998. Household Strategies and Rural Livelihood Diversification. The Journal of Development Studies. 35(1): 1-38.

Finan, J. 1999. Drought and Demagoguery: A Political Ecology of Climate Variability in Northeast Brazil. Workshop Proceedings from Public Philosophy, Environment, and Social Justice. October 21-22 1999. Carnegie Council on Ethics and International Affairs. New York, USA.

Hatch, J. 1983. Our Knowledge: Traditional Farming Practices in Rural Bolivia Vol:1 Altiplano Region Edited and Illustrated by John K Hatch.

Jovel, R., A. Tapia and I. Thomson. 1998. Bolivia: Assessment of the Damage Caused by the EL Niño Event in 1997-1998. (Bolivia: Evaluación de los Daños Originados por el Fenómeno de El Niño en 1997-1998). Corporación Andina de Fomento. La Paz, Bolivia. December 18 1998.

Kusterer, K. 1989. Small farmer attitudes and aspirations. USAID Program Evaluation Discussion Paper No. 26. USAID, Washington, DC.

Le Tacon, Ph., J. Vacher, M. Eldin, and E. Imaña. 1991. Los Riesgos de Helada en el Altiplano Boliviano. D. Morales and J. Vacher (eds) Actas del VII Congreso Internacional sobre Cultivos Andinos. La Paz, Bolivia, Febrero. Pp. 287-291.

Markowitz, L. and C. Valdivia. (Forthcoming). Patterns of technology adoption at San José de Llanga: Lessons in Agricultural Change. Chapter 7 in Coppock D. L. and C. Valdivia (eds). Sustaining Agropastoralism on the Bolivian Altiplano: The Case of San Jose de Llanga. Global Livestock Collaborative Research Support Program and International Livestock Research Institute.

Morduch, J. 1995. "Income Smoothing and Consumption Smoothing" Journal of Economic Perspectives. 9 (Summer): 103-114.

Muchna, P and A. Iglesias. 1992. Vulnerability of Maize Yields to Climate Change in Different Farming Sectors in Zimbabwe. Chapter 11 in Peterson G., D. Kral and M. Viney (eds) Climate Change and Agriculture: Analysis of Potential International Impacts. American Society of Agronomy.

Osunade, M.A. Adewole. 1994. Indigenous Climate Knowledge and Agricultural practices in Southwestern Nigeria. Malaysian Journal of Tropical Geography.1 : 21-28.

Pepin, N. 1996. Indigenous knowledge concerning weather: The example of Lesotho. Weather 51: 242-248.

Price, M. F. 1995 . Climate change in mountain regions: a marginal issue?". The Environmentalist 15: 272-280.

Stern, P. and W. Easterling (eds). 1999. Making Climate Forecasts Matter: Panel on the Human Dimensions of Seasonal-to-Interannual Climate Variability. Committee on the Human Dimensions of Global Change. Commission of Behavioral and Social Sciences and Education. National Research Council. National Academy Press. Washington D.C.

Stricherz,V. 1999. Long-Term Forecasting Cold Give Nations Tools to Survive Climate Change." January 23. Online Internet. November 11, 1999. Available http://www.eurakalert.org/releases/uw-lfn011999.html.

Reardon, T., C. Delgado, and P. Matlon. 1992. Determinants and effects of income diversification amongst farm households in Burkina Faso. The Journal of Development Studies. 28 (2) : 264-296.

Rosenzweig, M. R. and Binswanger, H. P. 1993. "Wealth, Weather Risk and the Composition and Profitability of Agricultural Investments" The Economic Journal 102 (January): 56-78

Valdivia, C., E. Dunn, and C. Jetté. 1996. Diversification, a risk management strategy in an Andean agropastoral community. American Journal of Agricultural Economics. 78(5) (December): 1329-1334.

Valdivia, C. and C. Jetté. 1997. Peasant Household Strategies Technologies and Diversification in Andean Fragile Lands. Agricultural Economics Working Paper. AEWP-1997-6. Pp.10. University of Missouri. Columbia, MO.

von Braun, J., Hotchkiss, D., and Immink, M. 1989. "Nontraditional Export Crops in Guatemala: Effects on Production, Income, and Nutrition". Research Rpt 73. IFPRI. Washington D.C. May.

Walker, T.S. and Jodha, N.S. 1986. How small Farm Households Adapt to Risk. Chapter 2 in Hazel, P., C. Pomareda, and A. Valdés (eds.) Crop Insurance for Agricultural Development. Baltimore: Johns Hopkins University Press.

Washington-Allen, R. 1993. Historical analysis of land use/land-cover change on the Bolivian Altiplano: a remote sensing perspective. MSc thesis. Range Science Dept. Utah State University.

Seasonal Rainfall Prediction and its Application in the Enhancement of Food Security

Diriba Korecha
National Meteorological Services Agency, Ethiopia

INTRODUCTION

Ethiopia is located along 3°N-15°N & 32°E-45°E with an area of about 1.02 million square km. The climate of the country varies from humid to arid zones. The mean annual rainfall ranges from 2400 mm (over southwestern) to not more than 500 mm (over north eastern & southeastern low lands). As a result, the inter-seasonal and inter annual rainfall is highly variable over the low rain belts of the country.

The country's economy is mainly dependent on rain-fed agriculture. Therefore, the importance of issuing accurate and timely seasonal weather forecast for the country on the likelihood of good or bad weather is indispensable.

Ethiopia has been affected by recurrent drought for a long time. The failure of seasonal rainfall adversely affects the country's socio-economy, in particular food production. Hence low performance in any rainy season would have a direct influence on agricultural production.

Since 1960 drought has been recurrent over many parts of the country. This situation had been exemplified by 1972, 1984 and 1999/2000 droughts. Drought, being common phenomena in the country, the National Meteorological Services Agency (NMSA) of Ethiopia had made concerted efforts to investigate the nature and impacts of this natural calamity. So far the very outset of the strategy followed by NMSA has been geared to the real objective which is related to the understanding of the cause & effects of this event. As such natural hazards originate from weather related phenomena, it is essential to prepare & provide seasonal rainfall prediction tailored to the enhancement of food securities. Therefore, the author is indebted to and pin points some achievements on the use of seasonal rainfall predictions in the enhancements of food securities over the country.

DATA SOURCES AND TECHNIQUES OF PREDICTION

Since 1988 NMSA has been issuing seasonal weather prediction for the three seasons. These are locally known as: *Bega* (dry season) from October to January, *Belg* (small rainy season) from February to May and *Kiremt* (main rainy season) from June to September. Various meteorological parameters, which were collected on real-time bases or archived in NMSA have been used to prepare the seasonal rainfall. For the present study historical rainfall data were standardized and examined for the coherence exists between seasonal rainfall anomalies and various oceanic and atmospheric parameters.

A number of research works have been done at NMSA to improve the status of seasonal rainfall prediction. In particular, attempts have been made on the coherence of warming or cooling of global SSTs and seasonal rainfall anomalies. The studies have shown that the emerging of cooling/warming SSTs over the equatorial Pacific Ocean result in excess/deficient rains during the main rainy season. Conversely, the warming of these portions of the ocean usually favor the presence of excessive rains during the dry and small rainy seasons. This can be inferred from figures I, II, III and IV. Evidently, this study contributes much for the success of year to year seasonal climate outlook capability.

The demand for seasonal rainfall prediction from policy makers, government & non-government organizations & others are ever increasing. In particular, the agricultural sectors give high attentions to perform their national & regional agricultural planning. Thus in order to meet these needs, National Meteorological Services Agency has adopted and developed many forecasting techniques, that enabled us in maximize the performance of the forecasting capabilities

VERIFICATION AND APPLICATIONS OF SEASONAL RAINFALL PREDICTION

Seasonal rainfall predictions were issued for the first time in 1987 on national basis. However, as the probabilistic and other forecasting techniques had been implemented, the predictions were issued for five to seven homogeneous rainfall zones of the country. As a result, the skill of the forecast is increasing from time to time. Nevertheless, no attempts have been made to quantify the performance of the seasonal rainfall predictions. But in recent years, as various forecasting techniques were being consolidated to predict the likelihood of the seasonal rains, it performed moderate to reasonably accurate.

In Ethiopia seasonal rainfall predictions are given before the onset of each season. In order to harmonize the seasonal rainfall predictions, the forecast forum is arranged by inviting representatives from user communities from agricultural, health, water resources, disaster prevention and NGO's. In the forum, seasonal rainfall prediction is thoroughly discussed. Moreover, various issues related to the techniques adopted in prediction, likelihood impact of anticipated weather on various socio-economic activities are usually interacted among the participants. In particular, the issues of food securities are highly sounded.

Seasonal rainfall prediction, which is given by NMSA is interpreted and applied mainly for food securities. The past experiences have shown that when the rainfall activity is anticipated to be deficient, the government has given priorities in maximizing food securities over the drought prone areas of the country. These situations are still exercised by the governments & the aligned NGO's to minimize the drought risks and the consequences. These are carried out to:

- formulate various action plans in mitigating the adverse effects of drought on various socio-economic sectors;
- minimize the drought risks by planting short-cycled or drought-resistant crop varieties ;
- monitor the severity of drought over the drought prone areas;
- enhance food securities by advising farmers in order to save their crop productions;
- appeal for donors in order to secure the problems related to hunger;
- avoid the extra expenses that would be allocated for agricultural inputs when the when the season is anticipated to be undermine the agricultural activities; and
- advise on the use of irrigation schemes over potentially water resources areas of the country.

Many international forecasting centers are issuing experimental seasonal rainfall prediction for different parts of the globe. Noticeably, NCEP (NOAA), IRI, ECMWF, Hadley Center (UK Met. Office) and some regional drought monitoring centers, had disseminated their prediction for the Sahel and equatorial African sub regions. These products are extremely very useful for those countries lack of international cooperation, historical data, communication facilities, and trained manpower. The National Meteorological Services Agency of Ethiopia has been used the product issued from the aforementioned centers when issued the seasonal rainfall predictions. However, the recent assessments have shown that much of these forecasting models had less skill scores or no signal over the horn of Africa. Thus, the centers should take in to account the general geographical and the near by water bodies to make the performance of the model modest.

RESULTS AND DISCUSSION

The use of seasonal rainfall predictions in Ethiopia is still in its infant stage. However, currently the policy makers, government and non-government organizations pertained to food securities use seasonal rainfall prediction as a major instrument and appreciate the existing forecasting progress. The drought legends have shown that many severe droughts have recurrent along the East Africa sub region. In particular, as the inter-seasonal of intern annual rainfall variability is very high along the drought prove areas of the country; the drought and its consequence are not uncommon. Since 1960, the severe droughts of 1972, 1984 and the current are commonly noticeable. These droughts were resulted in extremely food shortages in the country. Consequently, many people were persisted and displaced. However, since the emerging issue of seasonal rainfall prediction these catastrophe here been minimized by developing early warning systems. For instance, the seasonal rainfall prediction that had given for 1995/96 have shown the likelihood of better rainfall activities over much of the country. As a result, the governments hove done by importing agricultural inputs and distributed to the farmers in time. This, excess crop-production have been produced which maximize the food securities of the country. Likewise, the drought of consequences of 1997/98 El Niño where also under control by taking into account the different rainfall prediction issued before hand.

The current drought was also anticipated, in particular over the portions of the country affected by the recurrent droughts, before the beginning of the seasons. As a result, the government said some attention in monitoring the extent of drought coverage. Moreover, the food securities of the country were also assessed and made appeal for the areas had the food shortage. These are in fact showed the Ethiopian lessons with reference to the use of seasonal rainfall prediction in the enhancement of food securities.

CONCLUDING REMARKS AND RECOMMENDATIONS

In this paper the author is trying to assess the development of seasonal rainfall prediction and its application with reference to the enhancement of food securities. The assessments have shown how the timely and accurately given seasonal rainfall prediction assisted policy makers in minimizing the likelihood occurrence of weather related risks. The performance of seasonal rainfall predictions, which rendered by prominent global climate prediction centers, were also assessed with reference to the Ethiopian case. However, in order to furnish good service for the user communities and maximize the skill score of seasonal rainfall prediction, the following should be taken care of:

- proper forecasting techniques should be designed by taking into account the regional geophysical feature,
- the seasonal predictions should be provided along with accompanied by their limitations so that the users can able to take appropriate cautious,
- the seasonal rainfall predictions should be verified frequently and publicize the failure if any, for the user communities,
- the forecast values should be assessed in terms of its contribution for various social and economic activities,
- seasonal climate predictions and their applications for the benefit of human beings should be exercise,
- the global climate prediction centers should assist those countries under frequent weather related hazards like; drought, floods, in developing their respective national climate prediction scheme. Moreover, they should have to examine the performance of their forecasting models,
- the nature of SSTs over the Indian Ocean should also be equally treated as Pacific and Atlantic Oceans for the benefit of African and South Asian countries and
- some pilot projects should be designed in order to assess and formulate the impact of weather related hazards of the world.

Water Efficiency Using Seasonal Forecasts Climate Risk Management Strategies

***Robert A Young*[1], *Jozef Syktus*[1], *Lawrence Lau*[2], *Norman L. Miller*[3], *Jack Katzfey*[4], *John McGregor*[4], *and Lisa Goddard*[5]**

[1]Queensland Department of Natural Resources (QDNR), Australia
[2]University of Queensland (UQ), Australia
[3]Lawrence Berkeley National Laboratory (LBNL), University of California, USA
[4]CSIRO Atmospheric Research, Australia
[5]International Research Institute for Climate Prediction, USA

SUMMARY

Problems faced in Queensland are more exacting and difficult than in other places around the world, because of the character of the Australian continent and its extremely variable climate. For QDNR the emphasis is the water resource and environmental applications rather than on the supporting technologies (e.g. climate models).

Water is now a prime limiting resource and is inseparable from environmental and natural resource management. Economic planners often neglect to account for water requirements for certain forms of development. Through an independent consultant QDNR has undertaking a review aimed at improving water use efficiency by the development of a plan for an integrated climate/water resources program which will complement and draw on existing resources within a full spectrum of the our earth science and climate program.

The six month nation wide investigation addressed specific issues such as beneficial decisions that storage managers, irrigators, local councils, catchment managers, and environmental flow managers would be able to make using seasonal forecasts as well as clearly identifying potential benefits to different stakeholders in some preliminary quantitative terms.

QDNR's operational climate risk management decision support systems exploits the Southern Oscillation Index (SOI) forecasts for spatial environmental monitoring as well as the integration of seasonal climate forecasts from coarse resolution global climate models (GCM) (300 km) enhanced by regional climate models (RCM) (10-75 km) and other techniques.

Preliminary investigations indicates that seasonal climate forecasting based on the El Niño—Southern Oscillation phenomenon has considerable potential for forecasting stream and river flows in Australia. QDNR wishes to develop a program that realizes the economic and environmental management implications of that potential, including the development and integration of longer lead forecasts than those currently available.

Advanced planning for Natural Resource Management is on a decadal time-scale and consequently the need for decision support tools is now rather than later this century. This initiative by DNR will expand our capability to address water resource requirements.

SESSION VI. ECONOMICS AND VALUATION OF SEASONAL CLIMATE FORECASTS

A Perspective on the Economics of Climate Prediction for Mozambique

Channing Arndt
Purdue University, USA

INTRODUCTION

Farming in many parts of Africa is a risky business in large part because an essential input, climate, is highly variable and, up until very recently, completely unpredictable beyond very short time frames. Since very little land in Africa is irrigated, variable rainfall translates into variable production levels. Since rural inhabitants are often very poor and depend upon agriculture for subsistence, the consequences of natural disasters such as drought for rural inhabitants can be devastating. For Mozambique, drought in 1991-92 and flooding in 2000 represent recent examples of climate induced crises.

The impacts of extreme climate can extend well beyond the farm. For most of the countries in Africa, agriculture represents a large share of the total economy. In these countries, production variability in agriculture has immediate and important macroeconomic impacts. For example, inadequate or poorly timed rainfall implies not just low yields, wasted inputs, and high prices, but also, in many cases, food insecurity or famine, large scale imports of food, balance of payments deterioration, substantial government spending on drought relief programs, depressed demand for non-agricultural goods, and rural-urban migration. In addition, industries that depend upon agricultural commodities for raw materials, such as food processing and textiles, are often adversely affected, particularly if alternative supply sources have not been secured. Indirectly, drought can have substantial impacts on the environment as affected populations increasingly focus on survival rather than preservation of natural resources for the future. For all of these impacts, a sufficiently early and reliable signal of impending adverse climatic conditions could potentially permit actions on the part of economic actors and government to devise coping strategies and thus palliate the negative economic and environmental impacts of a poor agricultural season.

Though perhaps less obvious, early warning of impending favorable climatic conditions (or not drought/flood) could also be valuable. For example, under favorable climate conditions, input use is often too low and cropping patterns overly skewed towards drought resistant crops. Thus, the full benefits of the favorable climate conditions are not captured. In addition, following a bountiful agricultural season, prices often tumble due to problems in the marketing system such as inadequate storage infrastructure and institutional rigidities in export markets. A timely seasonal forecast of either high probability of very favorable conditions or low probability of very unfavorable conditions could permit farmers to adjust cropping patterns and input use in order to profit more fully from a good season. Also, a timely forecast could give the marketing system and downstream users, such as the food processing industry, time to prepare for a bountiful harvest.

This paper discusses the potential role of climate forecasts for Africa and presents methods for assessing the economic value of reliable forecast information. Section two highlights the potential importance of reliable climate information for development strategy in Africa and reviews economic assessments of the value of forecast information undertaken to date. Section three discusses methodological approaches for assessing the value of climate information with an emphasis on general equilibrium approaches. Finally, in section four, a general equilibrium model of Mozambique is employed to estimate gains from potential reactions to a drought forecast.

POTENTIAL ROLE OF CLIMATE FORECASTS AND REVIEW OF ECONOMIC WORK IN AFRICA TO DATE

The appearance of reasonably reliable climate forecasts for many parts of Africa coincides with the need to make important strategic decisions regarding the character of development in Africa generally. Some argue that Africa should follow in the footsteps of the successful economies of East Asia. In this case, development is driven by the growth of labor-intensive industries, such as textiles and footwear, with an export orientation. Bloom and Sachs (1998) are primary proponents of this strategy. In their view, the agricultural resource base in Africa is, generally, of such low quality that agriculture cannot effectively function as a leading sector. High rainfall variability is explicitly fingered as a major detracting factor.

On the opposite side, Hertel, Masters, and Elbehri (1998) and Wood and Mayer (1998) argue that Africa's natural resource base is simply too large to ignore. Africa's massive land endowment combined with dismally low educational attainment across the continent point to a development strategy focused on primary products. In this view, Africa should follow in the development footsteps of, for example, the Americas (both Latin and North) and Australia. Exports of primary products play the prominent role in the development process for at least the medium term. As educational attainment improved, processing of primary products, both agricultural and mineral, would figure more prominently in the export mix.1

The choice between these competing visions of Africa's economic future has profound implications for economic policy. The policies and institutions necessary to catalyze success in primary products and primary product processing are not the same as those necessary to catalyze success in labor intensive manufactures.

Climate forecast information has the potential to play a role in this debate. If the adverse effects of drought/flood can be mitigated and the benefits of favorable climate accentuated through climate information, then the value of the enormous African land resource base could be substantially leveraged. The gains from climate forecasts to economies with significant primary product processing capacity could be even larger than for early stage of development economies that are basically dependent on primary products. Benson and Clay (1998) find that the impacts of drought are relatively stronger in economies with substantial agro-processing. These results indicate that climate forecasts might be more valuable to Zimbabwe than Mozambique despite a lower share of primary agriculture in GDP in the former. Furthermore, should a primary products and primary products processing strategy be opted for, forecast information could play a significant role in African development for the medium term.

While the above paragraphs hint that the role of climate forecasts could be large, there is very little firm economic analysis to support the notion. This is especially true for Africa. Benson and Clay (1998) conducted the study, referred to above, examining the impact of drought on sub-Saharan African economies. Piesse, Thirtle, and van Zyl (1996) examine the effects of drought on productivity in South Africa. Boudreau (1998) examines the effects of climate variation on rural households in Zimbabwe.

The work done in Africa to date highlights the central problem with evaluating the economics of climate forecasts. Specifically, a clear counter-factual is necessary. As has been pointed out many times, it is not enough to say that climate variability has strong economic impacts. The key is whether the presence of climate forecasts can allow agents to take actions, which palliate the negative effects of unfavorable climate and/or enhance the positive effects of favorable climate. To evaluate economic impact of a forecast, the same climate scenario must be evaluated twice: one time with and one time without the forecast. The difference between these two scenarios is the economic benefit of the forecast (or the cost of ignorance).

LEVELS OF ECONOMIC ANALYSIS AND METHODS FOR EACH LEVEL

Levels of Economic Analysis

The preceding section emphasized the reaction of agents to forecast information. However, agents may react at many levels in the agricultural system. Due to very limited empirical experience with farmer reaction to forecast information, simulation modeling of farming systems must play a primary role in assessing value. The need for a counter-factual also highlights the central role of simulation models. They allow focus on the impacts of forecast information assuming a rational response by agents to the information provided.

Farmer reaction is the first, most micro-economic level in the agricultural system. Incorporation of climate information into existing bio-economic models of village economies presents a promising avenue for analysis. Advanced versions of these models already explicitly incorporate climate risk (e.g., Barbier and Hazell, 1998). Collaboration with climate forecasters might fruitfully refine the current, unconditional, specification of climate risk and provide additional climate outcome distributions conditional on either available or projected climate prediction technology. These bio-economic models are of strong potential interest in that the structure of the agricultural system is allowed to change over time. Sufficiently reliable climate forecasts might permit an evolution in the structure of farm operations. A shift to a different structure, with coping mechanisms more dependent on the climate forecast, might allow farmers to reap much greater benefit from forecast information than the benefit that could be realized from adjustments to farm management plans within the confines of existing structures. In addition, in the bio-economic models, environmental impacts are explicitly treated. This allows one to gain insight into the character and pace of environmental degradation as well as farmer reactions to a diminished resource base. These effects can be examined with and without forecasts.

The next level of analysis is the agricultural sector level where the marketing system, in particular, is in focus. Reaction of marketing sector participants could result in large benefits. For example, in the Sahel, aggregate demand for agricultural inputs such as fertilizer and herbicides varies dramatically with the aggregate level of rainfall. In years with above average rainfall, total demand for agricultural inputs (at a constant price) is well above total demand in years with below average rainfall. Unfortunately, Sahelian countries are both highly dependent on imports of either the inputs themselves or their component chemicals and distant from ports. This implies that it is very difficult to modulate aggregate agricultural input availability with realized climate outcomes. Total input supply is effectively determined prior to realized demand. Climate information could play a key role in more efficiently matching aggregate input supply with aggregate input demand, at least on average.

A focus on the marketing system is also consistent with current climate prediction capabilities. Harrison and Graham (1998) point out that: "Analyses of the skill of atmospheric models frequently show that the skill is maximized at regional scales typically on the order of 105 [square] kilometers or more." While a host of marketing participants operate at very localized levels, others, such as major input suppliers, focus at the regional level where climate prediction skill is highest. Furthermore, these marketing agents are relatively limited in number, thousands of marketing agents compared with millions of farmers. Finally, these marketing sector participants are likely to have higher educational attainment on average than farmers. Hence, in these early stages of experience with climate forecasts, the task of communicating the forecasts to marketing agents is likely to be substantially less daunting than the task of communicating the forecast to millions of subsistence farmers. In this light, the magnitude and distribution of benefits when marketing agents are the primary users of forecast information is a relevant question.

Lastly, the marketing sector will serve as a conduit of information to farmers. For example, following the drought forecast issued in 1997 for Zimbabwe, substantial anecdotal evidence indicates that the banking sector tightened credit to the agricultural sector. The direction of this move is likely correct. A credible drought forecast implies greater risk in agricultural loans. Tighter "credit communicates" to farmers the signal to reduce use of purchased inputs. This is also the correct signal in terms of direction. An elevated risk of drought should logically imply reduced input use. However, dialogue with the banking system (and other marketing participants) is necessary in order to assure that the response to a forecast is also correct in magnitude and composition. Excessive credit tightening across the board might prevent farmers from taking prudent steps such as purchase of appropriate plant varieties.

An economy-wide approach is the final most aggregate level of analysis. Using an economy-wide approach, one can examine the linkages and feedback effects between farmer reactions, reactions of the marketing sector, and reactions of the non-agricultural sector. As emphasized above, poor or favorable climate can have a major macroeconomic impact when agriculture/agricultural processing represents a substantial portion of economic activity. The economy-wide approach permits one to account for these spillovers and interactions.

These levels of analysis should be mutually reinforcing. For example, recent work in Mozambique by Arndt, Jensen, Robinson, and Tarp (forthcoming) illustrates that improvements in the efficiency of the marketing system leads to large aggregate benefits. As illustrated by the example of input supply for the Sahel, forecast information could lead to greater efficiencies in the marketing system. In this vein, synergies might well exist between studies of the marketing system and the economy-wide analyses. The marketing sector studies should indicate the nature and potential magnitude of any efficiency gains in the marketing system from forecast information. The studies should also provide information on institutional structure (e.g., nature of competition and policy environment). With this information in hand, the CGE model would provide both a quantitative assessment of the aggregate impact of adaptation to forecast information by marketing system agents and information regarding the distributional implications of these changes.

Across these levels of analysis, accounting for uncertainty will play a key role. As Faragò, Wilhite, and Glantz (1997) point out, "a forecast is just a forecast: it's not a guarantee." The failure of drought to materialize in much of southern Africa during the 1997-98 agricultural season is a case in point. For the foreseeable future, forecasts will enable one to reweight the probability of the various possible outcomes, i.e., to change the perceived distribution of possible climate outcomes. However, a distribution of outcomes still exists.

More Detail on Economy-Wide Approaches

The capacity to analyze economy-wide effects of a variety of shocks, such as drought or major policy reform, has increased dramatically in recent decades. Models of this sort are known as computable general equilibrium (CGE) models. The attraction of the CGE approach lies in its completeness. Many of the relationships defined in a CGE model are accounting relationships, which must hold both in the model and in an economy. For example, CGE models include the following basic accounting constraints:

- the value of imports cannot exceed total foreign exchange availability (including foreign borrowings),
- total consumption of a good cannot exceed total supply,
- investment cannot exceed total savings,
- government spending must be financed,
- households must respect their budget constraint,
- production requires real resources,
- factors of production (land, labor, capital) are in finite supply,
- demand for factors of production by productive activities cannot exceed factor supply, and
- households (and potentially government) own and receive revenue streams from factors of production.

None of the above propositions are in dispute. However, they serve to circumscribe, sometimes surprisingly tightly, the range of feasible economic outcomes.

On top of this accounting framework, the economic behavior of agents in the system is specified. In classic CGE models, consumers maximize utility of consumption and producers maximize profit in a perfectly competitive environment. Government behavior is exogenous and time is not explicitly treated (e.g., the model is static). Current applications often depart significantly from this classical standard. Recent CGE applications explicitly incorporate risk-reducing behavior, specify labor-leisure tradeoffs, examine the implications of imperfect competition, and/or incorporate dynamics.

The CGE approach is particularly valuable for assessing the impacts, including distributional effects, of large shocks. When shocks are large, interaction and feedback effects can be very important. Drought falls under this category of shock for almost all African economies.

DROUGHT IN GENERAL EQUILIBRIUM: AN APPLICATION TO MOZAMBIQUE

Arndt and Bacou (2000) undertake a first step analysis of the impacts of drought and the potential role of drought forecasts using a CGE approach for Mozambique. They undertake three experiments using the CGE model. The first simulates an unanticipated drought. In this experiment, farmers adopt an agricultural production plan; and this plan cannot be changed based on realized climate outcomes. In experiment two, farmers, but no other agents, are assumed to receive a perfect forecast and react to it. In experiment three, both farmers and marketing system agents react to a perfect forecast. The following responses might be expected of marketing agents under a credible drought forecast: grain stocks are maintained, ports prepare for increased grain import volumes, the transport systems prepares to distribute inland towards normally surplus regions, input suppliers reduce sales expectations, and agricultural processors seek alternative sources of supply. These forecast reactions, as well as the risk reductions inherent in a reliable forecast, are assumed to yield a 2% gain in marketing system efficiency. These simple experiments are designed to illustrate where the largest benefits to forecast information are likely to reside.

Drought is simulated through reductions in a technology parameter. Under drought conditions, we suppose that the same amount of inputs (labor, seed, etc.) gives less output. The exact technology declines employed for the eight agricultural activities in the model are: grain and other export crops- 0.67; basic food crops (primarily vegetables)- 0.75; cassava, raw cashew, raw cotton, and livestock- 0.85; and forestry- 1.00. From a value-of-forecast perspective, the crucial element is not so much the average productivity decline in agriculture as the dispersion across activities. This dispersion allows the possibility to reallocate resources from drought intolerant to more drought tolerant activities (price changes will also influence cropping choice as discussed below).

Table 1 shows the implications of drought for household welfare across the three experiments. As one might expect, rural households bear the brunt of the burden of a drought. More surprisingly, gains to households from even perfect crop reallocation (experiment 2) are small and skewed towards urban households. In contrast, gains from marketing sector improvements (compare experiments 2 and 3) are much more significant and more evenly distributed. While urban households reap more than 60% of the value of the (small) gains from crop reallocation, urban and rural households evenly share the (larger) incremental gains from marketing sector improvements.

Price changes and differential consumption patterns underpin the distributional effects. The producer agricultural terms of trade is defined as the producer price of agricultural goods relative to the producer price of non-agricultural goods. As net sellers of agricultural goods, increases in the agricultural terms of trade benefit farmers since the products they sell buy more non-agricultural goods. In experiment one, agricultural terms of trade on a producer price basis increase by more than 28%.

Table 1. *Household welfare (equivalent variation on consumption).*

	Percentage form base			
	Base Run	Exp 1	Exp 2	Exp 3
Urban	0.0	–4.9	–4.6	–4.0
Rural	0.0	–10.9	–10.7	–10.1
Total	0.0	–7.7	–7.4	–6.9

With drought forecast information, farmers consider both drought tolerance and price in their resource allocation decisions. They will allocate resources to crops whose prices are expected to be high. This increase in supply of crops that would otherwise have particularly high prices reduces the improvement in agricultural terms of trade to 24%. These relative price declines in agriculture pass much of the benefits of crop reallocation based on forecast information to urban groups.

In contrast, when efficiencies gained from reactions of marketing sector agents are included (experiment 3), agricultural terms of trade improve by 24.7% relative to the base implying a small improvement relative to pure crop reallocation (experiment 2). This gain in terms of trade favors rural households. In addition, since marketing costs are particularly important for agricultural commodities and agricultural commodities are particularly important to rural households both as income sources and as part of their consumption basket, rural households gain from marketing sector improvements.

CAVEATS, CONCLUSIONS, AND SUGGESTIONS FOR FURTHER RESEARCH

Stern and Easterling (1999) assert that economic analysis of the implications of climate information is still in its "infancy" (p. 128). While this analysis represents a step forward, prominent omissions remain in the approach employed here. First, both the forecasts and the reactions to the forecasts are, in fact, imperfect. In addition, incorrect forecasts will occur and these are likely to imply costs. From this perspective, the benefits of forecasts are overstated in the above analysis. Second, the model takes a national focus when a regional approach would be more appropriate. In Mozambique, impacts of drought vary substantially by region. Drought in the South is frequently accompanied by good rains in the North. These regional effects and potential negative correlations in climate outcomes are interesting even in the absence of forecasts. With forecasts, they imply enhanced opportunity to move resources from low productivity regions to high productivity regions in response to forecast information. The drought shocks employed are nationally representative. The differences in productivity effects across activities are certain to be much greater on a regional basis. Third, more empirical evidence on likely reactions of marketing sector participants and the benefits of those reactions is required.

Despite these limitations, the key role of the marketing system in gaining value from seasonal forecasts might be a robust result. Even under the impossible hypothesis of a perfect forecast and a perfect farmer reaction, gains from crop reallocation following a drought forecast appear to be small. If, on the other hand, seasonal forecasts induce relatively small marketing sector efficiency improvements (a more plausible proposition), the economy-wide gains appear to be significant and evenly distributed between urban and rural zones.

From a realization of forecast value perspective, this is probably good news. It is far easier to extend a probabilistic seasonal forecast to thousands of relatively sophisticated marketing sector participants than to millions of subsistence farmers. In addition, many marketing sector participants operate on the larger spatial scales where forecast skill is highest. Finally, marketing sector participants have the potential to communicate forecast information to farmers via the price system (higher interest rates on operating loans to farmers following a drought forecast, for example). Overall, the results merit cautious optimism on the eventual value of forecast information to Africa and continued economic and climatological research.

NOTES

1 Africa is far from monolithic and exceptions to these generalizations exist.

REFERENCES

Arndt, C. and M. Bacou. 2000 "Economy-Wide Effects of Climate Variability and Climate Prediction in Mozambique." Am. J. Agric. Econ., forthcoming.

Arndt, C., H.T. Jensen, S. Robinson, and F. Tarp. forthcoming "Marketing Margins and Agricultural Technology in Mozambique." J. . Develop. Studies.

Barbier, B. and P. Hazell. 1998 "Implications of population growth and declining access to trans-humant grazing areas for the sustainability of agro-pastoral systems in the semi-arid areas of Niger; A bio-economic modeling approach." Paper presented at an International Symposium on Property Rights, Risk and Livestock Development, September 27-30, DSE, Feldafing, Germany.

Benson, C., and E Clay. 1998 "The Impact of Drought on Sub-Saharan African Economies—A Preliminary Examination." World Bank technical paper no. 401.

Bloom, D.E. and J.D. Sachs. 1998 "Geography, demography, and economic growth in Africa." Brookings Papers on Economic Activity, 2.

Boudreau, Tanya. 1997 "Combining Risk Map Analysis and Climate Prediction: The Possible Effects of El Niño on Rural Households in Zimbabwe." Save the Children Fund.

Fargò, T., D. A. Wilhite, and M. H. Glantz. 1997 "A Forecast Is Just a Forecast: It's Not a Guarantee." Internet Journal for African Studies. http://www.brad.ac.uk/research/ijas/ijasno2/ijasno2.html.

Harrison, M.S.J. and N.E. Graham 1998 "Forecast Quality, Forecast Applications and Forecast Value: Cases from Southern African Seasonal Forecasts." United Kingdom Meteorological Office. unpublished.

Hertel, T., W. Masters, and A. Elbehri, 1998 "The Uruguay Round and Africa: A Global, General Equilibrium Analysis." J. African Economies, 7(2), July, pp. 208-36.

Piesse, J., Thirtle, C. and Van Zyl, J. 1996 "Effects of the 1992 Drought on Productivity in the South African Homelands: An Application of the Malmqusit Index." J. Agricu. E., 47(2), pp. 247-254.

Stern, P.C., and W.E. Easterling. 1999, "Making Climate Forecasts Matter." National Research, New York: National Academy Press.

Wood, A. and J. Mayer. 1998 "Africa's export structure in a comparative perspective." Institute of Development Studies and United Nations Conference on Trade and Development, Study No. 4.

The Potential Value of ENSO Forecasts to the Rice Sector in Taiwan

Ching-Cheng Chang[1], Chi-Chung Chen[2], and Bruce A. McCarl[2]
[1]Academia Sinica, Taiwan, ROC
[2]Texas A&M University, USA

SUMMARY

The most pronounced ENSO anomalies appear in the monsoon-Pacific trade wind regions. The impact of ENSO on agriculture has been studied extensively for various crops at many different scales (e.g., Adams, et al., 1995; Solow, et al., 1998, Chen and McCarl, 1999, Hill et al., 1999). However, very few of them have focused on rice, the most important staple food crop in the Asian-Pacific countries. Of 25 major rice-producing nations, 17 are located in this region, which together produce 92% and consume 90% of the world's total rice production. Throughout the region, much of the present interannual variability in precipitation is linked to the ENSO phenomenon (Matthews et al., 1995). As a consequence, any major change in ENSO behavior would have considerable impacts on the rice yield, which devastate the food security and welfare of farmers in this region.

In a previous study, we find that the production impacts from ENSO on the exporting countries are fairly diversified. The negative impacts in the cold events are likely to be more severe than they are in the warm events. Such domestic production changes due to the ENSO events alter the trade volume as well as the marketing share. The overall trade impact of a warm event is in general larger than that of a cold event. Nevertheless, both the cold and warm events could trigger strong changes in trade volume as well as in the market shares due to the highly diversified regional impacts on production. Furthermore, using the yield regression results in the 1991/92 El Niño event, we predict that the 1997/98 event will increase total rice trade volume by 31 per cent which is fairly close to the observed data. In this regard, further research on how to incorporate forecast information into the decision making process to improve our societal welfare is obviously needed.

The purpose of the present study is to investigate the value of improved ENSO forecasts to rice producers and consumers around the world as well as in Taiwan. A world trade model is constructed and linked with Taiwan agricultural sector model. The uncertain rice yield around the world and Taiwan will be estimated through the econometric method. A three-phase decision-tree approach will be adopted. In the first phase, we add a balance constraint that insures average storage additions equal average withdrawals into a synthesized base model. In the second phase, we assume knowledge of ENSO phase is available when the crop mix is chosen. In the third phase we have knowledge of event outcomes, thus prices, consumption levels and trade activities are set accordingly. The base model is an extension of McCarl et al's U.S. sector model (ASM) as used in Chang et al and Lambert et al unified with a rice world trade model.

This model could simultaneously examine the consequences of ENSO impact and information in the world and in Taiwan. The welfare implications and economic value of the ENSO forecasts will be evaluated based on a straightforward application of the Bayesian decision theory. The value of an ENSO forecast is measured by the expected increase in economic benefits arising from the use of the forecast in decision making. The policy implications in terms of investing in improving climate forecasts will also be examined.

Political-Economic Uncertainty and Climatic Hazards: The Utility of Climate Forecasts for Small-scale Farmers in Tlaxcala, Mexico

Hallie Eakin
University of Arizona, USA

INTRODUCTION

Farmers' vulnerability to climatic variability and extreme climate events is often measured by estimating the direct and indirect impacts of climatic variables on farmers' production (Parry, 1990). From this understanding of vulnerability, particular strategies can be designed to help farmers mitigate those impacts. In this conception of vulnerability, climate forecasts appear to be an ideal mitigation solution: by using forecasts to select appropriate crops, time farm activities and investments, farmers can take advantage of good weather conditions and mitigate the impact of extreme events.

It is, however, difficult to address the vulnerability of farmers to climate events and variability without simultaneously addressing the implications of political and economic factors that also influence farmers' decisions (Brklachich *et al.*, 1997; Chiotti *et al.*, 1997; Smit *et al.*, 1996; Smithers and Smit, 1997). Farmers' exposure to climatic variability, and their capacity to cope with its impacts is a function of the policy environment and economic context in which they pursue their livelihoods (Watts and Bohle, 1993). While this may be obvious, if the broader risk environment is not taken into consideration in the development and promotion of climate forecasts as a risk-mitigating technology, some farmers, particularly those who are struggling with dramatic changes in agricultural policy and market conditions, may be excluded from the potential benefits of forecast use. For many farmers, typically it is not a drought or frost alone that devastates production, but the occurrence of these events under circumstances of severe financial, technical and environmental constraints often exacerbated by changes in agricultural and economic policy.

This point was illustrated in a 1997 study of the potential utility of climate forecasts for small-scale maize producers in Tlaxcala, Mexico (Eakin, 1998). The study was undertaken in collaboration with the National Autonomous University of Mexico and the Autonomous University of Tlaxcala who together are producing experimental forecasts for the agricultural sector of Tlaxcala (Magaña R., 1999). Over the course of two months, over 70 farmers were interviewed in the municipalities of Apizaco and Terrenate in an effort to identify the impacts of extreme events on their livelihoods, their climate information needs, the adaptation strategies they would follow with advance knowledge of climate anomalies, their perceptions of climate events and forecasting, and their recommendations on the ideal mode of forecast delivery.

RESEARCH SITE

Tlaxcala is the smallest state in Mexico with an area of just over 4,000 km2. The state is situated in the uplands of the Neovolcanic Axis and the Sierra Nevada mountains, making it particularly susceptible to frosts and enhancing the spatial variability of rainfall. Its climate ranges from semi-arid to sub-humid, with average annual rainfall from as low as 400 mm to over 900 mm in the agricultural zones. The rainy season typically begins in May and lasts through September permitting a growing season of no more than 180 days, although the growing season length is heavily dependent on the timing of the first and last frosts. The area is often affected by a mid-season (July/August) decline in rainfall, locally called the canicula.

More than 80 percent of the state's agriculture is rainfed, and the majority of Tlaxcala's farmers have less than five hectares of land on which they primarily produce maize, beans and barley. Maize is by far the dominant crop in Tlaxcala, and typically more than half of what is produced is used for home consumption. Maize yields are highly variable in Tlaxcala as a result of both climatological constraints and uncertainties in the political and economic environment. In one of every three or four years, it is not uncommon for farmers to receive half their normal yield in maize, or to be left with a frost-damaged harvest that is unmarketable.

TRADITIONAL FORECASTING

As in many parts of the world where agriculture plays a central role in culture and economy, Tlaxcaltecan farmers have developed their own methods of climate forecasting based on generations of experience, local religious beliefs and close observations of their environment. The farmers interviewed in this study relied on interpretations of cloud color and form, ambient temperature and animal behavior as indications of approaching weather conditions. Many of these observations have been coded in local sayings and rhyme and are thus passed through generations. With respect to longer-term climate forecasts, farmers reported using an almanac called El Galván, and the cabañuelas, an elaborate system of observing daily weather patterns over the course of two months, and relating these observations to average monthly climate conditions in the months to come (see Eakin, 2000).

While the use of these prediction methods varied from farmer to farmer, the existence of traditional forecasting methods in Tlaxcala illustrated the strong demand for climate information and the importance of climatic variability in farmers' planning. The fact that farmers were accustomed to the concept of long-lead forecasting has provided an opportunity for the UNAM climatologists to use the traditional methods as analogies in introducing the new ENSO-based forecasts. Instead of conflicting with local knowledge, the new forecasts appear to be received by Tlaxcaltecan farmers as complementary to their traditional practices. Many of the local environmental indicators that farmers relate to climatic events are also being noted by the climatologists of the project as possible indications of linkages between local environmental processes and larger-scale climatic factors such as ENSO.

INFORMATION NEEDS

The traditional forecasts and in-depth interviews with the farmers about their farming practices also provided concrete illustrations of what climatic factors were most important to the Tlaxcaltecan farmers at what points in the production process. Because of the severe constraints on crop production imposed by the relatively short growing season, farmers emphasized the importance of being able to take advantage of the first rains of the season. They also were preoccupied by problems relating to irregularities in the timing and distribution of rainfall throughout the rainy season. Knowing when during the season they could expect adequate or deficient rainfall would help plan their input applications, allocation of labor and other activities. The timing of frost events was also a high information priority, along with the timing and severity of the *canicula.* The number of expected hurricanes approaching the Gulf of Mexico was also information farmers desired because of their association of local convective activity with rainfall on Mexico's eastern coast.

These information needs are now setting the parameters of the climatological research. While precipitation is more predictable than frost, the projects' climatologists are now trying to link the frequency of frost events to ENSO in an effort to better serve Tlaxcala's farmers. Importantly, the interviews with the farmers also illustrated the particular language that farmers are accustomed to using to describe particular climatic features such as the *canicula,* providing a vocabulary that can be used to convey the new forecast messages.

ADAPTATION STRATEGIES

The farmers interviewed in Tlaxcala had, through generations of experience, developed a substantial repertoire of strategies to confront the frequency of frost and drought in their localities. Despite the existence of local forecasting methods, most of the farmers interviewed suggested that they rarely acted on the basis of a climate prediction. Instead, they indicated that they typically responded to the development of poor climate conditions as they occurred. An exception to this is the practice of deep tillage as a moisture conservation strategy, which needs to be implemented well in advance of dry conditions. Farmers claimed that a particularly deep tillage immediately following the end of the rainy season (December) enhanced seed germination in the following spring, and thus augmented the subsequent growing season, should there be inadequate or late spring rainfall.

Farmers suggested substituting the more commercial corn varieties with local, faster-maturing varieties that were more resistant to irregular rainfall and drought. In the event of a heavy spring drought, farmers recommended substituting forage crops such as barely, oats or wheat in place of corn, given the hardiness of these crops and their relatively faster maturation. Wheat, barley and some varieties of hybrid maize are also thought to be more frost resistant by some farmers. If a greater frequency of frost events is anticipated, farmers suggested that one might adjust one's crop choice, or, if possible, one might choose the land less susceptible to frosts for the more sensitive crops.

Those farmers who used chemical fertilizers or herbicides stressed the importance of adjusting the timing of these applications to moisture conditions. Recognizing the risk of "burning" their crops by applying fertilizer in dry conditions, many of the interviewed producers said that they would not invest in chemicals should there be inadequate rainfall. Weeding, plowing, or other activities that disturb the topsoil are also often abandoned during dry spells in an effort to conserve the available moisture beneath the soil surface.

These and other strategies could conceivably be used as preventative or proactive measures with advance warning from a climate forecast. Indeed, in many countries, one of the more effective uses of forecasts is thought to be through crop substitution and input adjustments (Adams *et al.,* 1995; Betsill *et al.,* 1997; Glantz, 1996; Golnaraghi and Kaul, 1995; Lagos and Buizer, 1991; Miller, 1993). However, despite the existing knowledge of a range of possibly appropriate production strategies, few of the Tlaxcaltecan farmers interviewed said that they consistently followed these strategies even as poor weather conditions developed.

In general, farmers reported being constrained by lack of resources—credit, equipment, inputs—as a result of dramatic changes in Mexico's agricultural policy over the last decade and a half. After the debt crisis of 1982, the Mexican government began to rapidly withdraw from its previous heavy involvement in the provision of credit, inputs, technical assistance and other support for small-scale producers. As Mexico joined the General Agreement on Tariffs and Trade in 1986, and eight years later signed the North American Free Trade Agreement with the United States and Canada, the government privatized the seed and fertilizer and began a process of price liberalization for the majority of Mexico's commodities. Farmers have had to adjust from the security of guaranteed prices and subsidized inputs to a highly uncertain environment in which the prices all commodities except maize and beans are unregulated and variable. Perhaps most significantly for Mexico's small-scale producers, the public agricultural bank, BANRURAL, was restructured, resulting in the exclusion of over 80 percent of Mexico producers from access to affordable credit.

These factors have put farmers in a very difficult position—the cost of production has escalated, the basic consumer staple, maize, is no longer profitable, yet farmers feel they have few alternatives. For example, in order to perform a deep tillage, farmers have to have the financial liquidity to rent the tractor and pay for the petrol immediately after the harvest—a time when many farmers are using their available resources to engage in seasonal migration to work as day laborers in the northern states or in the United States. Thus this task is often left for the spring, far too late to aid in moisture conservation, and sometimes too late to take advantage of the first rains.

Similarly, in order to take advantage of ideal (in climatic terms) planting dates, farmers must have the financial resources available to purchase necessary inputs. Without agricultural credit, the timing of planting is often more a factor of the farmers' finances than of weather, or for that matter, of recommendations accompanying a forecast. While the Mexican government has implemented a program to provide a direct cash support to farmers to assist in input purchases, this assistance has been criticized as being too little, and for often arriving too late in the season to be of use.

Perhaps most importantly, although barley, oat, and wheat have been determined by the state's agricultural service to be more appropriate for Tlaxcala's variable climate, farmers are also reluctant to substitute crops that have purely commercial uses for maize, which serves as a food stable, a commercial good, a household savings and insurance account, as well as a source of seed and feed for livestock. With only 1-2 hectares available for planting, farmers who switch maize for barley or another grain are putting at risk their family's food security. With the liberalization of most commodity prices, the prices of alternative grains are perceived as very unreliable, as are the prices of production inputs. Should the price of barley or oat fall unexpectedly, or should a frost make the harvest un-sellable, finding the financial resources to provide one's family with the basic staple, maize, would be problematic. Thus the vast majority of the small-scale producers interviewed indicated that would only abandon their maize crop as a last resort.

From these findings, it is clear that the potential of climate forecasts to alleviate a farmers' vulnerability to climatic hazards is interdependent with the farmer's ability to cope with changing political and economic conditions. The uncertainties relating to poor market conditions, escalating input costs and severely restricted finance appear to exacerbate farmers' vulnerability to climatic events by inhibiting the flexibility of their responses. In relative terms, the constraints of the political-economic environment may be having a far greater impact on farmers' livelihoods, their crop yields and production decisions than the highly variable climatic conditions in which they operate. Under these circumstances, climate forecasts alone cannot address the vulnerability of farmers' livelihoods to climatic variability. If successfully using a climate forecast entails switching crops or new input strategies, the risks associated with these decisions must also be addressed. If not, it is possible that for some farmers, using a climate forecast will be perceived as exacerbating rather than mitigating the vulnerability of their livelihood.

Since this study was completed, the climate forecasts prepared by UNAM and UAT have been distributed widely in the state. Several farmers have reported using the forecasts to adjust their production strategies, despite the significant obstacles of resource access and availability described above. Why some farmers appear to be in a better position to make use of the information and the particular contexts of successful forecast use will be the subject of the next phase of the ongoing project on climate forecast use in Tlaxcala. While the data has not yet been collected, we are working with the hypotheses that those farmers who appear to be in the best position to use the forecasts are those who are able, through participation in producer associations or cooperatives, through differential access to credit or technical assistance, or through participation in particular government or private sector programs that offer crop insurance, or market and price guarantees for alternative crops.

The results of the study detailed above suggest that there needs to be coordination between the incentives and structural constraints imposed by agricultural policy, the cultural and economic reality of agricultural households, and mitigation strategies proposed on the basis of climate forecasts. Climate forecasts are not, in this perspective, substantially different from other agricultural tools and services. As such, this information must be provided with full recognition that it will not be on the basis of climate information alone that a farmer will make a production decision but rather on the basis of a synthesis of numerous factors which together structure the farm household's vulnerability.

REFERENCES

Adams, R., Bryant, K., McCarl, B., Legler, D., O'Brien, J., Solow, A., and Weiher, R. (1995). Value of Improved Long-Range Weather Information. *Contemporary Economic Policy* XIII, 10-19.

Betsill, M. M., Glantz, M. H., and Crandall, K. (1997). Preparing for El Niño: What role for forecasts? Environment 39.

Brklachich, M., McNabb, D., Bryant, C., and Dumanski, J. (1997). Adaptability of agriculture systems to global climate change: a Renfrew County, Ontario, Canada Pilot Study. *In* "Agricultural restructuring and sustainability: a geographical perspective" (B. Ilbery, Q. Chiotti and T. Rickard, eds.). CAB International, Oxon and New York.

Chiotti, Q., Johnston, T., Smit, B., and Ebel, B. (1997). Agricultural response to climate change: A preliminary investigation of farm-level adaptation in southern Alberta. *In* "Agricultural Restructuring and Sustainability" (B. Ilbery, Q. Chiotti and T. Rickard, eds.), pp. 201-218. CAB International, Oxon and New York.

Eakin, H. (1998). Adapting to climate variability in Tlaxcala, Mexico: Constraints and opportunities for small-scale maize producers. Masters, University of Arizona, Tucson.

Eakin, H. (2000). Seasonal Climate Forecasting and the Relevance of Local Knowledge. *Physical Geography* forthcoming.

Glantz, M. (1996). "Currents of change: El Niño's impact on climate and society," Cambridge University Press, Cambridge.

Golnaraghi, M., and Kaul, R. (1995). The Science of policymaking and responding to ENSO. Environment 37, 16.

Lagos, P., and Buizer, J. (1991). El Niño and Peru: A nation's response to interannual climate variability. *In* "Natural and technological disasters: Causes, effects and preventable measures" (G. S. Forbes, E. W. Miller and R. F. Schmaltz, eds.). Pennsylvania Academy of Science.

Magaña R., V. O., ed. (1999). "Los Impactos de El Niño en México,". SEP—CONACYT, Districto Federal.

Miller, A. (1993). Applications of seasonal to interannual climate prediction: the Nordeste Region of Brazil. *In* "Usable Science: Food Security, Early Warning and El Niño" (M. Glantz, ed.). NCAR, Budapest, Hungary.

Parry, M. (1990). "Climate change and world agriculture," Earthscan Publications Limited, London.

Smit, B., McNabb, D., and Smithers, J. (1996). Agricultural Adaptation to Climatic Variation. *Climatic Change* 33, 7-29.

Smithers, J., and Smit, B. (1997). Agricultural system response to environmental stress. *In* "Agricultural Restructuring and Sustainability: A Geographical Perspective" (B. Ilbery, Q. Chiotti and T. Rickard, eds.). CAB International, Oxon and New York.

Watts, M. J., and Bohle, H. G. (1993). The space of vulnerability: the causal structure of hunger and famine. *Progress in human geography* 17, 43-67.

Agricultural Economic Responses to Forecasted Climate Variation: Crop Diversification, Storage, and Trade

Daniel G. Hallstrom[1] and Daniel A. Sumner[2]
[1]North Carolina State University, USA
[2]University of California, Davis, USA

INTRODUCTION

The role of climate variation in agriculture is well known, but the economics of climate variations is less developed. This is particularly true of the potential role of improved climate forecasts in the economics of agricultural markets. Recently the literature has been expanding rapidly (Adams, et al., Mjelde, et al., Sumner, Hallstrom and Lee). This brief paper summarizes the scope and main results from recent research reported more fully to the economics audience in Hallstrom (2000).

The focus of this research is on how responses (adaptations) such as crop choice, storage and interregional trade modify the effects of climate variation and how these responses interact with improved climate forecasts. Motivation for our focus is that as climate forecasts improve in the coming years, agricultural markets are also expected to become increasingly open to international trade. The broad economic questions we address are as follows. How do climate variations affect prices, costs and economic welfare of crop producers and consumers? How do forecasts affect these relationships and outcomes and what is the value of forecasts in this system? How do responses such as crop choice, storage and trade influence effects of climate and forecasts? And, conversely, how do forecasts affect crop choice, storage and trade?

APPROACH

Our basic approach has been to study the behavior of aggregates in a highly abstract and stylized economic equilibrium model that captures a few key ideas. In this research, we do not attempt to examine statistical evidence about responses to climate forecasts. Instead, we examine changes that occur in response to climate shifts and climate forecasts by simulating the behavior of competitive grain markets. The focus is on market-level (aggregate) response, but underlying the market response are dynamic optimization decisions by individuals and firms. These market participants have Bayesian responses to climate information and reoptimize their production, consumption, storage and trade behavior each time new information becomes available.

There are three main economic responses to climate variation and forecasts. Cropland allocation responds to ex ante comparative advantage and to the benefits from crop diversification. For simplicity, there are no post-planting variable input adjustments and so cross-crop arbitrage determines how much land is devoted to each crop. Consumption and storage respond to prices, and price expectations, and provide inter-temporal arbitrage. Finally, trade between regions that experience different climate shocks and weather realizations provides spatial arbitrage. Market prices across crops, time and space adjust to equilibrate the system.

Given these important dimensions, the model is as parsimonious and simple as possible. There are two production regions that have the same size, consumer preferences and base productivity. However, climate variations apply only to region 1. For simplicity, transport costs are zero so that with free trade prices in the two regions must be equal. There are only two potential crops, which have the same base yield and price. However, the yield of crop 1 is more sensitive to climate variation and thus is more variable. Climate directly affects only the expected crop yields in region 1. There are three equally likely climate states: unfavorable (lower mean yield), neutral (cross-climate average mean yield) and favorable (higher mean yield). The climate forecasts provide correct information, prior to planting, about which yield distribution the realized yield is drawn from, but of course actual yield depends on the realized draw from that distribution.

The mathematical specification of the model and some more precise definitions of the economic logic are provided in Hallstrom (1999). Key analytical results of main interest to economists are in Hallstrom (2000). Here we lay out some results on interest to the climate research community.

In order to proceed with quantified simulations we must specify economic, agronomic and climatic parameters. Since we are implicitly considering ultimate applications to the global staple grain markets the parameters we chose are calibrated to approximate those market data. The global own elasticity of demand is –0.25 at the mean for each grain. Each region has an initial market share of 0.5 so with open trade demand facing each national industry remains inelastic. Therefore, revenue/outlays necessarily fall when yield rises (because prices fall by more). Total area (550 million hectares), base market prices, mean yields and variances and co-variances are approximately equal to recent data from world wheat and coarse grain markets. Climate variation shifts the mean yield of both crops in region 1 either up or down. More important in the decision-making process is the relative mean yield of crop 1 compared to crop 2. In region 1 the relative yield symmetrically increases with favorable growing conditions and decreases with unfavorable growing conditions. Percentage shifts in relative yields considered in the overall research range from two percent to eight percent. Here, however, we focus on only the results from the two- percent and the four- percent relative yield shifts. Finally, the discount rate, including storage costs is set at 10 percent.

RESULTS

A large variety of simulation results may be derived under alternative parameter specifications and assumptions about storage and trade. None of the simulations are realistic, but the pattern of results illustrates principles of methodological interest and some rough quantitative guidance. The simulations provide thousands of individual calculations and results are best displayed graphically. We provide a few sample results in figures 1 through 6 and more complete results in Hallstrom (1999). The results displayed deal with changes in expected prices and land allocations as well as some welfare measures showing the importance of economic linkages to the value of climate forecasts. Variances of the relevant distributions, especially production and prices are also of interest and we discuss those below without providing detailed results.

Figure 1 shows the relationship between the world market price of crop 1 and the availability of both crop 1 and crop 2 in million metric tons (mmt). Given interregional trade there is a single price in the two regions, and given substitution in consumption and production, the stocks and current harvest of both crops are needed to understand the pattern of expected price for crop 1. Figure 2, illustrates how crop specialization and diversification responds to the availability of each crop. Figure 2 applies to the case in which there is no climate forecast. The more of crop 1 available the lower the current share of land devoted to crop 1. Note that over the relevant range, the share of land planted to crop 1 varies up to 60 percent and down to about 40 percent.

Now we introduce a forecast about the climate affecting region 1. Figure 3 shows how the expected price for crop 1 depends inversely on crop 1 availability and the forecast of upcoming climate. These functions also depend on the availability of crop 2, here held constant at a relatively low 590 million metric tons. When availability of crop 1 is low or very high, below 700 mmt or above 850 mmt, a climate forecast is of little relevance. For intermediate levels of crop 1 availability, say, 750 mmt, we may discern a higher price for the unfa-

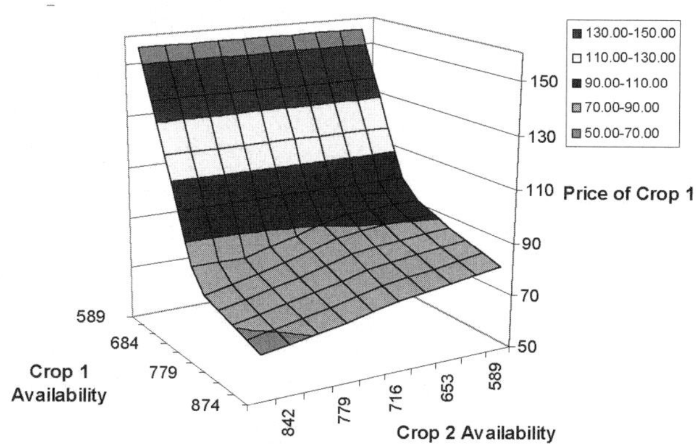

Figure 1. *Rational expectation price function for crop 1.*

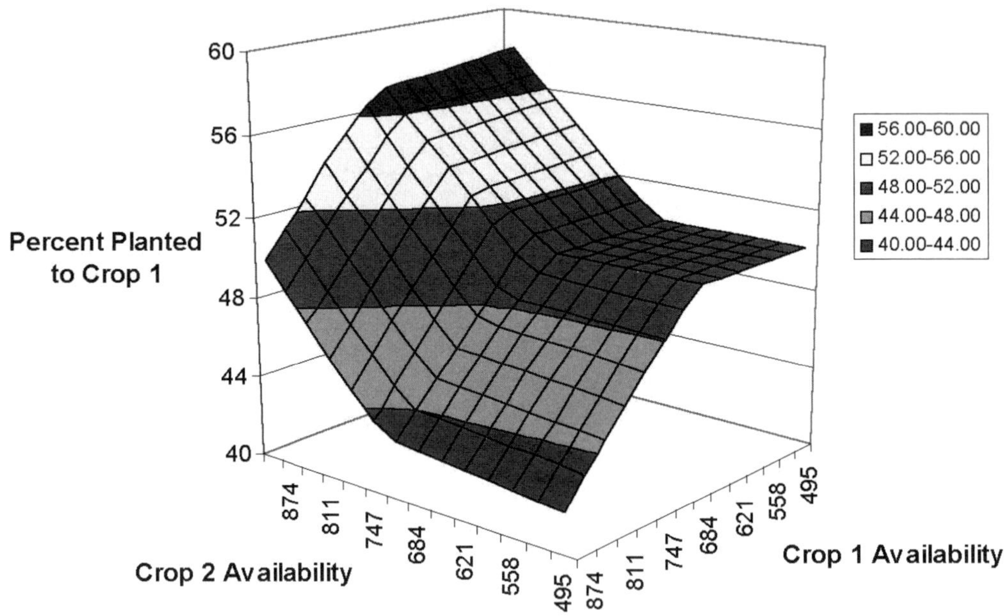

Figure 2. *Percent of available land planted to crop 1.*

vorable climate forecast. Figure 4 shows how farmers respond to the climate forecast in the model. In this case a favorable climate forecast causes farmers in country 1 to plant between 70 and 100 percent of their land to crop 1 depending on the prior availability of the two crops. Comparing Figure 2 to Figure 4 allows us to see dramatically how the forecast affects farmer behavior in this model, and how the response to the forecast depends on current crop availability, which is itself the cumulative outcome of the past history of planting, trade and storage decisions.

Figure 5 provides some results on the flow value of the forecast (under the two- percent relative yield shift) as a function of the crop availability. Net value is an increasing function of crop avail-

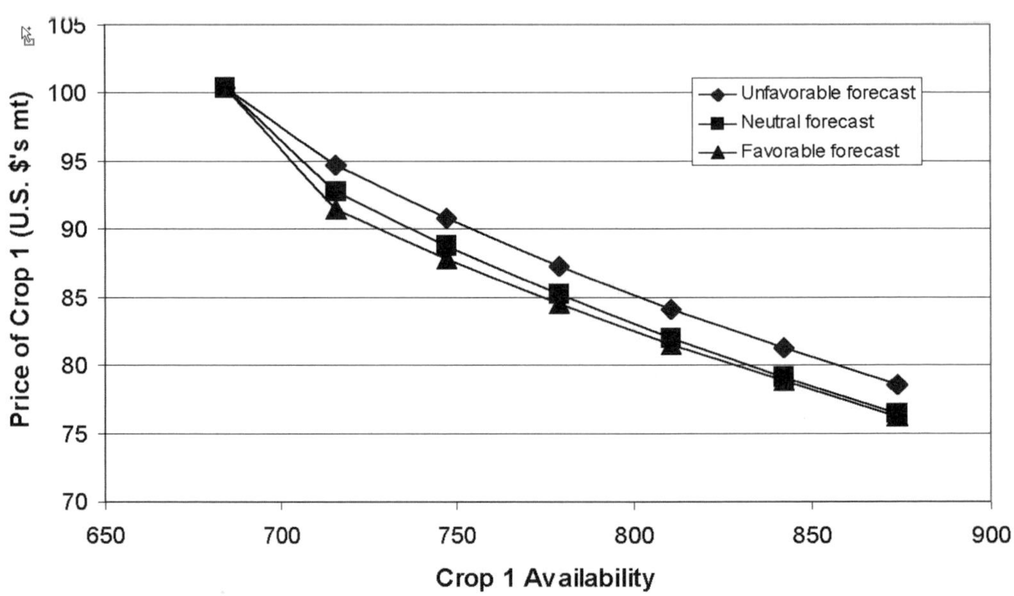

Figure 3. *Cross section of the rational expectation price function for crop 1 with improved climate information (Crop 2 availability at 590 mmt).*

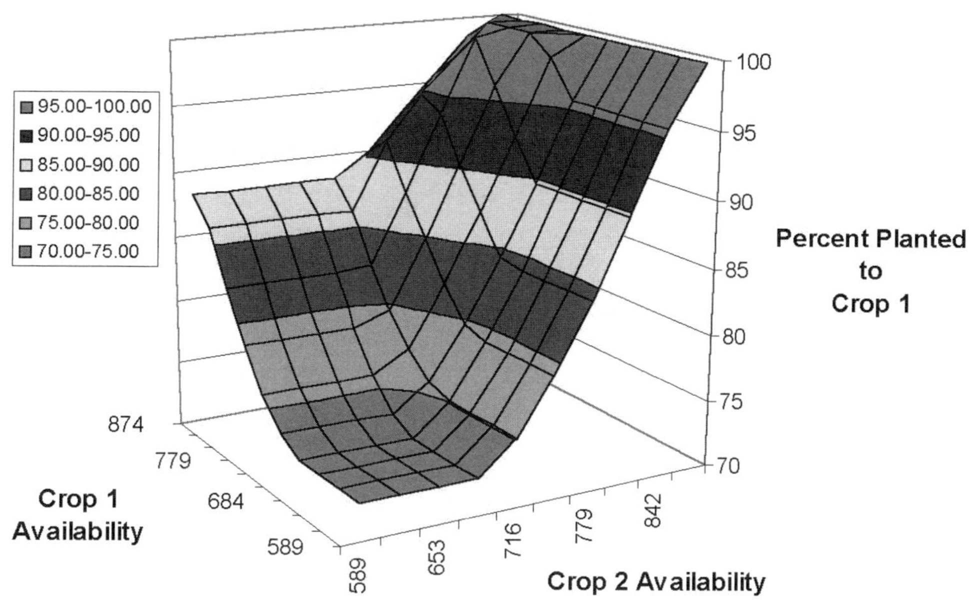

Figure 4. *Percent of land in Region 1 planted to Crop 1 (favorable forecast).*

ability. When stocks are very low, farmers continue to plant both crops in more equal proportions and gains from specialization in response to the forecast are limited. But, with large stocks on hand there is scope for profitable use of the temporary comparative advantage afforded by the climate variation and forecast. The value of a forecast doubles between the low availability cases shown in the front of the three- dimensional figure and the back of the figure.

Finally, Figure 6 shows the land response to the forecast and crop availability when no interregional trade is allowed. Now only region 1 data is relevant because the two regions are isolated by policy or transportation costs. In Figure 6, crop 2 availability is held fixed at 295 mmt (half of 590) and

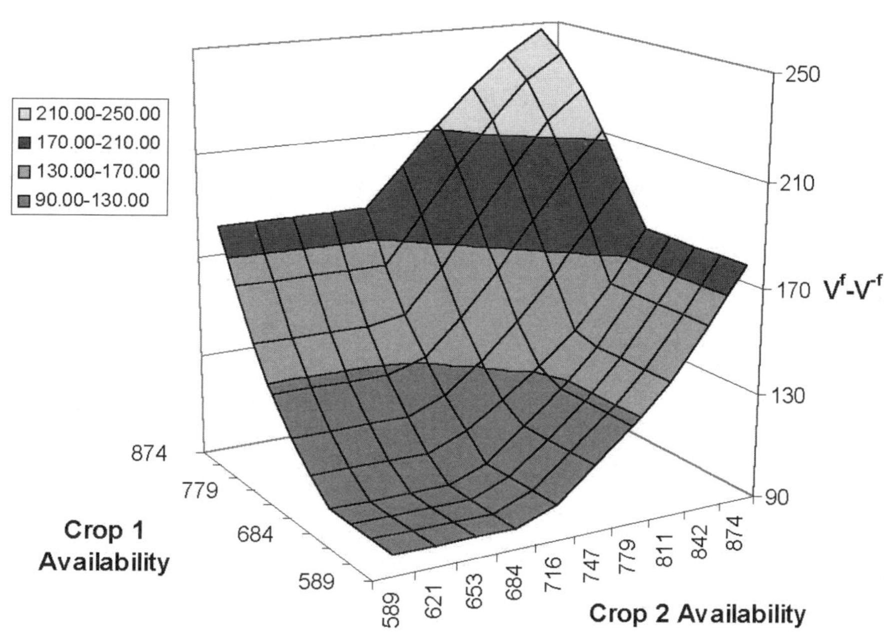

Figure 5. *Dollar value of the current forecast.*

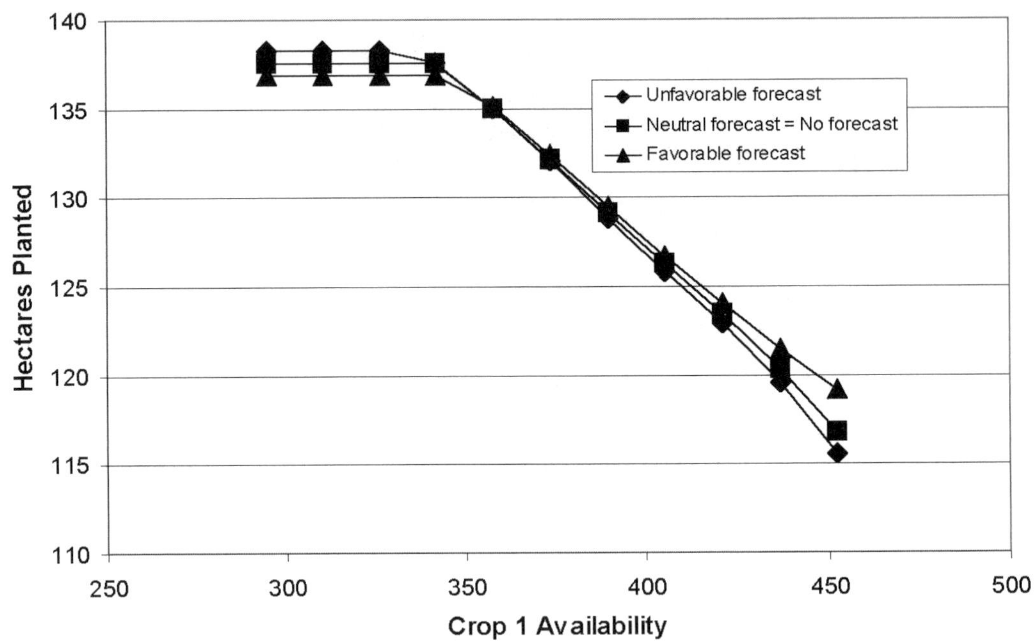

Figure 6. *Hectares planted to Crop 1 with autarky (Crop 2 availability at 295 mmt)*

hectares planted to crop 1 depend on availability and the forecast. Region 1 has a total of 275 million hectares to allocate between crops, and Figure 6 show how that land is allocated with low initial availability of crop 2. When trade is not an option the amount planted to crop 1 remains between 42 percent and 50 percent of total area available. Further, the forecast makes very little difference to the share of land allocated to crop 1. Even with large stocks, the difference in the land allocated to crop 1 is no more than three million hectares or less than two percent points.

Table 1 summarized these quantitative results in terms of the value of the forecast (under a four-percent relative yield shift) to producers, consumers and the economy as a whole. Consumers are the big winners from climate forecasts. As with other innovations, climate forecasts allow more production and thus lower prices, and since the demand for grain is inelastic, the percentage price decline exceeds the quantity increase and producer revenue falls. Producer welfare thus falls. This is a familiar result in the economics of agricultural innovation and follows directly from the inelastic demand. Note that with trade, consumers in both regions gain even though the forecast applies only in country 1 and that producers in country 2, who cannot make use of the forecast directly, lose more than producers in county 1. Overall, the world gains $484 billion from a relatively modest fore-

Table 1. *The value of a climate forecast to consumers and producers in each country, with and without interregional trade**

	Consumers	Producers	Net
	——— $ millions ———		
Region 1 (open trade)	1,729	1,460	269
Region 2 (open trade)	1,729	1,514	215
Total World (open trade)	3,458	2974	484
Region 1 (autarky)	10.3	1.3	12

*These calculations apply to the case in which forecasted climate shifts cause a 4% deviation in the ratio of expected yields in country 1.

cast applied only to the mean of grain yield distributions.

Contrast these results with those in the bottom row of Table 1. Now no trade is allowed and so behavior in region 2 is unaffected by the forecast. In region 1 there is no opportunity to import or export so the scope of response to the forecast is limited. Storage is still an option, but now, consistent with Figure 5, the annual value of a forecast is tiny. Consumers gain a total of $10 million compared to $3,458 million when trade is allowed. Producers do not lose, but the gain is trivial. The net gain is now only about 2.5 percent of the gain with free trade.

SUMMARY OF MAJOR FINDINGS

Major findings of the study may be summarized in a series of points:

- Climate forecasts increase specialization and trade.
- Forecasts decrease expected price, but (due to specialization) increase price variability.
- Related to the above, forecasts increase world output, but increase output variability.
- Response of planted area to forecasts depends on current market conditions; specialization increases with stocks.
- Net welfare in each region increases with climate forecasts.
- More storage (lower storage cost) significantly increases the value of forecasts.
- Inelastic demand means that climate forecasts benefit consumers and decrease producer profit (quasi-rent).
- Distributional effects are much larger than net effects.
- Ad valorem (%) trade restrictions/subsidies that are equal across crops reduce he net value of forecasts, but now producers benefit.
- Quantitative trade barriers (such as import bans) drive the value of a forecast near zero.
- Modeling research from here must explore robustness of these findings.
- Models should be generalized and expanded to more crops, regions and with parameters more closely tied to market and climate data.

REFERENCES

Adams, R.M., K.J. Bryant, B.A. McCarl, D.M. Legler, J.J. O'Brien, A. Solow, and R. Weiher. 1995. Value of Improved Long Range Weather Information. Cont. Econ. Pol. 13:10-19.

Hallstrom, D.G. 2000. Interannual Climate Variations and World Grain Markets: The Costs of Surprise Versus Variability. Mimeo, Dept. of Agric. and Res. Economics, North Carolina State University.

Hallstrom, D.G. 1999. Surprise Versus Variability: Climate Fluctuations and Improved Climate Information in International Grain Markets. Unpublished Ph.D. Dissertation, University of California, Davis.

Mjelde, J.W., S.T. Sonka, B.L. Dixon and P.J. Lamb. 1988. Valuing Forecast Characteristics in a Dynamic Agricultural Productions System. Amer. J. Agr. Econ. 70:674-84.

Sumner, D.A., D.G. Hallstrom, and H. Lee. 1998. Trade Policy and the Effects of Climate Forecasts on Agricultural Markets. Amer. J. Agr. Econ. 80:1102-1108.

Seasonal Climate Forecasts and the Wheat Trade, What are the Economic Implications?

Harvey Hill, A. Love, J. Mjelde, S. Fuller, G. Hammer, D. Rosenthal, and D. Butler
Texas A&M, USA

SUMMARY

Currently, little agricultural economic research has addressed the international trade welfare implications of wheat producers' aggregate responses to seasonal forecasts. This research addresses this deficiency by modeling the effect of producers' responses to seasonal forecasts on the international wheat trade. The study provides insights into the expected welfare implications of the use of seasonal forecasts. The imperfect nature of seasonal forecasts information is explicitly acknowledged by incorporating stochasticity into the models. The wheat (*Triticum aestivum* L.) trade is studied because: a) it is vulnerable to climate variability; b) many wheat producing regions in the world are teleconnected to El Niño/Southern Oscillation (ENSO); c) wheat is the largest traded agricultural commodity; d) wheat is grown in areas were few substitute crops exist; and e) few substitutes exist for wheat use.

The model is a stochastic-dynamic model representing wheat production, consumption, inventories, and trade for the major exporting countries: United States, Canada, and Australia. Argentina, Europe, and importing countries are modeled in less detail. Biophysical simulation models estimate regional yield changes resulting from the use of seasonal forecasts. Aggregate yield estimates with and without improving climate forecasts are obtained. Price expectations are estimated by combining historic seasonal forecast information with a quasi-rational price expectations model. Price expectations are obtained with and without seasonal forecasts. The model simulates wheat production and trade for a 25-year time horizon. Monte Carlo techniques based on random draws from an empirical distribution of climate conditions are used to estimate changes in producer and consumer surpluses. Seasonal forecasts are based on the five phase method developed by Stone and Auliciems.

Expected percentage changes in consumer surplus for Australia, Canada, and the United States are 2.8%, 2.2% and 1.8% and for producers 13.0%, -7.6%, and 4.8%. The standard deviations of the percentage changes for Australia, Canada, and the United States are 3.2%, 2.8%, and 2.4% for consumer surplus, and 22.7%, 19.3%, and 4.2% for producer surplus. On average, Australian producers and consumers benefit more in percentage terms than for U.S. and Canadian producers and consumers. Canadian producers receive the smallest benefits and have the widest distribution of gains. The probability of a negative percentage change in either producer and/or consumer surplus occurring over the 25 year is non-trivial. These results reinforce the notion of benefits to producers and consumers are dependent on the realized climate outcomes over the relevant time frame. Consumers and producers in different countries and even within countries can be affected differently. Studies presenting only a long-run expected value of seasonal forecasts ignore this issue which is important as most policy decisions are made on shorter time frames than long-run expected value.

SESSION VI. THE SOCIAL CONTEXT: UNCERTAINTIES, RISKS, AND CHALLENGES

Opportunities and Constraints to Using Seasonal Precipitation Forecasting to Improve Agricultural Production Systems and Livelihood Security in the Sahel-Sudan Region: A Case Study of Burkina Faso, CFAR—Phase I

Carla Roncoli[1], Paul H. Kirshen[2], Keith Ingram[1], and Ian Flitcroft[1]
[1]University of Georgia, USA
[2]Tufts University, USA

INTRODUCTION

Several national and international meteorological services produce seasonal or climate precipitation forecasts for the Sahel-Sudan region of West Africa. These forecasts are based on the effects that oceanic sea surface temperature (SST) anomalies (for example, those associated with El Niño-Southern Oscillation, ENSO) have on regional weather patterns. The following countries are totally or partly in the Sahel-Sudan; Senegal, Mauritania, Mali, Burkina Faso, Niger, Chad, Nigeria, and Sudan (see Figure 1).

These forecasts, which are of the expected deviation of seasonal rainfall from the long-term average, may have sufficient precision that governments, organizations, businesses, and farmers can use them to improve decision making in agricultural production

While these new methodologies have generated a great deal of enthusiasm in the scientific and development communities, some have also pointed to the risks entailed for potential users, and in particular by resource-poor farmers, in making decisions based on probabilistic information (Hulme et al. 1992b). In response to these concerns, research efforts have been directed to understanding the framework of risks and costs faced rural producers and identifying points of entry and appropriate modalities for integrating forecasts in production and livelihood strategies in ways that are culturally acceptable and positively affect quality of life, food security, and environmental sustainability.

This was the primary goal of Phase I of the Climate Forecasting and Agricultural Resources (C FAR) project, a multidisciplinary research project funded by the US National Oceanic Atmospheric Administration (Office of Global Programs) and jointly implemented by the University of Georgia and Tufts University in the US and the National Meteorological Service and the NGO Plan International in Burkina Faso between the years 1997 and 2000. The research scope included:

- an assessment of the state of the art in climate forecasting;
- an evaluation of information priorities, communication networks, and potential application by institutional actors, such as early warn-

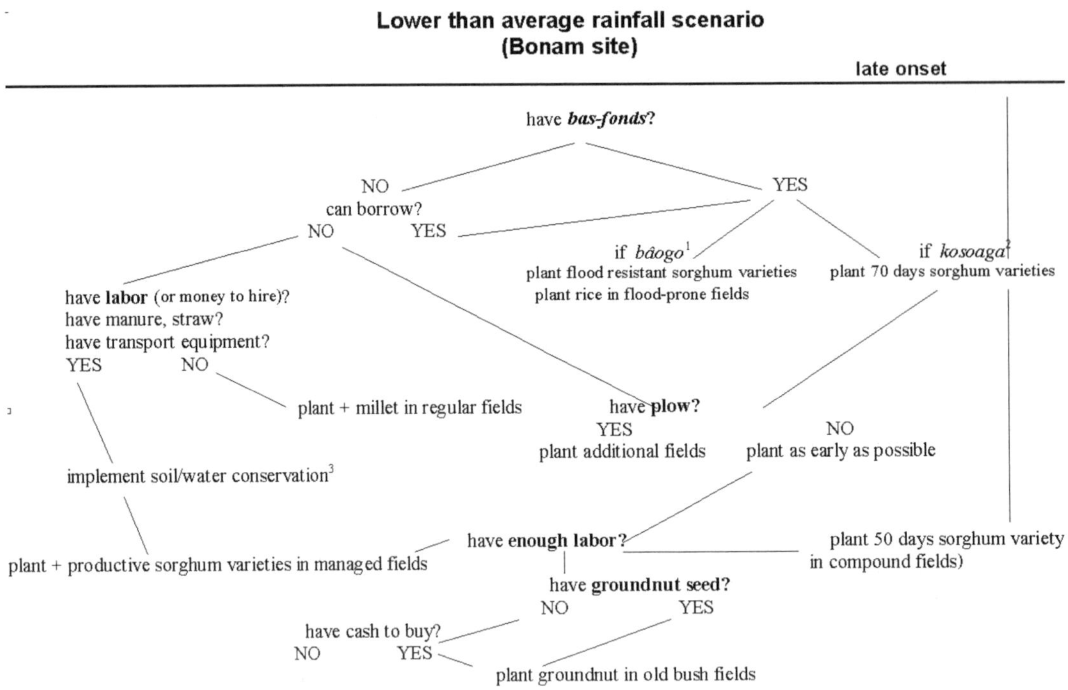

[1] *Baogo* are valley bottoms or areas near water courses where water stagnates for up to a week after a rain.
[2] *Kosoaga* are lowlands at the bottom of slopes where water flows, but does not remain for more than a day or two after a rain.
[3] Straw mulching, stone barriers, and *zai* (depressions filled with organic matter in which seed is planted).

***Figure 1.** Lower than average rainfall scenario.*

ing systems, development organizations, extension services, government agencies, and commercial enterprises;.

- an ethnographic study conducted in three agro-ecological zones of Burkina Faso on local knowledge pertaining rainfall, the role of rainfall information in farmers' decisions, the key forecast parameters needed by farmers, and the optimal ways for their dissemination.

Burkina Faso was selected as a case study because of the University of Georgia's long-term involvement in field research and institutional relationships there and because the country encompasses three agro-ecological zones that are representative of climate patterns and agro-ecological zones of the Sahel-Sudan region (including agro-pastoralism in the North, and commercial cotton production in the South, and subsistence-oriented grain farming in the central part of the country). It also exemplifies production systems characterized by dependence on rainfed farming, extreme poverty of productive resources, low levels of agricultural investments and infrastructural development, and few non-farm income earning opportunities, combined with high levels of rainfall-related risk.

This mix of factors means that farmers are very interested in information that can assist them in anticipating and mitigating risk, but also rather limited in their ability to optimally respond to rainfall forecasts or to recover from the impact of decisions made on the basis of inaccurate forecasts. As such, the Burkina case study presents an instance of the significant potential and the daunting challenges of applying rainfall forecasts in developing countries.

As reported by the Economist Intelligence Unit (1999), Burkina Faso is a poor country with a 1998 Gross Domestic Product per capita of $210. The range for the Sahel-Sudan region is $200 to $710; for comparison, the value for the United States of America is $31,000 (Economist Intelligence Unit, 1999). The economy is based mainly upon agriculture with income supplemented by remittances from migrant labor in neighboring countries. There are few rewarding and reliable cash earning opportunities. The Economist Intelligence Unit (1999) reports that 90 percent of the population in Burkina

Faso depend upon agricultural and livestock production for livelihoods and 80 percent of the population depend upon subsistence agriculture. Most of the cereals produced (millet, maize, and sorghum) are for household consumption and are grown under rainfed conditions with low levels of off-farm inputs such as fertilizer or mechanized cultivation. Cotton is the main export crop and, to a less extent, peanuts, sesame , and rice are also grown for the market. There are also some relatively small areas where vegetables are irrigated from small catchment reservoirs during the dry season for local consumption and export. Livestock management is an important component of the agricultural system, particularly for the agro- pastoralist groups in the drier Northern areas where cattle and other livestock are the principal source of income.

CFAR-I findings are summarized in this paper. After descriptions of the climate and institutional analysis, the village level results are presented

SAHEL-SUDAN CLIMATE

The Sahel-Sudan climatic zone is characterized by a strong South to North gradient of decreasing annual rainfall. Rain falls during a single wet season consisting of short intense storms over a three month (Sahel) to five month (Sudan) period varying from 100 mm to below 650 mm of annual rainfall in the semi-arid north to 650 mm to over 1000 mm in the semi-humid south (there are many definitions of the climate of these regions, these are from Sivakumar and Gnoumou, 1987). The native vegetation ranges from dwarf shrub grasslands in the driest zones to woodlands in wetter zones. There are several major rivers flowing through the region fed by the wetter regions in the south. Few of these have major reservoirs and sufficient flows to support large scale and widespread irrigation. Within the region there are also ephemeral streams, which are too undependable for major irrigation. For example in Burkina Faso, the total cultivated land is 3.27 million hectares, but only 15,000 hectares are irrigated (Economist Intelligence Unit, 1999).

The region's rainfall is characterized by extreme seasonal and annual temporal and spatial variability (Nicholson and Palao, 1993). In general, the less the rainfall, the greater the variability (Jackson, 1989). Large inter-annual fluctuations in total rainfall are common: for example, in Boulsa (central Burkina Faso) the rainfall was close to the long-term mean (727mm) in 1992, but it registered a record high in 1994 (1,010mm) and a record low in 1997 (491mm) (DPA, Boulsa 1998). Annual rainfall also differs significantly between neighboring villages or even farms: during the 1985 wet season, Flitcroft et al (1989) observed an almost twofold difference in total rainfall between gauges 14 km apart. This patterns is further exacerbated by the persistence of wetter and drier periods, which may result in prolonged droughts and cause famines such as those that occurred in the 1970s and 1980s (Nicholson 1986).

CLIMATE FORECASTING

Seasonal rainfall predictions are based on the realization that the SSTs influence the global atmospheric circulation and this has led to a search for regions of the world where such an influence might enable seasonal forecasting (Palmer and Anderson, 1994). The atmosphere is more strongly influenced by the SSTs in the tropics than in the extra-tropics. Moreover, these SSTs boundary conditions tend to persist for months making climate forecasting in some regions of the tropics feasible on these time scales. Such forecasts have been developed for a number of regions in the tropics, including Northeast Brazil, South Africa, West Africa, India and others (Mason et al., 1996)

In the Sahel-Sudan, the forecast research community has been most interested in seasonal prediction of the total rainfall during the wet season, often defined as the July-August-September total. In recent years, seasonal forecasts have been produced and transmitted to national meteorological services (NMS, but not internally disseminated) by the US Center for Climate Prediction, the United Kingdom Hadley Meteorological Center, the European Center for Medium Range Weather Forecasting, and the International Research Institute. To further the development and application of seasonal forecasts for the Sahel-Sudan and the rest of West Africa, the African Center for Meteorological Applications and Development (ACMAD) have organized fora to develop consensus probabilistic seasonal forecasts for the region for 1998 (PRESAO-1) and 1999 (PRESAO-2) wet seasons based upon the individual forecasts of

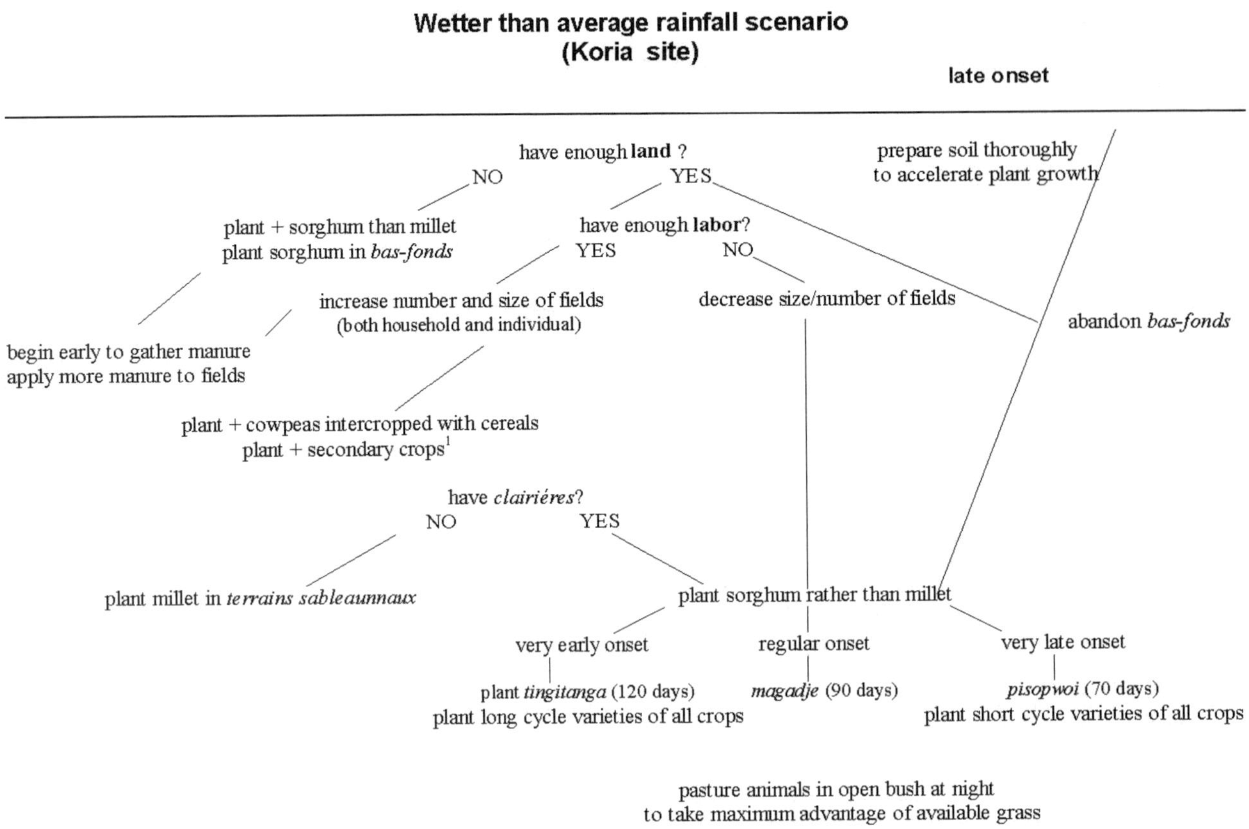

[1] Groundnuts, Bambara groundnuts, sesame, sorrel, okra, calabash, etc.

Figure 2. *Wetter than average rainfall scenario.*

organizations both inside and outside of West Africa. As shown in Figure 2, the 1999 forecast displayed, for each broad forecast area, the probability of the total rainfall being in the high, low, or average range compared to long-term historical rainfall. Forecasts were distributed at the fora and were available on ACMAD's World Wide Web page. Another outcome of the fora was that most West African meteorological services used their own models in July 1999 to update the seasonal forecasts. Hulme et al. (1992b) believe that the skill of these types of forecasts is insufficient for use at the village level to make direct farming decisions. They feel that a response to an incorrect forecast could result in "a waste of resources…at worst destitution and even threaten survival" (Hulme et al. 1992b, page 116). Part of CFAR's later research is to address how to best incorporate the probabilistic forecasts into farmers' decision-making processes and the other issues raised by Hulme et al (1992b) on forecast use by rural farmers.

INSTITUTIONAL ANALYSIS

The institutional analysis conducted by the CFAR project included organizations and agencies with missions related to agro-meteorology and agricultural and livestock production. Its primary purpose was to determine how to disseminate forecasts both to organizations and to farmers. Another purpose was to obtain recommendations from institutions for forecast improvements, even if they were beyond the state of the art. The analysis included interviews with African and non-African forecast producers, regional organizations such as ACMAD, Burkinabe institutions such as NMS, Group for Coordination of Food Security (known as CCI), commercial agencies that market and/or produce crops, agricultural extension services, development organizations, and farmers groups.

Overall it appears that an effective information flow system from forecasters to agricultural organizations and farmers is feasible. For example, the

NMS and CCI between them regularly contact, on at least a monthly basis through their newsletters, most of the government or semi-government organizations that either directly operate agriculture projects or have interaction with farmers through extension services and cooperatives. There is also the potential to distribute the bulletins the few relevant private and public institutions that presently do not receive any bulletins, but all of whom expressed interest in interviews in receiving forecast bulletins. The present speed of the information flow (ranging from real-time via radio to monthly via bulletins distributed within one week) seems adequate for seasonal forecasts. Given the proper incentives, the speed of dissemination could probably be increased if necessary.

In 1999, the Burkina Faso NMS broadcast the consensus climate forecast on national radio and also distributed it directly to SOFITEX at their request (see further description in ETHNOGRAPHIC STUDY). Since the Burkinabe national agricultural extension services are reducing their presence at the village level, radio broadcasts, particularly in local languages, may become important means to disseminate information to farmers and herders in Burkina Faso. The potential for successful forecast dissemination by national and local rural radio stations to farmers and herders is very high as demonstrated in Mali in 1998. The Mali National Meteorology Service used radio to transmit regional forecasts to farmers. The forecast information was widely used and harvests were higher than they would have been without the forecast (M. Konaté, personal communication, Bamako, Mali, December 1998). ACMAD also plans to start a pilot project throughout Africa to test the concept of using battery-less, windup radios to help rural farmers quickly receive time sensitive climate information and forecasts (ACMAD RANET Project, funded by US Agency for International Development, US AID).

Institutions in Burkina Faso have requested many improvements in the forecasts to be more useful to farmers and herders. Major requests included higher skill and spatial precision, the start of rains, and the distribution during the season. These requests have also been made by institutions in other Sahelian nations (ACMAD, 1998) and were also identified by Hulme et al (1992a) as being potentially useful to forecast users in the Sahel. Unfortunately, many of these are beyond the immediate state of the art (Hulme et al, 1992a).

ETHNOGRAPHIC STUDY

Research Sites

The CFAR project selected three villages as research sites, Bwahoun in the Southwest, Bonam in the Central Plateau, and Koria in the Sahel. These areas represent r the three main climatic conditions of the country, with the Southwest site benefiting from 5 to 6 months of rain with a long-term average of 1,0??mm, the Central Plateau site receiving 4 to 5 months with a long term average of 617 mm, and the Sahel site only 3 to 4 months with a long term average of 476 mm (see Figure 1).

Research was conducted between January 1998 and October 1999 in the course of several 4 to 6 weeks fieldwork periods during the dry season and during the rainy season. It encompassed a range of significant climate events, such as a major drought in the Central Plateau and equally devastating heavy rains in 1999 which washed off crops, houses, and roads in the Southwest. The time span enabled us to observe how these events impacted production strategies and outcomes, how local households coped with the resulting shortages, and how they shaped expectations and adaptations relative to the subsequent farming season.

Production remains largely dependent on seasonal rainfall, and therefore a high risk endeavor, in all three sites, although each differs considerably from the others in terms of livelihood configurations, degrees of integration into the market economy, and levels of capital investment and technological development in agriculture. The expansion of cotton production has catalyzed greater commoditizaton in the Southwest, labor migration to Côte d'Ivoire plays key roles in the livelihood of farming households in the Central Plateau, and livestock production and trade continuing to provide the main source of cash in the Sahel.

Each of the three communities count about 3,000 residents and are situated at 15 to 30 km from a provincial administrative center. Ethnic make-up vary, including a predominantly Mossi farming population in the Plateau, an indigenous Bwa community increasingly outnumbered by the influx of Mossi immigrants in the Southwest, and a majority of agro-pastoral Fulbe in the Sahel, subdivided between 'free' Fulbe and caste groups, including formerly enslaved Rimaibé. Sahelian Fulbe rigorously observe Islam precepts and condemn anything that deviates from them. Mossi of the Plateau integrate elements of Islam and Christianity with an enduring allegiance to ancestor-based spirituality. Among Bwaba, who have resisted the penetration of monotheistic religions, traditional beliefs are being eroded by the socioeconomic changes triggered by cotton production.

Export agriculture has enabled Bwahoun farmers to have far greater access to technology, inputs, and credit, which are channeled through SOFITEX, a semi-private company (a partnership of the Burkina government and the French textile industry) that purchases, processes, and export cotton. A few wealthy farmers have tractors and the large majority of farming households owns at least one plow and pair of oxen, while less than one fourth of households in Bonam and none in the Sahel does (where soils are too thin to allow for deep tillage). Unlike fields in the Plateau and in the Sahel which receive only manure and organic matter, Bwahoun farmers use large quantities of chemical fertilizers, herbicides, and pesticides, which SOFITEX provides on credit to farmers' groups, subtracting their cost from what they pay for cotton delivered by the group after the harvest. This system means that farmers carry not only all the costs but also all the risks of production since, if the value of farm output is less than their debt, farmers have to pool their resources and/or sell assets to repay the company for the inputs purchased on credit.

The importance of cotton in both colonial and post-colonial national economies (to which it accrues half of all export revenues) has also meant that cotton producing areas have been better served by agricultural extension in the past. In recent years, however, the outreach capacity of the government agricultural extension service was drastically reduced by IMF-induced structural adjustment policies that seek to reduce public spending. Consequently, SOFITEX has deployed a cohort of technical agents to carry out cotton-related extension in its zones of operation, while in the other sites foreign funded projects or international NGOs have intervened to provide resources for the specific areas and activities. These programs as well as current national policies of decentralized natural resource management have prompted the formation of viable organizational networks at village level which can be tapped into to disseminate forecasts.

Different levels of exposure and adherence to monotheistic religions, formal education, agricultural extension, and development programs affect the extent to which local knowledge, including indigenous forecasts, remains a viable basis for farmers' decisions. The influence of Islam in the Sahel and of cotton-related extension service in the Southwest have contributed to eroding interest and confidence in traditional knowledge, while a rich repertoire, based on environmental observations of tree fruit and flower production, duration and intensity of cold and hot periods, bird and insect behavior, and star and moon movements, was still found to be at work in the Central Plateau (although the credibility of these indicators is being challenged by increasingly erratic and anomalous climate patterns).

Information Priorities and Delivery

In all sites, producers concur that, in the last two or three decades, drawing a livelihood from the natural environment, whether by farming or by herding, has become increasingly arduous and risky. Climate variation, perceived as a higher frequency of water-deficit years, late onset of the rainy season, premature end of the rains, and anomalous rainfall distribution, is one of the most frequently mentioned causal factors increasing risk.

Given that livelihoods of most rural households depends on rainfed farming it is hardly surprising that most farmers expressed a great deal of interest in receiving rainfall forecasts. What follows is a synthesis of responses by farmers in the three sites concerning what kind of information would be most useful, when and how it should be delivered, while the next section will delineate possible adaptive responses.

What

Farmers in all sites stressed that, to be useful, forecasts on the quantity of rainfall need to be qualified according to the duration and distribution of rainfall over time and space. In all sites, the most salient parameters are the following: (a) the timing of the onset and end of the rainy season; (b) likelihood of water deficit periods; and (c) quantity of rainfall.

The relative importance of these parameters vary according to the different *agricultural strategies* and *crop requirements* in various zones. For instance, cotton and maize, the two major crops in the Southwest are vulnerable to water stresses at key points of their growth cycles, so Bwahoun farmers are particularly interested in knowing when water deficit periods may occur (to stagger planting in ways that avoid exposing crops to them in their vulnerable stages). Farmers in the Central Plateau, who grow mostly sorghum, are more interested in the duration of the rainy season, in order to select among varieties with different water requirements and growth cycles. In the Sahel, information on seasonal rainfall quantity can help farmers to know whether to plant millet in low or high water retention areas.

Information priorities are also affected by *recent experience.* For instance, in 1998 Bwahoun farmers stressed the occurrence of water deficit periods, which had affected cotton after planting and maize during head formation. They were not concerned with rainfall quantity, but they were in 1999 since both cotton and maize suffered from excessive rains (had farmers known about them in advance they would have replaced these crops with rice).

A *combination* of different parameters would also be more useful than one single parameter; for instance, Bwahoun farmers would have been able to prepare for flooding conditions in August had they known that the 1999 rains were expected to begin late but be heavy (with most seasonal rainfall being concentrated in July and August).

When

Most farmers make planting decisions based on: a) what had happened during the previous season (and the impact on their fields); and b) revisions on the basis of short term assessments of upcoming season. The planting period, which lasts between 30 and 90 days according to climatic zone and date of onset, is the most critical part of the farming season. The *timing and nature of onset* and the performance of *crop establishment* (how many times farmers had to replant) are considered the most reliable indicators for the rest of the rainy season. Farmers tend to associate late onset with drought and early onset with good rainfall, assumptions that can prove erroneous as they did in 1999.

By and large farmers in all sites requested that forecast be delivered from one to two full months before the expected onset of the rainy season (late April or early May). This would enable them to optimize labor and land allocation and to enact the most common responses to rainfall forecasts, such as procuring seed of different varieties and preparing fields in different locations. However, other adaptations, such as clear new fields, apply more or less manure, submit orders for inputs, and implement soil and water conservation require the information to be delivered early in the dry season. But seasonal forecasts provided at the end of the dry season could still contribute to adjusting farming decisions, in combination with farmers' own observations at the onset of the rainy season.

Seasonal forecast could also assist SOFITEX to decide what variety of cotton seed to distribute in different parts of the country, but to ensure that seed is in delivered to farmers in time for planting, they would need the information by February at the latest. The agency may also use forecasts to combat pests and diseases whose cycles may be related to rainfall conditions, as in the case of the 'white fly' infestation of 1998. Advance information on rainfall conditions that favor their occurrence is essential since chemicals need to be ordered and shipped from Europe.

How

Literacy levels remain very low in all sites, but particularly in the Sahel where families still resist sending children to school, as they are needed to tend to livestock.

In most sites farmers identified the best way to deliver forecasts to be by a combination of radio programs, followed up by visits by and meetings with agricultural extension agents. Radio broadcasting would ensure widespread coverage, while extension work would enable farmers to ask questions and get advice. This approach also reflects local learning styles, which are based on experience and interaction rather than verbal instructions. Popular gathering places, such as Friday mosques in the Sahel, markets in the Plateau, and beer drinking places in the Southwest can also be points of information delivery and discussion.

Radio programs in local languages, which are broadcasted by the national radio or by neighboring countries, are eagerly followed. All households in Bwahoun and about half of them in Bonam and Koria have radios but may not always have money to buy batteries (on the other hand, Bwahoun farmers buy them in bulk to operate spraying equipment). A few school teachers, extension workers, and wealthy farmers have generator-powered televisions, and villagers often gather to watch soccer matches.

Overall the ability of the government agricultural extension service is extremely limited, to the extent that agents do not even have fuel to get to the field, except when foreign funded project provide funds for specific areas and purposes. SOFITEX has deployed its own network of village-based agents, although being mostly young, poorly trained, and accountable to the company, they are not trusted by farmers. For instance, in 1999, SOFITEX disseminated the 1999 forecast produced by PRESAO-2 to farmers in order to convince them to continue planting cotton even if rains were late in starting. But its field agents were unable to interpret the information for farmers and were not taken seriously by them.

In order for forecasts to be understood by and useful to farmers, they need not only to provide relevant information, at the optimal time, and in the most appropriate form and language, but also by credible sources. This calls for a much closer *inter-institutional collaboration* between meteorological experts and all the players that intervene in rural areas, such as technical services, development projects, and local level institutions. For instance, extension and/or development agents may need to be trained in the interpretation of forecasts, while they may advise meteorologists about the information needs of rural producers.

ADAPTATION RESPONSES TO RAINFALL FORECAST

The main strategy whereby households attempt to reduce risk is by diversifying production and livelihood systems (Reardon et al. 1988, Ellis 1998), especially in the Sahel and in the Plateau, while in the Southwest cotton has rather induced a process of decreasing diversification. Farmers are becoming more involved in livestock production while agro-pastoralists are expanding their farmed acreages, and households are engaging in various non-farm activities (crafts, trade, labor migration). Agricultural diversification means carefully orchestrating what, where, when, and how to plant their crops in response to both concurrent and expected conditions, resulting in a constellation of fields, with different soil types and in different locations, planted in staggered sequence with various combinations of crops and crop varieties having different growth cycles and water (Carter 1997).

Most potential applications of seasonal rainfall responses envisioned by farmers interviewed pertain to this decision-making process, whereby farmers decide what to plant, how much, where, when, and how. For instance, they may:

a) choose certain crops and crop varieties (with different growth cycle, water requirements)
b) increase/reduce areas planted by crop, by location, by topo-sequence
c) adjust planting time and planting order among crops, locations, topo-sequence
d) change row orientation, with respect to slope
e) implement water conservation techniques
f) apply more or less inputs
g) store animal feed or sell livestock

In addition to farming options, they may also undertake a range of consumption, investment, or income-generation strategies to cope with expected shortages, such as:

a) Migrate to seek employment
b) Undertake non-farm economic activities
c) Disinvest household and personal assets
d) Avoid selling remaining food stocks
e) Reduce expenditures
f) Ration food

However, a study on farmers' creativity and resilience in responding to anomalous rainfall conditions should not neglect to also acknowledge that each adaptation entails trade-offs, as in the case of practices that mitigate some risks but expose farmers to others, or in the case of options that reduce risk at the cost of lower productivity, poorer nutrition, environmental degradation, or erosion of social relationships (Davis, 1996, Adam et al. 1998).

Farmers of all sites strongly emphasized how their ability to adequately respond to forecasts is hindered by poverty of resources and low risk tolerance. Availability of labor and land (in terms of both quantity and quality) are major constraints to agricultural adaptation in all sites. Lack of farm equipment that can expedite planting work and plant growth and enhance water utilization also hinders farmers' capacity to minimize the consequences of a forecasted drought, while dearth of agricultural capital or credit (except for input provision contingent on cotton production) limits their flexibility in resource mobilization and allocation. This means that household resource access profile largely defines which particular strategy is selected and, when and how it is enacted (Blaikie et al 1994).

Moreover, poverty of resources and lack of capital means that most farming are unable to withstand income shocks due to risky decisions, such as those based on faulty forecasts. This is because either they produce very limited surplus and operate at low levels above subsistence (as in the case of grain farmers in the Sahel and in the Plateau) or because they are encumbered by high debt burden (as in the case of cotton producers in the Southwest).

Summary accounts of the *complexities, constraints, compromises, and costs* inherent to agricultural decisions in each site are outlined here below. The attached decision diagrams exemplify farmers' responses to forecasts in situations of drier or wetter than average conditions (according what was actually obtained in each field site at the time of research, that is low rainfall in Bonam in 1997 and heavy rainfall in Bwahoun and Koria in 1999).

The Central Plateau

The principal mechanisms whereby Bonam farmers cope with climate variability is by adjusting field location, crop mixes, and seed varieties. Moisture retention capacity of soils and fields is a primary consideration in deciding what and when to plant.

If lower than average rainfall is expected, farmers will plant sorghum, the staple grain of choice, in valley bottom or near water courses. Those who do not have access to such fields may plant drought resistant varieties of sorghum or millet in their regular fields. Upland areas, characterized by gravelly, often eroded soils and poor infiltration may be planted with millet or abandoned. In case of higher than average rainfall, sorghum may be planted in regular fields and elevated areas instead of millet, while flood-prone valley fields may planted with rice. Even if heavy rain is predicted, farmers rarely abandon valley fields to avoid losing their use rights over them. In fact, because of their importance in drought years, valley fields are subject to increasing competition and those who own them are reluctant to lend them out to those who do not.

Since millet is more drought resistant (but suffers in wet conditions) farmers will increase millet acreages in response to a forecast for drier conditions. But farmers with large households avoid planting millet on most of their fields because: a) it is less productive than sorghum per unit of land, and b) since it can be ground by hand and consumed uncooked, it lends itself to uncontrolled use so that same quantity of stored grain lasts less than sorghum.

Maize, which has a growing cycle of 70 days, is well adapted to a shortened farming season, but it needs regular rainfall, careful weeding, and greater soil fertility; so it is only planted in small manured plots around the compound to ensure some food before the harvest of staple grains. Likewise, a 50 days variety of sorghum which can be planted late, has similar requirements, besides being disliked for the poor taste and texture of its flour. It also entails the risk of being predated upon by birds or damaged if it ripens too early while it is still raining.

During the last 10 to 15 years there has been a marked shift from longer cycle (120-150 days) to shorter-cycle (70-90 days) varieties of sorghum, most of which have been imported from the drier northern provinces. Shorter cycle varieties, however, are said to be less palatable and nutritious, less productive per unit of land, and more vulnerable to weeds, pests, and water stresses than traditional varieties, which are still planted when the rains begin early as in 1999.

Water conservation techniques, such as straw mulching, stone barriers, and zai (planting in small depressions filled with organic matter) are also implemented to improve water retention in regular fields, especially those planted with sorghum which is more demanding of water and nutrients than millet. But their feasibility depends of availability of labor, transport equipment, and materials (grass, straw, manure, etc.) and work must begin early in the dry season.

Lack of labor is a major constraint on many other forms of adaptation, including diversification of farming systems over different crops, fields, and locations. This especially the case for poor households where there is out-migration of able bodied labor to the Ivory Coast, which intensifies in the aftermath of climate shocks such as the 1997 drought.

Access to labor saving technology, such as animal traction, is therefore a key factor in determining households' capacity to minimize rainfall-related risk by expanding and/or diversifying their holdings and by expediting the planting process, which is crucial to good crop performance. However, only about one fourth of Bonam households have plows, and significantly those are the wealthiest in the village and those that were better able to cope with the consequences of the 1997 drought.

Southwest

The first strategy whereby Bwahoun farmers would respond to a rainfall forecasts is by adjusting land preparation by tractor or plow, orienting ridges with regards to slopes in ways that either facilitate or prevent water flow, that is either parallel or perpendicular to the slope.

Rainfall forecast may also influence their decisions concerning the relative proportion of cotton and maize to be plant in their fields. Because of cotton's high demands of land and labor, other crops that were previously grown have largely given way to maize, whose productivity is higher and labor demands less incompatible with those of cotton. Very limited acreages of sorghum, groundnuts, sesame, beans, and rice are still cultivated with minimal investments of labor and inputs.

Availability of household labor and suitable land are key factors shaping productive decisions in response to forecasts. *If the household is short of labor,* high rainfall may require them to reduce acreages of both cotton and maize in order to weed properly (since weeds proliferate in high humidity conditions). Perhaps they may plant more sorghum, which is less vulnerable to weeds. *If a household has sufficient labor,* it may increase cotton acreages if high (but not excessive) rainfall is expected.

However, there are costs and risks associated with this strategy: a) households will still need to grow enough maize to provide for at least part of their consumption, since grain prices make it financially unwise to rely mostly on the market; b) expanding cotton also means needing more inputs (and therefore a greater debt burden); c) cotton needs a full four-to-five months of rain, so if the rains end prematurely it will not reach full maturation. Therefore in case of late onset (as it happened in 1999) farmers may plant more maize than cotton, although they still need to do enough cotton to obtain inputs on credit needed for both crops.

But maize is also vulnerable to waterlogging, so if wet conditions are predicted and farmers *have access to such land,* they will plant it on upland fields (see Figure 3, possibly displacing groundnuts), staggering planting dates to minimize the risk of vulnerable stages in plant growth coinciding with periods of water stress. If they *do not have access to such land,* they must plant as early as possible to ensure that plants are grown strong enough to withstand flooding when it begins raining heavily in July. But the ability to do so is predicated upon having enough household labor and farm equipment (tractors or several plowing teams, since at planting time oxen are still weak from the long dry season and unable to work at a fast pace).

High rainfall is favorable to both rice and sesame, which are grown respectively in lowland and upland fields, and which can be very profitable. However, both crops have high labor demands, especially at planting and harvesting, that conflict with those of maize and cotton, so that most farmers can only plant small plots.

If low rainfall is predicted, farmers will emphasize food crops versus cash crops, by reducing cotton acreages and shifting to a shorter cycle, drought resistant variety of maize, which, however, is considered less palatable, less marketable, and less durable. Maize and sorghum will replace rice in valley fields. Completing the planting process quickly is critical to accelerate plant growth, so that farm equipment such as tractors and mechanical seeders, which enable farmers to get a head start by

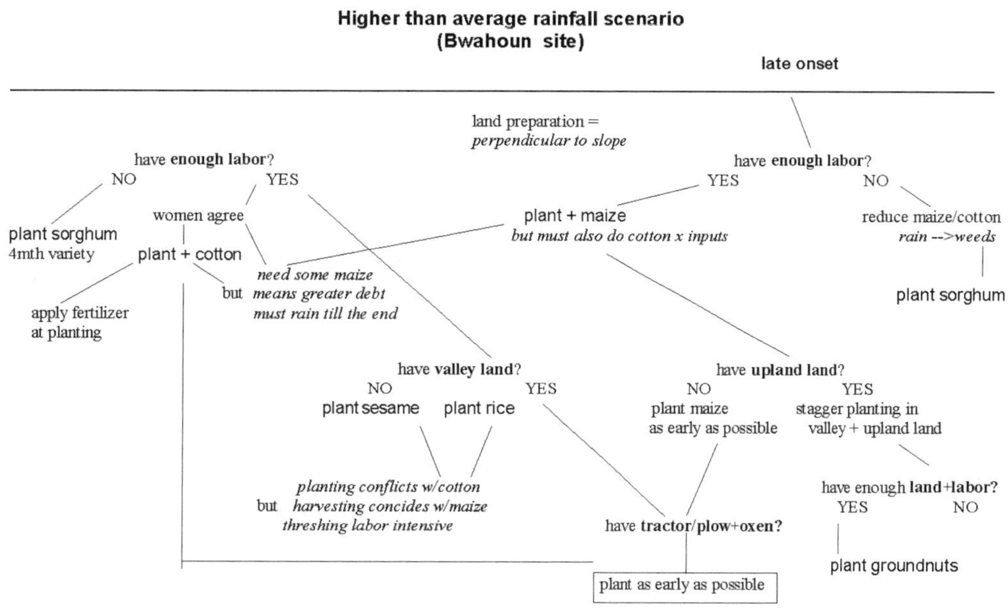

Figure 3. *Higher than average rainfall scenario.*

planting on dry ground before it rains, are key assets.

Use of chemical inputs is also adjusted according to rainfall conditions, so forecasts could help farmers save money by ordering and applying them in conservative ways. For instance, in drier situations, farmers may need less herbicides (the most costly of all inputs), while in wetter conditions, they may apply fertilizer at planting and pesticide in ways that minimize their being washed off.

The Sahel

Sahelian agro-pastoralists are somewhat differently affected by seasonal rainfall since they can move with their herds for at least part of the year. However, various ecological, demographic, and political conditions are making it increasingly difficult for the Fulbe to draw most of their livelihoods from herding. These include decreasing availability of water and grass, population growth and expansion of cultivation, and political restrictions and legal prosecution against trespassing in neighboring countries.

Potential livestock management responses to forecasts elicited from herders are limited to strategies that are enacted during the rainy season rather than in anticipation of it, such as stocking fodder and pasturing animals near fields in drier than average conditions and maximizing grazing time (by taking animals to the bush at night) to increase weight and, therefore, value, in high rainfall conditions.

Farmers (and agro-pastoralists who farm), on the other hand, can apply forecasts in multiple ways although land scarcity limits adaptive options, at it is also the case in the other sites (although here herders rather than farmers have customary control over the land whereby they endeavor to limit expansion of cultivation into pasture areas).

Forecasts can also help farmers direct their land preparation efforts: for instance, if late onset is predicted, they may have to reduce acreages in order to ensure that fields are thoroughly hoed to speed up plant growth and prevent weeds from choking seedlings. Farmers can also calibrate manure application, increasing inputs to maximize benefit to crops if high rainfall is predicted, and reducing them in case of drought to avoid scorching.

Choosing what to plant where is the main way farmers can respond to rainfall conditions. Farmers classify farm land in three categories: elevated clay areas (*clairières*), sandy soils (*terroirs sablonneux*), and valley bottoms (*bas-fonds*). As the last three decades have brought about a drier climate, farmers have moved away from clay areas into sandy soils (which remain moist and cool for longer periods) and even to the valley which were previously only used for grazing (competition for access to valley bottoms has intensified farmer-herder tensions). Movement away from clay areas to valley bottoms is also associated with a gradual shift from millet, the staple crop of choice in the Sahel to sorghum. The latter, in fact, grows well in higher water retention areas, such as valley fields and, its shorter growth cycle is better adapted to the increasing frequency of late onset or premature end of the rainy season.

Therefore, more sorghum, especially shorter cycle varieties, will be planted if late onset is predicted, but also in case of abundant rains since it is less susceptible than millet to water damage. Most millet grown is a longer cycle (120 days) local variety that is well adapted to the sandy soils and low rainfall prevalent in the Sahel, but there are three main sorghum varieties to select from, according to duration of growth cycle (which vary between 70 and 120 days) and water and soil fertility requirements, some of which have been imported from the Central Plateau. Forecasts for high rainfall also induce farmers to plant more secondary crops, such as sesame, sorrel, maize, cowpeas, and calabash.

In a water deficit season, farmers who have access to such land will maximize cultivation in valley bottom, while clay areas may be abandoned since they bake and crack when dry. In this case, sorghum will be largely planted in valley bottoms and millet in the regular fields. In high rainfall conditions, valley bottom may remain uncultivated as grazing areas, regular fields may be planted with both sorghum and millet, and clay areas will be brought back into cultivation and planted with sorghum. In this case, farmers may either increase acreages to maximize yields (if they have enough labor) or reduce them to ensure that fields are properly weeded and hoed (if they are short of labor).

CONCLUSIONS

We have presented in detail the incentives and constraints to the use of modern climate forecasting to improve agriculture in three agro-ecological regions in Burkina Faso representative of conditions in the Sahel-Sudan. We have found that:

The improvement and expansion of modern forecasts for the region is an active area of research. However, the *skill of present forecasts* of total seasonal precipitation may be insufficient for use at the village level to make direct farming decisions.

It appears that an effective *information flow system* from forecasters to institutions and farmers is feasible in Burkina Faso. The present speed of the information seems adequate for seasonal forecasts. Given the proper incentives, the speed of dissemination could probably be increased.

There is strong interest by farmers in climate information because of its *relevance to their livelihoods:* rainfall is one of the most salient topics of conversation during the rainy season.

There is a need for forecasts because, while farmers have used *local forecasting knowledge,* the latter are becoming less reliable due to more recent climatic, ecological, and agronomic changes. Cultural influences such as Islam and agricultural extension have also weakened this knowledge base in some areas.

Farmers can *benefit* from forecasts by enacting a variety of adaptive strategies at their disposal: in most areas production and livelihood systems are diversified to encompass a variety of response options.

Farmers mostly benefit from forecasts provided *before the onset* of the rainy season but they also can use forecasts provided after the onset since the planting season for certain crops lasts till early August.

Farmers need a *combination of information,* the relevance of which varies according to crop and area, but especially timing of the onset and end of the rainy season and distribution patterns of rainfall during the farming season.

The *context of information delivery* (particularly the credibility of the source delivering the information and the opportunity for farmers to discuss, ask questions, and learn by doing) is very important. This calls for greater inter-institutional cooperation between meteorological services on the one hand and technical services and development organizations on the other.

Risk tolerance is low for resource-poor farmers because: 1) in some areas surpluses are very marginal and farmers have few assets to fall back onto; and 2) in others they are burdened with heavy cash debts that need to be repaid after the farming season.

In applying forecasts farmers face a number of *resource constraints:* low risk tolerance and especially scarcity of land, labor, capital, and technology. This suggests that forecasts should be delivered in the context of an integrated development strategy to support i) sustainable resource use and ii) the viability of local production and livelihood systems. In particular, provision of agricultural credit and technology could help enhance flexibility and timeliness of response.

ACKNOWLEDGEMENTS

This research has been supported by the Economics and Human Dimensions of Climate Fluctuations Program, Office of Global Programs, United States National and Atmospheric Administration under Grant Number NA76GP0327 . We thank CFAR's institutional partner PLAN International, particularly Country Director Felipe Sanchez and Project Manager Eric Mamboue, for exceptional research support in Burkina Faso and team members Bahadio Salam, Boena Salifou, Boureima Fatimata, Dicko Hanafi, Ouattara Françoise, and Tiendrebeogo Frederic for their invaluable contribution to field research.

FOOTNOTE

1 A climate forecast is generally considered to be a forecast of meteorological conditions for a period of one month or more (a season). These forecasts can be prepared with advance or lead times ranging from the start of the forecasted period to several months to one year in advance. Climate forecasts are also often referred to as seasonal forecasts.

REFERENCES

ACMAD (1998) West African forum on climate variability and prediction, organized by ACMAD, AGRHYMET, ICRISAT, MEDIAS/START, Abidjan, 4-8 May.

Adams, A.M., J. Cekan, and R. Sauerborn. 1998. Towards a Conceptual Framework of Household Coping: Reflections from Rural West Africa." *Africa,* 68, 2, pp. 263-283.

Blaikie, Piers, Terry Cannon, Ian Davis, Ben Wisner. 1994. *At Risk: Natural Hazards, People's Vulnerability, and Disasters.* Routledge, London.

Carter, Michael. 1997. "Environment, Technology, and the Social Articulation of Risk in West African Agriculture." *Economic Development and Cultural Change,* 45, 3, pp. 557-590.

Davis, S. 1996. *Adaptable Livelihoods: Coping with Food Insecurity in the Malian Sahel.* Macmillan, London.

Economist Intelligent Unit (1999). *Country Reports, 1999-2000.* Great Britain.

Ellis, Frank. 1998. "Household Strategies and Rural Livelihood Diversification." *Journal of Development Studies,* 35 (1), pp. 1-38.

Flitcroft, I.D., Milford, J.R., and Dugdale, G. (1989) Relating point to area average rainfall in semi-arid West Africa and the implications for rainfall estimates derived from satellite data. *Journal of Applied Meteorology* 28, 252 –266.

Hulme, M., Biot, Y., Borton, J., Buchanan-Smith, M., Davies, S., Folland, C., Nicholds, N., Seddon, D., and Ward, N. (1992a) Seasonal rainfall forecasting for Africa. Part 1: Current Status and Future Developments, *International Journal of Environmental Studies* 39, 245-256.

Hulme, M., Biot, Y., Borton, J., Buchanan-Smith, M., Davies, S., Folland, C., Nicholds, N., Seddon, D., and Ward, N. (1992b) Seasonal rainfall forecasting for Africa. Part II : Application and Impact assessment. *International Journal of Environmental Studies* 40, 103-121.

Jackson, I.J. (1989). *Climate, water and agriculture in the tropics.* Longman.

Konate, M. (1999).Mali National Meteorological Service, Bamako.

Mason, S.J., Joubert, A.M., Cosijn, C. and Crimp, S.J. (1996) Review of seasonal forecasting techniques and their applicability to Southern Africa. *Water South Africa* 22, 203-209.

Nicholson, S.E .(1985) Sub-Saharan rainfall 1981-84. *Journal of Climate and Applied Meteorology* 24, 1388-1391.

Nicholson, S.E. (1986) Climate, drought and famine in Africa. In *Food in sub-Saharan Africa.* Eds. Hansen A. and McMillan, D.E.. Lynne Rienner, Boulder, Colorado.

Nicholson S. E. and Palao I. M. (1993) A re-evaluation of rainfall variability in the Sahel 1. Characteristics of rainfall fluctuations. *International Journal Of Climatology* 13 (4), 371-389.

Palmer, T.N. and Anderson, D.L.T. (1994) The prospects for seasonal forecasting: A review paper. *Quarterly Journal of the Royal Meteorological Society* 120, 755-793.

Reardon, Thomas, Peter Matlon, Christopher Delgado. 1988. "Coping with Household-level Food Insecurity in Drought-affected Areas of Burkina Faso." *World Development,* 16 (9), pp. 1065-1074.

Sivakumar, M.V.K. and Gnoumou F. (1987) Agroclimatology of West Africa: Burkina Faso. *Information Bulletin No. 23.* International Crops Research Institute for the Semi-Arid Tropics.

Strategies for Seasonal Climate Risk Management in Subsistence Agriculture

Ancha Srinivasan
Regional Science Institute, Japan

INTRODUCTION

Highly variable seasonal climate characterized by droughts, floods, typhoons, frosts, extremely hot or cold spells, etc. is the principal source of production, financial and environmental risks in both commercial and subsistence agriculture. However, the nature and magnitude of risks, and perceptions of and reactions to them in subsistence agriculture differ greatly from those in commercial agriculture. Subsistence farmers try to stabilize production by focusing on food crops that are usually combined or rotated with livestock (which provide draught power, income, dietary protein and improve soil fertility), while commercial farmers aim at maximizing the economic return by allocating more area to cash crops. Subsistence agriculture is labor-intensive and is characterized by small farm size, negligible assets, low and unstable production due to high dependence on rainfall, low input use, small and slow-evolving technology component, and a close association between household and production activities. Further, subsistence farmers have limited access to climate information networks. In contrast, commercial farmers have relatively big farms and use external inputs and modern technology in terms of new genotypes, mechanization, etc. They usually maintain cash reserves and can afford to pay for climate prediction and crop insurance services provided by the private sector. Climate risk can be very high in subsistence farming involving loss of life or out-migration but it is limited to debt or termination of farming activity in commercial farming. Subsistence farmers are, therefore, more risk-averse than commercial farmers. Further, due to displacement and discriminatory policies in many countries, subsistence farmers are forced to cultivate ecological areas where resource degradation is already high. This is especially true in Asia, where croplands are being converted into industrial zones, and land reforms often resulted in allocation of poor lands to subsistence farmers.

Insofar as subsistence pastoralists are concerned, they tend to maintain large herds even in periods of climate risk in the hope of having at least some animals alive while commercial farmers are more able and/or willing to reduce livestock numbers in time. The latter usually keep drought or snow-resistant livestock species and cope with climate risk better than subsistence herders with more vulnerable livestock. Subsistence farmers in some regions receive access to resources from village headmen, on whom they depend for all decisions about resource management. Complex land tenure patterns also restrict their ability to manage risks. Commercial farmers, on the contrary, have clearly defined land rights and receive information on risk management directly from Government sources.

Nearly 60% of world's arable land is still farmed by subsistence methods. Analysis of climate risk management options and their economic and environmental dimensions in subsistence farming is, therefore, important to help farmers make right decisions, to assist planners in evaluating the efficacy of risk management tools, and to aid governments in preparedness, prevention and planning activities. As the timing, frequency and extent of extreme events have close relationship with phenomena such as El Niño Southern Oscillation (ENSO), North Atlantic Oscillation and sea surface temperatures in the Indian Ocean, the ability to predict and effectively communicate climate anomalies may ultimately help subsistence farmers in minimizing the adverse impacts. This report attempts to identify strategies that are being, and can potentially be applied to manage climate risks in subsistence agriculture within the wider context of poverty alleviation and food insecurity.

CLIMATE RISK MANAGEMENT STRATEGIES IN SUBSISTENCE AGRICULTURE

Climate events are key external forces in the evolution of diverse, dynamic, complex and often conservative risk management strategies. The ultimate decisions, which may be tactical or strategic, vary among individuals, communities, and regions depending on the relative timing of a climatic event in relation to crop calendar. Tactical or short-term responses, which are more common in subsistence agriculture, are taken up during the cropping season while strategic decisions on system improvement are planned far in advance. Even within a region, different types of farms are known to respond differently to similar climatic events during a season (Dorward, 1996). Farmers who experience climate disasters more regularly adopt a wider range of strategies than those who do not. The choice also varies with individual's ability to bear risk, differences in goals and attitudes toward risk, and the economic and social framework including the availability of social security safety nets in a region. For example, the existence of a state emergency fodder fund in Mongolia was believed to have affected herders to lessen their own production of winter fodder, and to quit some risk-avoiding practices. When such fund was wound up, there was an immediate modification in management by herders towards more risk-averse practices, such as mixed-herding and collection of hay for themselves (Swift, 1999).

Farmers use both on-farm (structural) and off-farm (non-structural) mechanisms to cope with climate risk. On-farm practices include early planting, varying crop mixes, water harvesting, windbreaks, etc. Off-farm risk-avoidance measures usually include production contracting, credit and insurance, which are available only in selected regions. For example, production contracting in oilseeds and poultry is common in parts of the Indian subcontinent. Other off-farm measures such as marketing pools, contract selling, forward selling and future contracts are common among farmers growing vegetables and fruits. In certain extreme cases, farmers migrate to seek employment and undertake non-farm related economic activities. Other off-farm responses to climatic extremes include disinvesting household and personal assets, reducing expenditures, rationing food consumption, retaining some food stocks, etc.

Climate risk management strategies may be grouped into three categories: passive adjustments, anticipatory actions and deliberate reactions. Passive adjustments include biologically driven changes in communities. Anticipatory actions are limited in subsistence agriculture as most of risk management occurs on the farm and almost at a subconscious level. In contrast, commercial farmers consider climate risk in a formal and structured manner and take up actions such as production and marketing contracts, hedging, and leveraging. Insofar as deliberate reactions are concerned, subsistence farmers employ dynamic agronomic strategies as against single strategies where one prediction prior to cultivation destines a series of decisions.

Given the complexity and intensity of land use mixes, over-riding importance of social issues, and the diversity of climate scenarios in subsistence agriculture, it is not possible to exhaustively review all strategies. In view of this, an attempt will be made to draw on examples that are familiar and to search for broader patterns.

Enterprise Diversification

Despite the prospect of lower returns and the need for a wider range of management expertise than specialization, diversification and adjusting the proportion of enterprises is the most basic and robust approach used by subsistence farmers. Besides being able to compensate the loss from one enterprise by the gain from another, diversification results in timeliness of operations and in balancing labor needs. Diversification strategies, however, vary in different regions. Farmers in West Africa, where the short mono-modal rainy season favors annual grassland crop production, diversify by combining cattle and cropping. In such systems, livestock minimize the effects of adverse climates through improving soil fertility, reducing soil erosion and contributing to the efficient use of labor. However, the extreme seasonal variation in forage dynamics due to rapid crop wilting and deterioration forces herders to migrate and diversify through trade with farmers and agro-pastoralists. In contrast, the east African bimodal rainfall pattern favors natural pastures and livestock production.

As crop cultivation is limited, pastoralists diversify by keeping several species of livestock. For example, Turkana herders in northwestern Kenya maintain domestic herbivores, camels, cattle, sheep, goats and donkeys, each with species-specific production patterns and managerial needs, although they inhabit an arid climate which favors camels and goats (Galvin and Ellis, 1996). Indeed, goats are easier to maintain in periods of drought than cattle. Likewise, a herder who herds yaks or yak-cattle crosses and a few cattle can withstand risk from an exceptionally heavy snowfall better than a pastoralist who herds only cattle, although the latter is likely to earn more income in the event of normal seasons (Swift, 1999).

In subsistence cropping, it is customary to adopt intercropping or mixed cropping with several species and traditional cultivars. The composition of the mix is varied at each planting and often reflects the date of onset of rains and duration of rainy period. Despite their low yield potential in good climates, traditional cultivars survive better than modern ones in drought-prone regions, especially due to their low water and nutrient requirements. Improving availability of seeds of desirable crops and genotypes that are tolerant to climatic vagaries thus assumes great significance in subsistence agriculture. Crop diversification also stabilizes income. In Senegal, for example, cassava farmers integrate perennial forage crops or bissap. The abandonment of diversification strategies is considered a major cause for collapse of some poultry farms in Andhra Pradesh due to climatic disasters such as cyclones.

Economic diversification through off-season activities such as selling of free natural resources (fuel-wood, naturally grown vegetables, fruits, etc) and ability to switch to non-farming activities also help in stabilizing income of subsistence farmers. Indeed non-farm income is considered an important determinant of overall farm productivity and farmers' ability to manage climate risk in Burkina Faso (Pietola et al., 1998).

Selection and Diversification of Management Practices

Since production risk due to climatic variability varies with soil type, crop variety and other factors, farmers make changes to farm operations before, during and after cropping season. This is often referred to as "response farming". Rational selection of the growing season and cultural practices, and land allocation among alternative crops are two main options in response farming (Kingwell, 1994). Farmers in West Africa, for example, adjust crop area and inputs on the basis of timing and amount of early season rainfall. Tailoring land allocation among crops at a farm scale to ENSO Phase has been shown to increase farm income by 9 to 20% in the Argentine Pampas (Messina et al., 1999). As growing crops in seasons with a strongly positive Southern Oscillation Index (SOI) before planting is more profitable than in seasons with a strongly negative SOI (Clewett et al., 1991), varying agronomic practices depending on the level of climate risk is possible, if predictions are reliable. In livestock management, maintaining feed reserves, use of antibiotics and reducing the number of livestock (destocking) prior to the onset of climate disasters are some useful options. Destocking also reduces environmental degradation (McKeon and White, 1992).

Choice of cropping system, crop, and genotype

Judicious selection of crops, genotypes and cropping systems often determines the success of a farm in coping with climate risk. Many subsistence farmers adopt a "crop-cafeteria" approach wherein they seek to choose and mix crops and varieties to take advantage of, and cope with climatic variation. As the variation in crop response to climate events is largely due to differences in tolerance to water availability, high temperature, etc. careful selection of crops and genotypes with multiple traits assumes significance. As crop failure is often associated with duration and frequency of drought events, especially in regions receiving annual rainfall below 400 mm, growing heat and drought-tolerant crops or genotypes is important. Indeed, a wide range of genetic diversity exists for characteristics such as heat tolerance (Srinivasan et al., 1996). Planting drought resistant crops in Brazil following a reliable prediction of El-Niño in 1992 resulted in grain production of nearly 80% of the normal, as against only 20% in 1987, when no El-Niño-related action was taken (Source: www.pmel.noaa.gov/toga-tao/el-nino/impacts-brazil.html). Crops such as cassava are largely unaffected by El-Niño as it can tolerate poor soils, drought and floods due to its deep root system. Switching crop mixes depending on water regime is a common strategy adopted in most drought-prone regions. Farmers often increase area under drought-tolerant and short season varieties and crops. For example, sorghum is preferred to rice in Mali in drought while late-planted sweet potato serves as the ultimate survival crop in Zimbabwe. Farmers growing short duration corn instead of cotton in red soils of Telangana, and jute instead of groundnut in Rayalaseema regions have succeeded in coping with drought. Likewise, sowing more rice and less cotton in Peru has been reported to reduce losses in an El-Niño year. In this connection, development of simple decision rules on crop choice and crop sequence based on seasonal climate indices in different regions is helpful. For example, in many countries bordering west of the Pacific Ocean, planting drought resistant crop like sorghum is helpful if SOI is consistently negative. Planting remunerative crops like cotton must be tried only if SOI is rapidly increasing or consistently positive (Meinke et al. 1998).

Choice of cropping system and its components is also vital, as some systems are better in terms of both economic and environmental sustainability. Rotation between cereals and crops such as bissap or cowpea (West Africa), and shifting cropping patterns towards relatively drought tolerant crops (Mozambique) and towards crops whose prices tend to increase in periods of climatic stresses are common strategies. Introducing legumes like lucerne and lupins on sandy soils and of medics, serradellas and subterranean clover in pastures (Ewing et al., 1987) or direct seeding of *Sesbania sesban* into the maize crop preceding the fallow in dry areas of Kenya (Swinkels et al., 1997) are known to reduce drought risk. Avila (1996) reported that traditional maize-bean intercropping system in Nigeria performed best under low rainfall conditions, while the improved maize-bean system and alley farming with *Leucaena* hedgerows performed better under average and maximum rainfall conditions. However, alley farming is considered sustainable because of a 10% annual yield decline in the traditional system. Selection of a crop in a rotation is also important. Using the stochastic bioeconomic farming system Model of an Uncertain Dryland Agricultural System (MUDAS) in Mediterranean climates of Western Australia, Schilizzi and Kingwell (1999) reported that chickpea was more profitable than field pea or faba bean in rotation with wheat. Wheat-chickpea rotation was also associated with the higher revenues and lesser revenue variance as compared with the continuous wheat rotation.

Opportunity cropping

Opportunity cropping denotes the practice of sowing a crop whenever a planting opportunity occurs. Farmers in coastal Andhra Pradesh, for example, partially recoup losses of rice from tropical cyclones by growing pulses like mungbean soon after the floods recede. Seasonal forecasts may be used to decide on cultivating a specific crop for which the demand is expected to grow following an adverse climatic event. For example, the ability to grow crops such as short duration soybean in an El-Niño year can benefit farmers in Peru and Chile substantially, as sharp declines in fishmeal production following El-Niño are likely to result in a shift in favor of soybean meal.

Sowing time, rate, and method

As planting opportunities are often limited by the amount and timing of rainfall, varying sowing time can minimize climate risk. For example, late sowing of short-duration cultivars is identified as an effective strategy for rainfed wheat in the monsoonal climate of the Pampas (Savin et al., 1995). Farmers in West Africa plant twice or thrice to take advantage of rainfall and remove seedlings based on the observed rainfall (Blench, 1999). Such mid-season corrections of plant density also help retain soil moisture (when used as mulch) and improve fodder security for livestock. Varying seed rates with the amount of early rainfall is another common strategy. Likewise, dry seeding instead of transplanting minimizes the risk from drought and improves profitability of rainfed lowland rice. With this method, crop establishment is about 40 days earlier and soil water regime for the following crop is better than with transplanting (Saleh and Bhuiyan, 1995).

Fertility management

Adjustments in use of soil amendments (farmyard manure, lime, etc.) and fertilizers are important to stabilize crop yields in adverse climates. Applying small doses of fertilizer at critical stages can enable the crop withstand the effects of drought to some extent. However, careful attention must be paid to the form, timing and placement of fertilizer, as well as the total amount applied, especially in relation to the availability of soil moisture. Tactical adjustment of nitrogen fertilizer use or cultivar maturity based on a seasonal prediction available at sowing in the northern part of the Australian grain belt increased profits up to 20% and reduce risk up to 35% (Hammer et al. 1996).

Water management

Water management assumes great significance in periods of drought and floods. Although irrigation reduces the risk of drought, it is not often within reach of subsistence farmers in many regions due to high capital and operating costs. On-farm water storage and water harvesting methods allow management of irrigation timing in such areas. Water use efficiency is enhanced through methods such as altering the size and pattern of ridges (Broad Bed Furrow method practiced in the semi-arid tropics of India) and adopting more efficient systems such as pitcher irrigation. Minimum tillage, stubble retention, shrub/tree supports, windbreaks, use of soil conditioning compounds (e.g., Terracottem in Burkina Faso, Ivory Coast and China) and mulching the soil with leaves (e.g., water hyacinth leaves from Lake Victoria) or farm residues enhance water availability and permit flexibility in sowing time and crop choice. Other methods to accumulate the amount of moisture that is not brought to ground in rainfall (Occult precipitation) include agro-forestry technologies, hanging mosquito nets in foggy areas to catch moisture, brick constructions, etc. In regions where water is available, water allocation systems are modified to cope with climate risk. In modified "warabandi" system in India and Pakistan, for example, water allocation at tertiary unit is based on supply to every field for a fixed duration and at a fixed time on fixed days usually in line with climatic and production cycles. Community-based irrigation systems and monitoring systems to discourage water theft in periods of drought are also common in many Asian countries. Strategies to manage waterlogging from heavy rainfall or floods include provision of drainage, construction of mounds and levees, early planting, using flood-tolerant (e.g. rice, castor, banana and cassava) and storm-tolerant (e.g. cotton) crops, and drying the soil profile with crops such as sweet sorghum or lucerne.

Pest and disease management

Extreme climate events impose additional risk via triggering pest and disease incidence. Such risk is usually minimized through adoption of an integrated program of crop rotations, varietal resistance, biological and chemical controls.

Harvest management

Rain at harvest time can adversely affect crop yield and quality. Subsistence farmers manage rain disruption to harvest through either bringing in labor from other areas in exchange of their services at a later stage or engaging family members, who are otherwise dependent solely on non-farm related activities. Other strategies include cultivating genotypes with varying maturity, commencing early harvest, using mechanical driers, etc.

Geographic spread

Although geographic spread of farming activities is an ideal way of coping with climate risks, choice of a location is not usually an option in subsistence farming. In regions with excess land and low population density, however, location of cropping activities can reduce risk, by using the best soil types, or by spreading crops over a wide area.

Choosing Low Risk or Risk-Minimizing Activities

Cultivating low risk crops such as sorghum, millet and pulses and combining low and high-risk crops are two common strategies adopted by subsistence farmers. Specialty crops are considered to be of high risk due to the higher production costs. However, a crop enterprise that is considered a high-risk activity in one area may be considered a low-risk activity in another area, as yield variability is associated with soils, management, and other factors besides climate. Custom farming (e.g., hybrid seed production) and leaving land as a fallow are also considered low-risk activities, especially for farmers who can change employment. Insofar as livestock production is concerned, destocking in advance of drought to avoid overgrazing, stock losses and accelerated erosion is an important strategy. Multi-species herding, selection of pastures and new locations for herding, fodder conservation as hay, straw or grain, and transhumance are also considered risk-minimizing activities. For example, some Mongolian herders keep at least three species for effective risk reduction from heavy snowfall. Herders in Lesotho avoid mountain areas above a certain altitude in winter.

Maintaining Flexibility in Agricultural Production

Subsistence farmers usually tailor their management decisions to the season ahead. Some practices that are flexible in farm management are listed below.

Crop and soil management

- Choosing certain soils, crops and varieties, and varying their proportions in mixed/intercropping systems (e.g., enlarging the use of clay soils in periods of drought in Zimbabwe)
- Altering the area planted by crop and by location (e.g., reducing land devoted to drought-tolerant crops in the wetter seasons in Mozambique)
- Using genotypes of varying maturity (e.g., planting early maturing short duration varieties that can escape drought or early frost)
- Adjusting sowing time and sequence (e.g., staggered sowing to spread flowering so as to manage frost and rain damage and delayed planting to avoid early-season drought)
- Withholding or applying suitable soil amendments, fertilizers and pesticides
- Varying seeding rates, plant spacing, row orientation, etc. to match seasonal expectations
- Adjusting water conservation practices depending on soil moisture status and climate

Livestock management

- Buying resistant breeds of livestock in exchange of old stock, or rearing goats instead of cattle in periods of drought
- Keeping small herds by selling off the oldest and least productive livestock. It is important to note that cash can always be converted into cattle but not vice versa.
- Transferring livestock temporarily to other areas that have a more favorable season
- Maintaining a higher live-weight of animals prior to occurrence of drought through storing and feeding low-cost but nutritious feeds such as wheat (the cheapest energy source) and oat grain (high fiber content)

Apportioning Crop/Livestock Production Geographically

Geographic dispersion of farming activities also reduces the impact of climate risks. However, farming at several noncontiguous locations is not common in subsistence agriculture, as most farmers have small farms. However, a different form of geographic distribution is common in rainfed lowland rice in South India. For example, when a dry season is predicted, farmers grow rice nurseries in a distant location, which is more assured in terms of water supply. Farmers either pay for nursery-care by local people in that area or seek the services of relatives in exchange of the final produce. When climatic conditions are favorable for planting, farmers bring seedlings for transplanting in their field.

Subsistence herders tend to move from areas with an unfavorable dry climate to areas with a more favorable climate. Such mobility also improves overall land use by spreading the livestock pressure. Transhumant migrations are common in several regions of Africa (Baier, 1980). In Lesotho, which experiences a major drought every 10 years, livestock are moved to the cattle posts in summer and then moved to the lowlands in winter. In mountains, livestock are moved to the lower elevation areas or valleys in winter. While geographic dispersion is still possible in relatively less populated regions of Africa, it is nearly impossible in most Asian countries, which have reached the limit of arable land.

Vertical integration

Vertical integration, where the output from one stage of production and distribution is transferred to another stage, is an important structural change occurring in most of Asia, parts of Africa and Central America, especially in production of livestock and cash crops. For example, farmer-owned dairy, oilseed and vegetable cooperatives are common in the Indian subcontinent. Many poultry farmers in India form an association-managed enterprise mix that may include a feed mill, a hatchery, a slaughter facility and a packing plant. Vertical integration provides farmers with marketing flexibility to spread cash sales and to avoid selling all of the produce at the bottom of the market.

Planning for climate risk

Because of an increasing frequency and severity of climate risks, subsistence farmers in some regions are beginning to plan ahead on a regional basis, often with the guidance of local governments. Drought contingency planning including development of alternative livelihood strategies (e.g., food-for-work programs, cultivation of short season crops such as leafy vegetables, etc.) is common in many regions of Asia, but on-farm preparations on an individual basis are still far from satisfactory. Keeping reserves of seed, fodder, hay and other animal feeds to cope with climate risks is gradually becoming common. In the case of livestock farmers facing the risk of severe winters, planning activities include reinforcement of winter shelters, animal fattening during summer, accustoming animals to winter cold and optimizing the number of livestock in autumn.

INDIGENOUS KNOWLEDGE IN RELATION TO CLIMATE PREDICTION AND RISK MANAGEMENT

Subsistence farmers and herders in Asia, Africa and Latin America have always had their own strategies for coping with climatic risk, which are often based on local methods of predicting seasonal climate. Farmers in South America (Orlove et al., 2000), Lesotho (Pepin, 1996) and India, for example, have developed complex cultural models of weather. In Andean South America, the changes in the apparent brightness of stars in the Pleiades around the time of the southern winter solstice have been used for many centuries by potato farmers to forecast interannual variations in summer rainfall and in autumn harvests. Farmers minimize the risk from low rainfall during the growing season in an El Niño year by varying the planting date. In South India, many farmers make cropping decisions based on "Krishi Panchag" (Agricultural Vedic Calendar), which includes a climate forecast for an entire year based on planetary positions. Biodynamic agriculture, a system initiated by an Austrian philosopher, Rudolf Stiener, and practiced now in nearly 25 countries also depends on similar calendars. That farmers rely on local predictors of seasonal climate indicates that it is not difficult to integrate seasonal forecasts in their farm strategy. It is vital, therefore, to pay attention to local knowledge systems and analyze their relationship to seasonal predictions. Indigenous knowledge on risk management is, however, a function of an area's economy, climate and culture and it is dynamic rather than static. Fujisaka (1997) reviewed various strategies adopted by farmers in Ethiopia, rainfed lowland rice farmers in eastern India, and bean farmers in the Andes of Ecuador to minimize seasonal climatic risk. He concluded that it is important to (a) recognize merits of these systems by conducting careful diagnostic research, (b) explore if such strategies can be integrated with predictive functions more effectively, and (c) blend farmers' practices with modern innovations. A few examples of local systems that are useful in minimizing climate risk are given below.

Tlaxcalan farmers in Mexico practice several polyculture and agro-forestry designs for protecting maize from drought and hot spells. Many rows of trees are often planted as windbreaks to shelter crops.

Farmers in upland villages of Matalom, Philippines practice an indigenous soil conservation technique "cemento-cemento" or "kahun-kahun" to reduce the impact of heavy rainfall in areas with slopes ranging from 10 to 40%.

Farmers in Peru and Bolivia practice "raised field agriculture" to enhance productivity in regions with unpredictable weather patterns. Raised fields are large earthen platforms that prevent waterlogging and flooding, increase soil fertility, conserve moisture, insure nutrient production and recycling, and improve crop microclimates. Introduction of this technique to Asia has considerable potential to reduce the impacts of floods.

Farmers of Bali follow robust water management practices devised by priests in water temples to minimize damage from floods and drought.

Farmers in Hindu Khush region use indigenous systems (posilo-kam posilo and obano-chiso) for selecting and storing nutritious tree fodders to reduce drought risk on livestock.

CONSTRAINTS FOR EFFECTIVE APPLICATION OF CLIMATE FORECASTS

There are many impediments for effective application of seasonal climate predictions in subsistence farming. Analysis of such constraints can help in devising more effective climate risk management strategies.

Complexity of forecasts

Current forecasts are still too complex to use and are not available in local languages in some regions. They give information on mean rainfall but not much on variables such as start, duration, intensity, and temporal distribution. Short lead times and modest reliability are often incompatible with the timing and type of farm decisions. A considerable gap exists between the information needed by subsistence farmers and that provided by the meteorological services in Africa (Blench, 1999).

Poor integration of predictions with other aspects of infrastructure and input supply

Prediction of a climate hazard is rarely linked to increased availability of appropriate seeds. For example, aid agencies supplied bean seeds to meet short-term food needs in Rwanda, but many varieties produced 30% less than traditional ones in drought periods.

Transmission delays

Most developing countries have poor infrastructure for climate predictions, and ill-developed communication media (newspapers, radio, television and the Internet). This results in delivery of information that is too late to initiate any effective strategies. Weak linkages between meteorological services and media or end users, and lack of coordination among multiple channels are other reasons for transmission delays.

Technological gaps

- Inadequate understanding of correlation of seasonal climate indices and rainfall patterns in different geographic regions: For example, SOI is useful in spring only but not in autumn and winter (due to poor correlation with rainfall) in many parts of South Australia and Victoria. In contrast, the accuracy of the SOI in determining the three-month outlook for rain is highest between mid-winter and mid-summer in eastern and northern Australia.
- Lack of in-depth studies on associations between ENSO Phase, crop yields and net returns in subsistence conditions
- Lack of appropriate decision support systems to evaluate the consequences of alternative management actions in response to climate predictions, especially in mixed farming, and
- Lack of effective applications (which lead to altering a decision that results in either an economic benefit or a reduction in risk) appropriate to subsistence farming.
- Societal factors such as those related to land tenure and restrictions on access to arable land and labor prevent farmers in effectively responding to climate forecasts. For example, water harvesting structures have to be built in a larger area than the size of farm holding itself, and erosion control measures such as water control terraces need to be constructed over the entire slope.

ROLE OF COMMUNICATION AND INFORMATION TECHNOLOGIES, AND DECISION SUPPORT SYSTEMS

Recent advances in communication and information technologies can enable farmers receive forecast information in timely fashion so that corrective actions may be taken. It is expected that at least 40 new land viewing satellites will be launched within the next decade. Imagery from these satellites will be useful in more precisely determining the existence and spatial limits to extreme climate events (McVicar and Jupp 1998). High-resolution imagery from satellites such as IKONOS will be increasingly used for delineating temporal and spatial patterns of localized climate impacts and for devising location-specific risk management strategies. Geographic information systems (GIS) can be used to characterize climate-yield relationships, visually represent the spatial distribution of variability in climate risk, and integrate various climatic impacts. For example, a framework to quantify and map drought hazard in maize was developed in South Africa by running crop growth models in a GIS, based upon the phase of the SOI (de Jager et al., 1998). In the near future, developing countries will increasingly use channels such as satellite television and the Internet to fulfill the needs of their communities for up-to-date and precise information. Climate forecasts would obviously be a part of such information. Jensen et al. (2000), through an example of Pl@nteInfo, demonstrated that it is possible to build web-based decision support systems (DSS), where personalized advice can be given in real time. Opportunities to develop such tools, perhaps through collective action and access by farmer and other community-based associations, for helping small, resource-poor farmers must be explored.

Biophysical models and DSS are an effective means of determining the response of crops, livestock and management to climatic variability, and are especially useful in identifying risk management strategies that are economically viable, environmentally sustainable, and socially acceptable. In this connection, the relevance of risk analysis programs such as @RISK (Saunders et al., 1997) and other DSS including TACT (Abrecht and Robinson, 1996), PYCAL (French and Schultz, 1984), RANGEPACK (Stafford Smith and Foran, 1992), SOWHAT, HOWOFTEN, and HOWWET to subsistence farming systems must be examined. Using SOWHAT, farmers can make informed decisions on plant density, top-dressing of nitrogen, crop rotations, etc. HOWOFTEN and HOWWET are helpful in selecting crop type and plant density, determining the duration of fallow and the probable dates of flooding events, and in estimating expected yields. Customization of models such as RAINMAN, which allows determination and adjustment of the probability of rainfall based on SOI or sea surface temperatures (Clewett and Drosdowsky, 1996) to the needs of developing countries is necessary. Also, prospects for developing integrated services such as the Climate Risk and Yield Information Service (Balston et al., 1997) must be examined.

CONCLUSIONS

Diversification, flexibility, and opportunism characterize climate risk management by subsistence farmers and herders. Despite encouraging improvements in the reliability of climate forecasts, they have not been fully integrated yet into overall farm planning due to various constraints. Given the weak statistical base for climate systems in regions where subsistence agriculture is practiced, an understanding of the past frequency of climatic events, ecological system dynamics of the region, and village farm records is crucial in analysis, identification and dissemination of successful climate risk management strategies. Because the skill of seasonal climate predictions varies with season, location and situation, it is vital to develop region-specific climate risk management plans, which are appropriate to attitudes and circumstances of farmers and are based on customary cultures and local knowledge. A holistic strategy integrating production, marketing and financial responses is likely minimize climate risk more effectively than a series of separate and individual responses. While the specific mix of options used by an individual will depend largely on personal goals and conditions including past experience and attitudes to climate risk, advances in information and communication technologies can greatly facilitate risk management on a regional basis. Additional plans for diversifying risk alleviation in each region, and disseminating successful strategies to new areas must be formulated.

Besides enhancing the content, reliability and lead times of climate forecasts, additional pragmatic ways to communicate and integrate them with farm management must be pursued in each region by involving governmental advisory networks, the private sector and researchers in a participatory and farmer-centered systems approach. Input from farmers is vital in ensuring the relevance of forecast products and in developing DSS and management options. Besides the deployment of site-specific information delivery and extension systems, efforts to develop a range of profitable and climate risk-reducing technology packages (including development of resilient farming systems, breeding short duration drought-resistant varieties, etc.) relevant to subsistence farming must be accelerated.

REFERENCES

Abrecht, D.G., and S. Robinson. 1996. TACT: a tactical decision aid using a CERES-based wheat simulation model. Ecol. Model. 86:241-244.

Avila, M. 1996. Concepts and Methods for Economic Evaluation of Alley Farming. In B. R. Tripathi, and

P.J. Psychas (eds) The AFNETA alley farming training manual—Volume 2: Source book for alley farming research. Alley Farming Research Network for Africa, IITA Ibadan, Nigeria.

Baier, S. 1980. An economic history of Central Niger. Clarendon Press, Oxford, UK.

Balston, J.M., D.G. Abrecht, and J.P. Egan. 1997. Development of a climate risk and yield information service. p 187-188. In Proc. of Workshop on Farming Systems Dev. Waite Institute, Adelaide, Australia, March 1997.

Blench, R. 1999. Seasonal climatic forecasting: who can use it and how should it be disseminated? Natural Resource Perspectives No. 47. Nov. 1999. http://www.oneworld.org/odi/nrp/47.html Accessed on 1 March 2000

Clewett, J.F., and Drosdowsky, L. 1996. Study 1.: Use of Southern Oscillation Index as a drought management tool. pp. 120-124. In D.M. Stafford Smith, J.F. Clewett, A.D. Moore, G.M. McKeon, R. Clark (eds) DroughtPlan: Building on Participation. Full project report, Working Paper No. 10.

Clewett, J.F., S.M. Howden, G.M. McKeon, I.J. Partridge, and J.C. Scanlan. 1991. Use of systems analysis and the Southern Oscillation Index to optimise management of grain sorghum production from a farm dam irrigation system. pp. 307-328. In R.C. Muchow and J.A. Bellamy, (eds). Climatic Risk in Crop Production: Models and Management for the Semiarid Tropics and Subtropics. CAB International, Wallingford, UK.

de Jager, J.M., Potgieter, A.B. and van den Berg, W.J. 1998. Framework for forecasting the extent and severity of drought in maize in the Free State Province of South Africa. Agric. Sys.57:351-365.

Dorward, A.R. 1996. Modelling diversity, change and uncertainty in peasant agriculture in northern Malawi. Agric. Sys. 51:469-486.

Ewing, M.A., D.J. Pannell, and P.K. James. 1987. The profitability of lupin—cereal rotations. pp.82-90. In Kingwell, R.S. and D.J Pannell, (eds). MIDAS, A Bioeconomic Model of a Dryland Agricultural System. Pudoc, Wageningen, The Netherlands.

French, R.J. and J.E. Schultz. 1984. Water use efficiency of wheat in a Mediterranean-type environment. II. Some limitations to efficiency. Aust. J. Agric. Res. 35:765-775.

Fujisaka, S. 1997. Research: Help or hindrance to good farmers in high-risk systems? Agric. Sys. 54:137-152.

Galvin, K. and Ellis, J. 1996. Climate patterns and human socio-ecological strategies in the rangelands of sub-saharan Africa. pp.67-78. In Odada, E., O. Totolo, M. Stafford Smith, and J. Ingram (eds). Global Change and Subsistence Rangelands in Southern Africa: the impacts of climatic variability and resource access on rural livelihoods. GCTE Working Document No.20, 99 pp. Canberra, Australia.

Hammer, G.L., D.P. Holzworth, and R. Stone. 1996. The value of skill in seasonal climate forecasting to wheat crop management in a region with high climatic variability. Aust. J. Agric. Res. 47:717-737.

Jensen, A.L., P.S. Boll, I. Thysen, and B.K. Pathak. 2000. Pla@nteInfo—a web-based system for personalized decision support in crop management. Comp. Electr. Agric., 25:271-293.

Kingwell, R.S. 1994. Risk attitude and dryland farm management. Agric. Sys. 45:191-202.

McKeon, G.M. and D.H. White. 1992. El Niño and better land management. Search 23, 197-200.

McVicar, T.R. and D.L.B. Jupp. 1998. The current and potential operational uses of remote sensing to aid decisions on Drought Exceptional Circumstances in Australia: a review. Agric. Sys. 57:399-468.

Meinke, H., P.S. Carberry, G.L. Hammer and R.C. Stone. 1998. The use of seasonal climate forecasts in cropping systems management. In Proc. of the 9th Australian Agronomy Conf. Accessed at http://life.csu.edu.au/agronomy/papers/146/146.htm

Messina, C.D., J.W. Hansen, and A.J. Hall. 1999. Land allocation conditioned on El Niño-Southern Oscillation phases in the Pampas of Argentina. Agric. Sys. 60:197-212.

Orlove, B.S., J.C.H. Chiang, and M.A. Cane. 2000. Forecasting Andean rainfall and crop yield from the influence of El Niño on Pleiades visibility. Nature 403:68-71.

Pepin, N. 1996. Indigenous knowledge concerning weather. The example of Lesotho. Weather 51(7):343-348.

Pietola, K., T. Reardon, and K. Savadoga, 1998. Adoption of improved land use technologies to increase food security in Burkina Faso: relating animal traction, productivity, and non-farm income. Agric. Sys. 58:441-464.

Saleh, A.F.M., and S.I. Bhuiyan. 1995. Crop and rain water management strategies for increasing productivity of rainfed lowland rice systems. Agric. Sys. 49:259-276.

Saunders, L.S., T.H. Webb, and J.R.F. Barringer. 1997. Integrating economic data with spatial biophysical data to analyse profitability and risks of wheat production on a regional basis. Agric. Sys. 55:583-599.

Savin, R., E.H. Satorre, A.J. Hall, and G.A. Slafer.1995. Assessing strategies for wheat cropping in the monsoonal climate of the Pampas using the CERES-wheat simulation model. Field Crops Res. 42: 81-91.

Schilizzi, S.G.M., and R.S. Kingwell, 1999. Effects of climatic and price uncertainty on the value of legume crops in a Mediterranean-type environment. Agric. Sys. 60:55-69.

Srinivasan, A., H. Takeda, and T. Senboku. 1996. Heat tolerance in food legumes as evaluated by cell membrane thermostability and chlorophyll fluorescence techniques. Euphytica 88:35-45.

Stafford Smith, M. and B. Foran. 1992. An approach to assessing the economic risk of different drought management tactics on a South Australian pastoral sheep station. Agric. Sys. 39:83-105.

Swift, J. 1999. Pastoral institutions and approaches to risk management and poverty alleviation in Central Asian countries in transition, Part 1. Posted 30 June 1999. http://www.fao.org/WAICENT/FAOINFO/SUSTDEV/ROdirect/ROan0018.htm

Swinkels, R.A., S. Franzel, K.D. Shepherd, E. Ohlsson, and J.K. Ndufa. 1997. The economics of short rotation improved fallows: Evidence from areas of high population density in Western Kenya. Agric. Sys. 55:99-121.

Understanding and Improving Community Responses and Channels of Seasonal Forecast Dissemination in Zimbabwe

Gina Ziervogel, Thomas E. Downing, Randall V. Martin, and Richard Washington
Environmental Change Institute, University of Oxford, United Kingdom

SUMMARY

The potential for effective dissemination of climate or crop yield forecasts to individual farmers can no longer be overlooked. Seasonal forecasting techniques have improved greatly in the last decade and have the means to improve livelihoods for both commercial and small-scale farmers. This research focuses on the challenges that face both the users and providers of seasonal forecasts and provides a way for formalizing this flow by using an agent-based modeling approach.

Linking seasonal climate predictions directly to impacts is one way in which the noise of the forecast can be reduced and give users a more explicit knowledge of what the seasonal forecast translates into with regards to their livelihoods. A maize model does this by correlating potential evapotranspiration with historical rainfall data to produce a water stress index that can be used to produce a crop forecast (Martin et al, In press) .

An exploratory method for modeling social decision-making is proposed. The approach uses an agent-based language to model the interaction of individual, community and national institutions and the evolution of the forecast-response behavior. Key features include real-world stakeholders represented as software agents. Behavior is based on declarative rules, in response to actions of other stakeholders or signals from the environment.

The new research will focus on a livelihood approach and incorporate a dynamic process of moving between the output from the forecaster to the end-user and subsequent feedback to the forecasting community. The rationale for this is attributed to the large communication gap that exists between the information presented to the country as a whole and that needed by subsistence farmers (Washington and Downing, 1999). Small-scale farmers in rural areas do not necessarily have the means to process the probabilistic forecasts as issued by the meteorological services, as the lead-time is often not long enough or the information is not temporally or spatially specific enough .

Longitudinal household panel data of farmers in resettlement and communal areas of Zimbabwe has been collected by Kinsey for the last 8 years (Kinsey, 1999). This will contribute to the fieldwork and provide historical trends that will be integrated into the model. Participatory research will elicit behavioral response to risk and decision making in the context of small-scale Zimbabwean farmers. The application of such a framework will enable researchers and decision-makers to evaluate a wide range of integrated scenarios of streams of easonal climate forecasts.

REFERENCES

Kinsey, B. H. 1999. Determinants of rural household incomes and their impact on poverty and food security in Zimbabwe. Dept. of Rural and Urban Planning, University of Zimbabwe; Economics and Social Institute, Free University, Amsterdam.

Martin, R., R. Washington and T.E. Downing. In press. Seasonal maize forecasting for South Africa and Zimbabwe derived from an agroclimatological model. J. Appl. Meteoro.

Washington, R. and T.E. Downing. 1999. Seasonal forecasting of African rainfall: prediction, responses and household food security. Geogr. J. 165:255-274.

Potential Effects of Global Warming on Public Health: Vector-borne Diseases in Taiwan

Gen-Shuh Wang[1] and Guo-Hwa Wang[2]
[1]National Taiwan University, Taiwan, ROC
[2]Chang Gung Institute of Nursing, Taiwan, ROC

SUMMARY

The potential influence of global warming on public health has been a subject of study for decades; however, no clear conclusions have been made. The vulnerability of environmental changes on public health is hypothetical until the degree of change is known. Among the public health issues that are under direct or indirect influence from global warming, vector-borne diseases deserves more study. This presentation reviews the potential impact of environmental changes (global warming) on public health, and focuses on the trends of vector-borne diseases in Taiwan. Confounding factors for study of climate change and the emerging vector-borne diseases are discussed. Among the twenty-one reportable infectious diseases in Taiwan, Japanese Encephalitis, Malaria, Scrub Typhus and Dengue Fever are observed to be temperature dependent. The reported and confirmed cases for these diseases in summer and early fall are much higher than other reportable diseases. The increasing temperature has cause great concerns for the potential public health threatening from these vector-borne diseases. Based on the researches conducted by scientists in Taiwan, the trends of the vector-borne diseases (focus on Dengue Fever) and their correlation with environmental factors (temperature, rainfall, and humidity) were be reported. For example, the relationship between the number of suspected dengue fever cases and the average temperature, rainfall and relative moisture of previous month was significant. Also Dengue outbreak has been reported in Taipei (north Taiwan), as predicted that the distribution of Dengue fever will move north due to the global warming. Unfortunately, due to the limited Dengue cases, it is difficult to conclude the correlation between the Dengue Fever and the global warming. However, it was observed in southern Taiwan that the Breteau Index (density of A. Aegypt and A. Albopictus in household) was much higher during the ENSO period, indicating higher risk of vector-borne outbreaks. In order to evaluate the impact of unusual temperature changes on vector-borne diseases during ENSO period, this presentation also include the correlation between the Breteau Index, temperature and rainfall based on data collected in Taiwan between 1993 and 1998. It has been concluded that global warming will shift the vector's geographic distributions, increase the reproductive rates, and shorten the pathogen incubation period. The rising temperature could also lead to more rapid amplification of epidemic spread. A detail study concerning the ecological system and climate change will help to fully understand the consequences of global warming. At present, a more effective vector control technology is necessary to prevent the transmission of the infectious diseases.

Need for Indicators in Targeting Climate Forecasts: Climatic Risk and Income Diversification

Abigail Amissah-Arthur
International Research Institute for Climate Prediction, USA

INTRODUCTION

Risk from climatic variability is a major limiting factor to agricultural production in sub-Saharan Africa. This problem is more exacerbated in the semi-arid regions where climatic extremes like droughts can lead to famine and result in acute food insecurity for millions of people. Climatic risk in crop production has been defined as a probability of occurrence of unfavorable weather conditions affecting crop performance (Rotter, 1993). Risk reduction options usually employed have included adjustment of tactical and operational decisions such as crop and income diversification, cultivar selection, fertilizer input and timing of activities. These options are usually undertaken before and or during a particular season. For the resource-poor farmer in sub-Saharan Africa, the implications of outstandingly good or bad years can be profound thus, for these farmers, the impact of extreme weather conditions far outweighs its apparent probability of occurrence.

Increasingly, adaptation to climate change and variability has become a focus for research and policy in Africa. By using climatic forecasts of expected seasonal rainfall, it is possible to determine if the season would be a low, normal or high rainfall season (Russell, 1991) thus, enabling assessment of the cropping potential of the period or area and the risks of yield variability. Such information could also enable tactical intra-seasonal crop management planning, decision-making and increase the capacity to adapt to climatic variability by reducing vulnerability to climate risks. Thus, climate forecasts can support farmers in reducing risk by allowing them to alter their production strategies by taking advantage of opportunities in good years and reduce losses in poor years. However, the abilities to use climate forecasts could be constrained by the options (e.g., draft power, sources of income, crop types/cultivars, fertilizer use etc) available to the farmers. Identifying these options is essential in understanding how farmers may use and or benefit from climate forecasts. It has been proposed (Valdivia et al., 2000) that diversification and use of climate forecasts may be related and that access to off-farm income provides flexibility and the capability to take advantage of climate forecasts. Using Kenya as a case study the following section examines the spatial pattern of diversification of income—a risk reducing strategy that highlights the vulnerability of small-holder based agricultural systems—and the need to use such relationships as indicators for targeting climate forecasts.

CLIMATE AND PREDICTION

The NCEP nested regional spectral model (RSM) with a resolution of 20km has been successfully used to reproduce the observed interannual variability of the Kenyan short rains and of the timing of the onset of rains for a 27-year period (Mason et al., 1999; Sun et al., 1999). According to Ogallo (2000), most empirical forecasting studies have shown that maximum predictability skill for Kenya is in the region around the coast and lake Victoria where 70% of the variance in October-December total seasonal rainfall is accounted for during periods with strong ENSO signals.

Kenya thus presents an attractive study site where some skill in climate prediction has been demonstrated and where potential application efforts could be targeted. Kenya is mainly an arid to semi-arid country. The rainfall pattern is strongly related to the topography and defines the agroecology of the country (Figure 1). The highest mean annual rainfall totals (1700-2400mm) are

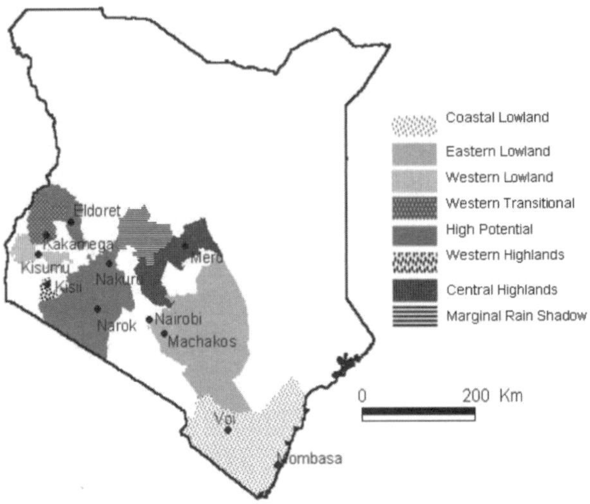

Figure 1. *Agro-ecological zones of Kenya.*

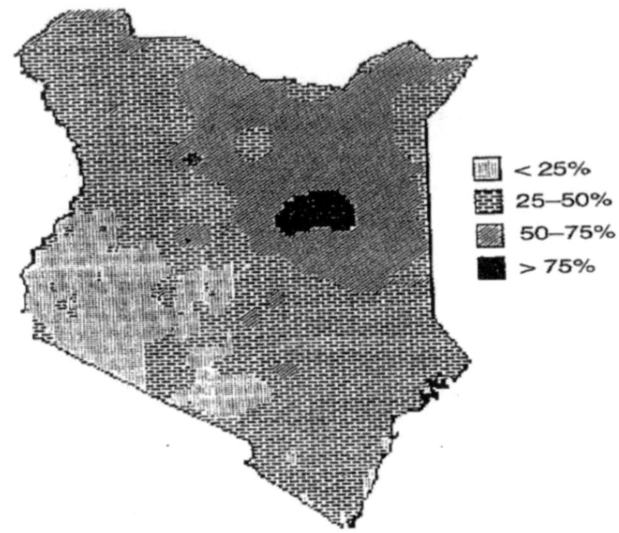

Figure 3. *Coefficient of variation (%) of annual rainfall. Adapted from Ogallo (2000).*

recorded over the highlands in the south-western and central parts of the country (Figure 2) while, the south-eastern and northern parts record a mean annual total ranging from 300-800mm. A high degree of inter-annual variability (Figure 3) is associated with the rainfall including frequent droughts (Figure 4). This situation constitutes a major source of risk and threat to sustainable agricultural production especially since only 20% of the land area is considered to be of high potential for cropping. Thus, crop production is a highly vulnerable enterprise in this region. Hence, the importance and need for timely and effective long-range prediction and early warning (Ogallo, 2000).

METHODOLOGY

Data and information was derived from a national household survey in 1995-97 of 1540 smallholder households by the Kenya Agricultural

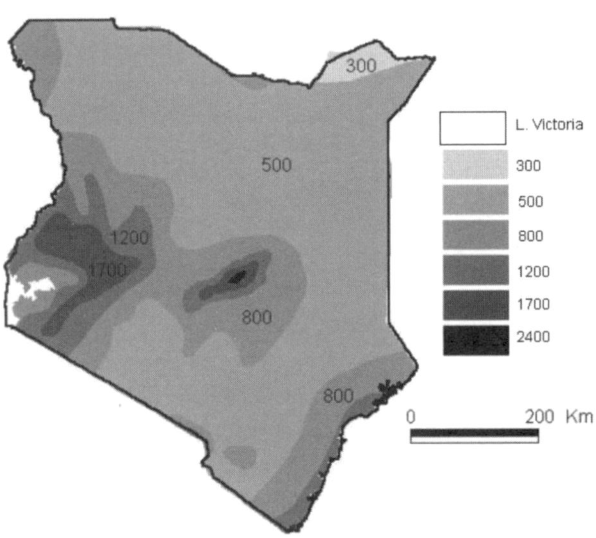

Figure 2. *Distribution of mean total rainfall (mm) across Kenya.*

Figure 4. *Annual drought probability (x 100). Adapted from Ogallo (2000).*

Research Institute (Argwings-Kodhek, 1998). The data was analysed and geo-referenced using the geographical information system. Three sources of household income were recognised: cropping, livestock and off-farm. Major source of income was determined to be either cropping or off-farm depending on which of the two accounted for the highest percentage of the total income. The livestock component was not considered for this purpose because in most parts of Kenya livestock is a low input-low output activity and is regarded as a store of value rather than a commercial enterprise (Argwings-Kodhek, 1998). Details of the approaches used are reported in Amissah-Arthur et al. (2000). By making use of existing body of literature that describes the historical background and socio-economic context of present day farming systems (Blench, 1999) indicators can be developed to understand how different groups will respond to climate forecasts.

CLIMATIC RISK AND INCOME DIVERSIFICATION

Small-scale small-holder farmers are known to produce the greater proportion of food consumed in sub-Saharan Africa (Odulaja and Kiros, 1996). In contrast to farmers in developed countries, the majority of the rural population in sub-Saharan Africa is engaged in farming. The crop and production practices of these farmers are based on their local knowledge and experience that has being developed over the years. Subsistence farmers in Africa generally aim at stabilizing their production, while those in developed countries aim at maximizing economic return (Gommez, 1999). Obviously, the strategies of these farmers, and their perception of risk, are very different.

Farmers in sub-Saharan Africa have developed several management and coping mechanisms to reduce risks of crop production and ensure productivity. The subsistence farmers' remedy for climatic variability in rainfed agriculture has often been to establish on a trial by error basis, a farming system that is conservative and provides a minimum return even in least favourable seasons. Faced with uncertainties caused by climate, economic and political events, these farmers have been rendered highly risk averse. Under these circumstances and in the absence of insurance markets, most farmers use ex-ante risk minimizing strategies to reduce vulnerability to climatic risks. Diversification in production and income alternatives is a strategy of drought-prone environments (Browman, 1987) to minimize the effects of climatic risk or maximize use of all resources available to the household (Ellis, 1993). The agricultural sector in Africa is not homogenous. Income diversification is a major risk reducing ex-ante strategy employed by many farmers. Figure 5 shows the spatial pattern of major source of income stratified for smallholders in 8 agro-ecological zones of Kenya. Crop production is the major source of income in the high potential zone. Both crop production and non-farm activities provide the major sources of income in the central highland region. However, the main source of income in the western highland, coastal, eastern and western lowlands is non-farm. Clearly, the situation in the coastal and eastern (Voi-Machakos axis) can be explained by (Figure 2) the low rainfall distribution in this region and the high inter-seasonal variability (Figure 3) of the rainfall makes crop production a risky enterprise. In general, more than 80% of Kenya can be classified as arid and semi-arid lands (Ogallo, 2000). Most areas in central and eastern Kenya receive less than 300mm rainfall in six out of 10 seasons (Downing et al., 1985) stressing the high risk of crop failure.

The total household income and income derived from cropping were similar for households in the

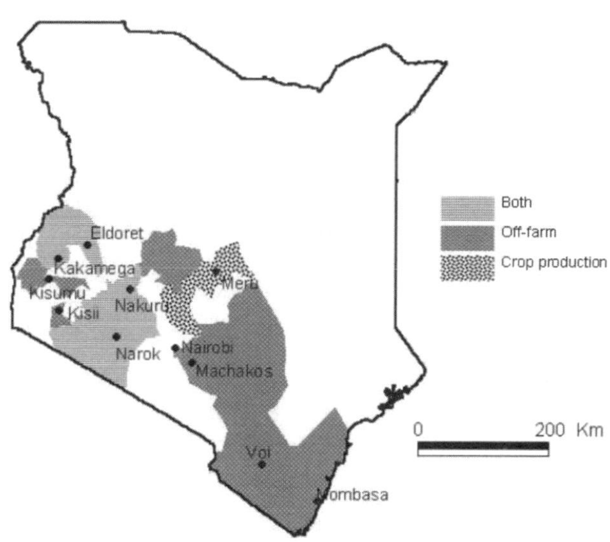

Figure 5. *Spatial pattern of household income composition*

western lowlands and highlands. However, non-farm derived income was higher in the western lowlands. The annual probability of drought (Figure 4) is in the range of 7-9% in the western lowlands as compared to 8-10% in the western highlands. For these two zones, the differences observed is less due to the climatic condition than to the fact that farm sizes are smaller in the western lowlands because of increasing population pressure and hence the need to derive income off-farm. Total household income is almost similar for western transitional and marginal rain-shadow, however annual probability of drought is about 25% in marginal rain-shadow compared with 8-10% of the western transitional zone. The major source of income in the higher drought prone zone is off-farm while it is on-farm for the lesser drought prone western transitional zone.

Figure 6 shows the total income and proportion derived from cropping, livestock and non-farm for the 8 agro-ecological zones. Cropping and livestock activities are two activities that are dependent and affected by the vagaries of weather and climate. The increasing importance of cropping in the total household income from the lowest in the coastal lowlands to the highest in the high potential zone is highlighted (Figure 6). Income from cropping alone in the high potential zone is about the size of the total income in the marginal rain shadow and western transitional zones while, it is about one and half times that of the western lowland and highlands. The two zones (eastern and coastal lowlands) with the highest proportion of their income

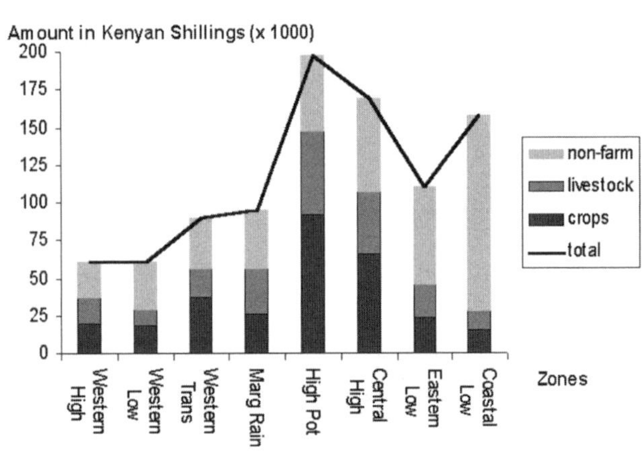

Figure 6. *Smallholder household income composition across eight agro-ecological zones. Data source: Argwings-Kodhek et al. (1998).*

from non-farm activities are mainly in the low rainfall arid and semi-arid parts where the annual drought probability (Figure 4) ranges from 14-23%. Interestingly, high potential zone and central highlands with the highest total income values and highest incomes from cropping are in the zone depicted as having a 20-26% probability of drought but lower rainfall variability of <25%. Clearly the lower the rainfall (Figure 2) the higher the variability (Figure 3), the higher the risk and the higher the dependence on other sources of income apart from farming. This is depicted by the situation observed in the coastal and eastern lowlands. In these zones, non-own farm, farm labor work is not an important source of off-farm income. This is a major difference in non-cropping activities between these two zones and the others in higher rainfall zones.

The major off-farm activities for all zones include salaried work, fishing, commerce and trading. A combination of increasing population pressure, low soil fertility and small farm sizes are the different reasons responsible for the need and high off-farm income observed for the marginal rain shadow and the western lowlands and highlands. Obviously, irrespective of the zone or the level of risk, income diversification is a well-entrenched practice by Kenyan farmers.

The patterns observed emphasize the fact that many small-holder farmers in the tropics utilize the diversity of their environments, and employ a wealth of techniques and means to get around the constraints imposed by climate. In most cases, farmers have developed a wide portfolio of income-earning non-cropping activities that offer some level of income stability in the face of looming crop failures. However, within these broad groups there are bound to be differences within the zones because farmer's access to resources and risk coping strategies determine the degree of individual diversification. For example Sheperd and Soule (1998) found that in the western highlands (Vihiga district), farm income made up only 7% of the total income for low resource endowed farmers. This they attributed to the small size of fields owned by the farmers leading to farmers seeking out off-farm activities as their major source of income. By contrast, high resource endowed farmers obtain more than 63% of their income from their farms. These farmers have larger farm sizes than their low resource endowed counterparts. Valdivia et al.

(1996) observed that variation in access to productive resources leads to different levels of diversification.

In an examination of a part of semi-arid eastern Kenya (Mbeere district), Tengberg et al. (1998) drew linkages between the sources of variability (environmental, social, economic) and the different diversity of farm practices. They examined different seasons which were normal and drought years and noted that there were differences in returns for the different years where some farmers had good returns in bad years and others bad returns in normal years. They concluded that the main sources of variability affecting cropping system and management practices in the zone are rainfall, soil fertility, and farmers' resource level and farm productivity.

Valdivia et al. (2000) proposed that diversification and use of climate forecasts may be related and that access to off-farm income provides flexibility and the capability to take advantage of climate forecasts. In a study of income diversification of Andean farmers and their adoption of climate forecasts, they hypothesized that specialized high risk farmers would find climate forecasts more useful than diversified farmers. However, their results concluded that the group of farmers who were most likely to alter their practices in response to climate forecasts, was that which had significant amount of off-farm income. If this finding were to be extended to the Kenyan smallholders, then it could be assumed that all the farmers in the different zones would alter their practices in some ways with the availability of climate forecasts since total income in each zone is quite highly diversified. According to Argwings-Kodhek et al. (1998) 50% of Kenyan smallholder households earn incomes from different off-farm activities. However, if Valdivia et al (2000) and Blench (1999) hypotheses were used here i.e., that farmers whose major source of income is cropping are regarded as the specialized farmers, then it would appear that farmers in the high-potential and central highlands would be those with the most tendency to adopt climate forecasts and alter their production practices. By contrast to Valdivia's Andean farmers, these Kenyan group of farmers also have significant amount of off-farm income. They thus, have access to resources from off-farm activities to shift strategies in response to the vagaries of climate. Farmers from the other zones especially coastal and eastern lowlands may tend to use the forecasts to alter their practices in good years and increase area cultivated. This study has provided an approach that could be used in site selection for more detailed fieldwork and for targeting climate forecasts to groups who are most likely to use it. By conducting farmer/field surveys at selected representative sites within the broader regions identified in this study, the recognized patterns and relationships can be validated.

THE NEED FOR INDICATORS FOR TARGETING CLIMATE FORECASTS

An indicator is usually used for a descriptor that is representative of a set of conditions or total effect of many variables. In order to ensure that climate forecasts are properly targeted for agricultural applications, it is essential that credible indicators be developed. Adoption of climate forecasts by smallholder producers is a function of various factors that vary in space and time. By conducting a critical review of the literature we (Amissah-Arthur et al., 2000) identified a set of indicators which characterizes the sub-Saharan Africa's smallholder farmers and their probability of adopting climate forecasts. With additional research and reviews other variables could be added to what is shaping out to be a minimum data set. An example of such an indicator smallholders sources of income- and its potential utilization in targeting climate forecasts has been examined in the present paper. However, there is a need to develop a research strategy to test and monitor these indicators in the field since it cannot be assumed that indicators can be transplanted easily from one culture or context to another. According to Stern and Easterling (1999) it is important to match the information provided in forecasts as much as possible to the coping strategies used by the forecasts recipients. Indicators as employed here, would be very useful in achieving this goal.

It will not be possible to conduct site specific surveys in every village and among all farmer groups to determine the use or potential adoption of climate forecasts in agricultural production. Indicators are a way of seeing "the big picture" while selected field studies ensures that these larger pictures are verified by looking at smaller pieces of it. Indicators are pointers that may also serve in identifying homogenous groups of farmers and assist in site selection for further field studies. By making use of existing body of literature that describes the historical background and socio-economic context of present day farming systems (Blench, 1999) indicators can be developed to understand how different groups will respond to climate forecasts. Used effectively, they reveal conditions and trends that would help scientists, researchers and decision-makers better focus efforts to increase the utilization of climate forecasts. However, before further progress can be made in using these indicators for targeting climate forecasts there is a need to develop a research framework to test the priority indicators.

REFERENCES

Amissah-Arthur, A., S. Jagtap, R. Chen, R. Basher. 2000. Indicators for Targeting Climate Forecast in sub-Saharan Africa. Research Report, IRI, Palisades, NY.

Argwings-Kodhek, G. T.S. Jayne, G. Nyambane, T. Awuor and T. Yamano 1998. How Can Micro-Level Household Information Make a Difference for Agricultural Policy Making?: Selected examples from the KAMPAP survey of smallholder agriculture and non-farm activities for selected districts in Kenya. Paper presented at the conference on Strategies for Raising Productivity in Agriculture, Egerton University/Tegemeo Institute of Agricultural Policy and Development, May 12, 1998, Nairobi Kenya.

Blench, R. 1999. Seasonal Climate Forecasting: Who Can Use it and How Should it be Disseminated? Natural Resource Perspectives, no. 47, November 1999.

Browman, D. L. 1987 Pastoralism in Highland Peru and Bolivia. Chapter 5. In: Arid Land Use Strategies in Risk Management. Westview Press, Boulder.

Downing, T.E., D.N., Mungai, H.R., Muturi 1985. Drought Climatology of Central and Eastern Kenya. Climatic Variability and Agricultural Production in Central and Eastern Kenya. Ministry of Environment and Natural Resources, Nairobi, Kenya, pp.1-37.

Ellis, F. 1993. Peasant Economics: Farm Households and Agrarian Development. Cambridge University Press, Cambridge, UK, p. 309.

Gommes, R. 1999. Agroclimatic concepts. Environment and Natural Resources Service (SDRN), FAO Research, Extension and Training Division. FAO, Rome.

Mason, S.J., L. Goddard, N.E. Graham, E. Yuleva, L. Sun and P.A. Arkin. 1999. The IRI Seasonal Climate Prediction System and the 1997/98 El Niño Event. Bull. Am. Meteorol. Soc. Meeting.

Odulaja, A. and F.G., Kiros. 1996. Modelling Agricultural Production of Small-scale Farmers in sub-Saharan Africa: A case study in western Kenya.

Ogallo, L.A. 2000. Predicting Drought in Kenya: Prospects and challenges. In Drought: A Global Assessment (ed. Donald Wilhite). Vol.1. Routledge, New York.

Ogallo, L.A., Janowiak, J., and Halpert, M.S. 1988. Teleconnections between Rainfall Over East Africa and Global sea surface temperature Anomalies. Journal of the Meteorological Society of Japan 66, 6:807-22.

Russell, J.S. 1991. Prospects for incorporation of Long-term Weather Forecasting into Crop Decision Support Systems. In (eds. R.C. Muchow and J.A. Bellamy) Climatic Risk in Crop Production: Models and management for the semi-arid tropics and subtropics. CABI, Wallingford.

Stern, P.C. and W. E., Easterling (eds.) 1999. Making Climate Forecasts Matter. Washington, DC: National Academy Press.

Sheperd, K.D. and M.J. Soule 1998. Soil Fertility Management in West Kenya: Dynamic simulation of productivity, profitability and sustainability at different resource endowment levels. Agriculture Ecosystems and Environment 71: 131-145.

Sun, L., F.H.M., Semazzi, F., Giorgi and L. Ogallo. 1999. Application of NCAR Regional Climate Model to Eastern Africa. Journal of Geophysical Research 104, D6: 529-62.

Tengberg, A., J. Ellis-Jones, R. Kiome and M. Stocking (1998). Applying the Concept of Agrodiversity to Indigenous Soil and Water Conservation Practices in Eastern Kenya. Agriculture Ecosystems and Environment 70: 259-272.

Rotter, R. 1993. Simulation of the Biophysical Limitations to Maize Production under Rainfed Conditions in Kenya. Evaluation and application of the model WOFOST. PhD Thesis, Universitat Trier, Trier, Germany.

Valdivia, C.E., E. Dunn, and C. Jette 1996. Diversification, a Risk Management Strategy in an Andean Agropastoral Community. American Journal of Agricultural Economics. 78(5): 1329-1334.

Valdivia, C.E., J.L., Gilles and S. Materer 2000. Climate Variability, a Producer Typology and the Use of Forecasts: Experiences from Andean semiarid smallholder producers. Proceedings of the International Forum on Climate Prediction, Agriculture and Development. IRI, New York.

AUTHOR INDEX

APPENDIX 1. LIST OF PARTICIPANTS

Adams, Richard
Resource Economics
Oregon State University
Corvallis, OR 97331
USA
Phone: +1-541-737-1435
Fax: +1-541-737-2563
E-mail: richard.m.adams@orst.edu

Adjez, Ahmed
National Climate Center
P.O.Box 45, Hacen Badi
El-Harrach, Alger 16200
ALGERIA
Phone: +213-2-66-01-34
Fax: +213-2-73-18-23
E-mail: ahmed_adjez@hotmail.com

Agrawala, Shardul
International Research Institute for Climate Prediction
PO Box 1000
Palisades, NY 10964-8000
USA
Phone: +1-845-680-4460
Fax: +1-845-680-4867
E-mail: shardul@iri.ldgo.columbia.edu

Amissah-Arthur, Abigail
International Research Institute for Climate Prediction
PO Box 1000
Palisades, NY 10964-8000
USA
Phone: +1-845-680-4461
Fax: +1-845-680-4864
E-mail: aaa@iri.ldgo.columbia.edu

Archer, Emma
Grad. School of Geography, Clark University
950 Main St.
Worcester, MA 01603
USA
Phone: +1-508-793-7333
Fax: +1-508-793-8881
E-mail: earcher@clarku.edu

Arkin, Phill
International Research Institute for Climate Prediction
PO Box 1000
Palisades, NY 10964-8000
USA
Phone: +1-845-680-4414
Fax: +1-845-680-4866
E-mail: parkin@iri.ldeo.columbia.edu

Arndt, Channing
Dept. Agricultural Economics
Purdue Univ.
West Lafayette, IN 47907
USA
Phone: +1-765-494-5837
Fax: +1-765-496-1224
E-mail: arndt@agecon.purdue.edu

Auer, Steve
NOAA/Office of Global Programs
1100 Wayne Avenue, Suite 1210
Silver Spring, MD 20910
USA
Phone: +1-301-427-2089 x153
Fax: +1-301-427-2222
E-mail: Stephen.Auer@noaa.gov

Baethgen, Walter
IFDC—Uruguay
Juan Ma. Perez 2917 Apt 501
Montevideo 11300
URUGUAY
Phone: +598-2-712-0838
Fax: +598-2-711-6958
E-mail: baethgen@undp.org.uy

Barlow, Matthew
International Research Institute for Climate Prediction
PO Box 1000
Palisades, NY 10964-8000
USA
Phone: +1-845-680-4473
Fax: +1-845-680-4864
E-mail: mattb@iri.ldeo.columbia.edu

Basher, Reid
International Research Institute for Climate Prediction
PO Box 1000
Palisades, NY 10964-8000
USA
Phone: +1-845-680-4455
Fax: +1-845-680-4864
E-mail: rbasher@iri.ldgo.columbia.edu

Bates, Bryson
CSIRO Land and Water
Private Bag
PO Wembley WA 6014
AUSTRALIA
Phone: +61-8-9333- 6330
Fax: +61-8-9333- 6211
E-mail: bryson.bates@per.clw.sciro.au

Berri, Guillermo J.
Dept. Ciencias de la Atmósfera, Univ. Buenos Aires
Pabellon 2, Ciudad Universitaria
1428 Buenos Aires
ARGENTINA
Phone: +54-11-4576-3356
Fax: +54-11-4765-9599
E-mail: berri@at1.fcen.uba.ar

Blumenthal, Benno
International Research Institute for Climate Prediction
PO Box 1000
Palisades, NY 10964-8000
USA
Phone: +1-845-680-4450
Fax: +1-680-4864
E-mail: benno@iri.ldgo.columbia.edu

Boonjung, Hatsachai
Suranaree University of Technology
111 University Ave.
T. Suranaree, A. Muang, Nakhon Ratchasima 30000
THAILAND
Phone: +66-44-224-012
Fax: +66-44-224-017
E-mail: boonjung@ccs.sut.ac.th

Broad, Ken
International Research Institute for Climate Prediction
PO Box 1000
Palisades, NY 10964-8000
USA
Phone: +1-845-680-4462
Fax: +1-845-680-4866
E-mail: kbroad@iri.ldgo.columbia.edu

Brumbelow, Kelly
Georgia Water Resources Institute, Georgia Tech University
790 Atlantic Dr.
Atlanta, GA 30332-0355
USA
Phone: +1-404-894-2329
Fax: +1-404-894-2677
E-mail: gt2105b@prism.gatech.edu, kelly@wrc1.ce.gatech.edu

Camargo, Suzana
International Research Institute for Climate Prediction
PO Box 1000
Palisades, NY 10964-8000
USA
Phone: +1-845-680-4416
Fax: +1-845-680-4865
E-mail: suzana@iri.ldeo.columbia.edu

Cane, Mark
LDEO, Columbia University
PO Box 1000
Palisades, NY 10964-8000
USA
Phone: +1-845-365-8344
Fax: +1-845-365-8736
E-mail: mcane@ldeo.columbia.edu

Cavalanti, Iracema
INPE/CPTEC
Rod. Predidente Dutra Km 40
12260-000
BRAZIL
Phone: +55-12-560-8423
Fax: +55-12-561-2835
E-mail: iracema@cptec.inpe.br

Centeno, Grace
International Rice Research Institute
PO Box 3127
1271 Makati City
PHILIPPINES
Phone: +632-845-0563 x643
Fax: +63-2-891-1292
E-mail: g.centeno@cgiar.org

Challinor, Andrew
Dept. of Meteorology, Univ. Reading
PO Box 243
Reading, Berks RG66BB
UK
Phone: +44-1189-316608
Fax: +44-1189-318316
E-mail: ajc@met.reading.ac.uk

Chang, Ching-Cheng
Inst. Economics, Academia Sinica
No. 128, Sect. 2, Yen-Chiou Yuan Rd.
Taipei 115
TAIWAN
Phone: +886-2-2782-2791 x201
Fax: +886-2-2785-3946
E-mail: emily@ieas.econ.sinica.edu.tw

Chen, Chi-Chun
Texas A&M University
USA
E-mail: chen@dust.tamu.edu

Cherry, Jessie
CIESIN
PO Box 1000
Palisades, NY 10964
USA
Phone: +1-845-365-8984
Fax: +1-845-365-8922
E-mail: jcherry@ciesin.org

Clarke, Allan J.
Oceanography Department (4320)
Florida State University
Tallahassee, FL 32306-4320
USA
Phone: +1-850-644-2240
Fax: +1-850-644-2581
E-mail: clarke@ocean.fsu.edu

Colman, Andrew
Met Office
London Road
Bracknell, Berkshire RG12 2SY
UK
Phone: +44-1344-854509
Fax: +44-1344-954499
E-mail: awcolman@meto.gov.uk

Conde Alvarez, Cecilia
Centro de Ciencias de la Atmosfera, UNAM
Ciudad Universitaria
Circuito Exterior, 04510 Mexico, D.F.
MEXICO
Phone: +52-5-622-4090
Fax: +52-5-622-4090
E-mail: conde@servidor.unam.mx

Crow, Michael
Vice Provost of Columbia University
205 LOW, 420 West 118th St.
New York, NY 10027
USA
Phone: 212-854-2330
E-mail: mc71@columbia.edu

Cullen, Heidi
International Research Institute for Climate Prediction
PO Box 1000
Palisades, NY 10964-8000
USA
Fax: 845-680-4864
E-mail: cullen@iri.ldeo.columbia.edu

de Serbinin, Alex
CIESIN
PO Box 1000
Palisades, NY 10964
USA
Phone: +1-845-365-8936
Fax: +1-845-365-8922
E-mail: adesherbinin@ciesin.org

DeWitt, David
International Research Institute for Climate Prediction
PO Box 1000
Palisades, NY 10964-8000
USA
Phone: +1-845-680-4415
Fax: +1-845-680-4865
E-mail: daved@iri.ldeo.columbia.edu

Diarra, Birama
Direction Nationale de la Meteo Rologie
BP 237
Bamako
MALI
Phone: +223-29-21-1
Fax: +223-29-21-01
E-mail: dnm@malinet.ml

Dilley, Maxx
The World Bank
Washington D.C.
USA
E-mail: mdilley@worldbank.org

Druyan, Leonard M.
Center for Climate Systems Research
NASA Goddard Institute for Space Studies
2880 Broadway
New York, NY 10025
USA
Phone: +1-212-678-5564
Fax: +1-212-678-5552
E-mail: LDruyan@giss.nasa.gov

Eakin, Hallie
Department of Geography, Univ. Arizona
Tucson, AZ 85721
USA
E-mail: hallicat@hotmail.com, heakin@u.arizona.edu

Farmer, Graham
Eritrea National Food Information System
PO Box 4908
Asmara
ERITREA
Phone: +291-1-181326
Fax: +291-1-181326
E-mail: fao@gemel.com.er

Ferrer, Rosa Maria
Centro de Ciencias de la Atmosfera, UNAM
Cuidad Universitaria, Circuito
Circuito Exterior, 04510 Mexico, D.F.
MEXICO
Phone: +525-6-22-40-92
Fax: +525-6-22-40-90

Freise, Amy
International START Secretariat
2000 Florida Avenue NW
Washington D.C 20009
USA
Phone: +1-202-462-2213
Fax: +1-202-457-5859
E-mail: AFreise@agu.org

Fuchs, Roland
International START Secretariat
2000 Florida Avenue NW
Washington D.C 20009
USA
Fax: +1-202-457-5859
E-mail: rfuchs@agu.org

Fulakeza, Matthew
Columbia University
2880 Broadway
New York, NY 10025
USA
Phone: +1-212-678-5572
Fax: +1-212-678-5552
E-mail: mfulakeza@giss.nasa.gov

Gadgil, Sulochana
Centre for Atmospheric and Oceanic Sciences
Indian Institute of Science
Bangalore—560 012
INDIA
Phone: +91-80-309-2505
E-mail: sulo@caos.iisc.ernet.in

Galvin, Kathleen
Natural Resource Ecology Lab
Colorado State University
Fort Collins, CO 80523
USA
Phone: +1-970-941-1642
Fax: +1-970-491-1965
E-mail: kathy@nrel.colostate.edu

Gameda, Sam
Agriculture and Agri-Food Canada
960 Carling Ave.
Ottawa, Ontario K1A 0C6
CANADA
Phone: +1-613-759-1532
Fax: +1-613-759-1924
E-mail: gamedas@EM.AGR.CA

Gathara, Simon
Drought Monitoring Centre
P.O. BOX 30259
Nairobi
KENYA
Phone: +254-2-567-864
Fax: +254-2-577-373
E-mail: simon.gathara@lion.meteo.go.ke

Goddard, Lisa
International Research Institute for Climate Prediction
PO Box 1000
Palisades, NY 10964-8000
USA
Phone: +1-845-680-4430
Fax: +1-845-680-4865
E-mail: goddard@iri.ldgo.columbia.edu

Grassl, Hartmut
Max Plank Institute for Meteorology
Bundesstrasse 55
D-20146 Hamburg
GERMANY
Fax: +49-40-41173-350
E-mail: grassl@dkrz.de

Haber, John
Columbia Earthscape
405 Low Library
New York, NY 10027
USA
Phone: +1-212-854-1434
Fax: +1-212-854-1307
E-mail: jh624@columbia.edu

Halide, Halmar
School of Mathematical and Physical Sciences, James Cook Univ.
Townsville, QLD 4811
AUSTRALIA
E-mail: halmar.halide@jcu.edu.au

Hall, Michael
NOAA/OGP
1100 Wayne Ave., Suite 1225
Silver Spring, MD 20910
USA
Phone: +1-301-427-2089
Fax: +1-301-427-2082
E-mail: hall@ogp.noaa.gov

Hallstrom, Daniel
North Carolina State University
USA
E-mail: dghallst@are1.cals.ncsu.edu

Hammer, Graeme
Queensland Dept. of Primary Industries
Towoomba, Qld 4350
AUSTRALIA
Phone: +61-7-46-881-379
Fax: +61-7-46-881-193
E-mail: HammerG@prose.dpi.qld.gov.au

Hansen, James
International Research Institute for Climate Prediction
PO Box 1000
Palisades, NY 10964-8000
USA
Phone: +1-845-680-4410
Fax: +1-845-680-4864
E-mail: jhansen@iri.ldeo.columbia.edu

Harris, Rebecca Lee
International Food Policy Research Institute
2033 K St., NW
Washington, DC 20006
USA
Phone: +1-202-862-5692
Fax: +1-202-467-4439
E-mail: r.l.harris@cgiar

Hill, Harvey
Dept. Agricultural Economics, Texas A&M Univ.
Blocker Building
College Station, TX 77843-2124
USA
Phone: +1-409-458-1506
Fax: +1-409-845-4582
E-mail: hsh@ag-eco.tamu.edu

Hillel, Daniel
Center for Environmental Studies
Rehov Hagalil 45
Karkur 37000
ISRAEL
Fax: +972-6-623-0861
E-mail: d_hillel@netvision.net.il

Hoogenboom, Gerrit
Biological and Agricultural Engineering
University of Georgia
Griffin, Georgia 30223-1797
USA
Phone: +1-770-229-3438
Fax: +1-770-228-7218
E-mail: gerrit@gaes.griffin.peachnet.edu

Hopp, Marianne
International Research Institute for Climate Prediction
PO Box 1000
Palisades, NY 10964-8000
USA
Phone: +1-845-680-4501
Fax: +1-845-680-4864
E-mail: mhopp@iri.ldeo.columbia.edu

Iglesias, Ana
Center for Climate Systems Research
Columbia University
2880 Broadway
New York, NY 10025
USA
Phone: +1-212-678-5626
Fax: +1-212-678-5648
E-mail: aiglesias@giss.nasa.gov

Ingram, John
NERC Centre for Ecology & Hydrology
Crowmarsh Gifford
Wallingford OX10 8BB
UK
Phone: +44-1491-692-410
Fax: +44-1491-692-313
E-mail: jsii@ceh.ac.uk

Jacobson, Barry
Cargill Central Research
PO Box 5699
Minneapolis, MN 55440-5699
USA
Phone: +1-612-742-3989
Fax: +1-612-742-7909
E-mail: Barry_Jacobson@Cargill.com

Jaeger, Jill
IHDP
Birneckergasse 10, A-1210
Vienna
AUSTRIA
Phone: +43-1-263-2104
E-mail: fuj.jaeger@magnet.at, jaeger.ihdp@uni-bonn.de

Jagtap, Srikant
Agricultural & Biological Engineering, Univ. Florida
PO Box 110870
Gainesville, FL 32611-0570
USA
Phone: +1-352-392-7719
Fax: +1-352-392-8476
E-mail: sjagtap@agen.ufl.edu

Jamieson, Peter D.
New Zealand Inst. Crop & Food Res.
Private Bag 4704
Christchurch
NEW ZEALAND
Phone: +64-3-256-400
Fax: +64-3-325-2074
E-mail: JamiesonP@crop.cri.nz

Johannsen, Chris
Purdue University/LARS
1202 Potter Hall
West Lafayette, IN 47907-1202
USA
Phone: +1-765-494-6248
Fax: +1-765-496-3216
E-mail: johan@purdue.edu

Jones, James
Agricultural & Biological Engineering, Univ. Florida
PO Box 110870
Gainesville, FL 32611-0570
USA
Phone: +1-352-392-8694
Fax: +1-352-392-4092
E-mail: jjones@agen.ufl.edu

Kane, Sally
NOAA/OGP
1100 Wayne Avenue, Suite 1210
Silver Spring, MD 20910-5803
USA
Phone: +1-301-427-2089
Fax: +1-301-427-2082
E-mail: kane@ogp.noaa.gov

Kermond, John
NOAA/OGP
1100 Wayne Avenue, Suite 1210
Silver Spring, MD 20910-5803
USA
Phone: +1-301-427-2089
Fax: +1-301-427-2082
E-mail:

Kiihl, Laura
Instituto Agronomico do Paraná
Rua Tomaz Antonio Gonzaga, 430
Jardim Lago Parque, 86015-320
BRAZIL
Phone: +43-376-2405
Fax: +43-376-2104
E-mail: laurar@pr.gov.br

Kirshen, Paul H.
Civil and Environmental Engineering Department
Tufts University
Medford, MA 02155
USA
Phone: +1-617-627-5589
Fax: +1-617-627-3994
E-mail: pkirshen@tufts.edu

Kitheka, Samuel
Kenya Agricultural Research Institute
KENYA
Phone: +254-145-20204
E-mail: eagletec@net2000ke.com

Korecha, Diriba
National Meteorological Services Agency
P.O. Box 1090
Addis Ababa
ETHIOPIA
Phone: +251-1-615779
Fax: +251-1-517-066
E-mail: dirbab@hotmail.com, nmsa@telecom.net.et

Kumar, Krishna
Indian Institute of Tropical Meteorology
Pashan Road
Pune 411 008
INDIA
Phone: +91-20-589-3600
Fax: +91-20-589-3825
E-mail: krishna@tropmet.ernet.in

Kushnir, Yochanan
LDEO, Columbia University
PO Box 1000
Palisades, NY 10964-8000
USA
Phone: +1-845-365-8669
Fax: +1-845-365-8736
E-mail: kushnir@ldeo.columbia.edu

Lall, Upmanu
International Research Institute for Climate Prediction
PO Box 1000
Palisades, NY 10964-8000
USA
Phone: +1-845-680-4421
Fax: +1-845-680-4864
E-mail: ulall@iri.ldeo.columbia.edu

Landman, Willem
International Research Institute for Climate Prediction
PO Box 1000
Palisades, NY 10964-8000
USA
Phone: +1-845-680-4525
Fax: +1-845-680-4864
E-mail: willem@iri.ldeo.columbia.edu

Lemos, Maria Carmen
Physics-Atmospheric Sciences Bldg., Rm 542, Univ. Arizona
P.O. Box 210081
Tucson, AZ 85721-0081
USA
Phone: +1-520-626-7242
Fax: +1-520-626-7248
E-mail: lemos@soar.atmo.arizona.edu

Leon, Alejandro
Office of Arid Land Studies, Univ. Arizona
1955 E. 6th St.
Tucson, AZ 85719
USA
Phone: +1-520-621-7895
Fax: +1-520-621-3816
E-mail: alexleon@Ag.Arizona.Edu

Lewis, John
US Agency for International Development
USA

Lin, Yeu-Woo
Central Weather Bureau
64 Kung Yuan Road
Taipei 100
TAIWAN
Phone: +886-2-2349-1113
Fax: +886-2-2349-1110
E-mail: linyw@cwb.gov.tw

Luby, Michael
Columbia Earthscape
Rm 405 Low Library, 535 W. 116th St.
New York, NY 10027
USA
Phone: +1-212-854-0423
Fax: +1-212-854-6309
E-mail: ml1047@columbia.edu

Lyon, Brad
International Research Institute for Climate Prediction
PO Box 1000
Palisades, NY 10964-8000
USA
Phone: +1-845-4475
Fax: 845-680-4864
E-mail: blyon@iri.ldeo.columbia.edu

Magalhaes, Antonio
The World Bank
SWS 315, Bloco A, Ap. 104
Brasilia, DF CEP 70384
BRAZIL
Phone: +55-61-329-1055
Fax: +55-61-329-1010
E-mail: Amagalhaes@worldbank.org

Malusalila, Percy
SeedCo
ZIMBABWE
Phone: +263-4-308-898
Fax: +263-4-308-898
E-mail: percyma@seedco.co.zw

Mearns, Linda
NCAR
PO Box 3000
Boulder, CO 80307-3000
USA
Phone: +1-303-497-8124
Fax: +1-303-497-8125
E-mail: lindam@atd.ucar.edu

Meinke, Holger
APSRU
PO Box 102
Toowoomba, QLD 4350
AUSTRALIA
Phone: +61-7-4688-1378
Fax: +61-7-4688-1193
E-mail: MeinkeH@prose.dpi.qld.gov.au

Meyer, Ray
Office of U.S. Foreign Disaster Assistance
OFDA, 8.06.044, RRB
Washington, DC 20523-8602
USA
Phone: +202-712-1078
E-mail: rmeyer@usaid.gov

Montecinos, Aldo
Dept. Geophysics, Universidad De Chile
Casilla 2777
Santiago
CHILE
Phone: +56-2-696-8790
Fax: +56-2-696-8686
E-mail: amonteci@dgf.uchile.cl

Morales-Acoltzi, Thomas
Centro de Ciencias de la Atmosfera, UNAM
Cuidad Universitaria, Circuito
Circuito Exterior, 04510 Mexico, D.F.
MEXICO
Phone: +52-5-622-4092
Fax: +52-5-622-4090
E-mail: acoltzi@ccaunam.atmosfcu.unam.mx

Moura, Antonio Divino
International Research Institute for Climate Prediction
PO Box 1000
Palisades, NY 10964-8000
USA
Phone: +1-845-680-4493
Fax: +845-680-4866
E-mail: amoura@iri.ldgo.columbia.edu

Mutter, Carolyn
International Research Institute for Climate Prediction
PO Box 1000
Palisades, NY 10964-8000
USA
Phone: +1-845-680-4428
Fax: +1-845-680-4866
E-mail: czm@iri.ldgo.columbia.edu

Nemani, Ramakrishna
School of Forestry
University of Montana
Missoula, MT 59812
USA
Phone: +1-406-243-4632
Fax: +1-406-243-4510
E-mail: nemani@ntsg.umt.edu

Nunes, Luci Hidalgo
Geological Inst. / Secretariat for Environment S. Paulo State
Av. Miguel Stefano, 3900
Sao Paulo, SP 04301-903
BRAZIL
Phone: +55-11-276-8288
Fax: +55-11-276-8572
E-mail: lhidalgo@uninet.com.br

O'Brien, James
Center for Ocean-Atmospheric Prediction Studies/FSU
2035 E. Paul Dirac Drive, R.M. Johnson Bldg.—Suite 200
Tallahassee, FL 32310
USA
Phone: +1-850-644-4581
Fax: +1-850-644-4841
E-mail: obrien@coaps.fsu.edu

Ogallo, Laban
Drought Monitoring Centre Nairobi
P.O. Box 30259
Nairobi
KENYA
Fax: +254-2-567-888
E-mail: laban.ogallo@meteo.go.ke

Olesen, Jørgen Eivind
Dept. Crop Physiology and Soil Science
PO Box 50
DK-8830 Tjele
DENMARK
Phone: +45-8999-1659
Fax: +45-8999-1619
E-mail: JorgenE.Olesen@agrsci.dk

Passalacqua, Gino
Calle Alfa Centavro No. 123
Urb. la Calera de Monterrico
Surquillo—Lima 34
PERU
Phone: +51-1-449-8490
E-mail: fampassa@inkanet.com.pe

Pfaff, Alex
Columbia University
420 W. 118th Street, Room 1306
New York, NY 10027
USA
Phone: +1-212-854-4190
Fax: +1-212-854-5765
E-mail: ap196@columbia.edu

Phillips, Jennifer
International Research Institute for Climate Prediction
PO Box 1000
Palisades, NY 10964-8000
USA
Phone: +1-845-680-4422
Fax: +1-845-680-4864
E-mail: jphillips@iri.ldeo.columbia.edu

Podesta, Guillermo
Univ. of Miami—RSMAS
4600 Rickenbacker Cswy.
Miami FL 33149
USA
Phone: +1-305-361-4142
Fax: +1-305-361-4622
E-mail: gui@rrsl.rsmas.miami.edu,
guipodesta@arnet.com.ar

Pozzi, Francesca
CIESIN
PO Box 1000
Palisades, NY 10964
USA
Phone: +1-845-365-8977
Fax: +1-845-365-8922
E-mail fpozzi@ciesin.org

Qian, Jian-Hua
International Research Institute for Climate Prediction
PO Box 1000
Palisades, NY 10964-8000
USA
Phone: +1-845-680-4429
Fax: +1-845-680-4865
E-mail: jqian@iri.ldeo.columbia.edu

Rajagopolan, Balaji
International Research Institute for Climate Prediction
PO Box 1000
Palisades, NY 10964-8000
USA
Phone: +1-845-680-4492
Fax: +1-845-680-4865
E-mail: rbala@iri.ldeo.columbia.edu

Rayner, Steve
International Research Institute for Climate Prediction
PO Box 1000
Palisades, NY 10964-8000
USA
Phone: +1-845-680-4423
Fax: +1-845-680-4864
E-mail: rayner@iri.ldeo.columbia.edu

Rebella, Cesar
Instituto de Clima y Agua, INTA
Las Cabañas y Los Reseros
(1712) Castelar, prov. de Bs.As.
ARGENTINA
Phone: +54-11-4621-125
Fax: +54-11-4621-5663
E-mail: thartmann@inta.gov.ar

Romero, Ricardo
INIA
Ruta 50 km 12
Colonia CC 39713—70.000
URUGUAY
Phone: +598-522-4060
Fax: +598-522-4061
E-mail: romero@inia.org.uy

Ropelewski, Chet
International Research Institute for Climate Prediction
PO Box 1000
Palisades, NY 10964-8000
USA
Phone: +1-845-680-4490
Fax: +1-845-680-4864
E-mail: chet@iri.ldgo.columbia.edu

Rosenzweig, Cynthia
NASA—Goddard Institute of Space Studies
2880 Broadway
New York, NY 10025
USA
Phone: +1-212-678-5562
Fax: +1-212-678-5648
E-mail: crosenzweig@giss.nasa.gov

Schmeltz, Alissa
Columbia University
535 W. 116th St., 401 Low Library, MC 4322
New York, NY 10027
USA
Phone: +1-212-854-7893
Fax: +1-212-854-3199
E-mail: as792@columbia.edu

Semenov, Mikhail
IACR Long Ashton Research Station
University of Bristol
Long Ashton, Bristol BS41 9AF
UK
Phone: +44-1275-549-421
Fax: +44-1275-394-007
E-mail: mikhail.semenov@bbsrc.ac.uk

Seth, Anji
International Research Institute for Climate Prediction
PO Box 1000
Palisades, NY 10964-8000
USA
Phone: +1-845-680-4419
Fax: +1-845-680-4864
E-mail: seth@iri.ldeo.columbia.edu

Shah, Kathy
NASA—Goddard Institute of Space Studies
2880 Broadway
New York, NY 10025
USA
Phone: +1-212-678-5625
Fax: +1-212-678-5552
E-mail: kshah@giss.nasa.gov

Simpsin, Jim
LDEO
PO Box 1000
Palisades, NY 10964-8000
USA
E-mail: simpsonj@ldeo.columbia.edu

Simpson, Caitlin
NOAA/OGP, Economics & Human Dimensions
1100 Wayne Avenue, Suite 1210
Silver Spring, MD 20910-5803
USA
Phone: +1-301-427-2089 x152
Fax: +1-301-427-2082
E-mail: simpson@ogp.noaa.gov

Singh, Bhawan
Dept. of Geography, Univ. Montreal
PO Box 612B, Succ. A
Montreal, Quebec H3C 3J7
CANADA
Phone: +1-514-343-8040
Fax: +1-514-343-8008
E-mail: bhawan.singh@umontreal.ca

Sivakumar, Mannava V.K.
World Meteorological Organization
7bix, Avenue de la Paix
1211 Geneva 2
SWITZERLAND
Phone: +41-22-730-8380
Fax: +41-22-730-8042
E-mail: sivakumar_m@gateway.wmo.ch

Small, Arthur
Columbia University
503 Uris Hall
New York, NY 10027
USA
Phone: +1-212-854-9016
Fax: +1-212-316-9214
E-mail: aas69@columbia.edu

Some, Bonaventure
AGRHYMET Center
PO Box 11011
Niamey
NIGER
Phone: +227-733-116
Fax: +227-732-435
E-mail: some@sahel.agrhymet.ne

Srinivasan, Ancha
Regional Science Institute
4-13 Kita 24 Nishi 2, Kita-ku
Sapporo 001-0024
JAPAN
Phone: +81-11-717-6660
Fax: +81-11-757-3601
E-mail: ancha@vtt.co.jp, ancha_s@yahoo.com

Stewart, Alan
PFRA-AAFC
Room 945 Canada Place, 9700 Jasper Ave.
Edmonton, Alberta T5J 4C3
CANADA
Phone: +1-780-495-5236
Fax: +1-780-495-4504
E-mail: stewarta@em.agr.ca

Stewart, Macol
NOAA/OGP, Economics & Human Dimensions
1100 Wayne Avenue, Suite 1210
Silver Spring, MD 20910-5803
USA
Phone: +1-301-427-2089
Fax: +1-301-427-2082
E-mail: mstewart@ogp.noaa.gov

Subbiah, Arjunapermal
Asian Disaster Preparedness Center
PO Box 4 Klong Luang
Pathumthani 12120
THAILAND
Phone: +66-2-524-5378
Fax: +66-2-524-5350
E-mail: kamal@ait.ac.th, ece@ait.ac.th

Sun, Liquang
International Research Institute for Climate Prediction
PO Box 1000
Palisades, NY 10964-8000
USA
Phone: +1-845-680-4433
Fax: +1-845-680-4865
E-mail: sun@iri.ldeo.columbia.edu

Susanto, Dwi
LDEO, Columbia University
PO Box 1000
Palisades, NY 10964-8000
USA
Phone: +1-845-365-8545
Fax: +1-845-365-8157
E-mail: dwi@ldeo.columbia.edu

Tilden, Scott
LDEO, Columbia University
PO Box 1000
Palisades, NY 10964-8000
USA
Phone: +1-845-365-8608
Fax: +1-845-365-8182
E-mail: tilden@ldeo.columbia.edu

Tourre, Yves
International Research Institute for Climate Prediction
PO Box 1000
Palisades, NY 10964-8000
USA
Phone: +1-845-680-4412
Fax:+1-845-680-4864
E-mail: tourre@iri.ldeo.columbia.edu

Unganai, Leonard
Zimbabwe Department of Meteorology
P.O. Box BE 150
Belvedere, Harare
ZIMBABWE
Phone: +263-4-778-172
Fax: +263-4-778-172
E-mail: advisor@weather.utande.co.zw

Uvo, Cintia
International Research Institute for Climate Prediction
PO Box 1000
Palisades, NY 10964-8000
USA
Phone: +1-845-680-4435
Fax: +1-845-680-4866
E-mail: cintia@iri.ldeo.columbia.edu

Valdivia, Corinne
Social Sciences Unit, Univ. Missouri-Columbia
200 Mumford Hall
Columbia, MO 65211
USA
Phone: +1-573-882-4020
Fax: +1-573-882-3958
E-mail: ValdiviaC@missouri.edu

Verdin, James
US Geological Survey
EROS Data Ctr, USGS
47914 252 St
Sioux Falls, SD 57198
USA
Phone: +1-605-594-6529
E-mail: verdin@edcmail.cr.usgs.gov

Vinocur, Marta
Fac. Agronomía y Veterinaria, Univ. Nacional de Río Cuarto
Ruta Nac. 36 Km 601
X5804 BYA Río Cuarto, Córdoba
ARGENTINA
Phone: +54-358-467-6191
Fax: +54-358-468-0280
E-mail: mvinocur@ayv.unrc.edu.ar, m_vinocur_grace_5800@yahoo.com

Wang, Gen-Shuh
Dept. Public Health, National Taiwan University
No. 1, Sec. 1, Jen-Ai Road, Room 1432
Taipei
TAIWAN
Phone: +886-2-2394-0612
Fax: +886-2-2394-0612
E-mail: gswang@ha.mc.ntu.edu.tw

Weber, Elke
Columbia University
406 Schermerhorn Hall
1190 Amsterdam Avenue
New York, NY 10027
USA
Phone: +1-212-854-1789
E-mail: euw2@columbia.edu

Wilcox, Brad
IAI, c/o INPE
Av. dos Astronautas, 1758
Sao Jose dos Campos SP 12227-010
BRAZIL
E-mail: bwilcox@dir.iai.int

Young, Robert
CINRS
PO Box 631 (Gate 3,Forestry Bldg, 80 Meiers Road)
Indooroopilly, Brisbane, QLD 4068
AUSTRALIA
E-mail: robert.young@mailbox.uq.edu.au

Yukins, Scott
Weather Services Corp.
420 Bedford St.
Lexington, MA 02420
Fax: 781-676-1001
Email: syukins@wx.com

Zebiak, Stephen E.
International Research Institute for Climate Prediction
PO Box 1000
Palisades, NY 10964-8000
USA
Phone: +1-845-680-4497
E-mail: steve@iri.ldgo.columbia.edu

Ziervogel, Gina
Environmental Change Unit, Univ. Oxford
la Mansfield Road
Oxford OX1 3TB
UK
Phone: +44-1865-281-185
Fax: +44-1865-281-181
E-mail: gina.ziervogel@geog.ox.ac.uk

Zubair, Lareef
International Research Institute for Climate Prediction
PO Box 1000
Palisades, NY 10964-8000
USA
Phone: +1-845-680-4424
Fax: +1-845-680-4864
E-mail: lareef@iri.ldeo.columbia.edu

Zullo, Jurandir
Cepagri/UNICAMP
Cidade Universitaria Zeferino Vaz
Campinas, Sao Paulo 13081970
BRAZIL
Phone: +55-19-289+3669
Fax: +55-19-289-3669
E-mail: jurandir@orion.cpa.unicamp.br

APPENDIX 2. COMPUTER DEMONSTRATIONS

Computer Demonstration of APSRU and APSIM—Australia

Graeme Hammer
Queensland Department of Primary Industries, Australia

APSRU is the Agricultural Production Systems Research Unit and APSIM is the Agricultural Production System Simulator. APSRU is a group of over 40 scientific and support staff based in Queensland, Australia. The group focuses on improving agricultural systems in NE Australia using systems approaches. A major area of research and development relates to application of seasonal climate forecasting in cropping systems. The cropping system simulation capability developed in APSIM is at the core of many research projects in APSRU. APSRU links closely with other researchers and with advisers and farmers in conduct of its research. More details are available at http://www.apsru.gov.au.

Information and Decision Support Systems for the Agricultural Sector: Monitoring NDVI to Determine Pasture Actual Status

Ricardo Romero[1], Walter E. Baethgen[2] and Agustín Giménez[1]
[1]INIA, Uruguay
[2]International Fertilizer Development Center, Uruguay

The Information and Decision Support Systems (IDSSs) are modern tools capable of processing large and complex information databases (weather, soil, initial conditions, etc.), and producing information that can be easily understandable. The IDSS link maps and databases through a GIS that allows spatial analyses at a regional level. Applications include: (a) impact studies and risk assessment; (b) land use feasibility studies; (c) impact studies of interannual and long-term climatic variability; and (d) crop and pasture production forecasts. NDVI data from AVHRR sensors of NOAA satellites can be used to estimate initial forage availability, and to calculate monthly normal values and anomalies, allowing comparison of current and expected values. The system can identify areas that are vulnerable to soil water deficit and, therefore, to forage crisis. IDSS is available online at http://www.inia.org.uy/disciplinas/agroclima/.

International Consortium for Agricultural Systems Applications—ICASA

Gerrit Hoogenboom
University of Georgia, USA

ICASA is a global network of international institutions and cooperating scientists who are interested in advancing the state-of-the-art in systems analysis and modeling. One of the goals of ICASA is to facilitate the application of computer models and systems tools for decision-making and policy development related to agricultural production systems, resource management, food security and sustainable growth. Current activities of ICASA include the promotion of data standards and file structures for experimental data documentation as well as crop model applications; the development of modules for crop model advancement; the organization of training workshops, conferences and symposia; and the promotion of open source code to facilitate sharing of models, software and data as part of the ICASA Toolbox. Additional information about ICASA can be found at the ICASA web site: http://www.ICASAnet.org.

Potential Impacts of Climate Change on World Food Supply: Interactive Internet Access to Data Sets from a Major Crop Modeling Study

Alex de Sherbinin
Center for International Earth Science Information Network, USA

A demonstration of a prototype tool to access and query data sets from a major EPA-funded crop modeling study. The web site includes explanatory text regarding the methodology utilized in the study and further development of scenarios based on the most recent general circulation models. It also provides links to other crop-climate web sites of interest. The tool will be available on-line by December 2000 at http://sedac.ciesin.org.

CLIMLAB2000

Yves Tourre
International Research Institute for Climate Prediction, USA

Primarily a statistical package, CLIMLAB2000 provides a set of tools that can be used to perform robust statistical analysis, data management and manipulation and display of results. More advanced statistical analyses such as Empirical Orthogonal Function (EOF) Analysis, Rotated EOFs and Canonical Correlation Analysis (CCA) are also included in the package. A database of global sea surface temperature (SST) is available and can be updated on-line. CLIMLAB2000 combines the power of the computational engine Octave for numerical and matrix calculations, and the ability of GrADS to perform spatial contour plots, coastlines and political boundaries. The program features a user-friendly interface, and is able to run on a PC with minimal processing power and memory. Its "friendliness" and instructions in three languages (English, French and Spanish) make it accessible to the international community working on Climate Applications. CLIMLAB2000 was distributed to Forum participants on CD-ROM, and can also be downloaded at http://iri.ldeo.columbia.edu/iri/programs/climlab2000/indexni.html.